Archimedes

New Studies in the History and Philosophy of Science and Technology

Volume 73

Archimedes has three fundamental goals: to further the integration of the histories of science and technology with one another; to investigate the technical, social and practical histories of specific developments in science and technology; and finally, where possible and desirable, to bring the histories of science and technology into closer contact with the philosophy of science.

The series is interested in receiving book proposals that treat the history of any of the sciences, ranging from biology through physics, all aspects of the history of technology, broadly construed, as well as historically-engaged philosophy of science or technology. Taken as a whole, Archimedes will be of interest to historians, philosophers, and scientists, as well as to those in business and industry who seek to understand how science and industry have come to be so strongly linked.

Submission/Instructions for Authors and Editors: The series editors aim to make a first decision within one month of submission. In case of a positive first decision the work will be provisionally contracted: the final decision about publication will depend upon the result of the anonymous peer-review of the complete manuscript. The series editors aim to have the work peer-reviewed within 3 months after submission of the complete manuscript.

The series editors discourage the submission of manuscripts that contain reprints of previously published material and of manuscripts that are below 150 printed pages (75,000 words). For inquiries and submission of proposals prospective authors can contact one of the editors:

Please find on the top right side of our webpage a link to our *Book Proposal Form.*

William R. Newman • Jutta Schickore
Editors

Traditions of Analysis and Synthesis

Springer

Editors
William R. Newman
Department of History and Philosophy of Science and Medicine
Indiana University
Bloomington, IN, USA

Jutta Schickore
Department of History and Philosophy of Science and Medicine
Indiana University
Bloomington, IN, USA

ISSN 1385-0180 ISSN 2215-0064 (electronic)
Archimedes
ISBN 978-3-031-76400-4 ISBN 978-3-031-76398-4 (eBook)
https://doi.org/10.1007/978-3-031-76398-4

This work was supported a Sawyer Seminar grant from the Andrew W. Mellon Foundation and by Indiana University

This Springer imprint is published by the registered company Springer Nature Switzerland AG
The registered company address is: Gewerbestrasse 11, 6330 Cham, Switzerland

Contents

About the Editors

William R. Newman is Distinguished Professor and Ruth N. Halls Professor in the Department of History and Philosophy of Science and Medicine at Indiana University. Most of Newman's work in the History of Science has been devoted to alchemy and "chymistry," the art-nature debate, and matter theories, particularly atomism. His most recent monograph is *Newton the Alchemist: Science, Enigma, and the Quest for Nature's "Secret Fire"* (Princeton: Princeton University Press, 2019). Newman is also General Editor of the *Chymistry of Isaac Newton*, an online resource combining born-digital editions of Newton's alchemical writings with multimedia replications of Newton's alchemical experiments. Newman is on the editorial boards of *Archimedes*, *Early Science and Medicine*, and *Ambix*.

Jutta Schickore is Ruth N. Halls Professor in the Department of History and Philosophy of Science and Medicine at Indiana University. Her research interests include philosophical and scientific debates about scientific methods in past and present, particularly about control experiments, (non)replicability, failure, and negative results; science and the public; and the relation between history and philosophy of science. Her publications include the monographs *Controlled Experiments* (Cambridge University Press, forthcoming 2024), *About Method. Experimenters, Snake Venom, and the History of Writing Scientifically* (University of Chicago Press, 2017), and *The Microscope and the Eye: A History of Reflections, 1740–1870* (University of Chicago Press, 2007). She has been a member of the Wissenschaftskolleg (Berlin, Germany 2024–2025), the Institute for Advanced Study (Princeton, NJ, 2007–2008 and 2017–2018), and the National Humanities Center (Research Triangle Park, NC, 2011).

Chapter 1
Introduction: Traditions of Analysis and Synthesis

William R. Newman

Analysis and synthesis are terms that bring to mind a host of scientific activities. Who has not heard of the analytical tests performed routinely by chemists? Even the most scientifically uninformed will at some point in their education have encountered humble litmus tests or their more sophisticated cousins, the pH test strips used to determine acidity and alkalinity. And we are daily bombarded by information, much of it disturbing, about analyses of the atmosphere and ocean that reveal growing levels of carbon dioxide, methane, and other gases that contribute to the present ecological crisis. But what do scientists do with the often refractory data that they collect by means of their analytical tests? In an effortless transformation of meaning, they "analyze" the data: yet here analysis has taken on a quite different sense, no longer referring to the physical determination of quantities and types of materials but instead indicating the mathematical and statistical techniques that allow one to screen out the otherwise intractable noise typically accompanying scientific research. As for synthesis, modern civilization provides us with examples at practically every moment of our waking lives in the form of synthetic products, ranging from the styrene polymer keyboard used to type the present text to the flavoring vanillin found in one's breakfast cereal. Both of these are materials that were first assembled—that is synthesized—in a chemical laboratory and then manufactured *en masse* in giant factories that form the basis of consumer culture. Moreover, entire branches of the pharmaceutical industry today owe their existence to synthetic organic chemistry, the field that recreates the molecules of the natural world and manipulates them in ways that are often unknown to nature itself. And yet in a fashion that parallels the use of analysis to refer not only to materials but to mental processes, we also combine disparate facts ranging from those acquired by scientific

W. R. Newman (✉)
Department of History and Philosophy of Science and Medicine, Indiana University, Bloomington, IN, USA
e-mail: wnewman@iu.edu

W. R. Newman, J. Schickore (eds.), *Traditions of Analysis and Synthesis*, Archimedes 73, https://doi.org/10.1007/978-3-031-76398-4_1

research to more mundane, casual thoughts into a "synthesis," typically meaning something like an overarching picture.

1.1 The Ancient Geometrical and Philosophical Traditions

Not only do analysis and synthesis refer to opposed physical or mental operations—the root sense of the Greek terms *analysis* and *synthesis* being "dissolution" or "coming apart," and "putting together"—but historically they have other senses as well, as the essays in this volume demonstrate.

The first of these historical traditions was primarily mathematical, and has a long history of scholarship devoted to it.[1] It is discussed in several of the chapters in this volume, especially Niccolò Guicciardini's, Alan Shapiro's, and Helen Hattab's. Already in the ancient world analysis and synthesis were paired terms referring to two complementary types of operation in Greek geometry. In simplest terms, geometrical analysis worked by first assuming that the proposition one was seeking to prove was true. One then proceeded by drawing logical inferences from it until one arrived at an independently known theorem. Once such certain knowledge had been attained, it could be used to prove the proposition that had only been assumed initially to be correct: this demonstrative stage constituted synthesis (Menn 2002, 198).[2] Although the method of analysis and synthesis had a rather technical sense in Greek geometry, this mathematical tradition also provided grist for ancient philosophers. Aristotle explicitly appeals to geometrical analysis in his *Nicomachean Ethics* (1112b15–27), where he compares it to the everyday experience of seeking a particular end. Just as the geometer first assumes the truth of a consequence and then reasons backwards to the conditions making it true, so a person seeking counsel can imagine a desired end and then work backwards, analytically, to the means of enacting it. By implication, reversing these steps to arrive at a practicable plan for executing the deed would then correspond to the synthesis of the mathematicians (Sweeney 1994, 228–9).[3]

But the Aristotelian tradition also described the processes of analysis and synthesis in ways that were less obviously built on geometrical tradition. In his *Rhetoric* and *Prior Analytics*, for example, Aristotle develops the idea that reasoning from effects to causes can be seen as a matter of associating "necessary signs" with specific events, such as a woman's production of milk with childbearing. Building on this observation, the ancient Aristotelian commentators Philoponus and Simplicius argued that such reasoning from signs to their concomitants constituted a form of

[1] See above all the classic study by Hintikka and Remes 1974 and the articles collected in Otte and Panza 1997. For a synoptic view, consult Oeing-Hanhoff 1971.

[2] Menn 2002 argues that the picture of the analysis—synthesis cycle given by Pappus of Alexandria and other later writers on Greek mathematics is simplistic and inadequate, but this has little bearing on the history of how the terms were received.

[3] See also Menn 2002, 208.

discovery of causes, and that once a cause had been discovered in this way, one could then reason from the cause to its effect. At least according to one modern scholar, this is the primary origin of the famous Renaissance method of *regressus*, whereby scientific knowledge was acquired by a two-fold process of reasoning from effects to causes and then back from causes to their effects.[4] Yet there are many other passages in the twin corpora of Plato and Aristotle and their commentators that also treat analysis (and to a lesser degree synthesis), contributing to the complexity of the issue.[5] For example, the ancient Neoplatonists, with their perennial goal of moving from the multiplicity and confusion of the material world to the simplicity of the transcendent One, saw this intellectual ascent as a form of analysis (or in Latin, *resolutio*). The understanding could strip away the pluralistic nature of the physical world in order to arrive at higher causes and principles. Here synthesis (in Latin *compositio*) implicitly preceded analysis, since the composite character of the world was a given, not a desideratum.

The "regressive" character of the geometrical analysis-synthesis cycle displayed by its movement in opposite directions found other, quite distinct applications in ancient philosophy and science as well. The famous second-century physician Galen employed paired analysis and synthesis in multiple fashions, as shown by Evan Ragland in this volume. Galen's *Ars medica* famously begins with the claim that all teaching begins either with the goal and works backwards by analysis, or begins with "the synthesis of those things discovered by analysis," or begins by breaking down definitions. But Galen did not view analysis and synthesis as mere pedagogical tools. His *Diagnosis and Cure of the Errors of the Soul* claims that analysis and synthesis provide the proper tools for devising and making mathematical instruments such as waterclocks: analysis leads to the necessary principles, which are then employed in a process of synthesis, resulting in the actual manufacture and testing of the device. Similarly, Galen spoke of medical diagnosis and anatomy as involving a type of analysis, and again, synthesis in the form of practice and testing. All of these cases appeal to analysis as a means of finding principles or points of origin; synthesis then employs those starting points in passing to practice or proof.

Most of the interpretations of the analysis-synthesis cycle that we have so far considered share a feature of considerable importance, namely the claim that it is analysis that yields discoveries and synthesis that in some sense confirms them. One can already see this in the geometrical method of antiquity, where synthesis was equated with formal proofs such as those given by Euclid, whereas analysis was linked with the informal, unpublished methods that led mathematicians to their actual discoveries. This has led Stephen Menn to note colorfully that an inevitable temptation resulted for philosophers and later scientists to view analysis as "the living core" of ancient mathematics, and synthesis as a sterile collection of "dead husks" (Menn 2002, 196). As Guicciardini shows in his chapter, such foundational

[4] See Morrison 1997.

[5] See, for example, the detailed article by Panza 1997 and compare also Byrne 1997.

mathematicians as René Descartes and Gottfied Wilhelm Leibniz were in an eager quest to rediscover or even outdo the ancient mathematical art of analysis. The culmination of this vaunting of analysis may perhaps be seen in the words of the Enlightenment philosopher Étienne Bonnot de Condillac, who lauded analysis but saw synthesis as a mere "uselessness and abuse of principles".[6] At first face, at least, it might seem that synthesis inevitably played at best a second-fiddle role to analysis.

1.2 Analysis and Synthesis as Paired Operations in Chemistry

There is at least one field, however, where synthesis acted as more than a mere restatement or confirmation of what analysis had already discovered. I refer to the domain of chemistry, where determination of the composition of materials was (and still is) of paramount importance. It is important to stress that this was already the case in discussions of premodern chymistry, where there were grave doubts that analysis could resolve materials into their discrete components at all, as opposed to creating new ones by the very act of decomposition. No less a scientist than William Harvey was willing to deny in the 1650s that "natural bodies are primarily produced or composed of those things into which they are ultimately resolved" (Harvey 1651, quoted in Frank 1980, 255–6). Similar concerns about the veridical nature of human-induced analysis underlay Robert Boyle's famous *Sceptical Chymist* of 1661. Boyle was understandably concerned that the analysis or *Scheidung* of Paracelsus, which the Swiss chymist viewed as the key to understanding nature in general (for which see the contribution of Kahn and Newman in this volume), was not a reliable means of arriving at the constituents of matter. Worried that the fire analysis of the Paracelsians yielded mere artifacts of combustion rather than actual pre-existent components of matter, Boyle cast doubt on the quest of finding ultimate principles of matter by any means. Yet he was willing to argue that chymical analysis could yield pre-existent ingredients of bodies as long as one did not make undue claims that these constituents were the ultimate materials into which more complex bodies could be dissolved. Hence at the beginning of the *Sceptical Chymist* Boyle points out that gold can be alloyed with copper and silver as well as other metals and metalloids, to yield a seemingly homogeneous body very unlike the original gold. By means of selective dissolution in acids, however, the gold can be retrieved intact. Similarly, mercury will combine with sulfur to yield the red solid vermilion, or with "saline bodies" to become a volatile white salt, both materials that are uniform to sight; yet in the end the mercury can be recaptured unchanged, in the case of vermilion merely by heating to the proper temperature. From these and other examples, Boyle concludes that the ingredients of the respective alloy or compound retained their robust existence throughout the synthesis and subsequent analysis.

[6] Condillac, as quoted in Shapiro's contribution to this volume at note 51.

Unlike the mathematical and closely related philosophical traditions, which typically required analysis to precede synthesis, Boyle's chymical examples could work in either direction. In his *Certain Physiological Essays* also published 1661, for example, Boyle "redintegrated," that is, resynthesized saltpeter that he had previously decomposed by burning it in the presence of charcoal. He knew from chymical and artisanal practice that saltpeter could be decomposed into spirit of niter (nitric acid) and a fixed salt by thermal decomposition. By recombining these ingredients he managed to regain saltpeter, thus providing evidence that these were pre-existent ingredients that retained their identity within the compound that we call saltpeter (or niter). Here the analysis preceded the synthesis, as opposed to the examples given at the beginning of the *Sceptical Chymist*.[7]

Again in contradistinction to mathematical analysis and synthesis and its philosophical offshoots, Boyle's analysis and "redintegration" were not only parallel processes that could function independently; rather, the probative force of the one operation depended on the other. While a mathematical proof of a proposition could exist very well without our knowing the steps employed in the discovery of the proposition (as is in fact historically the usual case), Boyle's syntheses required the mirror operation of analysis to have any significance at all. Similarly, he argued that without resynthesis, chymical analysis provided no certainty that its products were anything but artifacts. Exactly the same structure of argument would soon underlie what is perhaps the most famous analysis-synthesis pair in the history of chemistry over the *longue durée*. I refer to Antoine Laurent Lavoisier's decomposition of water into oxygen and hydrogen and its resynthesis from those elements in a famous experiment of 1785, discussed in Bernadette Bensaude-Vincent's chapter in this volume. By breaking water down into its components and then rebuilding it out of the same materials, Lavoisier was able to show indisputably that water was itself a compound, not a fundamental element that underlay the material world in general.

Neither Lavoisier's analyses and syntheses nor Boyle's decompositions and "redintegrations" originated out of nothing. In reality, both scientists were indebted to a long chymical tradition of paired analysis and synthesis that in the seventeenth century went by the name "reduction to the pristine state" (*reductio in pristinum statum*).[8] The major proponent of the reduction to the pristine state in the generation before Boyle was the Wittenberg medical professor Daniel Sennert, who figures prominently in Joel Klein's chapter. Sennert employed reductions to the pristine state extensively to attack the opinion, common among university professors of the time, that no pre-existing materials could subsist in "perfect mixtures" (seemingly homogeneous bodies, including what we would today call chemical compounds) other than either the Aristotelian prime matter or the four elements, fire, air, water, and earth. Like Boyle, Sennert synthesized alloys and chemical compounds from known ingredients to produce apparent perfect mixtures. And again like Boyle,

[7] For a recent treatment of Boyle's niter redintegration experiment and its significance, see Buyse 2024.

[8] For the history of the *reductio in pristinum statum* and its implications for atomism, see Newman 2006.

Sennert then analyzed these "mixts" into their ingredients, arguing that this showed the ingredients to have been present all the while, despite the common learned view that the "mixts" were actually purely homogeneous materials. Similar experiments were also carried out by another slightly later favorite of Boyle's, the Flemish chymist Joan Baptista Van Helmont, who added an important quantitative dimension to the *reductio in pristinum statum*.[9]

And yet if we push back the curtains of history a bit further, neither Boyle nor Sennert was the first to employ the reduction to the pristine state as a means of arguing for robust ingredients that persisted in so-called perfect mixtures. An alchemical tradition going back to the High Middle Ages had already used a similar approach to the reduction to the pristine state to argue explicitly against a theory of mixture that had been championed by Thomas Aquinas in the second half of the thirteenth century. Basing himself largely on Aristotle's *De generatione et corruptione*, Thomas interpreted a famous passage of the Stagirite's work, "mixture is the union of the altered miscibles" (328b22) in a very strong sense, where "union" was taken to mean that the four elements undergoing mixture lost their actual being and were reduced to the four elementary qualities, hot, cold, wet, and dry, acting on the undifferentiated Aristotelian prime matter. This position committed Thomas and his followers to claim that one should not be able to recapture the four elements or any other intermediate materials from a perfect mixture, once it had been formed. The "union" (*hēnosis*) of *De generatione et corruptione* was absolute, leaving no room for the original ingredients of the mixture to persist. A perceptive challenge to the Thomistic position emerged in the *Theorica et practica* of the little-known scholastic Paul of Taranto, a Franciscan alchemist of the High Middle Ages, whose work is discussed in Kahn and Newman's chapter. Paul used empirical evidence drawn from the laboratory to demonstrate the retrievability of intermediate principles from supposed perfect mixtures and squarely confronted the Thomistic position. Other alchemists of the period, even though most did not engage in this sort of head-on challenge presented by Paul of Taranto, implicitly rejected the Thomistic view when they maintained the robust existence of the alchemical principles sulfur and mercury within metals and minerals (Newman 2006, 23–44).

One can see, then, that the paired analysis and synthesis required for the reduction to the pristine state had a history rooted in a very specific debate. Boyle's 1666 *Origin of Forms and Qualities* was in fact still arguing against the Thomistic position (which had by then been adopted by the followers of John Duns Scotus and others) when he employed a reduction to the pristine state of camphor dissolved in sulfuric acid and then precipitated by adding water to make the following claim:

> This Experiment may serve to countenance what we elsewhere argue against the Schools, touching the Controversie about Mistion. For whereas though some of them dissent, yet most of them maintain, that the Elements alwaies loose their Forms in the mix'd Bodies they constitute. (Boyle 1666, 396)

---

[9] For Van Helmont's contribution to what would ultimately come to be known as "the balance-sheet method," see Newman and Principe 2002.

Just as in the cases of gold recaptured from an alloy or mercury reduced from vermilion, the recovery of the camphor showed that it had been present all along in the acid, even though its presence had been undetectable to the senses.

Let us now make some general points. First, unlike the "regressive" traditions anchored in ancient geometry, philosophy and medicine, it appears that the explicit interest of alchemy and early modern chymistry in analysis and synthesis originated in a polemical debate on the nature of mixture and homogeneity in general. The roots of this debate are not to be found in the tradition of ancient mathematical analysis and synthesis, nor in the parts of the Aristotelian or Galenic corpora that consciously model themselves on the practices of the geometers. They lie rather in a hybrid offspring of hands-on alchemical research and the Aristotelian commentary tradition building on passages where the Stagirite discusses mixture and combination, primarily *De generatione et corruptione* and the *Meteorology*, but bringing in other parts of the corpus as well. And second, the subsequent history of chemistry in the eighteenth century shows the continued power of this analytico-synthetic tradition, as the famous example of Lavoisier reveals.

1.3 Analysis as Testing

As several of the authors in this volume show, however, the multiple traditions that concerned themselves with the nature and character of matter also had other uses for analysis and synthesis, not merely as paired halves of a single process, but as independent actors with their own goals. This appears in Laurence Totelin's contribution, for example, in two ways. First, the ancient medical tradition of making compound medicines viewed this practice as synthetic. Sustained attempts were made to view compound medicines as aggregates of the so-called primary qualities, hot, cold, wet, and dry, which were in turn the immediate components of the four elements (in combination with prime matter). The manual blending of ingredients meant to provide a proper mix of qualities was explicitly seen as a synthesis, in the sense of a compounding. When medical practitioners were unable to acquire specific ingredients, many of which were rare and expensive, their practice allowed them to substitute one material for another. Indeed, the art of substitution (*quid pro quo* in Latin), became an established part of pharmacy, yielding ersatz ingredients that eventually came to be called *succedanea* in Western medical practice.

But ancient pharmacology also employed analysis, and in a very telling way. While the art of compounding could be an entirely innocent practice, it could also involve fraud or adulteration. How could one determine if a material purchased as verdigris (copper acetate) were genuine or simply some other green material of a similar appearance? The natural history writer Pliny says to expose the material to papyrus that has been soaked in oak gall, upon which an adulteration with atrament will cause it to turn black. More commonly, minerals were heated or burned and carefully observed to see if changes in color, weight, or texture occurred. These practices, along with the assaying tests of ancient metallurgists such as cupellation and cementation, were significant forerunners of modern chemical analysis. Yet this is "analysis" in a quite different sense from that of Boyle's paired analyses and redintegrations, for here analysis does not necessarily imply a physical decompounding of a material into its constituents, but rather a mere identification of them within it.

As Peter Ramberg points out in his chapter on the nineteenth-century development of the discipline called analytical chemistry, this second sense of analysis had long been present in the chymical tradition, whether implicitly or explicitly. Ramberg also notes that even the modern term "reagents" (*reagentia* in Latin) was already being used by the seventeenth-century chymist J. J. Becher to mean standard materials used to test for the presence of other materials (somewhat like Pliny's oak gall). The seventeenth century saw a rapid development of such tests, particularly in the form of color indicators, which Boyle famously made great use of. Another example of such interest appears in the work of the Kiel medical professor Johann Daniel Major, as described by Vera Keller in her chapter. Major experimented extensively with color tests among other indicators in the context of his theory that *taxis* or military grouping can be used to discuss matter at the micro-level.

Needless to say, analysis in the sense of material testing has gone on to form one of the principal backbones of modern chemical practice. But ironically, this sense of analysis not only does not necessarily involve chemical decompounding, it no longer need involve chemistry at all. Such modern analytical tests as X-Ray fluorescence, Raman, and Infrared spectrometry, as well as X-Ray crystallography all share the feature that they are non-destructive, and do not even employ chemical reactions to acquire their data, unlike the traditional wet analysis of older chemistry. Although such analytical procedures reveal some or all of the components making up a material, and can sometimes even determine the relative or absolute quantities in which they are present, they do not do this by physically decompounding their samples. They are remote descendants of the identification tests employed in ancient pharmacology and metallurgy, not scions of the *reductio in pristinum statum.*

1.4 Conflated, Transferred, and Inverted Traditions of Analysis and Synthesis

Even in the ancient world, the Greek and Latin terms for analysis and synthesis were being used to describe both physical and mental processes. Hence it comes as no surprise to learn that the mathematical, philosophical, and chymical traditions associated with these terms would evolve and combine to form interesting and influential new developments as early as the High Middle Ages. As Helen Hattab points out in her essay, Thomas Aquinas was already employing *resolutio* and *compositio* in both material and conceptual senses. René Descartes and Thomas Hobbes also blurred the distinction between varying meanings of analysis and synthesis, though in different ways, as Hattab also discusses. But the most famous of these hybrid uses of the terms is probably to be found in Isaac Newton's celebrated comments in Query 31 of the 1717 *Opticks* (Query 23 in the earlier Latin version), which form the starting point of Alan Shapiro's essay in this volume. In this passage, Newton explicitly presented natural philosophy as following the path of mathematical analysis and synthesis. Yet he managed to combine this with the decompositional language of chymistry:

> By this way of Analysis we may proceed from Compounds to Ingredients, and from Motions to the Forces producing them; and in general, from Effects to their Causes, and from particular Causes to more general ones, till the Argument end in the most general. (Newton 1718, 380)

As Shapiro argues, the ingredients making up Newton's "compounds" here in the published version of Query 31 are not chemicals, but light rays. And yet in the first draft of the Query, Newton framed the analysis of light as proceeding "from compound bodies to their ingredients."[10] The fact that Newton initially used the materialist language of "compound bodies" for light rays is no doubt a vestige of his personal conviction that light consisted of material corpuscles. But it supports another important point as well, namely that Newton's early prismatic analysis and resynthesis of light was very likely modeled upon, and perhaps even inspired by, Boyle's material analyses and resyntheses of niter, camphor, and multiple chemical compounds containing metals. Newton had read Boyle's principal chymical works by the late 1660s, and the language of the latter's *Certain Physiological Essays* emerges in Newton's early *Lectiones opticae* and *Optica,* where he speaks of the spectral rays reassembled to form white light as a "redintegrated whiteness" (*albedo redintegrata*) (Newman 2019, 132–3). This heuristic use of the traditional reduction to the pristine state represents a major transfer from the realm of chymistry to that of natural philosophy writ large. Newton's integration of chymical, mathematical, and philosophical analysis and synthesis was not a mere rhetorical flourish, but a fruitful path to scientific discovery.

Similarly complex uses of analysis and synthesis can be found in other authors of Newton's period and later. As Tawrin Baker shows in his chapter, early modern physicians ranged from thinking of analysis and synthesis as a method of teaching, in the Galenic fashion, to conceiving of anatomy as a whole as employing physical dissection (analysis) and mental reassembly (synthesis). Analysis and synthesis are present here both in the "regressive" sense of ancient mathematics and philosophy, and in the decompositional sense employed by chymists. Yet as the early modern period passed into the modern per se, the decompositional sense of the two processes came to be more pronounced, and even passed into discussions of human cognition. One sees this in the work of Condillac and his followers Lavoisier and Guyton de Morveau, for example, as stressed by Bensaude Vincent. Condillac joined the numerous mathematicians and philosophers of the early modern period (discussed in Guicciardini's chapter) who adopted the term "analysis" for algebra. The fact that Condillac conceived of mental operations as involving a process of

[10] See Shapiro's essay in this volume, note 19.

composition and decomposition made it an easy matter for Lavoisier and his colleagues to employ his system in their nomenclature reform for chemistry. Although there is still a debt to ancient mathematics here, the overall thrust was now along the lines of chymical analysis and synthesis, albeit transported to the realm of names and concepts.

The complex and evolving senses of analysis and synthesis appear in other fields as well. As Julia Kursell's chapter shows, the early nineteenth-century German musical writer Gottfried Weber is usually credited with creating the field of music analysis, which he described explicitly in terms of decomposition (using the German verb *zergliedern* and the Greek-based *Analyse*). Musicology was therefore seen as a straightforward dissection of music into its compositional parts. A generation later, however, analysis was being used in a radically different way by acoustical scientists in Germany. Hermann von Helmholtz was devising instruments for physically analyzing and synthesizing musical tones, as would his follower, the pioneer in experimental psychology Carl Stumpf. While the complexity of the analysis-synthesis tradition in the nineteenth century makes it hard to disentangle the various threads, one cannot help but wonder if these attempts found their ultimate inspiration in the Newtonian analysis and resynthesis of light. At any rate, the experiments carried out by these acousticians are remarkably similar in structure to those in the tradition of the reduction to the pristine state discussed earlier in the introduction. In Stumpf's case, the complex sounds of different musical instruments would first be decomposed into their constituents and then reassembled, piece by piece, with the help of sophisticated auditory apparatus in order to verify (with the help of trained listeners) that the synthetic product was identical to the original "sample."

Further ramifications of analysis and synthesis in the modern period appear in the electrical research described by Friedrich Steinle in his chapter. One sees in the work of French natural philosophers such as Edme Mariotte, Charles Dufay, and Jean-Baptiste le Rond d'Alembert a systematic attempt to arrive at the "principles" of electricity. According to d'Alembert, the multiple phenomena evinced by experimentation should be "reduced," if possible, to a single, basic principle. These principles were not underlying, hidden causes, but rather products of induction arrived at by means of systematic experiment over multiple cases. One can see something like the traditional, inductive passage from multiplicity to simplicity associated with the "regressive" tradition of analysis, though in a way where induction is equated with knowledge arrived at by experiment, and where one does not ascribe causality to the principles discovered.

The issues become even more complex in Jutta Schickore's chapter, where the Kantian reformulation of philosophy in late eighteenth-century and nineteenth-century Germany is examined alongside the tradition of "inductive science" and philosophy. Analysis and synthesis received a new formulation in Kant's terminology of analytic and synthetic propositions and in other parts of his philosophy, where the traditional meanings of the terms were inverted or reversed, but his innovations did not erase their more traditional signification. By examining introductory German logic books of the eighteenth century, Schickore reveals a striking mixture of the decompositional sense of analysis with the "regressive" one inherited from ancient philosophy, medicine, and mathematics. A vigorous tradition of connecting analysis with experimental induction along the lines discussed by Steinle also continued in Germany and in Britain, as Schickore shows. One of the most interesting features of this "inductive" tradition lay in its inclusion of hypothetico-deductive reasoning, a feature that modern philosophers such as Karl Popper sharply excluded from the Baconian program of induction. And yet by the beginning of the nineteenth century, the traditional pairing of analysis with synthesis was on the decline. The modern senses of the terms, which owe more to the decompositional model than to the "regressive" ones descending from ancient prototypes, had largely come to prevail.

The power that analysis holds for the modern mind is a result of the many divergent and yet mutually reinforcing tendencies described in the present book. The philosophical tradition associated with Kant, and the "Chemical Revolution" linked to Lavoisier, were both products of the late eighteenth century, and yet both had roots in the distinction between analysis and synthesis as it evolved over the *longue durée*. One significant feature of this book lies in its attempt to disentangle the various traditions involved, which can sometimes be achieved by following them back chronologically to their sources. This is apparent in the decompositional model, for example, where the analytico-synthetic probative ideal of the reduction to the pristine state, already present in the work of Paul of Taranto and developed much further by Sennert, Van Helmont, Boyle, Newton, and Lavoisier, ramified into areas as diverse as Newtonian optics and the nineteenth century science of acoustics practiced by Helmholtz and Stumpf. The chymical, pharmaceutical, and metallurgical traditions of the material test, on the other hand, provided fertile grounds for analytical testing in general, in a way quite distinct from the reduction to the pristine state (which was itself a test of a different sort). But for all the power and pervasiveness of analytical tests, analysis has remained only part of the equation. As one of the founders of the field of organic chemistry, Marcellin Berthelot, stated in a passage quoted at length by Bensaude Vincent, the power of organic chemistry to create the very products of its study gave it a unique status among the sciences: "This creative faculty, similar to that of art, distinguishes it essentially from natural or historical sciences." In short, while analysis came to represent probing, testing, and clarifying, synthesis had become the exemplification of makers' knowledge, according to which certainty is best found in the objects that we ourselves create (Pérez-Ramos 1988). It is precisely the synthetic makers' knowledge ideal to which a (possibly spurious) quotation attributed to Richard Feynman alludes in the following words—"What I do not create, I do not understand."

This volume—and its companion collection of essays on experimental control—originated in a Sawyer Seminar at Indiana University Bloomington titled "Rigor: Control, Analysis and Synthesis in Historical and Systematic Perspectives," which was funded by the Mellon Foundation. Mellon Sawyer Seminars are temporary research centers, gathering together members of faculty, postdoctoral fellows, and graduate students for in-depth study of a scholarly subject via reading groups, seminars, and workshops. During the course of our activities, we organized two international conferences, which brought together scholars in history, philosophy, and social studies of science, who examined the contemporary and historical dimensions of rigor in experimental practice. The contributors to this volume participated in the first Sawyer conference (September 2021), before reconvening in early 2023 for an author workshop, at which the draft chapters for the volume were intensely discussed.

Several institutions and individuals have helped to make our work possible. We gratefully acknowledge the Mellon Foundation's generous financial support—especially the Foundation's flexibility as we dealt with the challenges of pursuing collaborative scholarship during a pandemic. We are grateful to Cory Rutz, the Director of Foundation Relations at Indiana University's Office of the Vice President for Research, for his prompt and efficient assistance in administering the grant. The author workshop took place at the IU Europe Gateway (Berlin) and was funded by a combined grant from the IU College of Arts and Sciences and the College Arts and Humanities Institute. We very much appreciate this support. We are indebted to Jed Buchwald for including our work in the Archimedes series, and to Chris Wilby for his efforts at moving the publication along. We would also like to say a big thank you to our department manager Dana Berg (Department of History and Philosophy of Science and Medicine at IU), office assistant Maggie Herms (IU HPSC), as well as Andrea Adam Moore (IU Europe Gateway), all of whom helped to organize our conferences and workshops. Finally, we warmly thank the many participants at the two conferences and at the various other Sawyer events for their valuable inputs, comments, questions, and critiques.

References

Boyle, Robert. 1666. *The origin of forms and qualities*. In *The works of Robert Boyle*, ed. Michael Hunter and Edward Davis, vol. 5, 1999–2000. London: Pickering & Chatto.

Buyse, Filip Adolf A. 2024. Boyle, Glauber, and Newton: The redintegration experiment with saltpeter. *ACS Omega* 9 (14): 15727–15731.

Byrne, Patrick H. 1997. *Analysis and science in Aristotle*. Albany: SUNY Press.

Frank, Robert. 1980. *Harvey and the Oxford physiologists*. Berkeley: University of California Press.

Hintikka, Jaakko, and Unto Remes. 1974. *The method of analysis: Its geometrical origin and general significance*. Dordrecht: D. Reidel.

Menn, Stephen. 2002. Plato and the method of analysis. *Phronesis* 47: 193–223.

Morrison, Don. 1997. Philoponus and Simplicius on tekmeriodic proof. In *Method and order in Renaissance philosophy of nature: The Aristotle commentary tradition*, ed. Daniel A. Di Liscia, Eckhard Kessler, and Charlotte Methuen, 1–22. Brookfield: Aldershot.

Newman, William R. 2006. *Atoms and alchemy*. Chicago: University of Chicago Press.

———. 2019. *Newton the alchemist*. Princeton: Princeton University Press.

Newman, William R., and Lawrence M. Principe. 2002. *Alchemy tried in the fire: Starkey, Boyle, and the fate of Helmontian chymistry*. Chicago: University of Chicago Press.

Newton, Isaac. 1718. *Opticks*. London: W. and J. Innys.

Oeing-Hanhoff, L. 1971. Analyse/synthese. In *Historisches wörterbuch der philosophie*, vol. 1, 232–248. Darmstadt: Schwabe Verlag.

Otte, Michael, and Marco Panza. 1997. *Analysis and synthesis in mathematics*. Dordrecht: Kluwer.

Panza, Marco. 1997. Classical sources for the concepts of analysis and synthesis. In *Analysis and synthesis*, ed. Michael Otte and Marco Panza, 365–414. Dordrecht: Kluwer.

Pérez-Ramos, Antonio. 1988. *Francis Bacon's idea of science and the maker's knowledge tradition*. Oxford: Clarendon Press/Oxford University Press.

Sweeney, Eileen C. 1994. Three notions of *resolutio* and the structure of reasoning in Aquinas. *The Thomist: A Speculative Quarterly Review* 58: 197–243.

William R. Newman is Distinguished Professor in the Department of History and Philosophy of Science and Medicine, Indiana University (Bloomington). His main present research interests focus on early modern "chymistry" and late medieval "alchemy," especially as exemplified by Isaac Newton, Robert Boyle, Daniel Sennert, and the first famous American scientist, George Starkey.

Chapter 2
The Dark Side of *Sunthesis*? Fraud and Substitution in Graeco-Roman Pharmacology

Laurence Totelin

Abstract This paper examines one specific aspect of the compounding—*sunthesis* in Greek—of remedies in ancient pharmacy, namely the substitution of one or several of the ingredients stipulated in the original recipe. This aspect has both positive and negative facets. The positive is the art of substitution, which should theoretically rely on sufficient knowledge of the powers of ingredients to replace like for like. The negative facet is the implication of the potential for fraud or adulteration, which led to the development of tests to determine the authenticity of a given preparation or its ingredients. The paper examines the boundary between substitution and fraud and assesses the roles that these two phenomena played in the development of ancient pharmacy.

Keywords Ancient pharmacy · Remedy · Synthesis · Fraud · Substitution · Adulteration

2.1 Introduction

In his treatise *On Drunkenness*, the philosopher Philo of Alexandria (end of the first century BCE–middle of the first century CE) discussed at length the effects that wine exerted on sensory perception and the reasons why inebriation can produce different effects on the mind at different times (154–205). Within his discussion, he noted that two discrete drugs could affect the body in completely different ways, depending on the proportions in which they were mixed:

> τί δ' αἱ ἐν τοῖς σκευαζομένοις ποσότητες; παρὰ γὰρ τὸ πλέον ἢ ἔλαττον αἵ τε βλάβαι καὶ ὠφέλειαι συνίστανται, καθάπερ ἐπὶ μυρίων ἄλλων καὶ μάλιστα τῶν κατὰ τὴν ἰατρικὴν ἐπιστήμην ἔχει φαρμάκων· ἡ γὰρ ἐν ταῖς συνθέσεσι ποσότης ὅροις καὶ κανόσι μεμέτρηται, ὧν οὔτε ἐντὸς κάμψαι οὔτε περαιτέρω προελθεῖν ἀσφαλές— τὸ μὲν γὰρ ἔλαττον χαλᾷ, τὸ δὲ πλέον ἐπιτείνει τὰς δυνάμεις· βλαβερὸν δ' ἑκάτερον, τὸ μὲν ἀδυνατοῦν ἐνεργῆσαι

L. Totelin (✉)
School of History, Archaeology and Religion, Cardiff University, Cardiff, UK
e-mail: TotelinLM@cardiff.ac.uk

W. R. Newman, J. Schickore (eds.), *Traditions of Analysis and Synthesis*, Archimedes 73, https://doi.org/10.1007/978-3-031-76398-4_2

> δι’ ἀσθένειαν, τὸ δὲ βλάψαι βιαζόμενον διὰ καρτερωτάτην ἰσχύν—, λειότησί τε αὖ καὶ τραχύτησι, πυκνότησί τε αὖ καὶ πιλήσεσι καὶ τοὐναντίον μανότησι καὶ ἐξαπλώσεσι τὸν εἰς βοήθειαν καὶ βλάβην ἔλεγχον ἐναργῶς διασυνίστησιν.
>
> What then of the **quantities in compounded things**? For it is a larger or smaller quantity that brings about damages and benefits, as in many examples, but particularly in the case of drugs in the medical art. **For quantity in these compounds** is measured by limits and rules, and it is not safe either to stop short or to advance beyond them. For too little weakens the **properties** [of the drug], and too much overstrains them. Each one of these outcomes is harmful. In the former, the remedy cannot be effective because of its weakness; in the latter, its strength is damaging on account of its exceeding force. And again, through its smoothness or roughness, its thickness and compactness on the one hand, and its looseness and slackness on the other, it clearly displays its capacity to help or to harm.[1]

Here, Philo was referring to the interlinked notions of *posotēs*, quantity, and *sunthesis*, the act of mixing together several simple drugs (*sun-*: together; *thesis*: the act of putting) to produce a compound remedy. For it is not sufficient to know which ingredients should be combined and their individual powers (*dunameis*); one must know which precise proportions of each ingredient should be used to avoid harm, whether through inefficacy or excessive strength. In a very basic sense, then, pharmacological *sunthesis* functioned similarly to addition in mathematics—indeed, the Greek term was also used to refer to addition and multiplication.[2] So central was the notion of *sunthesis* to medical knowledge (*iatrikē epistemē*) that the titles of two of the most important ancient Greek treatises on compound remedies included the word: Galen's *On the Composition of Drugs according to Places* and *On the Composition of Drugs according to Kind* (late second century CE–early third century CE), the former listing remedies according to which part of the body they could treat and the latter listing them according to type (e.g., plaster; *kollurion*).[3] The Latin equivalent to the Greek *sunthesis* was *compositio* (*cum*: together; *positio*: the act of putting), which is also reflected in the title of Scribonius Largus' important *Compounding of Drugs* (*Compositiones medicamentorum*, first century CE).[4] This notion of *sunthesis* had a very long story, recently told in Paula de Vos' *Compound*

---

[1] Philo, *De ebrietate* 184–5. The edition followed is Wendland 1897 (repr. 1962). For an English translation, with facing Greek text, see also Colson and Whitaker 1930. Unless stated otherwise, all translations from the Greek and Latin are mine.

[2] The notion of *analusis* also appears in the works of the ancient physician Galen, but never in a pharmacological context.

[3] *On the Composition of Drugs according to Places* (*De compositione medicamentorum secundum locos*, henceforth: *De comp. med. sec. loc.*) is found in volumes 12 and 13 of Kühn's edition; *On the Composition of Drugs according to Kind* (*De compositione medicamentorum per genera*, henceforth *De comp. med. per gen.*) is found in volume 13 of Kühn's edition, which includes a translation into Latin (Kühn 1821–1833). No full translation in a modern language is available. Throughout the remainder of this article, for every Galenic and pseudo-Galenic passage, I shall provide a reference to Kühn's edition as well as to a more recent edition, if available. For a general introduction to Galenic pharmacology, see Vogt 2009.

[4] For an edition and translation into French, see Jouanna-Bouchet 2016. Ianto Jocks has produced an excellent English translation as part of his PhD thesis (Jocks 2020).

Remedies, which takes us from the ancient Mediterranean to eighteenth century New Spain, and beyond.[5]

In a more limited sense, the Greek word *sunthesis* also referred to a part of a written recipe: that in which the ingredients and their quantities were outlined. Indeed, the ideal written recipe contained four parts: the *prographē*, its heading; the *epangelia*, the indication of the compound's properties; the *sunthesis*, also referred to as *summetria*, its ingredients and their proportions; and the *skeuasia*, the mode of preparation used for the remedy.[6] The recipe below is a typical example:

> Ἑρμοφίλου θαλασσερὸς κολλύριον ἐπιτετευγμένον πρὸς ὑποχύσεις καὶ πᾶσαν ἀμβλυωπίαν, ποιεῖ καὶ πρὸς ἀρχὰς ὑποχύσεως. ἔστι δὲ εὐωδέστατον φάρμακον. ♃ Καδμείας δραχμὰς ιστ′. μέλανος Ἰνδικοῦ < ιστ′. πεπέρεως λευκοῦ δραχμὰς η′. ἰοῦ < δ′. ὀπίου Μηδικοῦ < δ′. ὀποβαλσάμου < δ′. κόμμεως < ιβ′. ὕδατι ἀναλάμβανε. ἡ χρῆσις μεθ' ὕδατος.
>
> *Kollurion* (eye-remedy) called *thalasseros* of Hermophilus, against cataract and any dim-sightedness. It also works against incipient cataract. It is a most fragrant drug: 16 drachms of calamine; 16 drachms of indigo (literally: Indian black); 8 drachms of white pepper; 4 drachms of verdigris; 4 drachms of Median opium; 4 drachms of opobalsamum; 12 drachms of gum. Mix with water. Use with water.[7]

The *prographē* states that the recipe is for a *kollurion*, a type of paste-like eye-remedy that was shaped like little breads and often stamped—we will encounter more *kolluria* in this paper.[8] The *prographē* also gives the name of the drug (*thalasseros*, linked to the sea) as well as its warrant, an otherwise unknown Hermophilus.[9] Then follows the *epangelia*, which lists the eye complaints against which the remedy is effective. The longest part of the recipe is the *sunthesis*, a list of ingredients presented in descending order of quantity until the final ingredient, gum, which is a form of excipient. This recipe, like many others for *kolluria*, involve a mix of vegetable (pepper, opium, opobalsamum, and gum) and mineral (calamine, indigo, for indigo was considered a mineral in antiquity, and verdigris) ingredients.[10] It also included two geographically qualified ingredients, "Indian black" (indigo) and Median opium, as was common in ancient medicine.[11] The recipe's *skeuasia* is extremely brief ("mix with water"), as many steps involved in the preparation of such a remedy would not have needed to be spelled out. As such, it was not necessary to specify in every single recipe that ingredients needed to be crushed finely.

---

[5] De Vos 2021.

[6] See Fabricius 1972, 24–30.

[7] Gal. *De comp. med. sec. loc.* 4.8, 12.781K.

[8] On ancient *kolluria*, see, e.g., Jackson 1996, 2228–31; Baker 2011; Pardon-Labonnelie 2011.

[9] See von Staden 1989, 584; Keyser 2008.

[10] Dioscorides, like other ancient authors, included indigo among metals: Diosc. 5.92, where indigo is described after lapis lazuli and before yellow ochre. Dioscorides, however, noted that some indigo was an excretion from Indian reeds, while other forms of indigo were scum that occurred in the processing of purple murex in copper cauldrons. The edition of Dioscorides followed throughout is Wellmann 1907–1914. For an English translation, see Beck 2020.

[11] On this phenomenon, see Totelin 2009, 2016.

The *sunthesis* in this recipe includes precise quantities, expressed in drachms, one of the most common ancient weights.[12] Other recipes did not give such detail, simply listing the ingredients and relying on the reader's knowledge of compounding.[13]

The compounding of remedies was a complex matter, which required care and was fraught with issues. Such issues could lead to the production of ineffective—even dangerous—remedies. As Heinrich von Staden noted, "The sources of pharmacological failure, errors, and inefficacy are numerous, according to Galen. Some are epistemological, some methodological, some logical, some linguistic, some ontological, some moral, some educational, some cultural."[14] In particular, issues with ingredient quantities could lead to the creation of ineffective remedies. Galen noted that it was common to find errors and willful alterations of the quantities of ingredients in pharmacological manuscripts.[15] This was exacerbated by the existence of numerous ancient metrological systems, between which exact equivalences could be difficult to establish. Galen spent much time explicating such systems.[16] More fundamentally, ancient pharmacology faced the issue of testing, which could only be done on human or animal bodies. This issue affected the testing of simple drugs, but it became even more complex in the testing of compound remedies, the effects of which were more (and sometimes less) than the sum of the individual ingredients. While ancient pharmacists did not hesitate to test new drugs on prisoners condemned to death, on friends, on non-human animals, or on themselves, they nonetheless expressed some concerns about this absolute limit of the medical art.[17] Thus, Galen wrote that,

> ὅτι ἡ πεῖρα ἐπισφαλής ἐστιν οὐδεὶς ἀγνοεῖ, τοῦτο δὲ πάσχει διὰ τὸ ὑποκείμενον περὶ ὅ ἡ τέχνη ἐστίν. οὐ γὰρ δέρματα καὶ ξύλα καὶ πλίνθοι, ὥσπερ τῶν ἄλλων τεχνῶν ὕλη τῆς ἰατρικῆς ἐστιν, ἐν οἷς ἔξεστι πειρᾶσθαι ἄνευ κινδύνου, ἀλλ' ἐν ἀνθρωπείῳ σώματι, ἐφ᾽οὗ πειρᾶσθαι τῶν ἀπειράστων οὐκ ἀσφαλές.

---

[12] For an introduction to numbering and measuring in the ancient world, see Richardson 2005.

[13] For a discussion of the ways in which quantities are expressed in the recipes of the Hippocratic Corpus, see Totelin 2009, 238–42.

[14] von Staden 1997, 61.

[15] See, e.g., *On Antidotes* (*De antidotis,* henceforth *De antid.*) 1.5, 14.31K. See von Staden 1997, 68; Hanson 2008, 49.

[16] See, e.g., *De comp. med. per gen.* 6.8, 13.893K for a discussion of the Roman *litra* and its equivalence in other metrological systems; see von Staden 1997, 71.

[17] On the experimentation of drugs on people condemned to death, see, e.g., Gal. *De antid.* 1.1, 14.2K (Attalus III and Mithradates VI tested remedies on people condemned to death); Gal. *De antid.* 2.7, 14.150K (the story of how the physician Zopyrus enjoined King Mithradates VI to test a remedy on a prisoner). On self-experimentation, see, e.g., Gal. *De sectis ad eos qui introducuntur.* 2, 1.67–68K = Helmreich 1893, 3–4). On giving remedies to test to friends, see, e.g., Gal. *De comp. med. per gen.* 3.2, 13.599K. On testing remedies on animals, see, e.g., Galen, *De theriaca ad Pisonem* (henceforth *De ther. ad Pis*) 2, 14.215K = Boudon-Millot 2016, 6. On the testing of drugs, see Grmek and Gourevitch 1985; Grmek 1997.

> Everyone is aware that experience (*peira*) is misleading, and everyone endures this because of the foundation on which the art (*technē*) [of medicine] is based. For the material of the medical art is not leather, wood, or bricks, as in the case of other arts. On these materials, it is possible to experiment (*peirasthai*) without danger, but not on the human body, on which it is not safe to experiment with untested things.[18]

In the remainder of this paper, I examine one specific aspect of compounding—namely, that of not using the ingredients listed in an original recipe. This aspect has both a positive and a negative facet. The positive is the art of substitution (*quid pro quo* in Latin), which—at least in theory—relies on sound knowledge of ingredients' powers (*dunameis*) to create a remedy that will remain effective. In ancient, particularly Galenic, pharmacology, the *dunamis* of a drug, its potential to act on the body, was a product of its qualities (i.e., hot, wet, dry, and cold). These qualities were "elementary" in that they were linked to the theory of the four elements (earth, water, air, and fire), which, in turn, was related to the theory of four humors (phlegm, blood, yellow bile, and black bile).[19]

The negative facet of the art of compounding is fraud or adulteration, which could lead to the production of ineffective or dangerous medicines and against which methods of detection—sometimes complex—were developed.[20] We shall explore the boundary between substitution and fraud and the roles that they played in further developing the pharmacological art of *sunthesis*.[21] We shall focus as far as possible on mineral ingredients, which will also lead us to make several observations on other ancient crafts that made use of them.

2.2 The Ancient Art of Pharmacological Substitutions

Ancient pharmacologists often discussed difficulties in procuring ingredients for their drugs. Many substances, especially vegetable ones, were only available at certain times of the year; others had to be transported over long distances.[22] Ancient transport was often unreliable: sea routes were inaccessible for long periods of the year; transport by road was slow; and war often disrupted it. Furthermore, ancient cities tended to siphon resources, which often made it difficult to source drugs in the

[18] Gal. *In Hippocratis de humoribus librum commentarii* 1.7, 16.80K.

[19] Vogt 2009, 307–10.

[20] For a definition of pharmacological fraud, see Stieb 1966, 3: 'Lacking defined standards, the term 'adulteration' may be considered to have included, always in association with intent or neglect: secret addition of extraneous substances, whether deleterious or merely to increase bulk and weight; the subtraction of constituents usually considered part of the substance; deterioration from an accepted standard of strength or quality. Adulteration may also include preparing a substance to conceal its defects and to make it appear better than it is.'

[21] Rudolf Schmitz noted that some substitutions could in fact be adulterations. He also noted the overlap between ancient lists of synonyms and lists of substitutes. Schmitz 1998, 394; see also Touwaide 2012.

[22] I explore these issues in Totelin 2025, forthcoming.

countryside, an issue acknowledged in the preface to the first book of Pseudo-Galenic *Remedies Easily Procured*.[23] To address these issues, ancient medical authors sometimes indicated how to replace ingredients that were not available. Examination of the use of the phrase "if this is not available" (*mē parontos*) in Galen's pharmacological works can shed some light on this process of substitution and when it was necessary.[24] Thus, Galen sometimes acknowledged that certain drugs might be readily available in some locations of the Roman empire but not elsewhere: for example, the compound remedy "with papyrus" (*dia chartou*) may not be available in the countryside, in which case the physician must devise an alternative; castor oil will be readily available in Egypt, but may need to be replaced by aged olive oil elsewhere; Pontic wax may be replaced with Tyrrhenian wax; a type of must (young wine) common in Galen's region of Asia could be substituted by honey wine.[25] While we have very little precise information about the prices of ancient drugs, it is clear that Galen often allowed for substitutions of expensive products with cheaper and/or more generic ones: cassia instead of the very expensive cinnamon; any bird fat in place of goose fat; a tawny wine rather than the famous Falernian wine; "Median juice," the sap of asafoetida, instead of Cyrenian juice—that is, the juice of the famous silphium plant, which may have been nearly extinct by the first century CE; or deer marrow instead of seal fat.[26] One of Galen's substitutions is linked to the weather: snow instead of very cold water.[27] Finally, Galen sometimes allowed one perfume to be replaced by another.[28] Many of these substitutions, then, were economically motivated, taking "economy" in a broad sense that encompasses pricing as well as availability. Galen did not explicitly outline the theoretical bases on which he made these substitutions, although they are typically self-evident to the modern reader: one oil replaces another; wax for wax; sweet wine for sweet wine; sweet-scented bark for sweet-scented bark; fat for fat; pungent sap for pungent sap; cold water for snow; perfume for perfume. The ingredient used as substitute, then, is understood to possess a *dunamis* similar to that of the original, if perhaps to a slightly lesser degree.

The types of comments that Galen and other ancient medical authors made regarding substitutions evolved into a Late Antique and Medieval genre of pharmacological treatise called *Peri antiballomenōn* in Greek and *Quid pro quo* in Latin.[29]

---

[23] Pseudo-Galen, *De rem. parab.* 1, pr., 14.313K.

[24] See also Nutton 2008, 213.

[25] Remedy with papyrus: *De comp. med sec. loc.* 1.8, 12.466K. Castor oil: *De comp. med. sec. loc.* 2.1, 12.510K; *De comp. med. per gen.* 6.8, 13.896K. Wax: *De comp. med. sec. loc.* 8.1, 13.119K. Must: *De comp. med. sec. loc.* 6.3, 12.915K.

[26] Cassia: *De comp. med. sec. loc.* 3.1, 12.606K. Bird fat: *De comp. med. sec. loc.* 1.2, 12.424K. Wine: *De comp. med. per gen.* 2.11, 13.513K. Median juice: *De comp. med. per gen.* 3.2, 13.567K; *De antid.* 2.15, 14.201K (this is in a poem of Damocrates). Deer marrow: *De comp. med. per gen.* 7.12, 13.1021K

[27] *De comp. med. sec. loc.* 2.1, 12.508K.

[28] E.g., *De comp. med. sec. loc.* 3.1, 12.604K; *De comp. med. per gen.* 4.6, 13.715K.

[29] For a list of medieval manuscripts including *Quid pro quo* treatises, see Riddle 1974, 175, n. 176.

These treatises, as John Riddle astutely observed, "assisted memory; [they] could not replace reliance on empirical observation."[30] An early example of the genre is the pseudo-Galenic *Peri antiballomenōn* (*On Substitutions*), which lists 369 substitutions of mostly vegetable, but also animal or mineral substances.[31] While the treatise is not authentically Galenic, it may be fairly early in date—it is summarized by the seventh-century CE medical author Paul of Aegina.[32]

The treatise opens with a relatively lengthy prologue, which serves as a dedication to a certain Diogenianus. It provides a definition of the topic—"we call substitutes the drugs that are used instead of others"[33]—and emphasizes that the art of substitution should rely on a sound knowledge of the powers (*dunameis*) of simples, as otherwise physicians may substitute drugs with others that are wholly different in their powers. It then gives an anecdote illustrating the utility of knowing appropriate substitutes:

> ἔσται δέ μοι ὁ λόγος πρὸς σὲ ἀληθὴς, ἀναμνησθέντι τοῦ ποτέ μοι συμβάντος ἐν Ἀλεξανδρείᾳ. εὐθέως παραγενομένου ἐκεῖσε γύναιόν μοι προσῆλθε μέλλον ἀποθνήσκειν, ἔχον διάθεσιν ἰσχυρὰν, ἧς τὴν διήγησιν οὐ πρόκειται νῦν εἰπεῖν. καὶ ζητοῦντός μου λυχνίδα, ἵνα τὸ δέον αὐτῇ προσάξω φάρμακον, εἰ μὴ εὐθέως εὗρον ἀκανθίου σπέρμα, ἔμελλεν ἀπόλλυσθαι παραχρῆμα τὸ γύναιον. ὡς δ' εὑρέθη τὸ ἀνάλογον τῇ λυχνίδι, εὐθέως ἐχρησάμην αὐτῷ καὶ συνῆλθεν εἰς ταὐτό. τῇ δὲ ἑξῆς ἡμέρᾳ παραγενόμενοί τινες τῶν θεωμένων αὐτὴν προτέρων ἰατρῶν ἠξίουν ἀκοῦσαι τὸ δέον φάρμακον. εὐθέως οὖν ἀκούσαντες παρεκάλεσαν γραφῆναι αὐτοῖς τὸν περὶ τῶν ἀντεμβαλλομένων λόγον.
>
> Let me tell you an authentic story, which as I recall happened to me in Alexandria. A little woman came to me, about to die, with a serious condition that I will not recount here. I sought to find *luchnis* so that I might provide her with the drug that she needed. The little woman would have died immediately had I not quickly found the seed of a thistle (*akanthion*). Since it was found to be the equivalent (*to analogon*) of *luchnis*, we used it immediately, and it gave the same result. The next day, some of the doctors who had examined her before came to me, wanting to hear what drug she had required. Having heard this, they entreated me promptly to write a book about substitutes.[34]

This story claims to be true, authentic (*alethēs*), and it is therefore an ideal introduction to a treatise on the art of substituting, an art that sets itself apart from that of fraud. While highly seductive, however, this story is, in fact, particularly unhelpful. Indeed, it offers no information as to the reasoning that led to the substitution—the

---

[30] Riddle 1992, 14.

[31] Fischer (2018) examines the manuscripts, editions and translations of this treatise. Touwaide (2012) provides an analysis of the types of substitutions found in the treatise: 1) simple substitutions; 2) bidirectional substitutions; 3) chains of substitutions.

[32] Paul's shortened version of the treatise includes the same introductory anecdote and 228 substitutions: Paul of Aegina 7.25 (Heiberg 1924, 401–08). For an English translation of Paul's text, see Adams 1947, 604–08. There exists an Arabic version (*Kitab Abdāl al-adwiya*): Ullmann 1970, 50, no 55.

[33] Ps.-Gal. *De succedaneis* (henceforth *De succed.*) Pr., 19.721K. This text is found in volume 19 of Kühn's edition.

[34] Ps.-Gal. *De succed.*, pr. 19.723K. The passage is translated and discussed by Gourevitch (2016, 257–8) and Touwaide (2012, 19).

reader never gets to know what accounts for the similarity between *luchnis* and *akanthion* with respect to their powers. It may be the case, however, that the author expected his reader to know this already. Galen described both the seed of *luchnis* and the root of *akanthion* as warming.[35] The absence of detail regarding the woman's condition further renders the anecdote pointless from a medical point of view.[36]

The list of substitutes that follows continues in the same vein: ingredients are presented as alternatives without any explanation as to what makes them suitable. No reference is made to quantities either, seemingly implying that the same quantity of the substitute will work effectively, or, at least, without deleterious effects.

Alain Touwaide suggested that this treatise might have been produced in an affluent context, in a place "with little direct contact with the natural environment," perhaps a city such as Rome or Alexandria.[37] A more detailed examination of the substitutions involving mineral products in the treatise will help us to refine that conclusion and provide further information on the principles of ancient substitutions (see Table 2.1). The translations suggested in Table 2.1 are based on those found in Lily Beck's indices to her translation of Dioscorides' *Materia Medica*, a key pharmacological text through antiquity and beyond; they are, however, tentative, as the identification of ancient minerals is fraught with difficulties.[38]

Minerals appear frequently in the pseudo-Galenic list of substitution, some in their natural state and some transformed through manufacture.[39] In total, 57 of the 369 substitutions in the treatise involve minerals. Most minerals mentioned in *On Substitutions* also appear in Dioscorides' *Materia Medica*, but some do not.[40] In particular, several stones (Table 2.1, nos. 26–31) and geographically qualified earths (nos. 12–14, 23) occur in *On Substitutions* but not in Dioscorides.

Most frequently, *On Substitutions* suggests mineral products as alternatives for other mineral products.[41] Furthermore, these mineral–mineral substitutions are usually like for like: that is, a salt replaces a salt (nos. 1–2); an earth an earth (nos. 10–15); a stone a stone (nos. 26–32); and an ore an ore (nos. 33–37). Given that the ancient theory of *dunameis* was based on an understanding of the elementary qualities of drugs, it seems reasonable that like would be substituted for like in this manner.[42]

---

[35] Gal. *De simplicium medicamentorum temperamentis et facultatibus* (henceforth *De simpl. med. temp. et fac.*) 6.1.15, 11.818K (*akanthion*) and 7.11.22, 12.65K (*luchnis*). This text is found in volumes 11 and 12 of Kühn's edition.

[36] Dioscorides recommended *akanthion* against tetanus; the *luchnis* that is used for wreaths against scorpion bites; and wild *luchnis* to draw out bilious matter and as a deterrent to scorpions (Diosc. 3.16 and 3.100–101).

[37] Touwaide 2012, 31–2.

[38] Beck 2020, 634–7.

[39] Touwaide 2012, 28.

[40] For a study of this key text, see Riddle 1985.

[41] Touwaide 2012, 30–31.

[42] For reasons that are not clear to me, *sandarachē* (arsenic sulfide) appears very frequently in the list: no. 3, 5, 8, 17, 25, 51, 54.

Table 2.1 Substitutions involving minerals in pseudo-Galen's on substitutions

Reference number	Greek text	Annotated translation	References in Dioscorides
1.	ἀντὶ ἁλὸς ἀμμωνιακοῦ, ἅλας Καππαδοκικόν.	Instead of ammoniac salt [mineral], Cappadocian salt [geographically qualified mineral]	5.109; not in Diosc.
2.	ἀντὶ ἁλὸς Καππαδοκικοῦ, ἅλας ἀμμωνιακόν.	Instead of Cappadocian salt [geographically qualified mineral], ammoniac salt [mineral]	Not in Diosc.; 5.109
3.	ἀντὶ ἁλὸς ἄνθους, σανδαράχη.	Instead of salt inflorescence [mineral], arsenic sulfide [mineral]	5.112; 5.105
4.	ἀντὶ Ἀρμενίου, μέλαν Ἰνδικόν.	Instead of azurite [geographically qualified mineral], indigo [geographically qualified mineral, as indigo was classified as a mineral]	5.90; 5.92
5.	ἀντὶ ἀρσενικοῦ, σανδαράχη.	Instead of yellow orpiment [mineral], arsenic sulfide [mineral]	5.104; 5.105
6.	ἀντὶ ἀσβέστου, ἡ εἰς τὰ βάφια ἄκανθα.	Instead of unslaked lime [mineral], thistle used for dyeing [vegetable]	5.115; not in Diosc.
7.	ἀντὶ ἀσβέστου ὃ λέγεται τίτανος, ἀδάρκης.	Instead of unslaked lime [mineral], *adarces* [mineral]	5.115; 5.119
8.	ἀντὶ Ἀσίου λίθου, λίθος γαγάτης ἢ ἅλες ἀμμωνιακοὶ ἢ σανδαράχη.	Instead of Assian stone [geographically qualified mineral], lignite [mineral], or ammoniac salts [mineral], or arsenic sulfide [mineral]	5.124; 5.128; 5.109; 5.105
9.	ἀντὶ ἀσφάλτου, πίσσα ὑγρὰ βρυττία ἢ γῆ ἀμπελῖτις.	Instead of asphalt [mineral], liquid Bruttian pitch [vegetable] or bituminous earth [mineral]	1.73; 1.72.5; 5.160
10.	ἀντὶ γῆς ἁπαλῆς ἢ ἀμπελίτιδος, μολυβδαίνα.	Instead of soft or bituminous earth [mineral], galena [mineral]	5.160; 5.85
11.	ἀντὶ γῆς ἀστέρος, γῆ κιμωλία.	Instead of *aster* (Samian earth) [mineral], Cimolian earth [geographically qualified earth]	5.123; 5.126
12.	ἀντὶ γῆς Ἐρετριάδος, τίτανος Θηβαϊκός.	Instead of Eretrian earth [geographically qualified mineral], Theban white earth [geographically qualified earth]	5.152; not in Diosc.

(continued)

Table 2.1 (continued)

Reference number	Greek text	Annotated translation	References in Dioscorides
13.	ἀντὶ γῆς Κρητικῆς, γῆ ἐρετριάς.	Instead of Cretan earth [geographically qualified mineral], Eretrian earth [geographically qualified mineral]	Not in Diosc.; 5.152
14.	ἀντὶ γῆς Μεγάρας, ἁλὸς ἄχνη.[a]	Instead of Megarian earth [geographically qualified mineral], salt froth [mineral]	Not in Diosc.; 5.110
15.	ἀντὶ γῆς Σαμίας, λευκογράφις Αἰγυπτία.	Instead of Samian earth [geographically qualified mineral], Egyptian pipe clay [geographically qualified mineral]	5.153; 5.134
16.	ἀντὶ διφρυγοῦς, μίσυ ὀπτὸν ἢ λίθος φρύγιος ἢ χαλκὸς κεκαυμένος ἢ λίθος πυρίτης.	Instead of pyrites [mineral], roasted copper ore [mineral], or calcinated copper [mineral], or copper pyrites [mineral]	5.125; 5.100; 5.76; 5.125
17.	ἀντὶ θείου ἀπύρου, σανδαράχη.	Instead of unburnt sulfur [mineral], arsenic sulfide [mineral]	5.107; 5.105
18.	ἀντὶ ἰοῦ σιδήρου, λιθάργυρος ἢ σκωρία σιδήρου.	Instead of iron rust [mineral], litharge [mineral], or iron slag [mineral]	5.80; 5.87; 5.80
19.	ἀντὶ ἰοῦ χαλκῆς, χολὴ γυπὸς ἢ πέρδικος.	Instead of copper rust [mineral], bile of vulture or of partridge [animal]	Cf. 5.79; Cf. 2.78.2, where bile of eagle (ἀετός) is mentioned; 2.78.2
20.	Ἀντὶ καδμίας, λευκογραφὶς Αἰγυπτία.	Instead of calamine [mineral], Egyptian pipe clay [geographically qualified mineral]	5.74; 5.134
21.	ἀντὶ κεραυνίου, λευκογραφίς.	Instead of truffle (?) [vegetable?], pipe clay [mineral]	Not in Diosc.; 5.134
22.	ἀντὶ κινναβάρεως, ῥοδοειδές.	Instead of cinnabar [mineral], rose-like remedy (?) [compound?]	5.94; occurs as an adjective qualifying the names of ingredients in Dioscorides but not as a single drug.
23.	ἀντὶ κισσήρεως, γῆ Κρητική.	Instead of pumice stone [mineral], Cretan earth [geographically qualified mineral]	5.108; not in Diosc.
24.	ἀντὶ κοραλλίου, σύμφυτον ἢ μῶλυ.	Instead of coral [mineral, as it was considered in antiquity], comfrey [vegetable] or *molu* [vegetable]	5.121; 4.9; 3.47

(continued)

Table 2.1 (continued)

Reference number	Greek text	Annotated translation	References in Dioscorides
25.	ἀντὶ λημνίας σφραγίδος, σανδαράχη.	Instead of Lemnian seal [geographically qualified mineral], arsenic sulfide [mineral]	5.97; 5.105
26.	ἀντὶ λίθου Φρυγίου, λίθος ἀργυρίτης ἢ πυρίτης.	Instead of Phrygian stone [geographically qualified mineral], stone containing silver ore [mineral] or copper pyrites [mineral]	5.123; not in Diosc.; 5.125
27.	ἀντὶ λίθου ἀχάτου, λίθος σαρδόνυξ.	Instead of the agate stone [mineral], the sardonyx stone [mineral]	Not in Diosc.; not in Diosc.
28.	ἀντὶ Χαλκηδονίου, λίθος κυάνεος.	Instead of the Chalcedonian stone [geographically qualified mineral], the deep blue stone [mineral]	Not in Diosc.; not in Diosc.
29.	ἀντὶ λίθου ὑακίνθου, λίθος βηρύλλιος.	Instead of the aquamarine stone [mineral], the beryl stone [mineral]	Not in Diosc.; not in Diosc.
30.	ἀντὶ λίθου σμαράγδου, λίθος ἴασπις.	Instead of the emerald stone [mineral], the jasper stone [mineral]	Not in Diosc.; 5.142
31.	ἀντὶ λίθου σπόγγου, λίθος ὁ ἐξουρούμενος.	Instead of the stone found in sponges [mineral], the stone that passes with urine [mineral]	5.144; not in Diosc.
32.	ἀντὶ μαγνήτου, λίθος Φρύγιος ἢ αἱματίτης.	Instead of the magnet stone [mineral], the Phrygian stone [geographically qualified mineral] or the hematite [mineral]	5.130; 5.123; 5.126
33.	ἀντὶ μίσυος ὀπτοῦ, διφρυγές.	Instead of roasted copper ore [mineral], copper pyrites [mineral]	5.100; 5.125
34.	ἀντὶ μίσυος Κυπρίου, ὤχρα Κύπρια.	Instead of Cypriot copper ore [geographically qualified mineral], Cypriot yellow ochre [geographically qualified mineral]	5.100; 5.93
35.	ἀντὶ μισυδίου, ὤχρα.	Instead of copper ore (?) [mineral], ochre [mineral]	5.100; 5.93
36.	ἀντὶ μολυβδαίνης, λιθάργυρον.	Instead of galena [mineral], litharge [mineral]	5.85; 5.87
37.	ἀντὶ μολύβδου κεκαυμένου, ψιμμίθιον.	Instead of calcinated lead [mineral], white lead [mineral]	5.81; 5.88

(continued)

Table 2.1 (continued)

Reference number	Greek text	Annotated translation	References in Dioscorides
38.	ἀντὶ νίτρου ἐρυθροῦ, ναρδόσταχυς.	Instead of red soda [mineral], spikenard [vegetable]	5.113; 1.7
39.	ἀντὶ νίτρου, ἀφρόνιτρον ἢ ἅλας ὀπόν.	Instead of soda [mineral], foam of soda [mineral] or roasted salt [mineral]	5.113; 5.113; 5.109
40.	ἀντὶ πίσσης βρυττίας ὑγρᾶς, ἄσφαλτος, πίσση ἐγχώριος περίσση.	Instead of liquid Bruttian pitch [geographically qualified vegetable product], asphalt [mineral], excellent native pitch [vegetable]	1.72.5; 1.73; 1.72
41.	ἀντὶ πομφόλυγος, καδμία κεκαυμένη.	Instead of zinc oxide [mineral], calcinated calamine [mineral]	5.75; 5.74
42.	ἀντὶ σηπίας ὀστράκου, κίσσηρις.	Instead of cuttlefish [animal], pumice stone [mineral]	2.21; 5.108
43.	ἀντὶ σκωρίας μολύβδου, ἕλκυσμα.	Instead of lead dross [mineral], silver dross [mineral]	5.82; 5.86
44.	ἀντὶ σκωρίας Κυπρίας, μελαντηρία Αἰγυπτική.	Instead of Cypriot slag [copper slag] [geographically qualified mineral], shoemaker's black [mineral]	5.76; 5.101
45.	ἀντὶ σποδίου, πομφόλυξ.	Instead of zinc oxide [mineral], zinc oxide [mineral]	5.75; 5.75
46.	ἀντὶ σποδοῦ Κυπρίας, σποδὸς φύλλων ἐλαίας.	Instead of Cypriot zinc oxide [geographically qualified mineral], the ashes of olive leaves [vegetable]	5.75; 1.105
47.	ἀντὶ στίμμεως Κοπτικοῦ, λεπὶς χαλκοῦ.	Instead of Coptic antimony [geographically qualified mineral], copper flake [mineral]	5.84 (but no mention of Coptos); 5.78
48.	ἀντὶ στυπτηρίας, ἅλας ὀρυκτόν.	Instead of alum [mineral], quarried salt [mineral]	5.106; 5.109
49.	ἀντὶ στυπτηρίας σχιστῆς, σίδιον.	Instead of split alum [mineral], pomegranate peel [vegetable]	5.106; 1.110
50.	ἀντὶ σηρικοῦ, λιθάργυρος.	Instead of red pigment [mineral], litharge [mineral]	Not in Diosc.; 5.87
51.	ἀντὶ σφέκλης, σανδαράχη.	Instead of wine tartrates [vegetable], arsenic sulfide [vegetable]	Not in Diosc.; 5.105
52.	ἀντὶ σώρεως, λιθάργυρος διφρυγὲς ἢ μελαντηρία.	Instead of melanterite [mineral], litharge [mineral], copper pyrites [mineral], or shoemaker's black [mineral]	5.102; 5.87; 5.125; 5.101

(continued)

Table 2.1 (continued)

Reference number	Greek text	Annotated translation	References in Dioscorides
53.	ἀντὶ τιτάνου, γῆ Ἐρετρία.	Instead of white earth [mineral], Eretrian earth [geographically qualified mineral]	Not in Diosc.; 5.152
54.	ἀντὶ φέκλης, σανδαράχη.	Instead of wine tartrates [vegetable], arsenic sulfide [mineral]	Not in Diosc.; 5.105
55.	ἀντὶ χαλκάνθης, λεπὶς χαλκοῦ.	Instead of vitriol [mineral], flake of copper [mineral]	5.98; 5.78
56.	ἀντὶ ψιμμυθίου, μόλυβδος κεκαυμένος ἢ σκωρία μολίβδου.	Instead of white lead [mineral], calcinated lead [mineral] or lead dross [mineral]	5.88; 5.81; 5.82
57.	ἀντὶ ὤχρας, μίσυ Κύπριον.	Instead of yellow ochre [mineral], Cypriot copper ore [geographically qualified mineral]	5.93; 5.100

Translations are based on those found in Beck's indices to her translation of Dioscorides' *Materia Medica* (Beck 2020, 634–7)

[a]The Kühn edition has the text 'ἀλόης ἄχνη', which I have emended here

Some cases are reported, however, in which plants were suggested as substitutes for minerals (e.g., comfrey for coral, which was considered to be a stone in antiquity, no.24; spikenard for red soda, no. 38; pomegranate peel for split alum, no. 49); minerals as substitutes for substances derived from vegetable products (arsenic sulfide for wine tartrates, no. 51); animal products as substitutes for mineral ones (bird bile for copper rust, no. 19); and mineral products as substitutes for animal ones (pumice stone for cuttlefish, no. 42).[43]

Given that *On Substitutions* is merely a list that offers no justification for what makes one product a suitable alternative for another, answers must be sought elsewhere, primarily in ancient treatises on simples, such as Dioscorides' *Materia Medica* or Galen's *On the Powers of Simple Medicines*. These treatises are not always as systematic as one might wish in their descriptions of drugs' *dunameis*, but they sometimes give us indications that explain some of the substitutions outlined above. Thus, Galen noted that both the pumice stone and the cuttlefish bone were drying; [44] and Dioscorides indicated that both pomegranate peel and split alum were warming.[45] It might be dangerous, however, to expect that a thorough reflection on

[43] We are faced here with several differences between ancient and modern classifications of natural products. For the ancients, coral was a plant that had transformed into stone. It is unclear how the ancients would have classified wine tartrates. The identification of ancient minerals is fraught with issues, and the identifications given here are only suggestive.

[44] Gal. *De Simpl. Med. Temp. et Fac.* 9.2.15, 12.205K (pumice) and 11.1.27, 12.347K (cuttlefish).

[45] Diosc. 5.106; 1.110.

dunameis was at play in all substitutions. Moreover, the way in which *dunameis* were determined might at times have been based on rather basic observations (see below for more on sensory observation). For example, one cannot help but notice a resemblance between pumice stone and cuttlefish bone. Did this resemblance give rise to the notion that they shared similar *dunameis*; or was this observation merely a starting point? Another example is the substitution of coral with comfrey. Ancient treatises on simples do not tell us whether these substances had similar *dunameis*. Dioscorides, however, noted that the root of comfrey was red, and gardeners might notice a slight resemblance between that root and coral. This visual resemblance may thus explain why comfrey was deemed a suitable alternative for coral.[46] While further research is required to determine this, it may be the case that color was often used to determine the *dunameis* of drugs and hence their appropriate substitutions.

As in the example of comfrey as a substitute for coral, the suggested substitutes in *On Substitutions* often appear to be more readily available than the original products. However, some of the proposed alternatives proposed may have been rarer—and presumably more expensive—than the original product, although it is important to note that no substance is intrinsically luxurious. For example, we can note that the rare spikenard was suggested as a substitute for red soda (no. 38) and silver dross for lead dross (no. 43).

Many of the mineral ingredients in the pseudo-Galenic list are geographically qualified. Thus, we find references to the "Armenian" [pigment] (azurite); Assian stone, Cappadocian salt, Chalcedonian stone, Cimolian earth, Coptic antimony, Cretan earth, Cypriot copper ore, Cypriot slag, Cypriot yellow ochre, Egyptian pipe clay, Egyptian shoemaker's black, Eretrian earth, "Indian" [pigment] (indigo), Lemnian seals, Megarian earth, Phrygian stone, Samian earth, and Theban white clay. With the exception of India, the regions whence these products came from—or allegedly came—were part of the Roman empire. The geographical epithets attached to ingredients in antiquity were not always an exact indication of their origin, but they still reflected the global nature of the ancient drug trade. Often a geographically qualified product replaced another one. For instance, instead of Samian earth, one could use Egyptian pipe clay. Depending on one's position in the Mediterranean, Egyptian products might have been easier to procure than Samian ones. In any case, the author placed greater value on sourcing products that allegedly originated from specific locations.

Several of the pseudo-Galenic substitutions involving minerals are particularly interesting for our purpose because they touch upon the boundary between substitution and fraud. For instance, *On Substitutions* suggested using either Bruttian pitch or a type of bituminous earth as alternatives for asphalt. Pliny the Elder mentioned that asphalt was often adulterated with vegetable pitch, likely because the two products were similar in appearance.[47] Another example is the suggested use of pome-

46 Diosc. 4.9.

47 Plin. *Historia naturalis* (henceforth *HN*) 25.180. The edition followed is André 1974. For an English translation, see Jones and Andrews 1956.

granate peel (*sidion*) in place of alum (*stupteria*). While, as mentioned above, both products were considered to be warming, we may note that, according to Pliny, pomegranate juice was used to test the authenticity of fluid alum from Melos:

> liquidi probatio ut sit limpidum lacteumque, sine offensis fricandi, cum quodam igniculo coloris. hoc phorimon vocant. an sit adulteratum, deprehenditur suco Punici mali; sincerum enim mixtura ea non nigrescit.

> The proof [of authenticity] of the liquid type is that it should be clear and milky, without grit when it is rubbed, and with a certain spark of color. They [the Greeks] call it *phorimon*. One can detect whether it is adulterated by means of the juice of pomegranate, for when pure it does not turn black when mixed with it [the juice].[48]

Such tests to determine products' authenticity may have played a role in determining the *dunameis* of ingredients. We now turn more fully to the topic of adulteration.

2.3 Adulteration

Beginning in the fourth century BCE with Theophrastus, who commented on the adulteration of Judean balsam, ancient technical writers expressed considerable concern over the problem of pharmacological fraud.[49] Galen and Pliny accused various people active in the ancient pharmacological trade of fraudulent behavior and secrecy,[50] with Pliny going so far as to say that physicians' workshops were manifestations of human frauds (*frauds hominum*)—that is, that the entire medical art was fraudulent.[51] He also singled out ready-made plasters and eye-remedies, sold in the "Seplasia" (the drug market), as manifestations of corruption and fraud.[52] To help their readers detect fraud, Pliny and Dioscorides recorded the many ways in which drugs could be adulterated and described various methods of detection (see

---

[48] Plin. *HN* 35.184. The edition followed is Croisille 1985. For an English translation, see Rackham 1952.

[49] Theophrastus, *Historia plantarum* 9.6.2; see also Diosc. 1.19.2; Plin. *HN* 12.119. The edition of Theophrastus' *Enquiry into Plants* followed here is Amigues 2006. That of book 12 of Pliny's *Natural History* is André 1970. See Rackham 1945 for an English translation. On ancient balsam, see Manolaraki 2015. On the topic of pharmacological fraud in antiquity, see Schmidt 1924, 120–5; Moulé 1920; Stieb 1966; Boudon-Millot 2003; Becker 2022; Totelin 2025, forthcoming.

[50] Gal. De *comp. med. sec. loc.* 3.2, 13.571K; *De Antid.* 1.1, 14.9–10K; 1.2, 14.7K; Plin. *HN* 21.144; 25.174; Ps.-Gal. *De ther. ad Pis.* 2.5, 14.216K = Boudon-Millot 2016, 7. The edition followed of book 21 of Pliny's *Natural History* is André 1969.

[51] Plin. *HN* 24.6. The edition followed is André 1972. See Jones and Andrews 1956 for an English translation.

[52] Plin. *HN* 34.108. The edition followed is Gallet de Santerre and Le Bonniec 1953. See Rackham 1952 for an English translation. The Seplasia was originally a market in Capoua where perfumers plied their trade, but the term was later applied to various other perfume and drug markets in Italy. See Allé 2010, 202–03.

below). Galen, by contrast, was more circumspect. He noted that, in his youth, he had been taught how to create imitation drugs under the mentorship of an unnamed teacher who was most motivated by monetary rewards. Galen preferred, however, "never to write down the compositions of the counterfeited goods (*tōn nothōn suntheseis*)."[53] Counterfeited drugs, then, were the products of *sunthesis*, of composition—Galen was not alluding here to simple forms of adulteration, where one product replaced another. Elsewhere, Galen recognized that some drug merchants were so skillful in creating imitation drugs that even those who were experts in the field could be fooled.[54] As solutions, he suggested that drugs be sourced from trusted friends or that travel be undertaken to procure lifetime supplies of ingredients, something that was possible especially for the long-lasting "metallic drugs," which are of particular interest to us here.[55] Galen's networks extended from the territories of the Iberians and the Mauritanians in the West to Syria and Pontus in the East.[56] In return, when he collected drugs on his extensive travels, he offered some to his friends. For instance, he made gifts of Cypriot calamine to his friends in Italy and Asia.[57] Few ancient healers, however, were as wealthy and well-connected as Galen, and they had no option but to contend with the realities of drug adulteration. As Julie Laskaris noted, while adulteration also occurred for plant (and animal) products in antiquity,

> one imagines that preparations containing metals would have been the more tempting to fake or adulterate in proportion to their greater value, which would have been derived from the costliness of the ingredients, the labour required for preparation, and their efficacy. On the other hand, the detection methods offered by Dioscorides and Pliny for metallic medicines were probably more accurate than any they could offer for botanical ones.[58]

Examination of some examples of ancient pharmacological fraud involving metals and other minerals will allow us to refine Laskaris' conclusion.

2.3.1 Types of Ancient Drug Adulteration Involving Mineral Ingredients

The most common method of adulteration in antiquity was simply to replace one product with another, or to mix the authentic product with its adulterant. We have already encountered this method in the case of pitch used as an adulterant for asphalt, as the two ingredients resembled one another in appearance. Another

[53] Gal. *De simpl. med. temp. et fac.* 9.3.8, 12.216K.

[54] Gal. *De antid.* 1.2, 14.7K.

[55] Galen travelled extensively as a young man and took the opportunity to collect vast amounts of drugs. For a summary of these travels, see Boudon-Millot 2012; Mattern 2013; Nutton 2020.

[56] Gal. *De antid.* 1.2, 14.8K.

[57] Gal. *De simpl. med. temp. et fac.* 9.3.11, 12.220K.

[58] Laskaris 2016, 159.

example was the use of Lemnian red ochre as an adulterant for cinnabar (*minium*), also red in color—both were used medicinally.[59] Now, as Pliny remarked, Lemnian ochre was a prized product that was sold under seal, wherefrom it derived its name *sphragis* (seal in Greek). Several generations later, Galen went to great lengths to procure Lemnian seals from the island of Lemnos, where he witnessed the production of the seals under the supervision of the priestess of Artemis and learned much about their medical qualities from locals.[60] As such, Lemnian seals may not always have been cheaper than cinnabar, but they may at times have been more readily available. Elsewhere, Pliny wrote that there existed two types of *minium*, one of better quality than the other. This second *minium* (perhaps red lead) could be found in almost all silver- and lead-mines, and was used to adulterate the better-quality one, bringing profits to the "workshops of the company."[61] A product called "Syrian" (*syricum*), which resembled *minium*, also served to adulterate it, although it appears that this was the case only among painters, as *syricum* is not mentioned in medicinal contexts.[62] We touch here upon the overlap between medicine and other ancient arts that required pigments, such as painting. This overlap is embodied in the figure of the *pigmentarius*, the pigment seller, who sometimes sold vegetable drugs. The legal compendium known as the *Digest* recorded a first-century BCE law, according to which *pigmentarii* who sold dangerous herbal and animal drugs inconsiderately would face punishment.[63] A century later, Scribonius Largus criticized *pigmentarii institores* who, motivated by greed, replaced true opium juice with cheaper poppy leaves when preparing a *kollurion* called *diaglaucium*.[64]

While the adulterations mentioned hitherto were relatively simple, others were considerably more complex. For example, Pliny described a method to create an imitation of true indigo, which came from India:

> qui adulterant, vero Indico tingunt stercora columbina aut cretam Selinusiam vel anulariam vitro inficiunt.

> Those who adulterate it [indigo] stain pigeons' excrements with true indigo, or they dye with woad the clay of Selinus or ring-earth [a white earth].[65]

Both methods consisted in tainting a whiteish product with a blue dye. In the first, a small amount of true indigo was still required, while in the other, another blue dye, woad, was used. The adulteration of hematite was even more complex:

---

[59] Plin. *HN* 35.34.

[60] Gal. *De simpl. med. temp. et fac.* 9.1.2, 12.170–175K. For an English translation of this account, see Brock 1929, 191–6.

[61] Plin. *HN* 33.120. See Moulé 1920, 220. The 'Syrian' was a manufactured product, made from Sinopian ochre and sandyx together. The edition followed is Zehnacker 1983. For an English translation, see Rackham 1952.

[62] Plin. *HN* 33.120; 35.24. See Moulé 1920, 221.

[63] *Digest* 48.8.3.

[64] Scrib. *Comp.* 22.

[65] Plin. *HN* 35.46.

δολοῦσι δέ τινες τὸν προειρημένον οὕτω· λαβόντες βῶλον σχιστοῦ λίθου πυκνόν τε καὶ στρογγύλον—τοιαῦται δέ εἰσιν αὐτοῦ αἱ λεγόμεναι ῥίζαι—εἰς κεραμεᾶν γάστραν σποδιὰν ἔχουσαν θερμὴν ἐγκρύπτουσιν, εἶτα διαλιπόντες μικρὸν ἀναιροῦνται καὶ ἐπ' ἀκόνης τρίβουσι δοκιμάζοντες, εἰ τὴν αἱματίτου χρόαν ἀπείληφε. κἂν μὲν οὕτως ἔχῃ, ἀποτίθενται, εἰ δὲ μή, πάλιν ἐγκρύπτουσι, συνεχῶς ἐφορῶντες καὶ δοκιμάζοντες· ἀφεθεὶς γὰρ ἐπὶ πολὺ ἐν τῇ σποδιᾷ μεταβάλλει τῷ χρώματι, ἔπειτα καὶ διαχεῖται.

Some adulterate this substance in the following way. They take a lump of talc that is firm and round—these are its so-called 'roots'— and bury it in an earthen belly-pot which contains hot ashes; then, after leaving it for a little while, they remove it and rub it against a whetstone to determine whether it has taken on the color of hematite. If it has acquired the color, they store it; if not, they bury it again, constantly checking and testing it. For if it is left too long in the ashes it changes color and then disintegrates.[66]

Powdered hematite featured frequently in ancient pharmacology—in particular, in the preparation of *kolluria*.[67] The stone, however, was most commonly used in the production of healing amulets, which took the form of intricately engraved gems. For example, uterine amulets were often made of hematite.[68] These amulets lead us to the overlap between healing and gem making, another area in which fraud was common. Thus, Pliny noted that it was extremely difficult to distinguish genuine stones from fakes.[69] As an example of such a successful imitation, he described a complex method used to produce fake sardonyx by sticking together three different stones. He then concluded that:

quin immo etiam exstant commentarii auctorum—quos non equidem demonstrabo—quibus modis ex crystallo smaragdum tinguant aliasque tralucentes, sardonychem e sarda, item ceteras ex aliis; neque enim est ulla fraus vitae lucrosior.

Indeed, there are even treatises by authorities, whom at least I shall not name, where they describe the ways in which they dye rock-crystals the color of emerald or other transparent stones, or make sardonyx from sard, and similarly various gems from others. Indeed, there is no fraud in life that is more profitable.[70]

While no such treatise is preserved, traces of such texts are discernible in the so-called Stockholm papyrus, which gave numerous recipes to make imitation gemstones.[71] The papyrus is usually classified as alchemical in nature, highlighting the links between medicine, gem making, and alchemy.[72]

---

66 Diosc. 5.126.3. See Moulé 1920, 219–20.

67 See, e.g., a remedy for the eye 'with hematite of the eye doctor Capiton' preserved in Gal. *De comp. med. sec. loc.* 4.7, 12.732K.

68 On these amulets, see, e.g., Aubert 1989; Hanson 1995; Dasen 2014.

69 Plin. *HN* 37.197.

70 Plin. *HN* 37.197–8. The edition followed is De Saint-Denis 1972. See Eichholz 1962 for an English translation.

71 For an edition of this papyrus, see Halleux 1981.

72 On fake gemstones, see Beretta 2009; Bol 2014. Several methods to detect false gems are recorded in the ancient lapidary treatises: *lithica kerygmata* 45, *Liber Damigeronis* 16 and 26 (Halleux and Schamp 1985, 174, 254, 265).

So far, we have discussed the adulterations of mineral drugs, which most often involved other minerals, but could also make use of animal and vegetable ingredients. Mineral ingredients were occasionally also used as adulterants for expensive plants. For instance, silver-foam and antimony were mixed with authentic spikenard to bulk it up.[73] Litharge, for its part, was mixed with myrrh and saffron to increase their weight.[74] If such adulterated spices were ingested, their effect on the patient's health would be deleterious. Spices and minerals, however, were also applied externally in medicine—for instance, in *kolluria*, which, as mentioned above, could be sold ready-made.

Rare, but significant, remains of ancient stamped *kolluria* have been discovered, including some stamped with the words *crocodes* (Latin) or *krokōdēs* (Greek) from a site near Kostolac (Serbia, ancient Viminacium) and Lyon.[75] *Crocodes/krokōdēs* can be translated either as "which contains saffron" or "saffron-colored." Chemical analyses, however, demonstrated that the *kolluria* contained no saffron. They did, on the other hand, contain metallic ingredients that would have imparted a saffron color: jarosite (deep yellow) and hematite (deep red) in the case of the Lyon *kollurion*, and cuprite (deep red) and zinc sulfide (yellow orange) in the case of the Kostolac *kolluria*.[76] Whether or not we are dealing here with a case of fraud remains unsettled.[77] Several ancient medical authors complained about the production of poor-quality *kolluria*, which could be attributed to fraud, poor skills, or an excess of legitimate substitutions of ingredients.[78]

2.3.2 Methods of Fraud Detection

It appears that ancient legislation was weak in the face of such widespread adulteration, emphasizing that the onus was on the buyer (*caveat emptor*) to verify the quality of potentially adulterated products.[79] To counteract the apparently widespread adulteration of metallic pharmacological ingredients, the ancients developed various tests. Stieb distinguished between organoleptic tests, which involved the senses, and physico-chemical tests, which assessed the physical or chemical qualities of a product.[80] While this distinction is useful, it should not be afforded too much weight in our examination of ancient pharmacological adulteration.

73 Plin. *HN* 12.43; Diosc. 1.7.3. See Moulé 1920, 213.

74 Plin. *HN* 12.71 (myrrh); Diosc. 1.26.2 (saffron). See Moulé 1920, 212–13.

75 Boyer et al. 1990, 240; Guineau 1991, 139; Pardon-Labonnelie, Spasic-Duric, and Uher 2020, 62–3; Pardon-Labonnelie 2021, 195–6.

76 Pardon-Labonnelie 2021, 196.

77 On the case for fraud, see Gourevitch 1998, 369, 2019, 149.

78 Plin. *HN* 34.108 (see above); Scrib. *Comp.* 38; Cassius Felix 29.11–12. See Pardon-Labonnelie 2010, 147.

79 Frier 1983; Bush 2002, 584–5.

80 Stieb 1966, 5–7.

The most common sensory tests for metallic products involved sight (observation of the color) and touch (including touch with the teeth). For instance, as mentioned above, true alum was recognizable by means of its milky color (sight) and did not feel gritty when rubbed (touch).[81] Such tests were by their very nature subjective, but as Galen noted, in the same way as someone who knows twins well will be able to differentiate between them, the expert pharmacologist will be able to distinguish substances and their properties.[82]

A common form of physico-chemical test for minerals was the flammability test. This was used, for instance, in the case of calamine, as some stones, found in particular at Cumae, resembled calamine but did not behave in the same way when subjected to fire:

> καὶ ἐκ τοῦ ἐπιτεθέντα μὲν τὸν λίθον τετριμμένον πυρὶ ἀποπηδᾶν καὶ τὸν ἐξ αὐτοῦ καπνὸν ὁμοειδῆ τῷ πυρὶ ὑπάρχειν, τὴν δὲ καδμείαν μένειν τε καὶ ἀναδιδόναι αἰθάλην μηλίζουσαν καὶ χαλκοφανῆ ὡσπερεὶ ζώνην τινὰ ποικίλην. ἔτι ὁ μὲν λίθος πυρωθεὶς καὶ ψυγεὶς ἀλλοιωθήσεται τῷ χρώματι καὶ κουφότερος ἔσται, ἡ δὲ καδμεία κατ' οὐδὲν μεταβάλλει, εἰ μή τις αὐτὴν ἐφ' ἱκανὰς ἐγκαύσῃ ὥρας.
>
> Further, crushed stone that is placed on the fire, leaps up, and the smoke it emits is of the same nature as the fire; but calamine stays still and emits quince-yellow soot, and it looks like copper, as if it were some sort of multicolored girdle. Moreover, the stone, when it has been burnt and cooled down will change color and will become lighter [in weight], but calamine will not change at all, unless someone burnt it for many hours.[83]

Sharp observation skills, which were based on the senses, were still required in this test of calamine. Perhaps because no test was failproof, it was common to have several different tests available for a single substance, although there was no indication of what test would work the best or whether one should deploy several tests on the same sample. For instance, in the case of verdigris, one could test it by sight; by biting it; by observing its color when placed on a hot fire-shovel; by burning it; and most originally by means of a paper test (the only description of such a test in ancient literature):

deprehenditur et papyro galla prius macerato, nigrescit enim statim aerugine inlita.
[The adulteration] is also discovered by means of papyrus that has previously been steeped in oak gall. For it blackens immediately when smeared with verdigris.[84]

To the modern reader, Pliny is clearly describing a chemical reaction, but he did not, of course, label it as such and did not explain how the test might have worked.

---

[81] Plin. *HN* 35.184. On notions of colors in analysis and synthesis, see Keller in this volume.

[82] Gal. *De comp. med. per gen.* 3.2, 13.570K.

[83] Diosc 5.74.5–6.

[84] Plin. *HN* 34.112. See Stieb 1966, 5–9; Bush 2002, 595–7.

2.4 Conclusions

In this paper, I have examined two important aspects of ancient pharmacological *sunthesis*: adulteration and substitution. Ancient and modern readers alike regard adulteration negatively and substitution positively. Treatises on pharmacological substitutions continued to be produced over the centuries. In the early modern period, these treatises acquired proper theoretical underpinnings, emphasizing the importance of thoroughly considering the question of drugs' *dunameis*, down to the degree to which a particular *dunamis* was present in a given drug and its substitute.[85] In the eighteenth century, as argued by Matthew Paskins, the ancient art of substituting was

> given additional support by the idea that the analytical procedures of chemistry itself can inform on what materials are really the same as each other; and hence on a rational reduction in the number of different things which the *medica* contains.[86]

Indeed, as Simon Werrett has pointed out, substitution was itself a way of experimenting and discovering the properties of drugs or other materials.[87]

This experimental dimension to the art of substitution may have been present from antiquity. As Philip van der Eijk has shown, Galen had developed a notion of "qualified experience" (*diōrismenē peira*), which sometimes comes very close to the modern notion of experimentation. Qualified experience, for instance, helped Galen determine whether a drug that worked well in some circumstances would work as well in others. Thus, a certain medicine might work well on an adult, but be dangerous when taken by a child.[88] Galen often criticized those who administered drugs without qualification (*adiorismōs*). We can therefore suggest that, when he recommended substitutes, Galen did so carefully, with qualification. His carefully considered substitution could then serve as an experiment to prove his theoretical reflection on the *dunameis* of alternative drugs.

By contrast, and although it is allegedly based on a knowledge of *dunameis*, the pseudo-Galenic treatise *On Substitutions* is a simple list of alternative products without any theoretical reflection or reference to qualified experience. We cannot even be certain whether all these substitutions were ever actually implemented. Indeed, some would have been highly dangerous, particularly if a highly toxic product was to be used in the same quantity as one that was not. We may speculate that some of these substitutions were forms of thought experiments, reflections on whether, for instance, a vegetable product of a certain color might serve as a substitute for a mineral product of the same color or vice versa.

Several of the substitutions listed in the pseudo-Galenic treatises may well be termed "fraudulent," as they do not appear to be based on sound reflection on

---

[85] Boumediene and Pugliano 2019.

[86] Paskins 2016, 59.

[87] Werrett 2019, 86. On the development of analytical chemistry, see Ramberg in this volume.

[88] On the concept of qualified experience, see van der Eijk 1997.

dunameis or seek to substitute an expensive product with a cheaper one. It is nigh on impossible to determine where the boundary between substitution and adulteration should be drawn. As Werrett observed, at times, substitution "shaded into adulteration, though judgements over the boundary were fraught." Indeed, the simplest forms of fraud consisted simply in substituting, or bulking up, one product for another. A slightly more complex form of adulteration involved the use of color—it was a very basic form of *sunthesis*. At its most sophisticated, however, adulteration involved intricate *sunthesis*, which developed through repeated practice, through experience. Further, creating fakes, or finding ways to detect them, were also based on forms of experimentation or, at least, repeated experience based on hypotheses.[89] As Galen had pointed out, it took time to learn how to create counterfeited drugs—it was an art, a *technē*. Indeed, in examining ancient drug adulteration, we may perceive several areas of overlap between ancient medicine and other ancient crafts, such as painting and gem making. To someone like Galen, however, such crafts were inferior to the medical craft as he practiced it. For not content to simply work with materials, he understood their *dunameis* and how these would combine in the practice of *sunthesis*. His art of *sunthesis* has clear theoretical underpinnings.

References

Adams, F. 1947. *The seven books of Paulus Aegineta.* Translated from the Greek with a commentary by F. Adams, vol. III. London: The Sydenham Society.

Allé, F. 2010. Travail et identité professionnelle: Analyse lexicographique des métiers du parfum dans l'Occident romain. *L'antiquité classique* 79: 199–212.

Amigues, S. 2006 *Théophraste. Recherches sur les plantes. Tome V. Livre IX.* Texte établi et traduit par S. Amigues. Paris: Les Belles Lettres.

André, J. 1969. *Pline l'Ancien Histoire naturelle. Livre XXI.* Texte établi, traduit et commenté par J. André. Paris: Les Belles Lettres.

———. 1970. *Pline l'ancien. Histoire naturelle. Livre XXII.* Texte établi, traduit et commenté par J. André. Paris: Les Belles Lettres.

———. 1972. *Pline l'Ancien. Histoire naturelle. Livre XXIV.* Texte établi et traduit par J. André. Paris: Les Belles Lettres.

———. 1974. *Pline l'ancien. Histoire naturelle. Livre XXV.* Texte établi, traduit et commenté par J. André. Paris: Les Belles Lettres.

Aubert, J.J. 1989. Threatened wombs: Aspects of ancient uterine magic. *Greek, Roman and Byzantine Studies* 30: 412–449.

Baker, P. 2011. Collyrium stamps: An indicator of regional practices in Roman Gaul. *European Journal of Archaeology* 14: 158–189.

Beck, L.Y. 2020. *Pedanius Dioscorides of Anazarbus. De materia medica.* Trans. Lily Y. Beck. Fourth, enlarged edition. Hildesheim: Olms–Weidmann.

Becker, H. 2022. *Caveat emptor*: The perils of shopping for medical products in the ancient marketplace. *Mouseion* 18: 321–349.

Beretta, M. 2009. *The alchemy of glass: Counterfeit, imitation, and transmutation in ancient glassmaking.* Sagamore Beach: Science History Publications.

[89] See Smith and Lores-Chavez 2023.

Bol, M. 2014. Coloring topaz, crystal and moonstone: Gems and the imitation of art and nature, 300–1500. In *Fakes!?: Hoaxes, counterfeits, and deception in early modern science*, ed. M. Beretta and M. Conforti, 108–129. Sagamore Beach: Science History Publications.
Boudon-Millot, V. 2003. Aux marges de la médecine rationnelle: Médecins et charlatans à Rome au temps de Galien (IIe s. de notre ère). *Revue des études Grecques* 116: 109–131.
———. 2012. *Galien de Pergame: Un médecin grec à Rome*. Paris: Les Belles Lettres.
———. 2016. *Galien. Tome VI: Thériaque à Pison*. Texte établi et traduit par V. Boudon-Millot. Paris: Les Belles Lettres.
Boumediene, S., and V. Pugliano. 2019. La route des succédanés: Les remèdes exotiques, l'innovation médicale et le marché des substituts au XVIe siècle. *Revue d'histoire moderne & contemporaine* 66: 24–54.
Boyer, R., V. Bel, L. Tranoy, et al. 1990. Découverte de la tombe d'un oculiste à Lyon (fin du IIe siècle après J.-C.). *Gallia* 47: 215–249.
Brock, A.J. 1929. *Greek medicine: Being extracts illustrative of medical writers from Hippocrates to Galen*. London: JM Dent & Sons.
Bush, J.F. 2002. 'By Hercules! The more common the wine, the more wholesome!' Science and the adulteration of food and other natural products in ancient Rome. *Food and Drug Law Journal* 57: 573–602.
Colson, F.H., and G.H. Whitaker. 1930. *Philo. On the unchangeableness of God. On husbandry. Concerning Noah's work as a planter. On drunkenness. On sobriety.* Trans. F. H. Colson and G. H. Whitaker. Cambridge, MA: Harvard University Press.
Croisille, J.-M. 1985. *Pline l'ancien. Histoire naturelle. Livre XXXV.* Texte établi et traduit par Jean-Michel Croisille. Paris: Les Belles Lettres.
Dasen, V. 2014. Healing images: Gems and medicine. *Oxford Journal of Archaeology* 33: 177–191.
De Saint-Denis, E. 1972. *Pline l'Ancien. Histoire naturelle. Livre XXXVII.* Texte établi et traduit par Eugène De Saint-Denis. Paris: Les Belles Lettres.
De Vos, P.S. 2021. *Compound remedies: Galenic pharmacy from the ancient Mediterranean to New Spain*. Pittsburgh: University of Pittsburgh Press.
Eichholz, D.E. 1962. *Pliny. Natural History, Volume X: Books 36–37*. Trans. D.E. Eichholz. Cambridge, MA: Harvard University Press.
Fabricius, C. 1972. *Galens Exzerpte aus älteren Pharmakologen*. Berlin: Walter de Gruyter.
Fischer, K.-D. 2018. Drugs to declare: Two pharmaceutical works attributed to Galen. Cuadernos de Filología Clásica. *Estudios griegos e indoeuropeos* 28: 225–241.
Frier, B.W. 1983. Roman law and the wine trade: The problem of vinegar sold as wine. *Zeitschrift der Savigny-Stiftung für Rechtsgeschichte: Romanistische Abteilung* 100: 257–295.
Gallet de Santerre, H., and H. Le Bonniec. 1953. *Pline l'Ancien. Histoire naturelle, Livre XXXIV.* Commentaire de Hubert Gallet de Santerre, Texte établi et traduit par Henri Le Bonniec. Paris: Les Belles Lettres.
Gourevitch, D. 1998. Collyres romain inscrits. *Histoire des sciences médicales* 32: 365–370.
———. 2016. Popular medicines and practices in Galen. In *Popular medicine in Graeco-Roman antiquity: Explorations*, ed. W.V. Harris, 251–271. Leiden: Brill.
———. 2019. *Pour une archéologie de la médecine romaine*. Paris: De Boccard.
Grmek, M.D. 1997. *Le chaudron de Médée: L'expérimentation sur le vivant dans l'Antiquité.* Synthélabo: Le Plessis Robinson.
Grmek, M.D., and D. Gourevitch. 1985. Les expériences pharmacologiques dans l'antiquité. *Archives internationales d'histoire des sciences* 35: 3–27.
Guineau, B. 1991. Étude physico-chimique de la composition de vingt collyres secs d'époque gallo-romaine. *Bulletin de la Société nationale des antiquaires de France* 1989: 132–140.
Halleux, R. 1981. *Les Alchimistes Grecs: Tome I. Papyrus de Leyde. Papyrus de Stockholm. Recettes.* Texte établi et traduit par R. Halleux. Paris: Les Belles Lettres.
Halleux, R., and J. Schamp. 1985. *Les lapidaires grecs. Lapidaire orphique. Kérygmes lapidaires d'Orphée. Socrate de Denys. Lapidaire nautique. Damigéron – Evax.* Texte établi et traduit par R. Halleux et J. Schamp. Paris: Les Belles Lettres.

Hanson, A.E. 1995. Uterine amulets and Greek uterine medicine. *Medicina nei secoli* 7: 281–299.
———. 2008. Galen: Author and critic. In *Editing texts: Texte edieren*, ed. G.W. West, 22–53. Göttingen: Vandenhoeck & Ruprecht.
Heiberg, J.L. 1924. *Paulus Aegineta, Libri V-VII.* Edidit J. L. Heiberg, CMG IX 2. Leipzig/Berlin: Teubner.
Helmreich, G. 1893. *Claudii Galeni Pergameni Scripta minora.* Vol. 3. Leipzig: Teubner.
Jackson, R. 1996. Eye medicine in the Roman empire. In *Aufstiege und Niedergang der Romischen Welt. Band 37/3*, ed. W. Haase, 2228–2251. Teilband Philosophie, Wissenschaften, Technik. Wissenschaften (Medizin und Biologie [Forts.]). Berlin: De Gruyter.
Jocks, I.T. 2020. *Scribonius Largus' Compounding of Drugs (Compositiones medicamentorum): Introduction, translation, and medico-historical comments*. PhD Thesis, University of Glasgow.
Jones, W.H.S., and A.C. Andrews. 1956. *Pliny. Natural History, Volume VII: Books 24–27.* Trans. W.H.S. Jones and A.C. Andrews. Cambridge, MA: Harvard University Press.
Jouanna-Bouchet, J. 2016. *Scribonius Largus. Composition médicales.* Texte établi, traduit et commenté par Joëlle Jouanna-Bouchet. Paris: Les Belles Lettres.
Keyser, P.T. 2008. Hermophilos. In *The encyclopedia of ancient natural scientists: The Greek tradition and its many heirs*, ed. P.T. Keyser and G.L. Irby-Massie, 380. London: Routledge.
Kühn, K.G. 1821–1833. *Claudii Galeni opera omnia.* Vol. 20. Leipzig: Car. Cnoblochius.
Laskaris, J. 2016. Metals in medicine: From Telephus to Galen. In *Popular medicine in Graeco-Roman Antiquity: Explorations*, ed. W.V. Harris, 147–160. Leiden: Brill.
Manolaraki, E. 2015. *Hebraei liquores*: The balsam in Judaea in Pliny's *Natural History. American Journal of Philology* 36: 633–667.
Mattern, S. 2013. *The prince of medicine: Galen in the Roman Empire*. New York: Oxford University Press.
Moulé, L. 1920. Les fraudes pharmaceutiques dans l'antiquité. *Bulletin de la Société française d'histoire de la médecine* 14: 199–226.
Nutton, V. 2008. Ancient Mediterranean pharmacology and cultural transfer. *European Review* 16: 211–217.
———. 2020. *Galen: A thinking doctor in imperial Rome*. London: Routledge.
Pardon-Labonnelie, M. 2010. Du savoir au savoir-faire: l'oculistique, une "spécialité" médicale gallo-romaine. In *Transmettre les savoirs dans les mondes hellénistique et romain*, ed. F. Le Blay, 133–153. Rennes: Presses Universitaires de Rennes.
———. 2011. Du κολλύριον au 'collyre'. In *La coupe d'Hygie: Médecine et chimie dans l'Antiquité*, ed. M. Pardon-Labonnelie, 33–49. Dijon: Éditions Universitaires de Dijon.
———. 2021. Les couleurs de la vue: Les propriétés thérapeutiques des couleurs dans l'ophtalmologie gréco-romaine. *Pallas* 117: 183–201.
Pardon-Labonnelie, M., D. Spasic-Duric, and E. Uher. 2020. Les collyres estampillés de Mésie supérieure: Un nouveau regard sur la tombe du "médecin et chirurgien oculiste de Viminacium". *Histoire des sciences médicales* 2: 55–84.
Paskins, M. 2016. One of these things is just like the others: Substitution as a motivator in eighteenth century chemistry. In *Theory choice in the history of chemical practices*, ed. E. Tobin and C. Ambrosio, 55–70. Cham: Springer.
Rackham, H. 1945. *Pliny. Natural history, Volume IV: Books 12–16.* Trans. H. Rackham. Cambridge, MA: Harvard University Press.
———. 1952. *Pliny. Natural History, Volume IX: Books 33–35.* Trans. H. Rackham. Cambridge, MA: Harvard University Press.
Richardson, W.F. 2005. *Numbering and measuring in the classical world.* Revised edition. Exeter: Bristol Phoenix Press.
Riddle, J.M. 1974. Theory and practice in medieval medicine. *Viator* 5: 157–184.
———. 1985. *Dioscorides on pharmacy and medicine.* Austin: University of Texas Press.
———. 1992. Methodology of historical drug research. In *Quid pro quo*, ed. J.M. Riddle, 1–19. Aldershot: Ashgate.

Schmidt, A. 1924. *Drogen und Drogenhandel im Altertum*. Leipzig: Verlag von Johann Ambrosius Barth.

Schmitz, R. 1998. *Geschichte der Pharmazie. Volume 1: Von den Anfängen bis zum Ausgang des* Mittelalter. Eschborn: Govi-Verlag.

Smith, P.H., and I. Lores-Chavez. 2023. Counterfeiting materials, imitating nature. In *The matter of mimesis: Studies of mimesis and materials in nature, art and science*, ed. M. Bol and E.C. Spary, 27–53. Leiden: Brill.

Stieb, E.W. 1966. *Drug adulteration: Detection and control in nineteenth-century Britain*. Madison: University of Wisconsin Press.

Totelin, L. 2009. *Hippocratic recipes: Oral and written transmission of pharmacological knowledge in fifth-and fourth-century Greece*. Leiden: Brill.

———. 2016. The world in a pill: Local specialties and global remedies in the Graeco-Roman world. In *The Routledge handbook of identity and the environment in the classical and medieval world*, ed. R.F. Kennedy and M. Jones-Lewis, 151–170. London: Routledge.

———. (forthcoming, 2025). *Selling pharmaka, buying health in ancient Greece and Rome: Retail therapy*. London: Routledge.

Touwaide, A. 2012. *Quid pro quo*: Revisiting the practice of substitution in ancient pharmacy. In *Herbs and healers from the ancient Mediterranean through the medieval west. Essays in honor of John M. Riddle*, ed. A. Van Arsdall and T. Graham, 19–61. Farnham: Ashgate.

Ullmann, M. 1970. *Die Medizin im Islam*. Leiden: Brill.

van der Eijk, P.J. 1997. Galen's use of the concept of 'qualified experience' in his dietetic and pharmacological works. In *Galen on Pharmacology, Philosophy, History of Medicine*, ed. A. Debru, 35–57. Leiden: Brill.

Vogt, S. 2009. Drugs and pharmacology. In *The Cambridge companion to Galen*, ed. R.J. Hankinson, 304–322. Cambridge: Cambridge University Press.

von Staden, H. 1989. *Herophilus: The art of medicine in early Alexandria*. Cambridge: Cambridge University Press.

———. 1997. Inefficacy, error, and failure: Galen on *dokima pharmaka aprakta*. In *Galen on pharmacology: Philosophy, history and medicine*, ed. A. Debru, 59–83. Leiden: Brill.

Wellmann, M. 1907–1914. In *Pedanii Dioscuridis Anazarbei De materia medica libri quinque*. Ed. Max Wellmann, 3 vols. Berlin: Weidmann.

Wendland, P. 1897, repr. 1962. *Philonis Alexandrini opera quae supersunt, vol. 2*. Berlin: Reimer.

Werrett, S. 2019. *Thrifty science: Making the most of materials in the history of experiment*. Chicago: University of Chicago Press.

Zehnacker, H. 1983. *Pline l'ancien. In Histoire naturelle. Livre XXXIII*. Texte établi et traduit par Hubert Zehnacker. Paris: Les Belles Lettres.

Laurence Totelin is Professor of Ancient History at Cardiff University, UK. She specializes in the history of Greek and Roman pharmacy, botany, and gynecology. Her recent publications include *A Cultural History of Medicine* (Bloomsbury, 2021) and, with Victoria Leonard and Mark Bradley, *Bodily Fluids in Antiquity* (Routledge, 2021).

Chapter 3
Spagyria, Scheidung, and Spagürlein: The Meanings of Analysis for Paracelsus

Didier Kahn and William R. Newman

Abstract Paracelsus is often lauded for having created a new disciplinary identity for alchemy by basing it on the twin operations of analysis and synthesis. Indeed, his neologism for the field, *Spagyria,* is often said to express this pairing by embodying the Greek terms for decompounding and compounding (σπάν and ἀγείρειν). The present article disputes both this etymological claim and the underlying belief that Paracelsus had an interest in synthesis that paralleled his very strong promotion of analysis (*Scheidung*). As the authors argue, there is good reason to think that Paracelsus actually modeled the word *Spagyria* on the early modern Swiss coin that went by the name *Spagürlein, Spagürli,* or *Spagürle.* This derivation was appropriate for a discipline based on *Scheidung*, since the coin was the product of numerous metallurgical processes involving separation. The claim that *Spagyria* was a fusion of σπάν and ἀγείρειν was actually a product of Paracelsus's followers, not of the Swiss chymist himself.

Keywords Alchemy · Paracelsus · *Spagyria* · Scheidung · Decompounding and compounding

> Capita nunc aliquot doctrinae in quibus cum medicis aliis non convenit, paucissimis ordine referam, ac rogo tuam Celsitudinem in primis, omnesque deinde lectores, ut velitis haec in judicando examinare per summos philosophos, hoc est, qui naturae immitatione callent corpora componere et resolvere Spagyrica arte in prima sua principia, quorum tria invenit verissima, videlicet, sulphur, salem et mercurium sive liquorem, quae non stulta imaginatione aut fallaci oratione narrantur ac animis infirmis obtruduntur contra naturae ordinem, verùm usu et experientia oculis subjiciuntur, verisque mentibus demonstrantur compositione

D. Kahn
CNRS, CELLF, UMR 8599, Sorbonne Université, Sorbonne, France
e-mail: dkahn@msh-paris.fr

W. R. Newman (✉)
Department of History and Philosophy of Science and Medicine, Indiana University, Bloomington, IN, USA
e-mail: wnewman@iu.edu

W. R. Newman, J. Schickore (eds.), *Traditions of Analysis and Synthesis*, Archimedes 73, https://doi.org/10.1007/978-3-031-76398-4_3

> et resolutione corporum, per ignem: Siquidem omnia corpora ex hisce tribus componuntur, in eademque resolvuntur, nec quicquam praeter haec tria in vilo corpore inveniuntur.

Let me now relate briefly and sequentially some chapters of a doctrine in which he [Paracelsus] does not agree with other physicians, and I ask your Highness especially, and all readers, that in judging these things you wish to have them examined by the greatest philosophers, that is, those who know how to compound and resolve bodies into their first principles by imitation of nature with Spagyric art, namely into sulfur, salt, and mercury or liquor, which are not related by fallacious rhetoric or by foolish imagination and thrust upon weak souls against the order of nature, but rather are brought beneath the eyes by use and experience, and are demonstrated to true intellects by composition and resolution of bodies through fire: Since all bodies are composed of these three and are resolved into them, nor are any things beyond these three found in common body.[1]

3.1 A Greek Origin for Spagyria?

In the passage above, the early follower of Paracelsus Adam von Bodenstein advised Cosimo de' Medici to imitate nature by means of the "Spagyric art," saying that like nature, the spagyrist employed a twofold process of synthesis and analysis. Bodenstein was stating a view that would soon become commonplace. Indeed, an etymology of the Paracelsian neologism *Spagyria* still accepted widely today derives the term by fusing the two Greek words σπάν and ἀγείρειν—to pull apart and to gather together. Thus, in 1844, the noted historian of alchemy and chemistry Hermann Kopp traced the phrase "die spagirische Kunst" back to the origins of alchemy, no doubt due to its Greek etymology, which he quoted.[2] Twenty-five years later, he confessed to his ignorance as to when the word "Spagiriker" first appeared, quoting Andreas Libavius extensively.[3] Seventeen years later, in his *Die Alchemie in älterer und neuerer Zeit* (1886), he admitted the word originated in the sixteenth century.[4] In the next century, James Riddick Partington, in his massive *A History of Chemistry*, approvingly mentioned the "probable" etymology given by Libavius, and referred to Kopp 1844 and 1869.[5] Recent scholars have emulated Partington's

[1] Bodenstein's 1563 dedication letter to Cosimo de Medici of the books on tartar by Paracelsus. See the edition of this dedication letter in Kühlmann and Telle (2001), 307.

[2] Kopp (1844), 160.

[3] Kopp (1869), 63–64. On Libavius, see below, n. 13–14. In this book, Kopp wondered how far the words of Carl von Prantl speaking of Plato's ideas on transformation of matter might be true (Prantl [1856], 138: "Es scheint nachweisbar zu sein (aus Philo Judaeus und Plotin), dass die Bezeichnung 'Spagiriker' gerade aus diesen Platonischen Ansichten betreffs des Trennens und Vereinigens (σπάω–ἀγείρω) floss").

[4] Kopp (1886), vol. 1: 4–5 *n*.

[5] Partington (1961), 134.

cautious but affirmatory statement on the Greek origin of *Spagyria*.[6] Less specialized authors, or incautious ones, have boldly stated that Paracelsus himself invented the word by relying on the Greek etymology.[7]

In reality, the derivation of *Spagyria* from σπάν and ἀγείρειν found its origin among the first generation of Paracelsian scholars, who were attempting to bring a modicum of order into the chaotic *Nachlass* of their master. The first author that we know of who hypothesized this etymology was the French Paracelsian Jacques Gohory, dedicated as he was to making obvious how deeply Paracelsus was rooted in the humanist tradition of the *prisca philosophia*.[8] Later in the same work, Gohory, among a series of conjectures on the Greek etymologies of Paracelsian words, suggested for *Spagyri* "παρὰ τὸ σπάω *traho* & ἄγυρις *concio*."[9] Two years later the German physician Martin Copus, in a book warning the reader against the dangers of antimony, suggested the soon-to-be standard etymology, no doubt normalized from Gohory's.[10] In 1576 another French Paracelsian, Joseph Du Chesne, claimed that:

> the learned named the spagiric art from two Greek words, ἀπὸ τοῦ σπᾶν και ἀγείρειν, because by this art a certain subtle, spiritual nature in which the power and effect of the remedy mainly resides, is extracted, and then brought together and condensed.[11]

One year later, Theodor Zwinger, at whose home Du Chesne had privately defended his doctoral theses in 1575, called the alchemists:

> those who can, through the action of fire, dissolve the heterogeneous bodies and coagulate the homogeneous. [...] Our popular Theophrastus Paracelsus, a man who excelled in this art to the point of being miraculous, called <alchemy> the spagiric art through an appropriate derivation, since it entirely consists ἐν τῷ σπᾷν καὶ ἀγείρειν, in extracting or separating, and compounding or coagulating.[12]

---

[6] See e.g. *Principe* (2013), 129; Kahn (2016), 192.

[7] See e.g. Blaser (1979), 92; Menten (2013), 86b, or even Hauck (2008).

[8] Gohory (1567), 196; Gohory (1568), 184: "Spagyros enim ubique suo vocabulo ac novo Chymicos intelligit, forte a σπάω quod 'abstrahere' significat, unde etiam dicuntur abstractores quintarum essentiarum [an unfortunate allusion to a satirical excerpt against alchemists by Erasmus], & 'gyro' propter anfractus quos nunc a poeta audiistis" [allusion to a verse by G. A. Augurelli he quoted just before].

[9] Gohory (1567), 356; Gohory (1568), 316.

[10] Copus (1569), sig. Ciij verso: "Diese *Tinctura* und andere dergleichen *praeparationes,* ist gefunden und wird zugerichtet/durch die Kunst der *separation* und *composition,* der nu offt gedacht/die itzund *corrupto vocabulo Graeco Spagirica a* σπάω *&* ἀγείρω, und auch sunst/*Artificium* ἀναλύσεως *&* συνθέσεως genandt wird."

[11] Du Chesne (1576a), 162: "eruditi viri ex Graecis duobus verbis sic <Spagiricam artem> nominarunt, ἀπὸ τοῦ σπᾶν και ἀγείρειν: quod per eam eliciatur, ac tum cogatur comprimaturque subtilis quaedam & spiritualis natura, in qua vis & effectus medicamenti praecipue consistit." Likewise in Du Chesne's own French trans.: Du Chesne (1576b), 189: "[...] d'autant que par icelle on tire, & puis on reserre & congele une substance plus subtile et spirituelle [...]."

[12] See the third edition of Zwinger's *Theatrum Humanae Vitae*, quoted by Gilly (Forthcoming), First part, Ch. VI, 78.

But it was the Saxon schoolmaster Andreas Libavius who exercised the widest influence in spreading the supposed Greek origin of *Spagyria* among subsequent chymists. In his *Commentariorum* of 1606, Libavius repeated the now popular Hellenic etymology in the following words:

> The moderns call it *spagiria* (σπαγειρία). Leo Suavius does not know from whence.... But most celebrated is that σύγκρισις καὶ διάκρισις of the old, called "coagulation" [and] "solution" by our artisans [i.e. chymists]. For the latter tear apart the structures of mixed bodies and break them up with their ingenious techniques and apparatus. Penetrating into the inner chambers of composite things, into the bedrooms and sanctuaries of their essences, they congregate and unite the homogeneous, while separating the heterogeneous. That is, in Greek σπάν and ἀγείρειν.[13]

What Libavius's etymology of *Spagyria* lacked in originality it would make up for in influence. His words would have a major impact on the mainstream of seventeenth-century chymistry, being taken up by influential writers such as the Wittenberg medical professor Daniel Sennert and his intellectual heirs.[14]

For all the impressiveness of these early modern attempts at framing a Hellenic genealogy for *Spagyria*, there is much more at stake here than mere etymology, for the derivation of *Spagyria* from these Greek terms would imply—if true—that Paracelsus himself saw the basis of his chymical art to lie in a twofold decomposition and synthesis (or resynthesis) of matter. In the present paper we aim to dispute that view and at the same time, to propose an alternative etymology for the puzzling term *Spagyria*. We will argue, in fact, that *Spagyria* does not come from a fusion of two Greek words, or even from Greek at all, but rather from a Swiss coin that Paracelsus himself refers to in one of his many writings on syphilis, namely the *Spagürlein*, also called the *Spagürli* or *Spagürle*.[15]

---

[13] Libavius (1606), 77: "Spagirian (σπαγειρίαν) appellant recentes. Nescit Leo Suavius unde.... Sed celebratissima est illa veterum σύγκρισις καὶ διάκρισις coagulatio, solutio nostris artificibus dicta. Divellunt hi, perfringuntque compages mistorum adminiculis & instrumentis ingeniosis; & in penetralia compositarum rerum, cubiculaque & adyta essentiarum penetrantes, homogenea congregant, uniunt, & ab heterogeneis separant. Id est Graecis σπάν καὶ ἀγείρειν [...]."

[14] Libavius's association of σπάν and ἀγείρειν with the terms that Democritus used for association and dissociation of atoms, σύγκρισις and διάκρισις, opened the door to an atomistic interpretation of *Spagyria*. For this development, see Newman (2006), chapter 3.

[15] See the online Grimm *Wörterbuch sub voce Spagürlein* at https://woerterbuchnetz.de/?sigle=DWB, consulted 1/25/23, for Paracelsus's use of the term in *Drey Bücher von den Frantzosen* (1529); ed. Johann Huser in Paracelsus, *Bücher und Schrifften* (1589–1591) and *Chirurgische Bücher und Schrifften* (1605), available online on the database THEO (www.paracelsus-project.org), referred to as H and H C, here H C:160b.

3.2 The Significance of "Spagyria" and "Scheidung" in Paracelsus's Work

There is no doubt that Paracelsus built much of his practice, as well as his cosmology, medicine, and philosophy, on the concept of analysis. "Scheidung" is at the basis of his work from the very beginning. One of the earliest occurrences of the word, perhaps even the earliest, appeared when Paracelsus, in *Das Sechste Buch in der Artzney* devoted to tartaric diseases, introduced what he called this "new" method, namely separation, as one that he recently discovered:

> As we intend to write about the incurable tartaric diseases, we want to [...] place before our eyes the perfect virgin *Experientia*, who, without any male seed, is a mother of all arts, and we want her to prove our whole writing beyond any doubt, so that her authority enable us to understand the origin of these diseases. [...] Moreover, we revel splendidly in the new discovery of separation, which has been, as far as we know, unknown to the ancients: judging by their writings, they came through the doors of experience as one-eyed people.[16]

As convincingly argued by Urs Leo Gantenbein, this treatise must date back at least to 1524: his arguments include among others the mention by Paracelsus of his "young blood,"[17] the explicit display of his highest reverence to the Virgin Mary—here in the clothes of the "virgin *Experientia*"—as in the theological treatises he wrote in Salzburg in 1524, and the mention quoted above of his recent discovery of "Scheidung."[18] What he means there is not, however, that he only recently became aware of the relevance of alchemy to medicine: he learned alchemical procedures much earlier in his youth, as stated in a famous excerpt from his *Grosse Wundartzney*.[19] His new discovery is, rather, the relevance of alchemical separation, i.e., analysis, to the understanding of the causes of diseases. Indeed, in the first chapter of the treatise he mentions the *tria prima*, i.e. mercury, sulphur and salt—for one of the first times ever in his writings[20]—as the first principles out of which all things are composed: a conclusion he only could reach through the means of

[16] *Das Sechste Buch in der Artzney von den Tartarischen Kranckheiten*, H 4:14–15: "Dieweil unnd wir von den unheilbarlichen Steinkranckheiten schreiben wöllen/wöllen wir [...] vor uns stellen/ die aller außbereiteste Jungfraw *Experientiam*, die ohne Männlichen Sahmen eine Mutter ist aller künsten/und wollend des ungezweiffelt sein/eine bewererin alles unsers schreibens: unnd also durch ihr ansehen von ihr entpfahen den ursprung diese Kranckheiten zu erkennen [...] Unnd fürtreffenlich ergetzet uns der newe fund der scheidung/der/als wir noch nit anders wissen tragen/ den Alten unbekannt ist gewesen/und nach ihrem schreiben als Einögig antretten sind die thüren diser Experientz."

[17] H 4:14 and 15.

[18] Gantenbein (2020), 15–18.

[19] H C:101c–102a. See Benzenhöfer (2002), 27–30.

[20] The *tria prima* are not named as mercury, sulphur and salt in this treatise, but only evoked—clearly enough—as "dreien materialischen dingen": see below, n. 23. Elsewhere in the treatise (H 4:17) they are named the *Tres Primae*. Their only other early mention, quoted by Gantenbein (2020), 20–25, is found in *De Genealogia Christi*, another of the theological treatises written in Salzburg in 1524.

analysis, as explained by himself in a number of his other writings.[21] His explanation here goes on to state that the three principles are a raw material, not without feces[22] hidden in their very substance: these feces are the cause of the tartaric diseases—at least of the kind which we develop by our very nature, as opposed to the other kind, which we develop through the means of external things such as what we ingest.[23] In other words, "Scheidung" is the means—only recently discovered by Paracelsus—to find out that every natural body is composed of the three principles, a knowledge which, in turn, enables him to understand how the tartaric diseases originate.

In the treatise supposedly following *Das Sechste Buch in der Artzney*, namely *Das Siebendte Buch in der Artzney* devoted to mental diseases, Paracelsus still mentions "the great art of separation," as if being in awe before its power—an attitude he will not display in later treatises. No powerful remedy, he states, against the sorts of mental illnesses he is discussing can be made without this art, which alone can prepare the wonderful quintessences that will perform this deed.[24] Here Paracelsus

---

[21] See e.g. *Opus Paramirum*, H 1:74–75: "Das so da brinnt/ist der *Sulphur*, nichts brenndt/allein der *Sulphur*: Das da raucht/ist der *Mercurius*/Nichts Sublimirt sich/allein es sey dan *Mercurius*: Das da in Eschen wirt/ist *Sal*, Nichts wird zu Eschen/allein es sey dan *Sal*. [...] Wiewol das ist/im lebendigen *Corpus* sicht niemandts nichts/dann ein Bawren gesicht: Die scheidung aber beweist die Substantzen. So red ich hie nit von der *prima Materia*: Dann ich will hie nit *Philosophiam* tractiren/sondern *Medicinam*. Also wie vom Saltz steht/so wissen vom Rauch/der beweist den *Mercurium*/der sich durch das Fewr auffhebt unnd Sublimirt: Unnd wiewol auch sein *prima materia* hie nit sichtbar ist/so ist doch sichtbar der ersten *Ultima Materia*: Also das der *Mercurius* da ist die ander Substantz des dings. Also was da brennt/vnnd den augen Fewrig erscheint/dasselbig ist der *Sulphur*/der verzeert sich/dann er ist *Volatile*. Nun ist das so da Fewr ist/auch ein Substantz/ vnnd ist die dritte/die das *Corpus* gantz macht. [...] Dan was in den Bawren augen nicht liget/dasselbige ligt inn der Kunst/das in die augen gebracht werdt/das ist *Scientia Separationis*."

[22] The term "feces" or "faeces," which is the plural of the Latin "faex," has the primary meaning of "*grounds, sediment, lees, dregs* of liquids (cf. sentina)," as recorded in the Latin-English dictionary of Lewis and Short. See the Brepols *Database of Latin Dictionaries, sub voce,* consulted 1/26/23: http://clt.brepolis.net.proxyiub.uits.iu.edu/dld/pages/QuickSearch.aspx. The term was often used in this sense in medicine and alchemy, for example during the processes of sublimation and distillation. See the *Oxford English Dictionary, sub voce,* consulted 1/26/23: https://www-oed-com.proxyiub.uits.iu.edu/view/Entry/67598?redirectedFrom=faeces#eid

[23] *Das Sechste Buch in der Artzney*, H 4:16: "das geschicht in zwen wege: Der eine ursprung ist in den dingen die uns solche kranckheiten zufügen: Der ander ursprung ist in der Natur/die eine solche kranckheit formiert [...] Alles das/so ausserhalb unsern Cörpern ist/des wir geniessen und gebrauchen/wirdt auß dreien materialischen dingen geformiert und in Eim geendet/als wir *de Generationibus* melden. Solche drey ding werden in der Natur also grob/daß sie nimmer ohn *feces* nit wachsen/und haben alle mal eine wildniß in inen/die ihnen in ihrer Substantz verborgen ligen/ auß deren die Tartarischen kranckheiten entspringen."

[24] *Das Siebende Buch in der Artzney De Morbis Amentium*, H 4:91: "Wollen wir uns fürsetzen ein Praeservatiff/das den Menschen behüt vor der Ersten Privatz der Sinnen *Caduci*, und deßgleichen von der *Mania*, und also auch vor der *Chorea*, und also auch vor der Suffocation/und also auch vor der Privatz *Sensuum*. So ist ein semlichs ohne grosse kunst der Separation nit zu machen/sondern allein durch die *Quintas Essentias* soll und muß ein solchs zuwegen gebracht werden/die do durch wunderbarliche krafft/diesen Privatzen allen fürkommen. Dann es bedarff nit einer kleinen krafft unnd tugent/wider solch groß und ubertreffenliche kranckheit vor zu bewaren und zu behüetten."

obviously alludes to his *Archidoxis*, a treatise certainly begun at roughly the same time as *Das Sechste Buch in der Artzney* but most probably continued, or perhaps completed, later.[25]

These two early mentions of the art of separation were followed by countless others throughout the years, either with the names "Scheidung," "*separatio*," "*spagyria*" or the like. The use of the word "*spagyria*" and its derivatives reached a peak in the Basel period in such works as *De Gradibus, De Vita Longa*, or the surgical works, and spread nearly everywhere in the other writings of Paracelsus. In the *Opus Paramirum* he even used the phrase "*scientia separationis*," and elsewhere, "*ars spagyrica*." Separation was sometimes the means by which the stomach-alchemist separated the good from the feces within the human body.[26] Elsewhere it was the tool used by the physician to separate a disease from the vital force,[27] or to segregate the poison of a natural product from its healing virtue.[28] The notion of separation had a number of possible applications.

Nothing of this says much, however, about the context in which Paracelsus came to praise separation beyond every other chymical procedure. A closer look at his writings hints at two different contexts, one religious, the other empirical. The former draws on Genesis: Creation is the model that the physician must emulate. As God separated the light from darkness, so the physician must learn to recognize the darkness in all created things that prevents them from becoming remedies.[29] As God created separation ("Scheidung") to separate everything contained within the great world from evil and death, so the physician must use it in the small world—i.e. the human body.[30] In cosmological writings, "Scheidung" is the means by which the

---

[25] Kahn (2024). *Das Siebendte Buch* contains a huge number of references to remedies described in the *Archidoxis*, whereas *Das Sechste Buch* only has very few references to them.

[26] A prominent feature in the early *Volumen Medicinae Paramirum* ("De Ente Veneni"), among others. See e.g. H 1:24–25, 30, 31.

[27] See below, n. 33.

[28] E.g. in *Volumen Medicinae Paramirum*, H 1:27; *Drey Bücher der Wundartzney Bertheoneae*, H C:361c.

[29] *Von Ursprung, Herkommen und Anfang der Frantzosen Acht Bücher* (H C:217b): "Von anbegin/das ist/von anfang der Welt/ist dz. Exempel der Artzney gesetzt worden/nach welchem wir Artzt uns richten sollend. [...] Erstlich in Beschaffung der Welt/ist Tag und Nacht ein ding gewesen/das war ein Dunckele/und zu nichten gut. Damit aber das es in nutz keme/ward das Liecht vom Finstern gescheiden/unnd also ward das Liecht der Tag/und die Nacht der Mond/und sein Gestirn. Das ist uns ein Exempel/das wir ein jegliche gewachsene Artzney/ein solche Dunckele zu sein sollen erkennen/unnd das mag nit widerredt werden."

[30] Ibid., 217b–c: "Dann der geschaffen hat die Scheidung/und die Scheidung gemacht/der hat gescheiden das jhenig/das die grosse Welt erhelt/vor ubel und tödtligkeit. [...] Auff solches wissend/so nuhn die Scheidung bey dem Artzt sein soll/das er zu gleicherweiß/ein Werck macht in der kleinen Welt zu schemen/Als Gott an der Sonnen gemacht hat. Und wie die Krafft der Sonnen und des Mons sind/also ist auch die Krafft seiner gescheidener Artzney/ein theil Sonn/ein theil dem Mongleich." See also e.g. *Von den natürlichen Bädern*, H 7:297.

Creator separated all natural products from the prime matter—to put it simply. Separation is the straightforward path in the process of *creatio ex nihilo*.[31]

The empirical context, on the other hand, is primarily that of mining and refining: for example, as gold separates ("sich scheidt") from its gangue, so the disease must be separated from life to be eliminated. Nor of course was gold the only metal that Paracelsus saw as undergoing "Scheidung" during its extraction. Building on an elaborate comparison in which he argues that subterranean mineral and metal veins form giant underground "trees," Paracelsus argues that iron, silver, and other metals undergo a separation from their earthly impurities during their refining in the same way that a chestnut can be removed from the parts of its shell and membrane:

> There are various trees that give their fruit not simply, but mixed, just as a chestnut and the like, which has externally a rough shell, after that another, then a membrane over the kernel. So there are also metals and genera of minerals that also exist in the form of such meats and skins, such as iron ore, silver ore, and other ores from which one must separate [*scheiden*] them, upon which one then finds the fruit within, which has been separated [*gescheidet*].[32]

Moreover, as it may happen that it is up to the artist to follow up a task left unfinished by nature, such as separating the pure from the impure, the metal from the ore, so it is up to the physician to complete what nature has left unfinished in the microcosm [i.e., man].[33] Were it not for separation ("Scheidung"), there would be neither good nor base metals brought to light. Similarly, in the human body there would be neither health nor disease without separation.[34]

Another "Scheidung"-comparison occurs twice, that of the distillation of wine. In a fragment from a book on mucilaginous substances within the body, Paracelsus states that such a substance is also found in wine as well as a residue from distillation of brandy and is called *Isop*.[35] Then, in his treatise *Von den Natürlichen Dingen*,

---

[31] See e.g. *Philosophia de Generationibus et Fructibus Quatuor Elementorum*, H 8:64, 66, 130 ff. See also Weeks and Kahn (2024), 32–33, and the relevant passages in this English translation of the *Philosophia de Generationibus et Fructibus* and other such texts.

[32] *Liber de Mineralibus*, H 8:340: "Es sind ettlich Beum/die geben ihr Frucht/und aber nit bloß/ sonder gemengt. Als ein Kesten/und ihrs gleichen/hatt am eussersten ein rauhe Schelffen/dornach ein andere/demnach ein Heuttli uber den Kern. Also sind auch Metallen/und *genera Mineralium*, die auch in solchen *Carnibus* und *Cutibus* ligend: Als Eisenertz/Silberertz/vnd ander Ertz/darumb mans muß darvon scheiden/so find man darnach die Frucht in demselbigen/so es gescheiden wirt."

[33] *Von den Blatern, Lähme, Beulen, Löcheren unnd Zittrachten der Frantzosen*, H C:269a: "Weiter auch wie sich in *Mineralibus Aquarum* begibt/dz. sich das Gold von Ertz scheidt/also scheidt sich auch der *Morbus* vom Leben. […] Und wie sich auch begibt/das der *Artifex* das/so der Natur ist uberbliben/vollenden soll/als das scheiden des Reinen vom unreinen/des Metallen vom Ertz: Also stehet neben im gleich der Artzt/das er solches/das der Natur in *Microcosmo* uberbleibt/vollende." This comparison is developed further on the following page (269b).

[34] *Drey Bücher der Wundartzney Bertheoneae*, H C:359b: "Wann wo die Scheidung nicht were/so würden weder gut noch böß Metallen an tag kommen: Also auch im Leib/so die Scheidung nicht were/weder Gesundtheit noch Kranckheit begegnete. Dieweil aber alle ding in die Scheidung geordnet seind/so ist der erste grund/das die scheidung soll erkennt werden bey einem Artzt." See also ibid., 362a.

[35] Paracelsus means viscous wine dregs. See *Ex Libro de Mucilagine* (H 5:216): "Also *de Pleuresi, Podagra* unnd *Arthetica* zu Judicieren. Item/also auch ist ein *Mucilago* im Wein: Aber sein *Mucilago* der ist viel/wirdt *Isop* genannt/der bleibt vom Brannten Wein/etc."

he explains that he who is either a physician or an alchemist must not use raw sulphur without a preparation: its *arcanum* must be separated from its "filth" until it becomes as white as snow. *Ysopus* belongs there, he adds, "that is, the art of separation, which has been called *Ysopaica* for ages in alchemy and every work of separation."[36]

From this partial selection, one can see the enormous significance that Paracelsus ascribed to "Scheidung" from his earliest years up to his maturity and beyond. Most striking is the discrepancy between the importance of "Scheidung", or analysis, to Paracelsus, and his relative lack of interest in synthesis. It is generally assumed that *spagyria* is the name used by Paracelsus for alchemy. However, the definitions of *spagyria* in Paracelsus are unequivocal: it is always defined as the art of separating the pure from the impure, excluding any mention of synthesis.[37] Even when Paracelsus spoke of "divine alchemy" ("Göttliche Alchimisterey"), it was only with reference to the separation of day from night, hot from cold, etc. in the creation of the world[38]—in other words separation, not synthesis.

Synthesis, however, does occur in a few instances in the works of Paracelsus. In *Von den ersten dreyen Essentiis* he began the treatise explaining, as he did in several other works, that every natural body produced by its own element, i.e. by one of the four elements, is composed of the three principles, salt, sulphur and mercury. But then he went on, explaining the process: from the three principles a conjunction occurs, yielding a body with a united essence. Unfortunately he did not pursue this line of argument, and explained instead which qualities each of the *tria prima* was endowed with, before turning to medicine.[39] Perfectly aware as he was that his the-

[36] *Von den Natürlichen Dingen* (H 7:164–65): "Der ein Artzt ist oder ein Alchimist/der soll den *Sulphur* nit brauchen/wie er an ihm selbst ist: Sondern Separirt in sein *Arcanum,* vom Unflat sauber geweschen und geschieden/dz. er werde baß geweschen/unnd werde in seiner Tugendt weisser dann der Schnee. Darzu gehört *Ysopus,* das ist/die Kunst *Separandi,* die dann von alter her *Ysopaica* heist/in der Alchimey und aller Sequestration. Roh aber ist [er] zu dem gemeinen Mann/ zu dem gemeinen Handel ein trefflich ding." Whether *Isop*/*Ysopus*/*Ysopaica* were derived from the plant hyssop (and why), or from any other source, is unknown.

[37] See e.g. *Quatuordecim Libri Paragraphorum* (H 3:360): "und solchs confortieren muß geschehen *per spagyricos gradus*, das *purum ab impuro* zogen werde." *Eilff Tractat: Vom Stul Lauff* (H 4:179): "gleich einem Spagirischen der die ding alle subtil aufftreibt/und scheidt/reinigt/unnd in viel weg/jetzt in dem weg/darnach in ein andern weg/so lang biß gefunden wirdt das jenig das er begert." *Scholia in libros de Gradibus* (H 7:363): "Spagyrus dicitur, qui singulas corporum substantias, purum corpus ab impuro, separare novit, habetque rerum experientiam." Ibid. (H 7:366–67): "Ubi namque desinit natura, Spagyricus incipit. Cum purum ab impuro separatur, feces quae remanent, nihil omnino valent. Primo igitur fiat separatio puri corporis ab impuro, estque tunc in secundo gradu, scilicet operativae virtutis. Deinde sublimatio separat corpus illud puratum a puriori per digestionem, scilicet, so du last digeriren/& supernatantia semper depone, quousque desunt: estque reliquum in tertio gradu: postea distilla in Sole, eritque in quarto gradu."

[38] *De Meteoris* (H 8:214). See Weeks and Kahn (2024), 32–33.

[39] *Von den ersten dreyen Essentiis* (H 3:15): "Ein jetlichs gewechs/daß sein Element producirt/wirt in drey ding gesetzt/dz. ist/in *Sal, Sulphur* unnd *Mercurium*: Auß den dreyen wirt ein Conjunction/ die gibt ein *corpus* und ein vereinigts wesen. Was hie dz. *corpus* antrifft/wirt nicht gemelt/allein das inner des *corporis*."

ory of the three principles necessarily involved their synthesis, he paid the latter very little attention.

Only in one other work, *De Renovatione et Restauratione*, did Paracelsus mention synthesis. Renovation and restoration of metals, he wrote, can be attained by reducing the metal to its three constituent principles, the *tria prima*. Then the "matter of the *tria prima*" may become the same metal again. On the contrary, human beings cannot be reduced to their prime matter—the *tria prima* or the human seed –, for this matter is beyond our grasp. Furthermore, it is not in our power to bring a human being up to a new birth and even create an immortal creature.[40]

These two occurrences are to be contrasted to the massive number of occurrences of separation and its synonyms in the works of Paracelsus. Thus, *spagyria* cannot be considered as a plain synonym for alchemy in the usual sense of the word: it lacks an essential component, namely synthesis, which is an integral part of traditional alchemical sources. To cite but one example, the late medieval *Theorica et practica* ascribed to the Franciscan alchemist Paul of Taranto explicitly states that alchemists can fabricate the known metals by conjoining the proper quantities of their principles, mercury and sulfur, as long as those principles also have the appropriate degree of volatility, the correct color, and the right purity. Paul even goes so far as to use the analysis of metals by fire and corrosives, followed by their resynthesis, to disprove the influential Thomistic theory that "perfect mixts" like the metals cannot be analyzed into their ingredients. This early example of the *reductio in pristinum statum,* a type of demonstration that would acquire great fame in the writings of Daniel Sennert and Robert Boyle, was a gift of alchemy, but not of Paracelsus.[41] Thus, the original model for "Scheidung" was not necessarily alchemy per se, if by that term one means the transmutation of metals, for the creation of a new metal, whether by direct combination of sulfur and mercury or by conversion of one metallic species into another, implied a type of synthesis. Hence transmutative alchemy typically included synthesis and, most of all, did not emphasize separation above all other operations.

---

[40] *De Renovatione et Restauratione* (H 6:101): "mag derselbig wol wider zu seinen dreyen Ersten kommen/daß sein Saltz/sein Schwefel/und sein *Mercurius* widerumb erscheinen/als in seiner ersten Geberung/und des Metallen Wesen gantz vergeht/unnd kein Metall mehr ist. Darnach mag auch wol beschehen/daß die *Materia trium Primarum* zu einem Metall wider wirt/als vor: Als auß deß Kupffers Ersten dreyen/widerumb ein Kupffer. Das ist auch wol *Restauratio* und *Renovatio* in den Metallen: dann er ist newgeboren/auß eim gemachten Metallen und perficirten. Aber diß ist kein *Restauratio* noch *Renovatio* hie/zu rechnen gegen dem Menschlichen: Dann auß ursachen/dz. wir nit mögen gebracht werden in die drey Ersten/oder in unser *sperma*, auß dem wir wider möchten Renovirt unnd Restaurirt werden/wie wir jetzt haben angezeigt von den Metallen: denn es wer darnach in unserm gewalt/das wir uns möchten besseren in der andern Geberung/dann die Erst gewesen were [...] Also wir auch auß uns möchten ein untödtliche Creatur Schöpfen/des wir nit Macht haben: unnd also einer solchen *prima materia* sind wir beraubt/unnd in ein unwiderbringliche gewandlet/die nit mag zuruck gezogen werden/sondern muß fürfaren/wie sie angefangen hat/und nicht gedencken dem wider zu zukommen/davon es außgangen ist."

[41] Newman (2006), 40–44.

To summarize the points that we have been making, Paracelsus's longstanding fascination with "Scheidung" or analysis, coupled with his relative lack of interest in synthesis, provides considerable reason to doubt that *Spagyria*, the synonym of "Scheidung", stemmed from a fusion of the Greek σπάν and ἀγείρειν. Aside from the fact that neologisms based on Greek words are rare in his works and awkwardly made up,[42] why would Paracelsus have built synthesis into the word (in the form of ἀγείρειν) only to ignore practically all discussion of synthesis in his corpus? From what source then did Paracelsus derive the term *Spagyria?*

3.3 Spagyria, Spagürlein, and Saigerprozess

As we stated earlier in this paper, an alternative possibility is that Paracelsus built the odd word *Spagyria* on the name of a contemporary Swiss coin, the *Spagürlein* (also called *Spagürli* or *Spagürle*). The word appears once in the edited work of Paracelsus, where he uses it to disparage learned physicians, whose fancy Greek terms are not worth "a *Spagürlein* more than [those of] the *Wallseer* who live in the high mountains and think no otherwise than that their language is that of the whole world".[43] As the vitriolic passage implies, the *Spagürlein* was a coin of low value. According to the Grimms' *Wörterbuch*, its value was set at three Lucerne *pfennige* in 1477.[44] Other sources, and indeed surviving specimens, reveal that the coin was made of silver, albeit not of the purest sort.[45] Initially stemming from mints in Northern Italy, the *Spagürlein* also began to be coined in the Swiss cantons in the first decade of the sixteenth century.[46] It was a widely dispersed coin in early modern Switzerland, as the quotation from Paracelsus suggests (Fig. 3.1).

[42] See e.g. *yliaster/iliastes, idechtrum, taphneus*… In his *Theophrasti Paracelsi* […] *Compendium* Gohory tried to divine Greek etymologies in such Paracelsian neologisms from *De vita longa* as *necrolii, scaiola, aniadus, adech,* and, less obscurely, *ilech.* Only his etymology of *spagyria* was found convincing by contemporary Paracelsians.

[43] Paracelsus, *Drey Bücher von den Frantzosen*, in Paracelsus (1605), 160b: "Nun so etliche Bücher der Artzney auff dem Kriechischen angefangen haben/vermeinen sie/dieweil die Sprach die Bücher regier/so regier sie auch die Krancken. Also lernen sie die Griechischen Bücher lesen/unnd so sie dieselbigen außlernen/so können sie nichts/unnd werden also *Doctores*/die heissen nicht Artzt/ sonder Kriechen. Kein Artzt soll sich beschirmen mit der Sprach/allein mit Practick. Nit mit einem Spagürlin ist Kriechen mehr begabet/dann die Walseer in den Hohen Bürgen/die doch auch nit anders meynen/jr Sprach sey die gantz Welt." As suggested by Urs Leo Gantenbein, whom we warmly thank, "Bürgen" is nothing more than an idiosyncratic way of writing "Bergen". Walser German is a dialect spoken by inhabitants of the Alps and other regions, known as the Walser people.

[44] See the online Grimm *Wörterbuch sub voce Spagürlein* at https://woerterbuchnetz.de/?sigle=DWB, consulted 1/25/23.

[45] See *Coin Archives* at www.coinarchives.com/w/openlink.php?l=5865420%7C6702%7C2309%7C8452851c0d7a6c67b919976e1716612f, consulted 6/2/24.

[46] Kunzmann and Luraschi (2000–2002).

Fig. 3.1 Obverse and reverse of a sixteenth-century *Spagürlein* from Lucerne. (Courtesy of Leu Numismatik)

But why then would Paracelsus have given his beloved pursuit of "Scheidung" a name derived from a low-grade silver coin? As we have already seen, one of the principal occupations that Paracelsus linked to "Scheidung" was the set of metallurgical operations required to separate commercially useful metals from their ores and to refine them once they had been smelted.[47] Coin manufacture in turn required metals of standard composition, which could only be acquired by means of refining. The mining and refining of silver had undergone a major innovation in the century or so before the birth of Paracelsus with the discovery of the *Saigerhüttenprozess* or *Saigerprozess*, a method of refining silver from copper ores that had previously been unprofitable for that purpose.[48] The *Saigerprozess* consisted of two principal operations, both of which followed the initial reduction of impure copper (often containing less than 2% silver) from its ore. The processes involved liquation of the reduced, impure copper after it had been enriched with lead in order to separate out a lead-silver mixture, followed by "drying" or oxidation of the exhausted metallic "cake" to remove residual lead in the form of molten litharge containing most of the remaining silver. The combined lead-silver that the liquation provided would then be cupelled in order to separate the silver in relatively pure form. As for the silver-rich litharge yielded by the "drying" process, it could be smelted to yield argentiferous lead, which could in turn be cupelled as well. This rather complicated but effective refining process helped set the stage for a boom in European mining during the fifteenth and sixteenth centuries. It was an example of contemporary high-tech that yielded immediate economic consequences.

At the same time, the *Saigerprozess* was also a striking example of mineral analysis, involving sequential stages of "Scheidung". First there was the smelting of the

[47]This point has also been made by Urs Leo Gantenbein, though without reference to the Swiss *Spagürlein* coin. See Gantenbein (2000).

[48]Most of the information found here on the *Saigerprozess* stems from L'Heritier and Tereygeol (2010).

copper ore, which involved roasting and then reduction of the metal. The nature of these operations as a "Scheidung" potentially involving the three Paracelsian principles would have been evident both from the sulfurous fumes given off during the roasting and from the slag that was left behind as a residue. Second, the liquation of the lead-copper-silver cake during the *Saigerprozess* itself had as its goal the separation of the lead-silver from the copper, hence a second "Scheidung." The "drying" process was also an example of *Scheidekunst*, since it resulted in the flow of molten, silver-rich litharge, from the metallic cake left behind by the liquation. Finally, the cupellation itself provided yet another stage of "Scheidung", since the lead and the porous walls of the bone-ash cupel provided a final separation of the silver from the lead.

One can see then that the sixteenth-century metallurgy of extracting silver provided a spectacular platform for observing entire sequences of "Scheidung." Even if such small change as the *Spagürlein* did not undergo the full panoply of refining processes available in the sixteenth century, and may even have been made of low-quality billon that had undergone surface enrichment, some of these sophisticated operations would have gone into its production.[49] But what did Paracelsus actually know of such practical metallurgy at firsthand? Here our knowledge is unfortunately vague, as indeed it is for most of Paracelsus's life. We know, for instance, that he moved with his father from the Swiss village of his birth, Einsiedeln, to the southern Austrian town of Villach when he was about 9 years old, in 1502. Villach had been a center for lead mining since the late Middle Ages, thanks to the immense *Bleiberg* located there, which was mined until 1993. But we have no idea how long Paracelsus remained in Villach before he took up a life of wandering at some point in the 1510s. His experience of Villach may well have been restricted to his childhood and early teenage years. Our first piece of relatively solid knowledge about Paracelsus's association with the mining industry stems from a famous passage in his *Grosse Wundarznei* of 1536. Here Paracelsus lists five famous churchmen whom he claims to have been teachers of his *von kintheit auf*, and from whom he claims to have learned *adepta philosophia.* At the end of this list, however, Paracelsus adds another teacher of quite different character, namely "the noble and steadfast Sigmund Füger of Schwaz, together with a number of his laborants."[50]

Since the 1880s it has been known that this Sigmund Füger was actually Sigmund Fieger of Schwaz, a member of the lesser Austrian nobility who controlled a famous silver mining industry in Carinthia.[51] Referred to in its day as "the mother of all mines" ("Aller Bergwerke Mutter"), Schwaz was perhaps the most famous center of

---

[49] Dr. Christian Weiss, Kurator Numismatik & Siegel at the Schweizerisches Nationalmuseum, kindly informs us that the *Spagürlein* was often made of debased alloy (billon), which then underwent a process of surface enrichment called *Weißsieden*, which worked by removing copper and other base metals from the surface of the coin flan by means of cooking in weak acids, such as found in a solution of tartar. Once the coin was put into circulation, the thin layer of purer silver would eventually wear off, revealing a more coppery color beneath.

[50] *Das Ander Buch der grossen Wundartzney* (H C:102a): "der Edel und Vest Sigmund Füger von Schwatz mit sampt einer anzal seiner gehaltenen laboranten." See Benzenhöfer (2002), 28.

[51] Schubert and Sudhoff (1889), 86–87.

mining in the Alps. According to one modern source, the primary ore available in Schwaz was a type of *Fahlerz*, a complex sulfide mineral containing numerous metallic elements in addition to copper and silver. The extraction technology employed in Schwaz came to be known as the *Abdarrprozess*, a local variant of the *Saigerprozess.*[52] Fieger not only directed the mining operation at Schwaz, but also employed the workers to whom Paracelsus refers in a laboratory probably devoted mainly to refining and *Probierkunst*. According to Eduard Schubert and Karl Sudhoff, Paracelsus was active in Schwaz at some point between 1510 and 1520; more recent scholarship has suggested the period 1522/23.[53] Unfortunately, we have been unable to uncover any evidence to support either set of dates, though Paracelsus's acknowledgement of Fieger's role in his education suggests that their interaction was early. At any rate, the first occurrences of the relevance of separation to Paracelsus date back, as we have seen, to writings from his Salzburg period (1524–1525). For now we can only say that it is entirely possible that the *Saigerprozess* in its Tyrolian incarnation was an impetus to Paracelsus's subsequent emphasis on *Spagyria* as the art of separation.

3.4 Conclusion

Although we have thrown light on the likely connection between the term *Spagyria* and the technology of metallic extraction and refining, particularly that of silver, obvious questions still remain. Given that Paracelsus's experience with the world of mining and metallurgy probably began either in Villach or Schwaz—both of them principalities in Austria—why would he have chosen the name of a small Swiss coin for his innovation? Many possibilities exist, given the incompleteness of our data on the life of Paracelsus. Yet one thing is clear: despite having abandoned Einsiedeln at an early age, Paracelsus was proud of being Swiss. As he says in his *Chronik des Landes Kärnten*, Carinthia may have been his "second fatherland," but Switzerland was the first.[54] It is entirely possible, though of course far from sure, that the term *Spagyria* may have been a reflection of Paracelsus's self-identification as a Schweizer. Furthermore, its derivation from the name of the Swiss coin may explain that the first occurrence of the word *spagyria*, or more exactly *spag*[*yrus*],[55] recently discovered by Urs Leo Gantenbein, is not in the Basel writings among Paracelsus's other numerous neologisms of this time period as was formerly

[52] Soukup (2007), 211 and 128, 132–133.

[53] Schubert and Sudhoff (1889), 87; Soukup (2007), 209.

[54] Paracelsus (1928), 4: "das ander mein vaterlant…" See likewise the *Grosse Wundarznei*, H C:56a: "Ich hab hierinn bißher ein Ländtlichen Spruch geführet/das mich keiner Rhetoric, noch Subtiliteten berühmen kan/sonder nach der Zungen meiner Geburt/und Landssprachen/der ich bin von Einsidlen/des Lands ein Schweitzer […]."

[55] As explained by Gantenbein in his apparatus criticus (see note 57 below), only the first four letters are legible due to the fold of the manuscript. The last four are the most plausible conjecture.

assumed,[56] but in one of the possibly earliest writings of Paracelsus, the *Super Salve regina explicatio*—a Latin treatise on the Virgin Mary.[57] Unlike his other neologisms, *Spagyria* seems to reflect Paracelsus's experience with the Switzerland of his youth, and the derivation of the term from a coin was probably meant to reflect the multiple processes of "Scheidung" that went into its manufacture, not the paired appearance of analysis and synthesis proposed by later Paracelsian commentators.

Finally, there is yet another possibility as to why the *Spagürlein* might have been associated with *Scheidung* in the mind of Paracelsus. In modern German, the term "Scheidemünzen," literally "division-coins," or "fractional coins" refers to small denomination coinage whose face value often exceeds the worth of its material. As we have pointed out, the *Spagürlein* was itself a low-value coin, and in Swiss numismatics it is in fact referred to sometimes as a "Scheidemünze."[58] Unfortunately, we have not been able to verify any occurrence of the term "Scheidemünze" before the seventeenth century, though the expression "Entscheidung der Oberwehr" (division of large coinage) already appears in the second half of the fifteenth century.[59] It remains an intriguing possibility whether Paracelsus was thinking of *Scheidung* in this sense as well when he engaged in his own coining of the expression *Spagyria*.

What further ramifications can we draw from this study? Above all, it appears that the tradition of paired analysis and synthesis already present in medieval alchemy (and no doubt earlier) did not play a significant role in the work of Paracelsus himself, whatever his early followers may have said. The pairing of analysis and synthesis, which supported the overthrow of the Thomistic theory of perfect mixture and the formulation of atomic and corpuscular theory in the seventeenth century, and which went on to serve as a basis for Antoine Laurent Lavoisier's famous "balance-sheet" method for relating input and output in chemical reactions, did not descend from Paracelsus. It was rather the product of an independent and earlier alchemical tradition that Paracelsus' followers integrated into his work. This is not the only case where earlier alchemical innovations have been absorbed by the reputation of Paracelsus, of course. One thinks of the medical alchemy pioneered in the fourteenth century by John of Rupescissa and pseudo-Ramon Lull, whose works Paracelsus knew both at first- and second-hand. Paracelsus invented iatrochemistry no more than he invented the paired process of analysis and synthesis. We say this

[56] Benzenhöfer (2005), 219–20 ("Kennwörter der in Basel gehaltenen Vorlesungen"). On his list of "Kennwörter der sicheren bzw. Wahrscheinlichen Frühwerke," Benzenhöfer listed as well one occurrence of "spagirisch." This occurrence, however, appears in the *Elf Traktat* (H 4:179), a work actually posterior to the Basel period, as demonstrated by Weeks (1997), 38–40.

[57] We warmly thank Urs Leo Gantenbein for indicating this finding. The text will be part of his *Neue Paracelsus-Edition*, vol. 2 (forthcoming). The occurrence is: "medicus ut peritus spagyrus, natura vera ut operatrix." As pointed out by Gantenbein, this is a paraphrase of a common medieval medical dictum: "natura omnium est operatrix, medicus vero minister."

[58] Kunzmann and Luraschi (2000–2002), 25.

[59] The earliest reference to "Scheidemünze" in the Grimms' *Wörterbuch* stems from 1691. See https://woerterbuchnetz.de/?sigle=DWB&lemid=S05968 *sub voce*. For the 1474 expression "Entscheidung der Oberwehr," see Maßmann (1911), 7. We thank Jutta Schickore for this reference as well as several others concerning "Scheidemünzen."

not in an attempt to belittle Paracelsus, however, but rather with the goal of determining the precise reasons for his undeniable success, beginning in the second half of the sixteenth century, and continuing even today. Obviously part of this success was due to the efforts of the first generation of Paracelsians to integrate Paracelsus into the learned tradition of the *prisca philosophia*, as we can see in both Gohory's and Bodenstein's writings, not to mention others of their contemporaries. The spurious Greek etymology of *spagyria* was part of this effort. Ironically, the sole mention of the *Spagürlein* made by Paracelsus himself occurred in the midst of a rabid attack against the humanist doctors who, he argued, thought of themselves as skilled physicians due to their knowledge of Greek, not their practical medical training.

Bibliography

Benzenhöfer, Udo. 2002. *Paracelsus*. Hamburg: Rowohlt Taschenbuch (1st edition, 1997).

———. 2005. *Studien zum Frühwerk des Paracelsus im Bereich Medizin und Naturkunde*. Münster: Klemm & Oelschläger.

Blaser, Robert-Henri. 1979. *Paracelsus in Basel*. Muttenz: St. Arbogast Verlag.

Copus, Martin. 1569. *Das Spissglas Antimonium oder Stibium genandt/in ein Glas gegossen/es sey Seel oder Rodt/das man* Vitrum Antimonii *nennt/ein warhafftige Gifft und gantz gefehrliche schedliche Artzney sey.* No place [Braunschweig]: no name.

de Menten, Pierre. 2013. *Dictionnaire de chimie: Une approche étymologique et historique*. Brussels: De Boeck.

Du Chesne, Joseph. 1576a. *Sclopetarius, sive De curandis vulneribus, quæ Sclopetorum & similium tormentorum ictibus acciderunt, Liber. Ejusdem Antidotarium Spagiricum, adversus eosdem ictus*. Lyon [Geneva]: Jean Lertout.

———. 1576b. *Traitté de la cure generale et particuliere des arcbusades, avec l'antidotaire spagirique pour preparer & composer les medicamens*. Lyon [Geneva]: Jean Lertout.

Gantenbein, Urs Leo. 2000. Die Beziehungen zwischen Alchemie und Hüttenwesen im frühen 16. Jahrhundert, insbesondere bei Paracelsus und Georgius Agricola. *Mitteilungen, Gesellschaft Deutscher Chemiker / Fachgruppe Geschichte der Chemie* 15: 11–31.

———. 2020. The Virgin Mary and the Universal Reformation of Paracelsus. *Daphnis* 48: 4–37.

Gilly, Carlos. Forthcoming. Theodor Zwinger e la crisi culturale della seconda meta del Cinquecento. https://doczz.it/doc/282880/theodor-zwinger-e-la-crisi-culturale-della-seconda-metà-del. Accessed 30 December 2022.

Gohory, Jacques (Leo Suavius). 1567. *Theophrasti Paracelsi Philosophiae et Medicinae Utriusque Universae Compendium.* Paris: Philippe de Roville.

———. 1568. *Theophrasti Paracelsi Philosophiae et Medicinae Utriusque Universae Compendium.* Basel: Pietro Perna.

Hauck, Dennis William. 2008. *The Complete Idiot's Guide to Alchemy*. New York: Alpha Books.

Kahn, Didier. 2016. *Le Fixe et le Volatil. Chimie et alchimie, de Paracelse à Lavoisier*. Paris: CNRS Editions.

———. 2024. L'Archidoxis et la Theophrastia. In *D'Uranie à Gollum: Mélanges en l'honneur d'Isabelle Pantin*, ed. Jean-Charles Monferran, Tristan Vigliano, and Alice Vintenon, 45–72. Paris: Champion.

Kopp, Hermann. 1844. *Geschichte der Chemie*. Vol. 2. Braunschweig: Vieweg.

———. 1869. *Beiträge zur Geschichte der Chemie*. Vol. 1. Braunschweig: Vieweg.

———. 1886. *Die Alchemie in älterer und neuerer Zeit*. Heidelberg: Carl Winter.

Kühlmann, Wilhelm, and Joachim Telle. 2001. *Corpus Paracelsisticum*. Vol. 1. Tübingen: Max Niemeyer.

Kunzmann, Ruedi, and Fabio Luraschi. 2000–2002. Bissoli aus der Münzstätte Bellinzona. *Schweizer Münzblätter* 50–52, Heft 202: 24–26.
L'Heritier, Maxime, and Florian Tereygeol. 2010. From Copper to Silver: Understanding the Saigerprozess through Experimental Liquation and Drying. *Historical Metallurgy* 44: 136–152.
Libavius, Andreas. 1606. *Alchymia, Commentariorum in librum primum alchymiae partis I. Lib. I*. Frankfurt: Peter Kopff.
Maßmann, Fritz. 1911. *Geschichte der Scheidemünze und Scheidemünzumlauf im Handelskammerbezirk Dortmund*. Heidelberg: Rössier & Herbert.
Newman, William R. 2006. *Atoms and Alchemy. Chemistry and the Experimental Origins of the Scientific Revolution*. Chicago–London: The University of Chicago Press.
Paracelsus, Theophrastus. 1605. In *Chirurgische Bücher und Schrifften*, ed. Johann Huser. Strasbourg: Lazarus Zetzner.
———. 1928. In *Sämtliche Werke*, ed. Karl Sudhoff, vol. 11, 1st ed. Munich–Berlin: R. Oldenburg.
Partington, James Riddick. 1961. *A History of Chemistry*. Vol. 2. London: Macmillan.
Principe, Lawrence M. 2013. *The Secrets of Alchemy*. Chicago–London: The University of Chicago Press.
Schubert, Eduard, and Karl Sudhoff. 1889. *Paracelsus-Forschungen*. Vol. 2. Frankfurt a. M: Reitz & Koehler.
Soukup, Rudolf Werner. 2007. *Chemie in Österreich: Bergbau, Alchemie und frühe Chemie*. Wien: Böhlau.
von Prantl, Carl. 1856. Die Keime der Alchemie bei den Alten. *Deutsche Vierteljahrs-Schrift* 1–2: 135–151.
Weeks, Andrew. 1997. *Paracelsus: Speculative Theory and the Crisis of the Early Reformation*. New York: SUNY Press.
Weeks, Andrew, and Didier Kahn. 2024. Introduction. In Paracelsus. *Cosmological and Meteorological Writings*. Leiden: E. J. Brill.

Didier Kahn is Senior Researcher at the CNRS (CELLF, UMR 8599, Sorbonne Université). His work focuses on the works of Paracelsus, the history of alchemy and its relations to science, literature and religion in early modern France and Europe. He is also a member of the editorial team of Diderot's complete works known as DPV (Paris: Hermann).

William R. Newman is Distinguished Professor in the Department of History and Philosophy of Science and Medicine, Indiana University (Bloomington). His main present research interests focus on early modern "chymistry" and late medieval "alchemy," especially as exemplified by Isaac Newton, Robert Boyle, Daniel Sennert, and the first famous American scientist, George Starkey.

Chapter 4
Chymistry Goes Further: Sensible *Principiata* and *Things Themselves* Over the Longue Durée

Joel A. Klein

Abstract This paper historicizes a constellation of interrelated ideas regarding the chymical principles as they developed and became resilient fixtures within a major chymical tradition. Focusing primarily on German chymists, several of whom have eluded sustained historical interest, it explores how experimental analysis was generally thought to produce sensible chymical principles, often conceived as *principiata*: bodies produced by combining or mixing fundamental elements or principles. These *principiata* allowed for the establishment of hierarchies of increasingly complex compounds and helped to define the most fundamental components of matter. Chymists' ability to separate tangible substances that were believed to be fundamental was considered so central to the chymical enterprise that it came to define or delimit chymistry itself and was often used to attack groups perceived as overly speculative or less empirical. This chymical tradition, which included Paracelsus as well as figures such as Andreas Libavius, Daniel Sennert, and Georg Ernst Stahl, significantly influenced later chemistry. Its approach to hierarchies of combinatorial principles was integral to the concept of chemical compounds and the delineation of chemistry as an autonomous discipline.

Keywords Chymistry · Analysis · Principles · Principiata

4.1 Introduction: Charging Down the Blind Alley?

In 1958, Marie Boas [Hall] wrote regarding the study of early modern elements and chymical principles that "historians are well advised to consider it a blind alley, and to look elsewhere for the theoretical problems which could and did aid in the advance of chemistry," concluding instead that the mechanical philosophy was the main precursor to modern chemistry (Boas 1958, 142). This general view that the principles were unrelated—or a hindrance—to the emergence of science has exerted

---

J. A. Klein (✉)
The Huntington Library, Art Museum, and Botanical Gardens, San Marino, CA, USA
e-mail: jklein@huntington.org

W. R. Newman, J. Schickore (eds.), *Traditions of Analysis and Synthesis*,
Archimedes 73, https://doi.org/10.1007/978-3-031-76398-4_4

a sustained influence. Robert Siegfried and Betty Jo Dobbs argued in an influential article that Aristotelian elements and Paracelsian principles were "*a priori* schemes… conceived more as metaphysical entities, than as specific substances to be handled in the laboratory" (Siegfried and Dobbs 1968, 276).[1] Hélène Metzger likewise concluded that Aristotle's elements did nothing but corrupt chemistry and that the Paracelsian principles were only to be "investigated by the metaphysician" (Metzger 1991, 18). She made the related points that chemistry's advance was deterred by its pre-modern subordination to medicine and that it was only able to progress when it abandoned alchemy and medicine and adopted a more mechanistic and theoretical purview. More recently, Ursula Klein has concluded that pre-modern elements and Paracelsian principles did not contribute to forming the concept of a chemical compound, which instead had its origin in eighteenth-century studies on chemical affinity, inspired by the practical operations of sixteenth-century metallurgy and seventeenth-century pharmacy (Klein 1994).

Historians of alchemy and "chymistry" have challenged these conclusions and have demonstrated, for instance, that the case for the metaphysicality and irrelevance of alchemy and Paracelsianism is based largely on several misconceptions: namely, that these traditions had adopted wholesale both the concept of the perfect mixture—which would mean that their analyses of compounds created new ones from homogenous materials—as well as the idea that elements and principles were not sensible bodies or substances but rather element matrices containing form-endowing principles that were noncorporeal or spiritual bearers of qualities. I have, for instance, discussed a tradition of learned chymistry in sixteenth- and seventeenth-century Germany that directly criticized the Paracelsian understanding of element matrices and quasi-spiritual principles, and William Newman has demonstrated that concepts of unchanging particles that remain intact beneath surface-level appearances were foundational in both medieval alchemy and the tradition of early modern chymical atomism that it inspired.

In a 2014 special edition of *Ambix*, Evan Ragland and I suggested that concepts of analysis and synthesis provide a convenient throughline for tracing changes and continuities over time, thus offering to bridge some of the disconnects between pre-modern *chymistry* and modern chemistry (Klein and Ragland 2014). This paper follows upon this diachronic ambition and historicizes a constellation of interrelated ideas regarding the chymical principles as they developed and became resilient fixtures within a major tradition over the *longue durée*. My focus here is largely—but not exclusively—on German chymists, several of whom have been largely overlooked by historians.

[1] Siegfried made the upshot of these claims clear in a later work that portrayed the history of chemistry as the victory of modern materialism over metaphysical speculation. He argued that both Aristotelianism and Paracelsianism were hobbled by their "emphasis on external properties and de-emphasis on the underlying matter," and the related idea that "observed properties were not generated so much from material composition as by spiritual presence," and that chemistry could thus only progress after these principles and elements were jettisoned in the eighteenth century. See Siegfried 2002, 5, 30.

In short, I maintain that within this tradition, experimental analysis was generally believed to produce sensible chymical principles or, similarly, *principiata*: that is, bodies produced by combining or mixing fundamental principles or elements. Such *principiata* allowed for the establishment of hierarchies of increasingly complex compounds and were often considered foundational at a "negative-empirical" level whereby the limits of laboratory analysis defined the principles or elements of nature.[2] This analysis of sensible materials that revealed hierarchical systems of composition was also routinely perceived as defining or delimiting chymistry itself. Indeed, it became the primary rhetorical weapon that chymists used to defend themselves and to attack other philosophies of nature. Taking aim at groups who were perceived as excessively speculative or detached from practice and experience, chymists trumpeted their ability to separate, identify, and sometimes recombine fundamental components of nature, which they often described as tangible "things themselves" in contradistinction to the ethereal intellection of Aristotelians and mechanists alike. Emerging from these points is the broader conclusion that portraying chymistry as an enterprise concerned with metaphysical or spiritual principles misses something essential about the nature of chymistry itself and chymists' self-understanding. In effect, I suggest that this tradition of analysis and synthesis was integral not only in the establishment of the concept of chemical compounds but also to the formation of an autonomous discipline of chemistry.

4.2 Background: From Paracelsian Spagyria to Sensual Philosophy

The Swiss–German medical reformer Theophrastus Paracelsus von Hohenheim (1493–1541) established a philosophy grounded on the practice of analysis that would be taken up by numerous followers, provide a new framework for medicine, and ultimately elevate European alchemy to a philosophy with cosmological and religious implications. Paracelsus emphasized the need to separate the constituents of matter from one another through *Scheidung* or *Spagyria*, which typically meant a dissolution by fire. Based on these processes, he argued that everything in nature was composed of salt, sulfur, and mercury, which he called the *tria prima* or the "three first principles." Paracelsus was not the first to prioritize chymical analysis, but his imperative to separate natural materials into their constituent principles such that they could be manifestly discerned was particularly influential. Following Paracelsus, other chymists, including Andreas Libavius (1555–1616), Jean Beguin (ca. 1550–1620), and Joan Baptiste van Helmont (1580–1644) expanded *spagyria* beyond separation to also include recombination. This method of analysis and

[2] On the negative-empirical principle, see Thackray 1970; Bensaude-Vincent and Stengers 1996, 37; Newman 2001, 324–5.

synthesis then became so central to the chymical enterprise that, for many practitioners, it was synonymous with chymistry itself.[3]

The "spagyric analysis" separating a given substance into sensible principles also became a primary weapon used against competing traditions which were perceived as overly speculative or unmoored from practice and experience. The English Paracelsian author, Richard Bostocke, for example, wrote in 1585,

> The Chymicall Phisition ... is ruled by experience, that is to say, by the knowledge of three substanties, whereof eche thing in the great world and man also consisteth, that is to say, by their several Sal, Sulphur and Mercury, y[t] by their several properties, vertues and nature, by palpable and visible experience....The right way to come to this knowleg is to trie all things by the fire: for the fire teacheth the science and arte of Phisicke.... So shall he knowe all things by visible and palpable experience, so that the true proofe and tryal shal appeare to his eyes & touched with his hands. So shall he have y[e] three Principia, ech of them separated from the other, in such sort, y[t] he may see them, & touch them in their efficacie and strength, then shal he have eyes, wherewith the phisition ought to looke and reade with al. Then shal he have that he may taste and not before. For then shall he know, not by his owne braines, nor by reading, or by reporte, or hearesay of others, but by experience, by dissolution of Nature, and by examyning and search of the causes, beginnings and foundations of the properties and vertues of thinges... (Bostocke 1585, Dv (v & r))

Joseph Du Chesne (1546–1609), Paracelsian author and physician-in-ordinary to the French King, Henry IV, similarly argued that chymistry's superiority over traditional Aristotelianism was rooted in the revelation of perspicuous principles by means of fire analysis. He wrote,

> The chymists or spagyrists, however, leaving those bare qualities of bodies, sought the foundations of their actions elsewhere, [in] tastes, odours, and colours. At last, by a wise inquisition, they knew there to be three diverse and distinct substances, which are found by a singular artifice in every natural, elemented body: that is, salt, sulphur, and mercury... For those aforementioned virtual and sensible qualities are to be found in these three hypostatical beginnings, not by imagination, analogy, or conjecture, but by the thing itself [*reipsa*] and the effect… (Du Chesne 1603, 90)[4]

Du Chesne concluded that tastes were caused by salt, odors arose from sulfur, and colors derived from diverse sources but primarily from mercury.

Although Du Chesne criticized Aristotelian elements and qualities heavily, he and many other Paracelsians retained a key place for these in their theory of matter. Both Du Chesne and German alchemical author Oswald Croll (1563–1609), for instance, both accommodated Aristotle to Paracelsus and concluded that the

[3] On Paracelsus' *Scheidung* and Paracelsian *spagyria*, see Principe and Newman 2002; Klein 2022, 55–6.

[4] "Chymici itaque, seu spagirici relictis nudis illis corporum qualitatibus, actionum, atque ipsorum etiam saporum, odorum, colorum fundamenta in alio quaesiverint. Tandem sagaci inquisitione cognoverunt illa esse tres diversas atque distinctas substantias illas, quas in omni corpore naturali elementato singulari artificio invenerunt: nempe sal, sulphur, atque mercurium. Haecque principia rerum interna, principia constituentia, virtualia, atque hypostatica nuncuparunt. In his quippe tribus principiis hypostaticis illae memoratae qualitates virtuales atque sensibiles, non imaginatione, analogia, aut coniectura, sed reipsa & effecte reperiuntur. Nempe sapores in sale potissimum: odores in sulphure, colores ex utrisque etiam, sed potissimum ex mercurio…"

chymists' sensible principles were active but obscured by the passive elements earth and water (Klein 2014). As Croll put it, such elements "are just the bodies and homes of the others and impede and obstruct their force" (Croll 1609, 21).[5] This view was taken directly from the rather complicated but influential philosophy of the Danish Paracelsian Petrus Severinus (1542–1602), for whom the passive elements were incorporeal "first receptacles" or "matrices" covering the active Paracelsian principles, which were held together by a strong mixture and only separable and made sensible through chymical analysis.[6]

This general understanding of active, sensible principles dulled by passive elements and revealed only by analysis was certainly influential, not least among the authors of the so-called French Textbook Tradition. The apothecary and chymical author Jean Beguin (1550–1620) stated that the chymist, an "*artifex sensatus*," could analyze and demonstrate three sensible bodies, which "might be proved by momentous reasons, but evident and ocular inspection far supersede these" (Beguin 1618, 56).[7] Beguin's successors adopted this view, and the author Nicaise Le Fèvre (1615–1669), for instance, turned it toward an explicit defense of chymistry against scholastics or those "who follow the schools." He wrote,

> if you ask from the [*Physicien Chymique*], what are the parts that constitute a body, he will not give you a naked answer, and will not be content to satisfy your curiosity with mere discourses, but he will endeavor to bring his demonstrations to your sight and also your other senses, by making you to touch, smell and taste the very parts which entered into the composition of the body in question, knowing very well that what remains after the resolution of the *mixte* was that very substance that constituted it.... You see then, that Chymistry rejects such arguments, staying close to visible and tangible things, as appears in the practice of this art: because if we affirm that such a body is compounded of an acid spirit, a bitter salt, and a sweet earth, we will see, touch, smell, and taste those parts which we extract, with all those conditions we attributed to them. (Le Fèvre 1660, 10–11)[8]

Le Fèvre explicitly coupled chymical analysis with the anatomist's scalpel, for just as the anatomist had found several similar parts constituting the human body, so too did the chymist endeavor to exhibit definitive principles to the senses, and thus he

[5] "…sunt aliorum saltem corpora & domicilia, & vim illorum impediunt & remorantur." On Croll's matter theory, see Hirai 2005, 295–323.

[6] On Severinus, see Shackelford 2004.

[7] "…etsi validis rationum momentis comprobari posset: tamen eas omnes evidentia longe superat inspectio ocularis…"

[8] "Voicy donc la difference qui est entre le Physicien Chymique & le Physicien qui suit la doctrine de l'Escole: Qui est, que si vous demandez au premier de quelles parties un corps est composé, il ne se contentera pas de vous le dire simplement, & de satisfaire à vostre curiosité par vos oreilles; mais il voudra vous le faire voir aussi & le faire connoistre à vos autres sens, en vous faisant toucher, flairer & goûter les parties qui composoinent ce corps, à cause qu'il scait que ce qui demeure apres la resolution du mixte, estoit cela mesme qui faisoit sa composition... Vous voyez que la Chymie rejette des arguments de cette nature, pour s'attacher aux choses qui sont visibles & palpables, ce que nous ferons voir dans la travail: car si nous vous disons qu'un tel corps est composé d'un esprit acide, d'un Sel amer & d'un terre douce, nous vous ferons voir, toucher, flairer, & goûter les parties que nous en tirerons, avec toutes les conditions que nous leur aurons attribuées."

concluded that the chymist has been justly called a "sensual philosopher [*Philosophe sensal*]" (Le Fèvre 1660, 10–11).

While this emphasis on tangible and sensible principles might appear straightforward, the French textbook authors had been influenced by Severinian philosophy and had concluded that the principles were, as Beguin put it, "neither bodies, because they are plainly spiritual … nor spirits, because they are corporeal" (Beguin, *Tyrocinium*, 54–55).[9] As I have demonstrated elsewhere, this understanding of quasi-spiritual principles was hardly the only understanding of chymical principles in the seventeenth century and was, in fact, the subject of extensive critique by multiple learned German chymists, including, for instance, head of the Coburg Gymnasium, Andreas Libavius, and the Wittenberg professor of medicine, Daniel Sennert (1572–1637) (Klein 2014).

4.3 Prima Mixta, Principiata, and Res Ipsae

By and large, these German chymists, when compared with their Paracelsian counterparts, were no less reliant on sensible principles separated by analysis, even though they accommodated the chymical principles to Aristotle in a manner that differed considerably from that adopted by the Severinian Paracelsians. William Newman has demonstrated that Libavius and Sennert both adapted the Democritean *syncrisis* and *diacrisis* (i.e., analysis and synthesis) of corpuscular matter to Paracelsian *spagyria*, having drawn extensively from the medieval alchemy of the *Summa perfectionis* of pseudo-Geber and the kindred *Meteorology IV* of Aristotle, especially as interpreted by the neo-Aristotelianism of Julius Caesar Scaliger (see. A major aspect of both individuals' thought was a hierarchical understanding of matter and mixture that allowed Aristotle's elements to co-exist with the Paracelsian principles. As later sections of this paper will demonstrate, even as explicit references to Aristotle faded, this hierarchical understanding of matter, its related terminology, and the central importance of analysis and synthesis of sensible components came together to exert extensive influence throughout the seventeenth and eighteenth centuries.

In the *Alchemia triumphans* of 1607, Libavius criticized certain Galenists for their myopic focus on bare elements and qualities while praising chymical analysis for its ability to discover three principles: liquid, oily fat [*oleosa pinguedo*], and salt, which were analogous to the principles mercury, sulfur, and salt. Libavius explained, however, that these did not contradict the existence of traditional elements, for God had produced *mixts* from the elements, and the elements were beyond the senses, so only these mixts could be subjected to alchemical study. Libavius concluded that such mixts should be conceived as "*principiata*" rather than *principia*—that is, themselves formed from more primitive principles (Libavius 1607, 716). This

[9] "…nec corpora; quia plane spiritualia … nec spiritus, quia corporei…"

general understanding of *principiata* as things produced from more fundamental principles was widely discussed in Aristotelian philosophical traditions, and it is worth pointing out that Libavius uses *principiata* as a means of harmonizing Aristotelian physics with alchemy, writing that it is not true, "that if Peripatetic physics posits elements as the primary sensibles, the principles of alchemy are mere fabrications. One does not invalidate the other..." (Libavius 1607, 716).

Sennert (1619, 294–5) similarly argued that the chymists' principles—salt, sulfur, and mercury—were the *prima mixta*, or first mixts of the Aristotelian elements, and that these were not only responsible for most of the sensible phenomena throughout the world, such as tastes and odors, but that they were observable after a chymical distillation. Sennert quoted Julius Caesar Scaliger (1484–1558), who wrote that "There is no taste in any element, as it is an element. Nor can taste be in a compound [*composito*] from the elements." Sennert noted that the elements *qua* elements have little power to act, except insofar as they are responsible for the sensible qualities of hot, cold, wet, and dry. Higher-order effects required different entities, and thus tastes, smells and colors were attributed to the *prima mixta*, which, although composed of the elements, were not merely mixtures of the elements but had their own forms given to them by God at the creation. To support this notion that new properties emerged with new mixtures, Sennert quoted Scaliger: "the form of every perfect mixture, even if it does not have a soul, like in diamond, is a nature of a fifth kind, very different from the four elements" (Sennert 1619, 244).

Sennert also directly quoted Du Chesne as an authority on the view that tastes, smells, and colors were caused by salt, sulfur, and mercury, and he argued that when the same effects and qualities were found in multiple substances, they required a common principle or "first subject," much the same as a quality such as hotness was explained by the presence of fire (Sennert 1619, 275–6). Sennert's experimental demonstration of atomism relied on reversible reactions that he called "reductions to the pristine state," where he dissolved a metal in a strong acid and eventually precipitated it, recovering the original ingredients (see Newman 2006, ch. 4). He employed these experiments to demonstrate the permanence of individual parts in a composition and to challenge medieval theories of mixture that required a resolution to the four elements and the destruction of their corresponding forms.

Sennert believed that multiple types of atoms existed and were governed by a hierarchy of forms, and while some atoms corresponded to the elements, there also existed others of higher-order substances, which had their own unique forms. Within this hierarchy, the *prima mixta* emerged as particularly significant because these were the limits of laboratory analysis.[10] In effect, by equating the chymical principles with the first mixts, Sennert was able to explain a large variety of phenomena in medicine and natural philosophy without recourse to the Aristotelian sensible qualities or four elements but, importantly, without rejecting these entirely or appealing to the incorporealities of the Paracelsians. In line with the seventeenth-century zeitgeist of reform, Sennert styled his chymical atomism based on *syncrisis*

[10] On the *prima mixta*, see Newman 2006, 127.

and *diacrisis* as an experimental investigation that sought to square the understanding not with the "notions of another man, but with the things."[11] Far from being inconsequential, Sennert's conception of hierarchically organized matter and his terminology of "*prima mixta*" had an important influence on the corpuscular philosophy of later naturalists such as Robert Boyle (1627–1691) (see Newman 2006).

The Italian-born Angelo Sala (1576–1637) spent most of his career in Germany and gained a certain notoriety for his atomist outlook as well as several experiments featuring *syncrisis* and *diacrisis*. In the 1617 *Anatomia vitrioli*, Sala announced a "reduction [of vitriol] to its pristine state" and delineated one of the earliest chymical syntheses confirmed by analysis (Hooykaas 1949, 77–8). In short, he made blue vitriol (i.e., copper sulfate) beginning with weighed amounts of copper, water, and spirit of sulfur (i.e., sulfuric acid); and he then decomposed the synthetic vitriol to recover the original reagents in their same proportions.[12] Sala concluded that vitriol was not a simple or essential substance but a collocation of particles of copper, water, and acid and, likewise, that no transmutation had occurred in the process.[13] Influenced by Libavius, Sala believed that all sulfuric acid, irrespective of its source of production (i.e., synthetic or natural vitriol), was identical and that the individual components of compound substances such as vitriol were fixed bodies. As he put it, "sulfur always remains sulfur and water always remains water, if their simple substances are regarded without admixtures" (Sala 1622a, b, 79). Hooykaas concluded that this understanding of material entities as having definite, constant composition and distinct properties approaches the modern concept of a "pure substance" (Hooykaas 1933).[14]

Sala defended his experiments and ideas against his would-be detractors, appealing to "the tribunal of fair and good judgment of Chymists and Naturalists" (Sala 1622a, b, 101).[15] He compared the "excessive talkativeness [*multiloquentia*]" of traditional philosophers with his arguments that were "confirmed by living examples" and challenged opponents to respond "with similar weapons, and establish their reasoning with living and evident examples … showing by the thing itself

[11] Sennert 1636: Sig. †† v. "Veritas enim est adaequatio rationum, quae sunt in intellectu, non cum alterius hominis nationibus, sed cum rebus." For what it is worth, the seventeenth-century English translation by Nicholas Culpepper and Abdiah Cole renders the final part of this quotation as "things themselves." See Sennert 1660, B2r.

[12] Sala's conclusion that the vitriol was 33% water by weight is rather close to the modern value for the percentage of hydration for copper sulfate, 36.08%

[13] Elsewhere Sala referred to such analytic and synthetic cycles using the Latin *redintegrare*, which, along with its English cognate "redintegration," was adopted by many later chymists. Sala 1622, 3r–v. In 1603, the Frenchman Nicolas Guibert (c. 1547 – c. 1620) discussed reactions in which metals were combined by plating or alloying, concluding that these were not transmutations (as they had long been portrayed), and that the original metals could be recovered. See Kahn 2016, 101–4.

[14] Translated by Hans Kubbinga as *The Concept of Element: Its Historical-Philosophical Development* (authorized translation, privately printed), 143.

[15] "…penes tribunal aequorum & boni judicii Chymicorum ac Naturalistarum."

[*reipsa*]...." (Sala 1622a, b, 101).[16] He continued in a passage worth quoting at length here:

> For in the arts, industries, or manual inventions which consist of evident examples and living demonstrations, such as the Art of Chymistry, it is not sufficient to be able to argue and produce reasoning, and to wastefully speak at length (so to speak), turning white into black and black into white. Rather, it is necessary to show by the thing itself [*reipsa*] and give something to be seen and touched, to prove effectively the thing itself [*rem ipsam*] which we wish to assert, such as we proclaim it to be. To do otherwise is nothing other than corrupting and adulterating the said art, which does not consist in empty phantasms and imaginations and truly chimerical speculations, but in live demonstrations, as stated. And this is not a field or contest for eloquence, but this is only a place for exercise, and a true gymnasium for demonstrating and actually producing the effect we allege; especially since the bodies and things which the Art of Chymistry is accustomed to dealing with are real bodies, and not empty and lightweight phantasms, but visible things that can be handled by human hands. (Sala 1622a, b, 101–03)[17]

Sala leaves the reader in little doubt as to his commitment to the separability and tangibility of fundamental chymical components and their centrality to his entire philosophy of nature. Likewise, elsewhere in his *Anatomia antimonii* (1617), he made it clear in a similar passage that chymistry's ambition was to use its ability to analyze and observe such "things themselves" to understand the composition of matter. He wrote,

> ... through the noble and excellent tool of the Art of Chemistry... we have been given the ability to learn, and to recognize with our own eyes, and distinguish substances of bodies completely unknown to the ancients, from which all things are naturally composed. (Sala 1617)

For Sala, the chymist's ability to produce physical bodies that could be presented to the senses was thus at the very core of the identity of chymistry and the first line of defense against foes.

Sala's son-in-law, physician to the Count of Oldenburg, Anton Günther Billich (1598–1640), similarly argued that *syncrisis* and *diacrisis* were at the heart of the definition of chymistry. Billich differentiated between chymistry's external end,

---

[16] "...si argumenta mea valent, quae tamen vivis exemplis confirmata in medium adduco, faciant id similibus armis; & vivis, ac evidentibus exemplis, rationes suas stabiliant, measque evertant: ostendetes reipsa..."

[17] "In artibus enim, industriis aut inventionibus manualibus, quae consistunt in evidentibus exemplis, & vivis demonstrationibus, qualis est Ars Chemica, non sufficit posse argumentari & ratiocinia proferre, & aërem (ut ita dicam) multiloquentia in vanum ferire, asferendo album esse atrum, & atrum album: sed necesse est ostendere reipsa, videndum & palpandum dare, comprobareque esse effectualiter rem ipsam quam astruere volumus, talem qualem eam praedicamus. Aliter enim facere, nihil aliud est quam dictam artem corrumpere & adulterare: quae ars non consistit in Vanis phantasiis & imaginationibus speculationibusque vere Chimaericis; sed in vivis demonstrationibus, ut dictum est. Et non hic est campus aut agon in quo certatur flosculis eloquentiae: sed hic solummodo locus est exercitatorius, ac vera palaestra demonstrandi & ipso opere effectum reddendi quod praetendimus: praesertim cum corpora & res de quibus Chymica Ars tractare consuevit, re ipsa corpora sint; & non vana leviaque phantasmata, sed res visibiles ac manibus humanis tractabiles."

which served medicine, and its internal end, which was primarily defined by *syncrisis* and *diacrisis*. This latter end was concerned with resolving compound bodies "into the parts from which they are proximately composed." In adopting this central focus on analysis and synthesis, he ridiculed Beguin and Du Chesne at length for their conception of quasi-spiritual chymical principles and concluded that these were best understood rather as "mixts, and not elements, but arisen from the elements, and not principles, but *principiata*" (Billich 1631, 22). Billich later went on to argue, however, that chymical analysis revealed the Aristotelian elements, that Aristotle himself could aptly be called a chymist, and that the object of chymistry is the same as the subject of Aristotle's *Meteorology IV*. These are among the reasons that Robert Boyle referred to Billich as "that fierce Champion of the Aristotelians against the Chymists" in his *Sceptical Chymist* (1661).

Boyle, we now know, was heavily influenced by another chymist who centered reversible reactions and the analysis and synthesis of chymical compounds withing his philosophy and also considered the chymists' principles to be *principiata*, but to very different ends (see Principe and Newman 2002). This was the influential Jan Baptist van Helmont (1579–1644), who believed that the Paracelsian principles were often merely artificial products of fire analysis. Helmont argued that the principles were produced rather than separated from substances in the same way that alcohol could be produced from diverse vegetables, concluding, "In like manner therefore those three things are *principiata,* but not principles" (van Helmont 1652, 333).[18] He continued to use the terms "mercury," "sulfur," and "salt" because these substances could be observed after a distillation of some bodies, though not all, and instead concluded that all substances were ultimately composed of water. He favored solution analysis over distillations by fire and sought a universal solvent that he called the "alkahest."

Helmont nonetheless heaped praise upon Paracelsian *spagyria* and boasted that such analyses and syntheses were superior to other methods—especially those of "the schools"—because they yielded tangible and perspicuous results.[19] It was precisely this understanding of chymical substances that inspired one of his primary arguments in support of the practice of chymistry. He wrote,

[18] "Sunt igitur similiter tria illa principiata; non autem principia."

[19] Much like Sala and Sennert, Helmont used cycles of analysis and synthesis to demonstrate that apparently uniform substances were actually compounds made from small particles. Helmont believed that compounds could be broken down into their initial components, which could be regained in their original quantity, and he demonstrated this in his synthesis of glass from salt of tartar and sand. After creating the glass, he ground it into powder and mixed it with more salt of tartar and then exposed the mixture to a humid environment causing it to form "oil of glass" or "waterglass" (i.e., potassium or sodium silicate). By adding acidic *chrysulca* (i.e., mostly nitric acid), he was then able to produce a nitrate salt and also separate out the same amount of sand that was used in the initial glass production. Newman and Principe have demonstrated that much of Helmont's chymistry was driven by this interest in gravimetry and, combined with his emphasis on analysis and synthesis, led to his explicit recognition of the concept of mass balance. See Newman and Principe 2002, 77–8.

> I praise my bountiful God, who has called me into the art of the fire, out of the dregs of other professions. For truly, Chymistry has its principles not through discourses, but those which are known by nature and evident by the fire: and it prepares the intellect to pierce the secrets of nature, and causes a further searching out in nature, than all other Sciences put together: and it pierces even unto the ultimate profundities of real truth: because it admits the operator unto the first roots of those things, with a pointing out of the operations of nature and the powers of art...

Scholastic ratiocinations and mechanical speculations were no match for the chymist, whose art allowed him to probe further into nature's foundations and operations. Helmont was explicit that chymistry's superior method was a product of its ability to disclose and exhibit the materials that constituted natural bodies, such that they could be seen and handled. As Helmont put it, chymical analysis revealed "things themselves [*res ipsas*]" in such a way that they "retire into a domesticated juice" and "become social [*socialia*] unto us."

Other passages throughout Helmont's works provide context that allows us to understand how he conceived of this special intimacy with matter afforded by the art of chymistry. In his text *De Lithiasi*, he wrote,

> Wee read in our Furnaces, that there is not a more certain kind of Science in Nature, for the knowing of things by their radical and constitutive causes; than while it is known, what, and how much is contained in any thing. So indeed, that the knowledge, and connexion of causes are not more clearly manifest, than when thou shalt so disclose things themselves [*res ipsas*], that they bewray themselves in thy presence; and do as it were talk with thee. For truly, real Beings, standing onely in their owne Original, and succeeding principles of seeds; and so, in a true substantial entity, do afford the Knowledge, and produce the cause of knowing the nature of Bodies, their middle parts, and extremities or utmost parts. (Van Helmont 1648, 10–11)[20]

To support this general understanding of the disclosure of things themselves, Helmont immediately quoted an extended passage from Pseudo-Lull's *Testamentum*, which criticized the "Logician [*Logicus*]," who, despite having profound intelligence and rhetorical abilities, approached nature only superficially and without direct knowledge. Commenting on this passage, Helmont claimed that it was only the art of "*Spagyria*" that "shews how to touch, and see the truths of those things in the clear Light" (Van Helmont 1648, 21).[21]

While Helmont, as we have seen, appealed to the concept of *principiata*, German alchemist and cameralist Johann Joachim Becher (1635–1682) made the hierarchical organization of principles and higher-order mixts a central feature of his

[20] "In nostris furnis legimus, non esse in natura certius sciendi genus, ad cognoscendum per causas radicales, ac constitutivas rerum; quam dum scitur quid, quantumque in re quaque, sit contentum. Ita quidem ut cognitio, & connexio causarum, non constent clarius, quam cum res ipsas ita recluseris, ut coram prodeant, ac velut tecum loquantur. Siquidem entia realia, duntaxat stantia in suis primordialibus, & succedentibus seminum principiis, adeoque in vera entitate substantiali, dant notitiam, & proferunt causam cognoscendi naturam corporum, mediorum, & extremitatum." Translation from Van Helmont 1664, 839.

[21] "Spagyria enim sola, est speculum veri Intellectus: monstratque tangere, & videre veritates earum, in claro lumine." Translation from Van Helmont 1664, 840.

philosophy. Indeed, this was among the primary features of Becher's work that were taken up and celebrated by Georg Ernst Stahl (1659–1734) and his many followers. In his 1664 *Oedipus chimicus*, Becher concluded that chymistry, being a practical science, must deal with material and practical subject matter, and thus it only "considers as first matter what first comes within the senses and the hands." This ruled out the intangible elemental first matter of the Aristotelians, which could only be understood by reason, leaving instead "the second [matter] of the Aristotelians, which is the first of the chymists, namely the accidents of the Aristotelian first matter" (Becher 1664, 14).[22]

Later, in the *Physica Subterranea*, Becher described earth and water as the "most singular principiated principles [*principia principiata & singularissima*]," noting that as "all things have come from earth and water, all things can ultimately be reduced to earth and water," and that these "most remote principles" acted as "specific seeds" responsible for the generation of higher orders of matter when variously mixed (Becher 1669, 129). Later in this text, he continued,

> I hope that no one will be so absurd as to interpret the three aforementioned [Paracelsian] principles in any other way than as proximate and *principiata*: namely, matter already disposed for action in the closest way possible. And even though they may be considered in this way, they are still improperly called salt, sulfur, and mercury, whatever way they are explained. (Becher 1669, 123)[23]

Becher concluded instead that minerals were resolved into water and three earths—*terra lapidea, terra pinguis, and terra fluida*—whereby the earths generally corresponded with salt, sulfur, and mercury. Within this schema, he also conceived of these principles as combined within a hierarchy to form *composita* at the first level of composition, followed by more complex *decomposita* and, finally, *superdecomposita*. As Newman has argued, Becher's hierarchical theory and his somewhat confusing terminology actually came from Robert Boyle, demonstrating another avenue for the extended influence of chymical atomism (Newman 2014, 63–77).

Becher's influence on later chymistry is evident in several analytic experiments that he believed illustrated his understanding of composition and sensibly demonstrated his diverse principles. In the *Oedipus chimicus*, he maintained that anyone who witnessed an analysis and "saw with their eyes and touched with their hands that vitriol consists of sulfur and salt" should be convinced of this (Becher 1664, 44).[24] In the *Physica Subterranea*, Becher described having performed a

[22] "Cum Chimica scientia practica sit, subjectum etiam habet materiale & practicum, quare id pro prima materia statuit, quod primum ei sub sensum & manus cadit, tale autem Aristotelicorum prima materia esse nequit, cum illa tantuum ratione comprehendi, oculis vero manibusque apprehendi non possit, alia merito nobis quaerenda erit, nempe Aristotelicorum secunda, quae Chimicorum prima est, puta primae materiae Aristotelicae accidentia, haec enim tractationi Chimicae inserviunt."

[23] "Neminem autem spero, ita absurdum fore, ut praefata tria principia aliter quam propinqua & principiata intelligat: nempe pro materia iam proxime ad actum disposita: & licet hoc modo considerentur, tamen quomodocunque explicentur, improprie sal, sulphur & Mercurius dicuntur."

[24] "…praesertim si vitriolum ex sulphure & sale constare oculis viderent, manibus tangerent."

"deflagration" or burning of sulfur that he believed had produced an acid salt and the sulfurous principle, *terra pinguis*. This experiment influenced Stahl profoundly, but another experiment was reported more widely: Becher recorded that he had melted some jasper in a crucible and, upon cooling it, had noticed that the brightly colored mineral had turned white but retained its hardness. The parts of the crucible that were not in contact with the jasper, however, had been tinged by the jasper's natural color and had assumed the appearance of the mineral. Becher believed that the crucible had been colored "by the soul of jasper [ab *anima Jaspidis*]," and thus he concluded that he had separated the characterizing substance in which there existed "a certain immortal form [*immortalem quandam formam*]."[25] Becher believed this substance to be his "subtle earth," which the chymists improperly called "mercury," and this experiment, as we shall see, exerted a considerable influence during the eighteenth century. The experiment was conveyed to an international audience in 1671 in *The Philosophical Transactions of the Royal Society*, and the representative of the Royal Society wrote that the experiment was consequential to the extent that "We cannot forbear giving the reader ... one very considerable experiment, said to have been actually made by the author himself."[26]

4.4 Philosophical and Rational Chymistry

In his outline of the history of chemistry, Antoine-François de Fourcroy (1755–1809) marked the beginning of "philosophical chemistry" with Becher and two other Germans: Johannes Bohn (1640–1718), professor of practical medicine at Leipzig, and Jacob Barner (1641–1686), who was likewise professor at Leipzig and eventually physician to the king of Poland (de Fourcroy 1782, 18).[27] Fourcroy praised Bohn and Barner's works for "the clearness of the ideas contained in them, and the order and method of their arrangement," remarking that "The publication of these two first philosophical works on our science coincides with the origin of experimental philosophy, and must be considered as the birth of true chemistry" (de Fourcroy 1800, 19).[28] In the following, I shall trace how earlier ideas about the analysis and synthesis of hierarchical *principiata* paved the way for the development of this self-styled philosophical chymistry.

In the preface to a 1685 series of "Dissertations of Chymico-Physics," Bohn presented a case for the notion that the chymical Art was better suited to natural

---

[25] Becher's understanding that the separation of the color of jasper signified the separation of its *anima* is highly likely to have been influenced by Johann Rudolph Glauber (1604–1667). On Glauber's separation of the soul of gold, see Smith 2004, 172.

[26] *The Philosophical Transactions of the Royal Society*, 6 (1671), 2233.

[27] Translation from Fourcroy 1796, 30–1.

[28] "La publication de ces deux premiers ouvrages philosophiques sur notre science coincide avec la création de la physique expérimentale, et doit être regardée comme la naissance de la véritable chimie." Translation from Fourcroy 1804, 29.

philosophy than to medicine. He complained that chymistry had been "entrusted to physicians or alchemists, as if it pertained only to them, while investigators and doctors of natural things have avoided the smoke, coals, and other annoyances of the chymical laboratory and have believed these unworthy of their speculations..." (Bohn 1685, *2v–*3r).[29] While the alchemists "boast[ed] of being concerned with the principles and elements of things and are the patrons of what are called the chemical principles," Bohn suggested that their principles of salt, sulfur, and mercury were "products of the fire ... lacking in the simplicity of elements" and that the alchemists, therefore, "obstruct the evolution of natural things." Physicians who addressed chymistry had not fared any better, however, for they "relegated chymistry to the final part of medicine and its minister, pharmacy," while those who taught chymistry as part of a medical curriculum "only treat[ed] chymistry as a subsidiary aspect of Medicine." Instead, Bohn suggested that chymistry ought to be learned "in the middle of a course of Philosophy...before even considering medicine," concluding that "chymistry thus pertains to Philosophy: for the Philosopher's task is to observe the diverse phenomena of this art, in order to recognize the nature of things from various resolutions and mixtures of concretes..." (Bohn 1685, *3v).[30] In effect, Bohn distinguished between chymistry's secondary end and its proximate or internal end, writing:

> The first of these [secondary ends] could be called philosophical, as it seeks only to extract the theories of the principles and affections of natural bodies and their etiologies through mere contemplation; the second, pharmaceutical or medical, which aims at the preparation of beneficial remedies; the third, mechanical or industrial, which, for example, salt-makers, brewers, etc. pursue; and finally, the fourth, alchemical, whose goal is the solitary transformation and exaltation of metals. If we nevertheless consider the proximate or internal end, there will be only one chemistry, which, without regard to the reason for the secondary ends, primarily works in the way that, using certain instruments and applying them in different ways, it separates mixed natural things into parts, combines these parts with each other and with other concretes in various ways, but in such a way that it investigates the reasons and causes of all the phenomena that emerge from this; in short, its end is the work itself, and when this is perfected, the operation of the chemist, as such, ceases. (Bohn 1685, *4r–*4v)[31]

---

[29] "...harum minimam haud esse suspicor, quod vel Medicis, vel Alchymistis, quasi ad hos solos spectaret, Ars illa concredita fuerit, rerum naturalium vero Investigatores atque Doctores fumos, carbones coeterasque Chymicorum Laboratorium molestias detrectarint ac speculationibus suis indignas crediderint..."

[30] "Feliciori sane successu atque cum uberiore emolumenta in medio Philosophiae curriculo, ut quidam Genuina Medicinam instituendi rationis Suasor disserit, antequam ne quidem de medicina cogitatur, Chymia doceretur, adeoque ad Philosophiam pertinet: cum Philosophi sit artis huius diversa spectare phaenomena, quo ex variis concretorum resolutionibus atque mixturis, quae fermentationis, quae effervescentiae, quae praecipitationis, natura sit, dignoscat."

[31] "Quarum prima dici posset Philosophica, tanquam merè contemplativa ac corporum naturalium principiorum & affectionum theorias harumque aetiologias tantum eruere gestiens: Altera Pharmaceutica seu Medica, quae remediorum commodorum praeparationem intendit: Tertia Mechanica seu opificiaria, quam v.g. Salitores, Cerevisiarii, Tinctores, Vitriarii, Saponarii, Metallurgi, Aurifabri similesque Opifices exercent: Quarta tandem Alcyhmistica, cuius scopus metallorum transmutatio & exaltatio solitarius est. Si nihilominus finem proximum, seu internum,

For Bohn, the ultimate end of chymistry lay squarely within the realm of natural philosophy, and the means to that end were analysis and synthesis.

Bohn developed these ideas further in a dissertation from this same volume on the "Dissolution of Bodies" (Bohn 1685, A1r–B4v).[32] He wrote that instead of exploring changes from states of solidity to fluidity, "Rather, I will be speaking about the analysis of the bonds [*compagis*] of mixtures in general, by which what was once one and continuous is divided into the smallest possible parts, either homogeneous or heterogeneous, either by the wet (as they say technically) or dry method" (Bohn 1685, A2r).[33] Following earlier chymists, Bohn bypassed the issue of the ultimate elements, writing, "I will not move any controversy about whether the elements themselves also obey [analysis], since the essence of the elements is still sufficiently hidden in the well of Democritus" (Bohn 1685, A2r).[34] Instead, Bohn maintained that chymical analysis only dealt with mixtures and concrete bodies and that "sounder philosophy" had demonstrated that "elements, because they are simple bodies, are not the subject of *diacrisis*." Explicitly referring to works by Boyle, Helmont, and Billich, Bohn answered the question of whether chymical analysis was able to reduce mixts into elements or principles in the negative. Bohn wrote that the "trivial chymists" who believe that the end products of their analyses are elements or principles are "not so much proving the existence and essence of these elements as merely supposing them through a sufficient degree of credulity" (Bohn 1685, A3v). In effect, Bohn questioned whether the substances separated by the instrument of the chymists' analytic fires actually existed when they were a part of the concrete whole, concluding, instead, that "reason requires us to suspect that they were produced by fire" (Bohn 1685, B2r).[35] Likewise, he explained why his chymistry, situated in natural philosophy, dealt only with mixts:

> Therefore, assuming with Philosophers of a more accurate mind that there are two kinds of *minima*, the first and the second, the first of which are the smallest particles of matter in the whole universe, which, although they have a determined shape because they are material, are nevertheless imperceptible to our senses because of their smallness; but the latter are the

---

intueamur, una tantum erit Chymia quae nulla habita finium secundorum ratione primario in eo laborat, quo intervenientibus certis instrumentis, diversi mode applicatis, mixta naturalia in partes divellat, has & invicem, & cum aliis concretis variè combinet, ita tamen, ut cunctorum inde emergentium phaenomenorum rationes & causas inquirat uno verbo, finis eius est ipsum opus, quo perfecto, operatio Chymici, ut talis, collimant."

32 "De Corporum Dissolutione."

33 "Sed de compagis mixtorum analysi tali & omni, qua, quod era unum atque continuum, in partes minimas, modo homogeneas, modo heterogeneas, dividitur, sive per viam humidam (ita technice loquuntur) sive siccam, hoc contingat." Bohn wrote that he used the words "dissolutio" and "discontinuatio" interchangeably and as synonyms, "in order to make it more clear that the subject of this dissertation is the analysis of mixed substances, in which the continuity of these substances is destroyed, while the contiguity of the atoms constituting their texture is maintained."

34 "An elementa quoque ipsi pareant, nemini movebo litem, cum eorum essentia in Democriti puteo satis adhuc abscondita lateat..."

35 For Bohn's understanding of the instruments of chymistry and his influence on Hermann Boerhaave, see Powers 2007, 2012.

first mixts [*prima mixta*], coagulated from the preceding ones, which, separately existing, likewise escape our sensory perception, but when combined in more concrete bodies, e.g. in earth, water, salt, sulfur, etc., they affect them under different patterns: I shorten this discourse in such a way that just as no one, except nature, reaches the first *minima*, so the power of art is limited to the second *minima*, or the first mixts… (Bohn 1685, B2r–B2v)[36]

The "Philosophers of a more accurate mind" to whom Bohn referred undoubtedly included Boyle as well as the German chymists, such as Sennert, who had developed this understanding of sensible *prima mixta* from earlier alchemical and Aristotelian traditions (Newman 2001, 2006). In the sentence that followed, Bohn referred explicitly to Billich and his distinction between confused and distinct analysis, whereby the former reduces concrete bodies into more composite particles while the latter reduced bodies into substances closer to the *minima secunda*.

According to Fourcroy, Stahl knew Barner's *Chymia Philosophia* "by heart at the age of fifteen years," and here we see another instance of how this new "Philosophical Chymistry" that addressed questions of natural philosophy was built primarily and explicitly on the *syncrisis* and *diacrisis* of earlier chymists.[37] Barner defined chymistry, simply, as "the art of separating pure from impure bodies by means of fire, and then combining them, and thus producing effective medicines," but his ambitions for his endeavor were extensive. He wrote, "And this is what I attribute to myself, this is what I want to be credited with, that I have revealed the causes of all operations for the first time, and have brought out true philosophical chymistry…." (1689, 4r).[38] Likewise, Barner asserted that chymistry's primary end was "to separate the parts of mixtures, so that a more accurate demonstration can reveal what they consist of and into what they are reduced" (Barner 1689, 6–7).[39] On this basis, he concluded, "Therefore Chymistry properly belongs to Physics [*Physica*], since it demonstrates the composition, constituent parts of those mixtures, sulfurs, salts, and the diversity that they themselves and their nature have from the union of mixture." Following Boyle, Helmont, and "other men of great name," Barner wrote that chymistry thus rightfully deserves the names "*Naturae Clavis*, *Scientia ac* τέκμαρσις [i.e., judging from sure signs]" (Barner 1689, 7–8).

Barner had greater confidence than Bohn in chymistry's ability to separate "the very constituent principles themselves as they previously existed in mixtures,

[36] "Supponens proin cum accuratioris genii Philosophis minima duplicia, prima sc. & secunda quorum illa primae totius universi materiae particulae sunt, quae utut, quia materiales, determinata sua figura gaudeant, propter exiguitatem nihilominus summam sensibus nostris haud patescunt; haec vero corpuscula seu prima mixta sunt, ex praecedaneis [sic] coagmentata, quae separatim existentia pariter sensoria nostra fugiunt, combinatae vero in corporibus magis concretis, v.g. in Terra, Aqua, Sale, Sulphure &c. sub diverso schemate eadem afficiunt: discursum hunc ita contraho, quod sicut minima prima nemo, nisi natura, attingit, ita artis potentiam minima secunda, seu prima mixta, terminentur…"

[37] Fourcroy, 3.23–24. Barner 1689, 124.

[38] "Atque hoc est, quod mihi adscribo, hoc mihi laudis tribui volo, quod primus operationum omnium causas tradiderim, & Chymiam vere philosophicam…produxerim."

[39] "… Chymiae sit finis primarius, quemve illa ex se habet, est mixtorum partes separare, ut accuratiori demonstrationis genere, ex quibus illa constent, in quae redigantur, innotescat."

without changing them" but followed Helmont in rejecting the Paracelsian *tria prima* of salt, sulphur, and mercury as principles *per se*. He wrote, "it is clear that it is more correct to think of these three as Helmont does, who calls them *principiata,* that is, arising from others, and uniquely water" (1689, 18).[40] Beyond Helmont, however, it is clear that Barner's understanding of chymical composition owed a significant debt to Sennert. In addition to the hierarchically organized *principiata*'s close affinity with the Sennertian *prima mixta*, Barner published a book titled *Prodromus Sennerti Novi.*[41] As the historian of chemistry Theodor Gerding quipped, Barner was "ein Schüler Sennert's und ein Anhänger Helmont's" (Gerding 1867, 94). Barner also concluded a short *Exercitium Chymicum* appended to the *Chymia Philosophica* with praise for the "Experimental Philosophy" of Boyle and other members of the Royal Society, which he regarded as exemplary of how the examination of natural bodies and their principles and union of mixture could serve the non-medical "second end of chymistry...pertaining to the natural sciences" (1689, 559).

What is especially striking here is that this hierarchical understanding of *principiata* governed by the limits of analysis was adopted by such a wide array of authors with divergent views on other questions about matter theory and chymistry. The Jena professor of medicine and chemistry, Georg Wolfgang Wedel (1645–1721), for instance, defended alchemical transmutation and supported the use of fire analysis to separate chymical principles. Wedel discussed *principiata* in a variety of contexts, suggesting, for instance, that salts, because they are not absolutely simple, were "not principles but rather *principiata*" (Wedel 1686, 410). Similar to Sennert and Billich, he believed that Aristotelian elements combined to form the chymical elements, which he described as the "first matter...in composition, which is the last in resolution" (Wedel 1715, 5). In a 1685 treatise entitled *De clave principiorum chimicorum*, he argued that the chymists' *syncrisis* and *diacrisis* demonstrated the principle established by philosophers "that some things are made from the combination of others, and others are made by separation" (Wedel 1685).[42] Remarking that these twin notions were the foundation of the chymical art, he likewise suggested that analysis and synthesis were the means by which chymistry moved beyond mere "labor or practice" to instead "cultivate a theory that arose from practice."

Leipzig professor Michael Ettmüller (1644–1683) was more enamored of mechanical ideas and explicitly followed Boyle, Helmont, and David von der Becke (1648–1684) in rejecting both the Aristotelian elements and Paracelsian principles

[40] "Ex dictis huc usque patet, rectius de tribus hisce sentire Helmontium, qui ea. principiata, hoc est, orta ex aliis, & unice aqua ..."

[41] See Barner 1674, where he promised to "examine that ancient teaching of Sennert in light of the more recent principles of anatomy and chymistry... and present those new dogmas of the more recent authors, brought back under the hammer, in a single systematic way, as Sennert did in his time."

[42] "Quod alii quoque Philosophi stabiliverant axioma ...: alia ex aliis combinatione, alia disjunctione fiunt, seorsim opere ipso praestant & demonstrant Chimici, quorum operationes in σύγκρισις et διάκρισις consistent."

as "the primary principles of composition and the ultimate principles of resolution." Rather, he believed that the chymists' principles were

> not so much primary as secondary, produced by seeds through the transmutation of proximate matter into a special body, and thus they are not so much the material principles of bodies as *principiata*, born from matter that is immediately prone to seed action. (Ettmüller 1685, 24)[43]

In one instance, he referred to a quaternary of *principiata*: acidic, alkaline, fatty, aqueous, and earthy particles; elsewhere, he limited these to saline-acidic, aqueous, and earthy. Nevertheless, he was clear that the differences between these *principiata*, which he described as having different "textures," arose from changes in the composition of elementary particles (i.e., through *syncrisis, diacrisis*, etc.) and via the related action of *semina*. Ettmüller also noted that this understanding of *principiata* yielded the conclusion that they were "mutually transmutable" and, for instance, that a compound body that is "deprived of the power of a seed" would return to water and that the water particles could then be transformed into another body.

Finally, the German mathematician and physician Joachim Jungius (1587–1657) brought logic to bear on the relationship between the chymical principles and the Aristotelian elements. In the *Doxoscopia physicae minores*, he criticized a tenet of traditional philosophy writing, "That Axiom is utterly false, that *Principiata* are just as the *Principia*" (Jungius 1662, Sig. Ee 2).[44] Instead, Jungius maintained that compounds [*Composito*], as *principiata*, could possess distinct properties that were not present in or could not be inferred from their simpler, foundational elements or principles.

4.5 Stahl and the Stahlians

Georg Ernst Stahl (1659–1734) has been remembered largely for his influential phlogiston theory, but it is worth pointing out that Stahlian chymistry's central emphasis was on analysis and synthesis and that this grew directly from the tradition under consideration (Chang 2015). Indeed, Stahl and his followers used much of the same terminology that we have encountered here, but beyond this, the influence of earlier traditions is clearly perceived in their definition and delimitation of the boundaries of chymistry, which gave them their primary defense against competing philosophies—namely, mechanistic physics.

---

43 "…seque videtur, non tam primigeneas esse particulas illas, quam secundogeneas, per semina sub materiae proximae in Corpus speciale transmutatione productas, adeoque non tam sunt Corporum Principia materialia, quam Principiata, ex materia, proxime quae seminis actioni subest pronate…"

44 "Falsißimum eft hoc Axioma quod talia sint Principiata qualia Principia."

In the first pages of the *Fundamenta Chymiae Dogmatico-Rationalis & Experimentalis* (1732), after defining chymistry and describing its instruments and objects, Stahl wrote,

> The subject of chemistry is mixed bodies: the principles of mixture are earth, water, and air; from these emerge the principiated principles [*principia principiata*] or concretes, which contribute to the composition of bodies and are called the proper principles of the chemists. (Stahl 1732, 3)[45]

These *principia principiata* were limited to salt and sulfur, whereas mercury was related to water or air, and thus the variety and composition of bodies depended entirely upon the mixture of the *principia* of earth, water, and air as well as the *principiata* that arose from these. As Stahl put it, "whoever wants to give the causes of effects and phenomena occurring in chymical operations, must not only know the principles, but also the exact mixture of those principiated principles" (Stahl 1732, 3)[46] These different mixtures led to the hierarchical schema in which principles combined to form mixed bodies, which combined to form compounds, which, in turn, combined to form aggregates. As he concluded elsewhere, "all the darkness and disputes about Principles arise from a neglect of that real distinction between original and secondary Mixts, or Mixts consisting of Principles and Bodies compounded of Mixts" (1723, 4).[47]

Stahl's chymical principles were imperceptible to the senses when separated from bodies, and even mixts and compounds were so small that they were similarly elusive. It was only when a larger aggregate was formed that it could be seen or touched (Stahl 1715, 227–9). Nonetheless, Stahl was clear that his "sulphurous principle" or "inflammable principle" of phlogiston was sensibly present in mixts and was revealed by experiments. He wrote, "All mixed physical things, more or less, noticeably have a share of this essence: namely, in all three so-called realms; the vegetal, animal, and mineral" (Stahl 1718, 82).[48] While this principle of inflammability was certainly material and existed in physical matter, it was not isolable and escaped the senses when separated from its original mixture. Stahl revealed the existence and nature of this principle in two experimental exemplars: a synthesis of sulfur and a deflagration of sulfur, which together have been called Stahl's "analytic

[45] "Subjectum Chymiae sunt corpora mixta: Principia Mixtionis sunt, terra, aqua, & aether; ex hisce emergunt principia principiata seu concreta, quae ad corporum compositionem concurrunt." N.B. that this 1732 text is different from the earlier and more readily available Stahl 1723, which bears great similarities to Shaw 1730.

[46] Ibid. "Quicunque itaque vult reddere causas effectuum ac phoenomenorum in chymicis operationibus occurrentium, non tantum principiorum, sed & principiatorum illorum mixtionem exacte nosse debet."

[47] "Totam videlicet de Principiis litem & obscuritatem fovet omissio realis illius distinctionis inter Mixta prima & secunda, seu Mixta ex principiis & Composita ex Mixtis." Translation from Shaw 1730, 5.

[48] "Alle vermischte cörperliche Dinge, mehr oder weniger, mercklich von diesem Wesen Antheil haben: und zwar in allen dreyen sogenannten Reichen; dem vegetablisichen, animalischen, und mineralischen."

cycle." Essentially, Stahl decomposed sulfur in the deflagration, wherein he burned sulfur under a bell jar, which ostensibly separated the inflammable principle, and he then collected the acidic residue that had combined with water from the air. However, unsure as to whether this acid was merely separated from the sulfur or produced in the fire, he turned to a synthetic experiment and combined sulfuric acid with phlogiston, creating sulfur once again.[49]

Stahl took such experimental analyses and syntheses of hierarchically organized principles, mixts, and compounds and used these to attack competing philosophies perceived to be overly speculative or less grounded in tangible experimental results. Much the same as his forebears had attacked Aristotelianism, he targeted the "Mechanical philosophy," writing,

> Although it prides itself on explaining all things with the utmost clarity, it has presumptuously applied itself to the contemplation of Chemico-physical matters. For even though I do not despise a sober use of it, no one sees any light brought from it unless they are blinded by prejudiced opinions. And this is not surprising. It often remains in doubt, merely skimming the surfaces and the bark of things without touching the core, content with deriving very abstract and extremely general explanations from the shapes and motions of particles, neglecting what a mixture is, what a composite and aggregated body is, and what the nature, properties, and distinctions of these are. And from this indeed, so many unhappy, fantastical chimeras, so many vain and incomplete applications in Chymistry have appeared to exist. (Stahl 1723, Sig. (2v))

Stahl, as other chymists had done previously, thus adopted an agnostic philosophical position on the nature of ultimate particles, preferring instead to focus on higher-order aggregates and mixts that could be perceived and subjected to chymical experiments. As Stahl's English interpreter Peter Shaw clarified, Stahl's subject of chymistry, the mixt, was understood to be "certain Corpuscles of such a degree of smallness, with regard to our senses, as not to be cognizable by them, unless in a numerous parcel" (1730, 7 n. *.).

Stahl's ideas were adopted, clarified, and in certain cases expanded by his students and colleagues at the University in Halle (Saale). Michael Alberti (1682–1757) and Johann Juncker (1679–1759), for instance, both identified the analysis and synthesis of sensible mixts and aggregates as chymistry's unique niche, and they continued to use the same terminology from earlier centuries. Alberti wrote,

> chymistry is according to its own and real sense the art of Synthesis and Analysis *[ars Syncriseos & Diacriseos*], by means of which suitable subjects are resolved and the resolved things are combined again ... and this description of chymistry agrees with that famous designation of the art of *Spagyria*, which is nothing other than what is concerned with the resolutions and combinations of bodies. (Alberti 1721, 5)[50]

[49] See Eklund 1971, 23–39.

[50] "...itaque Chymia juxta proprium & realem sensum ars Syncriseos & Diacriseos, mediante qua apta subjecta resolvuntur & resoluta iterum combinantur ... & cum hac Chymiae descriptione consentit illa famigerata appellatio artis Spagyricae, quae non alia est, quam quae circa resolutiones & combinationes corporum versatur."

Both Alberti and Juncker likewise described the hierarchy of matter with reference to *principiata,* concluding that chymistry's domain was restricted to what could be sensed and that it ought to remain agnostic about the nature of the ultimate elements. Juncker wrote,

> Formal and ultimate causes, no less than the forms of bodies, are hidden from us in many cases or cannot be discovered. Therefore, we [chymists] are content to investigate and propose the more proximate, material causes, which demonstrate the effect as well as the instrumental causes more clearly. For example, if the chymist is asked about cinnabar and what causes its bright red color, he demonstrates that it depends on the closer union of sulfur with mercury as a material cause. If further asked, from where sulfur produces this coloring effect, he proves that it mainly originates from the inflammable earth mixed in it, and would mostly be satisfied with this. He leaves it to speculative physicists to explain, how sulfur reflects such a beautiful color; what is the figure or position of its molecules, etc. (Juncker 1730, 7)[51]

Alberti similarly argued that because the foundational sulfur, saline, and mercury principles were elusive and immediately combined with other principles if separated from a mixt, "it is nowhere within the power of the artist to collect elementary materials outside of mixture, much less to offer them to external senses." In effect, he argued that "chymistry is concerned only with mixed and composite bodies, which are subject to future dissolutions and combinations." Alberti likewise concluded that analysis and synthesis revealed "the relationships and mutual habitudes of principles," but he departed from many of his chymical predecessors and went so far as to conclude that chymistry "does not so directly aid the art of healing ... but rather looks more toward practical physics" (Alberti 1721, 4).[52]

Alberti reiterated his agnostic outlook with respect to the fundamental principles, arguing that it was not the office of chymistry to "resolve, separate, collect, and investigate" the primary essences of bodies, because observable fundamental particles would invariably elude both physical and chemical efforts. Instead, he concluded that "it is more correct to search for chemical subjects in mixtures and coarse aggregates, in which state they are somewhat more susceptible to the senses and use..."[53] Juncker similarly argued that even attempting to gain direct knowledge of the essence of principles was futile and that we should instead aspire to obtain

[51] "...Formales & ultimae causae non secus ac forma corporum in plerisque nos latent, vel erui nequeunt. Itaque conteni sumus, propiores materiales, & quae proxime effectum edunt clariusque demonstrantur, nec non instrumentales investigare & proponere. Sic si de cinnabari quaeratur, quid in ea. colorem illum vivide rubentem efficiat, Chemicus ipsum a sulphure arctius cum mercurio tamquam causa materiali socia juncto pendere ostendit; Si ulterius rogetur, unde sulphur hunc colorantem effectum edat, probat, eum ab inflammabili terra immixta potissimum oriri, atque in his fere acquiescit; altiorem autem quaestionem, quomodo, qua figure, quo situ sulphur cum mercurio sub vario lucis accessu tam rubicundam superficiem constituat, speculatoribus physicis extricandam relinquit."

[52] Alberti, "Fundamenta Chymiae," 4.

[53] Ibid., 7: "Frustraneum etiam erit conamen operationes chymicas ad exquirendas, resolvendas, separandas, collegendas & indagandas primas corporum essentias, aut quod adhuc magis est ad rerum seminia presequi, cum tam physico, quam chymico conatui semper hi conceptus occulti erunt, dum ad primas materias seorsim observandas non facile aditus patet, unde rectius subjecta

"practical knowledge, so that we may learn to know the affections of principles as they manifest themselves in slightly larger or sensible molecules of bodies" (Juncker 1730, 92).[54]

Even so, Juncker explained why chymistry, while powerless to separate, isolate, and perceive ultimate principles, was able to arrive at conclusions regarding these. He wrote,

> In any mixture, the principles or constituent parts are all connected as one, yet each retains its own essence and original qualities in this nexus. That is why we can be sure that the resolution of concrete things into their constitutive principles, which exhibit their original nature and properties, has been well carried out, when, by taking these same principles again, the new and same thing is synthesized. For example, if mineral sulfur is destroyed by separating its phlogiston from the acid, and if, by adding phlogiston from charcoal or another source by a just operation, then it is again the same sulfur. (Juncker 1730, 104)[55]

The analytic cycle was thus a key aspect of what made Juncker's chymistry "philosophical." He explained that this chemistry was "truly scientific" by virtue of its concern with "the matter from which things cohere, the mode and motion of cohering, and the various respective properties with respect to both concretion and dissolution," and that these could be studied via experiments "in such a way that progress is visible from simplicity to mixtures and compositions, and conversely from these to resolutions into a simpler state" (Juncker 1730, 36).

Juncker thus built a case for the superiority of chemistry in his *Conspectus chemiae* on the analysis and synthesis of compound bodies, which he believed would provide the deepest understanding of nature and, in particular, "certainty about the constitutive parts of these bodies, their mixture, and the reason for the many qualities and phenomena that depend on this." However, while it was often impossible to discover the primary elements directly or to subject these to experiment, "at least the secondary and proximate elements of any body" could be revealed "to the senses, and when these are observed, through chemical transposition and many other effects, the primary elements are also most likely to be recognized" (Juncker 1730, a2r–a2v).[56]

---

chymica in mixtionibus & crassis aggregationibus perquirenda erunt, in quo statu paulo magis sensui & usui obnoxia sunt..."

[54] "Frustra etiam quis laborabit, specialissimam principiorum essentiam ... Unde hac relicta aspirandum potius est ad practicam illam scientiam, ut affectiones principiorum, quemadmodum in paullo grandioribus aut sensibilibus moleculis corporum sese exserunt, pernoscere discamus."

[55] "In quavis mixtione principia, seu partes constituentes connexae, *pro unto stant*, singula tamen essentiam suam & affectiones pristinas in hoc nexu retinent. Hinc quoque resolutio concretorum in principia constitutiva, *antiquam* naturam ac proprietates suas exhibentia, bene succedit, itemque *nova eademque syncrisis*, si eadem principia assumantur. E.g. sulphur minerale, e mixtione sua destruitur, si [phlogiston] expulso, acida illius pars separetur; at si huic iterum [phlogiston] ex carbonibus aut aliunde addatur justa operatione, denuo sit *idem* sulphur."

[56] "quandoquidem saepe in promtu est, si non prima, tamen secunda & proxima corporis alicuius compositi elementa sensibus subiicere; perspectis autum secundis e transpositione chemica plurimisque effectibus, prima quoque verosimillime cognoscuntur." Stahlian transposition refers to the replacement and recombination of chemical corpuscles. See Chang 2002, 41 n. 28.

4.6 Conclusion: Chymistry Goes Further

The power and extent of this tradition's influence on later chemistry is made clear from Stahl's French followers, who continued to support conceptions of the analysis of mixts into sensible *principiata* late into the eighteenth century. Furthermore, some took the sensible analysis of mixts beyond Stahl's own conclusions and defended chemistry against speculative physics even more stridently.

Pierre-Joseph Macquer (1718–1784), famous for his *Dictionnaire de chymie* (1766), praised Becher's theory of principles as interpreted by Stahl as "the source of the most important discoveries in chemistry," and maintained that most bodies could not be reduced into their primary principles or elements and that analysis could only demonstrate simple substances that "although compounded of a certain number of principles do themselves the office of principles in the composition of bodies less simple than in themselves," and were thus called "*principiate principles* [*principes principiés*]" (Macquer 1766, 325–31).[57] Macquer opined that principiated things had greater claim to the name of principle because they subsisted in their state upon separation from a body, were "characterized by peculiar properties," and were "capable of reproducing by their union a compound entirely like that from which they were originally separated." He likewise explained that there were *principiate principles* of different degrees of simplicity, and that substances that could not be further decomposed were thus primary principles; these then combined to form secondary principles; and secondary principles combined to produce tertiary principles, and so on (Macquer 1766, Dd2v).

Gabriel Francois Venel (1723–1775) is remembered primarily for his contributions to the *Encyclopédie*, in which he, likewise, distinguished between primary and principiate principles, but he turned Stahlian chymistry toward a particularly aggressive assault on physics. He argued that chemistry penetrated "the interior of certain bodies about which Physics knows only the surface and the exterior figure" (Venel 1753, 409).[58] It revealed internal chemical properties inherent to bodies and found that the cause of such qualities was located in the elements themselves or in the nature of the mixture. While physics regarded such qualities merely as modes or accidents, the chemist viewed them as physically manipulable substances, which included color, the principle of inflammability, taste, and odor. Whereas a physicist would say that fire is light that is thrown off when a body is heated, "the chemist is able to remove the principle of inflammability, that is to say, fire, just as he is able to squeeze water out of a sponge and collect it in another vessel" (Venel 1753, 419).[59] Color was no different:

---

[57] Translations from *Dictionary of Chemistry* Vol. II (London: Cadell and Elmsley, 1778), translated by James Keir.

[58] "...l'intérieur de certains corps dont la Physique ne connoît que la surface & la figure extérieure..."

[59] "...car le chimiste peut aussi bien enlever au charbon, & montrer à part le principe de l'inflammabilité, c'est-à-dire le feu, qu'exprimer l'eau d'une éponge & la recevoir dans un vaisseau."

> The color seen in a colored body, for the Physicist, is a certain disposition of the body's surface, which allows it to reflect specific rays; but for the Chemist, a plant's greenness is inherent to a specific resinous green body, and he is able to remove it from the plant. The blue coloring of clay is due to a metallic material that he is also able to separate. Even the blue of jasper, which seems so closely united to the fossil substance, has been extracted, according to the famous experiment of Becher. (Venel 1753, 419)[60]

Venel's appeal to Becher's jasper experiment—completed over 80 years earlier—is particularly striking, for while Stahl had extensively discussed analyses that revealed phlogiston, he did not believe that the mercurial principle, *terra fluida*, had been revealed, and to my knowledge, never made mention of Becher's experiment with jasper. Nevertheless, Venel persisted and suggested that physicists and chemists approached natural phenomena in ways that were different but not contradictory. Even so, he opined that "the Chemist simply goes one step further" (Venel 1753, 419).[61]

Finally, Martin Fichman has concluded that Antoine Laurent Lavoisier's (1743–1794) chemistry owed a considerable debt to the French Stahlians, most notably their distinction between the properties of individual particles and the properties of mixts and aggregates (Fichman 1971, 94–122). J. B. Gough argued that French Stahlian chemists' greatest contribution—constituting what he styled the "Stahlian revolution"—arose from their attempt to distinguish chemistry from physics and was, in essence, that they "intellectually isolated and defined the chemical molecule and made it the unique subject of the chemical discipline" (Gough 1988, 23). Gough pointed to their understanding that indivisible *parties constituantes* combined to form *parties intégrantes*, which were thought to be "the smallest particle into which a homogenous chemical substance may be divided without decomposing it" (Gough 1988, 24). He concluded that the *partie intégrante* was the forerunner of the modern chemical molecule and served as a central focus of chemistry, "helping to define it and to maintain its autonomy" (Gough 1988, 31). I have shown here the influence of a tradition of earlier chymistry focused on principles and *principiata*, which, far from being a metaphysical dead end, was an archetype of the combination of theory and practice grounded upon the experimental analysis and synthesis of bodies. It led to the formulation of hierarchical models of chymical combination and provided chymists with a defense that impelled their burgeoning discipline further in the direction of autonomy. In effect, this earlier tradition must be considered within the broader history of chemistry's development.

---

[60] "La couleur considérée dans le corps coloré est, pour le physicien, une certaine disposition de la surface de ce corps, qui le rend propre à renvoyer tel ou tel rayon; mais pour le chimiste, la verdure d'une plante est inhérente à un certain corps résineux verd, qu'il sait enlever à cette plante; la couleur bleue de l'argille est dûe à une matiere métallique qu'il en sait aussi séparer; celle du jaspe, qui semble si parfaitement un avec cette substance fossile, en a pourtant été tirée & retenue, selon la fameuse expérience de Becher."

[61] "Le chimiste fait seulement un pas de plus…"

References

Alberti, Michael. 1721. Fundamenta Chymiae. In *Introductio in Medicinam Practicam ... Therapia Medica...cum additamento Fundamentorum Philosophiae Naturalis usui medico accommodatae et Chymiae*. Halae: Orphanotrophei.

Barner, Jacob. 1689. *Chymia Philosophica Perfecte Delineata....* Nuremberg: Andreae Ottonis.

Beguin, Jean. 1618. *Tyrocinium chymicum e naturae fonte et manuali experiential deprompt um.* Königsberg.

Bensaude-Vincent, Bernadette, and Isabelle Stengers. 1996. *A history of chemistry*. Trans. Deborah van Dam. Cambridge, MA: Harvard University Press.

Billich, Anton Günther. 1631. *Observationes et Paradoxica Chymiatrica*. Leiden: Joannes Maire.

Boas, Marie. 1958. *Robert Boyle and seventeenth-century chemistry*. Cambridge: Cambridge University Press.

Bohn, Johannes. 1685. Praefatio ad Lectorem. In *Dissertationes Chymico-Physicae*. Leipzig: Gleditschii.

Bostocke, Richard. 1585. *The difference Betweene the Auncient Phisicke... and the latter Phisicke.* London: Robert Walley.

Chang, Ku-ming (Kevin). 2002. Fermentation, phlogiston and matter theory: Chemistry and natural philosophy in Georg Ernst Stahl's Zymotechnia Fundamentalis. *Early Science and Medicine* 7: 31–63.

———. 2015. Phlogiston and chemical principles: The development and formulation of Georg Ernst Stahl's principle of inflammability. In Bridging traditions: Alchemy, Chymistry, and Paracelsian practices in the early modern era, ed. K. Parshall, M. T. Walton, and B. T. Moran, 101–130. Kirksville: Early Modern Studies.

Croll, Oswald. 1609. *Basilica Chymica continens Philosophicam propria laborum Experientia.* Frankfurt.

de Fourcroy, Antoine-François. 1782. *Leçons élémentaires d'histoire naturelle et de chimie*. Paris.

———. 1796. *Elements of chemistry*. Transl. by R. Heron, vol. 1, 30–31. London: J. Murray & S. Highley.

———. 1800. *Système des Connaissances Chimiques*. Paris: Baudouin.

Du Chesne, Joseph. 1603. *Liber de Priscorum Philosophorum*. S. Gervasii: Vignon.

Eklund, Jon. 1971. *Chemical analysis and the phlogiston theory 1738–1772: Prelude to revolution*. PhD dissertation. Yale University.

Ettmüller, Michael. 1685. *Medicus Theoria et Praxi Generali*. Frankfurt: Michael Günther.

Fichman, Martin. 1971. French Stahlism and chemical studies of air, 1750–1770. *Ambix* 18: 94–122.

Gerding, Theodor. 1867. *Geschichte der Chemie*. Leipzig: Grunow.

Gough, J.B. 1988. Lavoisier and the fulfillment of the stahlian revolution. *Osiris* 4: 23–31.

Hirai, Hiro. 2005. *Le concept de semence dans les théories de la matière à la Renaissance*. Turnhout: Brepols.

Hooykaas, Reijer. 1933. *Het Begrip Element: In Zijn Historisch-Wijsgeerige Ontwikkeling*. Ph.D Dissertation. Utrecht University. Trans. Hans Kubbinga as *The concept of element: Its historical-philosophical development*. Authorized translation, privately printed.

———. 1949. The experimental origin of chemical atomic and molecular theory before Boyle. *Chymia* 2: 65–80.

Juncker, Johann. 1730. *Conspectus Chemiae Theoretico-Practicae*. Halae Magdeberg: Orphanotrophei.

Jungius, Joachim. 1662. *Doxoscopiae physicae minores*. Hamburg: Johann Naumann.

Kahn, Didier. 2016. *Le fixe et le volatil: Chimie et alchimie, de Paracelse à Lavoisier*. Paris: CNRS Éditions.

Klein, Joel A. 2014. Corporeal elements and principles in the learned German chymical tradition. *Ambix* 61 (4): 345–365.

———. 2022. Practice and experiment: Cultures of chymical analysis. In *A cultural history of chemistry in the early modern age*, ed. Bruce T. Moran, vol. 3, 41–65. London: Bloomsbury Academic.
Klein, Joel, and Evan Ragland. 2014. Introduction: Analysis and synthesis in medieval and early modern Europe. *Ambix* 61 (4): 319–326.
Klein, Ursula. 1994. Origin of the concept of chemical compound. *Science in Context* 7 (2): 163–204.
Le Fèvre, Nicaise. 1660. *Traicté de la Chymie*. Paris: Thomas Jolly.
Libavius, Andreas. 1607. *Alchymia triumphans de iniusta in se collegii Galenici spurii in Academia Parisiensi censura; et Ioannis Riolani manographia, falsi convicta, et funditus eversa*. Frankfurt.
Macquer, Pierre-Joseph. 1766. *Dictionnaire de Chymie*. Paris: Chez Lacombe.
Metzger, Hélène. 1991. *Chemistry*. Trans. Colette V. Michael. West Cornwall: Locust Hill Press.
Newman, William R. 2001. Experimental corpuscular theory in Aristotelian alchemy: From Geber to Sennert. In *Late medieval and early modern corpuscular matter theories*, ed. Christoph Lüthy, John E. Murdoch, and William R. Newman, 291–329. Leiden: Brill.
———. 2006. *Atoms and alchemy: Chymistry and the experimental origins of the scientific revolution*. Chicago: University of Chicago Press.
———. 2014. Robert Boyle, transmutation, and the history of chemistry before Lavoisier: A response to Kuhn. *Osiris* 29 (1): 63–77.
Newman, William R., and Lawrence M. Principe. 2002. *Alchemy tried in the fire: Starkey, Boyle, and the fate of Helmontian Chymistry*. Chicago: University of Chicago Press.
Powers, John C. 2007. Chemistry without principles: Hermann Boerhaave on instruments and elements. In *New narratives in eighteenth-century Chemistry*, ed. Lawrence M. Principe, 45–61. Dordrecht: Springer.
———. 2012. *Inventing Chemistry: Herman Boerhaave and the reform of the chemical arts*. Chicago: University of Chicago Press.
Principe, Lawrence M., and William R. Newman. 2002. *Alchemy tried in the fire: Starkey, Boyle, and the fate of Helmontian Chymistry*. Chicago: University of Chicago Press.
Sala, Angelo. 1617. *Anatomia antimonii: id est, Dissectio tam Dogmatica quam Hermetica Antimonii*. Leiden: Basson.
———. 1622a. *Anatomia vitrioli*. Hamburg: Carstens.
———. 1622b. *Chrysologia, seu Examen Auri Chymicum*. Hamburg: Carstens.
Sennert, Daniel. 1619. *De Chymicorum cum Aristotelicis et Galenicis Consensu ac Dissensu*. Wittenberg: Schürer.
———. 1636. *Hypomnemata Physica*. Frankfurt: Clemens Schleich.
Shackelford, Jole. 2004. *A philosophical path for Paracelsian medicine: The ideas, intellectual context, and influence of Petrus Severinus (1540–1642)*. Copenhagen: Museum Tusculanum Press.
Shaw, Peter. 1730. *Philosophical principles of universal chemistry*. London.
Siegfried, Robert. 2002. *From elements to atoms: A history of chemical composition*. Vol. 92, i. Philadelphia: American Philosophical Society.
Siegfried, Robert, and Betty Jo Dobbs. 1968. Composition, a neglected aspect of the chemical revolution. *Annals of Science* 24: 275–293.
Smith, Pamela. 2004. *The body of the artisan: Art and experience in the scientific revolution*. Chicago: University of Chicago Press.
Stahl, Georg Ernst. 1715. *Opusculum Chymico-Physico Medicum*. Halae Magdeburgicae: Orphanotrophei.
———. 1718. *Zufällige Gedanken und nüzliche Bedencken über den Streit von dem sogennanten Sulphure*. Halle: Waysenhaus.
———. 1723. *Fundamenta Chymiae Dogmaticae & Experimentalis*. Nürnberg: Wolfgang Mauritius.
———. 1732. *Fundamenta Chymiae Dogmatico-Rationalis & Experimentalis*. Nürnberg: B. Guolfg.

Thackray, Arnold. 1970. *Atoms and powers*. Cambridge, MA: Harvard University Press.
Van Helmont, Jean Baptiste. 1648. *Opuscula Medica Inaudita*. Amsterdam: Elzevier.
———. 1652. *Ortus medicinae, id est, initia physicae inaudita*. Amsterdam: Ludovicum Elzevirium.
———. 1664. *Van Helmont's workes*. London: Lodowick Lloyd.
Venel, Gabriel Francois. 1753. Chymie, ou Chimie. In *Encyclopédie ou Dictionnaire raisonné des sciences, des arts et des métiers*, ed. Denis Diderot and Jean-Baptiste le Rond d'Alembert, vol. 3, 409–419. Paris.
Wedel, Georg Wolfgang. 1686. *Pharmacia Acroamatica*. Jena: Johannis Bielckii.
———. 1715. *Compendium Chimiae*. Jena: Johannis Bielckii.

Joel A. Klein is the Molina Curator for the History of Medicine and Allied Sciences at The Huntington Library in San Marino, CA. He earned his PhD from Indiana University and has had postdoctoral fellowships at Columbia University and the Science History Institute.

Chapter 5
Philosophical Methods of Analysis and Synthesis from Medieval Scholasticism to Descartes and Hobbes

Helen Hattab

Abstract Drawing on scholarship that traces the medieval appropriation of ancient methods of analysis and synthesis, I demonstrate that the intermingling of differing senses of analysis and synthesis, resolution, and composition predates the early moderns. Section 5.1 maps out five distinct ancient senses of resolution and composition and illustrates that several are conflated in the works of St. Thomas Aquinas, whose philosophy witnessed a resurgence with the rise of the Jesuit order in the sixteenth century. Section 5.2 examines how earlier methods of analysis and synthesis were taken up and developed by scholastic logicians influential in Descartes' and Hobbes' contexts. Sections 5.3 and 5.4 clarify Descartes' and Hobbes' claims about philosophical analysis and synthesis in light of this background. I demonstrate that the incoherencies generated by linking their philosophical methods solely to Zabarella's *regressus* may be resolved by reinterpreting these claims in their wider context. The question as to whether/how their own and prior philosophical methods of analysis and synthesis shape the methodical procedures that Descartes and Hobbes employ to tackle scientific problems is shelved, but the philosophical texts examined indicate that they need not be directly connected.

Keywords Descartes · Hobbes · Zabarella · Regressus · Resolution-composition

In the seventeenth century, various traditions that had long employed methods of analysis and synthesis or resolution and composition were conflated, leading to confusion and controversy among scholars regarding the type(s) of method that philosophers such as Descartes, Hobbes, and Spinoza employed and how such methods might be linked. One is the ancient mathematical tradition revived in the Renaissance and discussed in this volume in the chapter by Niccolo Guicciardini. One scholarly debate about the influence of this tradition on early modern philosophy concerns whether the Euclidean geometrical method of synthesis, which derives philosophical conclusions from definitions and axioms and is most clearly evident in Spinoza's

---

H. Hattab (✉)
Department of Philosophy, The University of Houston, Houston, TX, USA
e-mail: hhattab@Central.UH.EDU

W. R. Newman, J. Schickore (eds.), *Traditions of Analysis and Synthesis*, Archimedes 73, https://doi.org/10.1007/978-3-031-76398-4_5

Ethics, is primarily a didactic method of presentation suited to gaining the average reader's assent, as Descartes suggests, or also a scientific method of discovery, as Hobbes implies (Hattab 2020; Sacksteder 1980).[1] The other influential tradition, with its methods of resolution or division and composition, is ancient philosophy. Such methods were linked to Aristotle's works, which had predominated university curricula since the institutions' medieval origins. I explore whether/how Descartes and Hobbes appropriate prior philosophical uses of analysis and synthesis in their theoretical reflections on method.

Drawing on scholarship that traces the medieval appropriation of distinct ancient methods of analysis and synthesis, I demonstrate that the intermingling of differing senses of analysis and synthesis, resolution, and composition predates the early moderns. Section 5.1 establishes that several ancient senses of resolution and composition are conflated in the works of St. Thomas Aquinas, whose philosophy witnessed a resurgence with the rise of the Jesuit order in the sixteenth century. Section 5.2 examines how earlier methods of analysis and synthesis were taken up and developed by scholastic logicians influential in Descartes' and Hobbes' contexts. Sections 5.3 and 5.4 clarify Descartes' and Hobbes' claims about philosophical analysis and synthesis in light of this background. Owing to time and space constraints, I do not address the question of whether/how their own and prior philosophical methods of analysis and synthesis shape the methodical procedures they employ to tackle scientific problems. Other chapters in this volume examine the practical implementation of such methods across various early modern domains.

5.1 Resolution and Composition in Medieval Scholasticism

Recent studies have identified at least five distinct senses of analysis or resolution that medieval commentators inherited from ancient sources and often conflated. Some are paired with a corresponding sense of synthesis or composition.

1) Physiological analysis into elements, mentioned in Alexander of Aphrodisias' and Ammonius' commentaries on the *Prior Analytics* and explained as follows by Calcidius in his commentary on Plato's *Timaeus*:

> If by means of our intellect, we wish to take away these qualities and quantities, these shapes and figures, and then consider what keeps all these things inseparably together and contains them, we shall find that there is nothing else than that which we are looking for, i.e., matter, and herewith we have found the material principle. This then is one of the two possible methods of arguing, called *resolutio*. (Van Winden 1959, 132)

"The opposite movement, *composition*, which "follows *resolutio* as union follows separation," works by reconstructing the object, by adding back in, if you will, the genera, qualities and forms which have been separated from it" (Sweeney 1994,

---

[1] Although Hobbes clearly embraces synthesis as a philosophical method of discovery, Sacksteder has shown convincingly that philosophical analysis and synthesis are distinct from the processes of analysis and synthesis that comprise the mathematical method Hobbes calls 'logistica'.

206–207).[2] Eileen Sweeney connects this sense of analysis/resolution (which she calls Calcidian resolution, also called 'dissolution' by medieval commentators) with Aristotle's argument in *Metaphysics* Bk VII, Ch. 3 that when all "affections, products and capacities of bodies" are taken away, including length, breadth, and depth, only matter remains, and, therefore, it seems to be substance (Aristotle 1985, 1625, 1029a11–20). In Bk VIII, Ch. 4, Aristotle uses "analysis" in this way while contrasting between two senses in which things are constituted: they are constituted from proper matter and also from the same primary component(s). In the first non-resolutive, developmental sense of being constituted, if phlegm comes from the fat, it also comes from the sweet, provided the fat came from the sweet. This is because the sweet is there at an earlier stage of development, later giving rise to the fat and eventually the phlegm. In the second resolutive sense, phlegm comes from bile "by analysis of the bile into its ultimate matter" and "it is produced if the other is analysed into its original constituents" (Aristotle 1985, p.1648, 1044a23–24). Sweeney traces how Calcidius pairs this physiological sense of analysis with a process of composition whereby the complex (e.g., phlegm) is reconstructed out of its parts, which are the same parts that constituted the dissolved bile. She argues for Calcidius as the likely source of the key conflations that Aquinas makes between this sense of analysis and another procedure in Aristotle (sense 2 below), which differs from physiological analysis and which Aristotle, unlike Aquinas, does not explicitly pair with a reverse process of composition.

2) Conceptual analysis, as described by Aristotle in *Physics* I, 184a25. Aristotle highlights that the physicist must begin with things that are less known by nature but that are better known to us and progress from these to the things that are better known by nature—that is, the elements and principles. That which is better known by perception is the whole or composite, and this is likened to the universal, which, since it also embraces many things as parts, resembles a whole. On this basis, Aquinas distinguishes between the following two ways of knowing the truth, adding a reverse compositive procedure that is not in Aristotle's text:

a) "....the method of analysis, by which we go from what is complex to what is simple or from a whole to a part, as it is said in Book I of the *Physics [184a21]* that the first objects of our knowledge are confused wholes. Now our knowledge of the truth is perfected by this method when we attain a distinct knowledge of the particular parts of a whole" (Aquinas 1961, 108, sec. 278).
b) "....that of synthesis, by which we go from what is simple to what is complex; and we attain knowledge of truth by this method when we succeed in knowing a whole" (Aquinas 1961, 108, sec.278).

Sweeney identifies another merging of senses (1) and (2) in Aquinas' first lecture on Aristotle's *Politics*, wherein Aquinas links Alexander of Aphrodisias' *Prior Analytics* illustration that analysis divides a sentence into letters and syllables with the physiological analysis of bodies to make a parallel claim that understanding how

[2] Van Winden's translation cited by Sweeney 1994, 206–207.

political constitutions differ likewise requires the state to be divided into the basic units from which it is constituted:

> Just as in other things to know the whole, it is necessary to divide the composite until one arrives at incomposite things, i.e., until one arrives at indivisibles which are the smallest parts of the whole: for example, in order to know sentences, it is necessary to divide until [one arrives] at letters, and to know natural, mixed bodies, it is necessary to divide them until [one arrives] at elements. (Sweeney 1994, 212)[3]

According to Aquinas, resolution in a broad sense that encompasses the division of any whole, physical or conceptual, into its elemental parts requires a subsequent process of composition that enables us to make judgments about the things caused by principles using the indivisible principles already known.

3) Analysis as order, found in the preface to Galen's *Ars medica.* Don Morrison describes this as a method by which one structures a treatise: that is, this method organizes "an entire body of already acquired knowledge" (Morrison 1997, 18). Unlike the first phase of a syllogistic proof known as the *regressus*, in which one reasons syllogistically first from effect back to cause to syllogistically demonstrate the effect from the proper cause after a mental consideration, analysis in this sense is not a method of discovery. However, medieval commentators on Galen, beginning with Pietro d'Abano, read the *regressus* back into Galen's text, confusing analysis as order with a demonstrative resolution from effect back to cause (Morrison 1997, 18). Section 5.2 discusses how Jacopo Zabarella uses the term "analysis" in Galen's original sense, carefully distinguishing it from the resolutive phase of the *regressus.* Descartes also exhibits familiarity with analysis as order, although his use differs from Zabarella's in ways that resemble the innovations of seventeenth-century logicians.

4) Resolution to first principles, as found in the Neoplatonic works of Proclus and Plotinus. Plotinus describes this dialectical sense of analysis as follows: "Our dialectic makes great use of division and *analysis* as the principal means of knowledge and as *imitating* the *procession* of beings from the One and their *reversion* back again…" (Plotinus 1969, 158). The procedure that Plotinus has in mind is clarified by Proclus, who describes dialectic as first employing division to reach the forms, then weaving the intelligible universe together from the first genera arrived at through division, and concluding with a process of resolution or analysis back to the metaphysical starting point (Sweeney 1994, 216). Resolution in this sense thus differs from conceptual analysis (sense (2)), even though both move from complex to simple. Sweeney highlights that the key difference from Aristotelian senses of resolution lies in the fact that while senses (1) and (2) both move down the ontological ladder to arrive at the constitutive elements of wholes, Neoplatonic resolution moves up the ontological chain back to the highest and simplest cause/principle. She explains that the Neoplatonic process of reason moves from that which is better known, both to us and in itself, to lower complex objects and then returns, through

[3] Sweeney's translation of this passage from Aquinas' first lecture in his commentary on Aristotle's *Politics*.

analysis/resolution, to the simpler and higher causes. Through this process, the simplest causes are not discovered (they are known at the start) but rather known with greater understanding. Scotus Erigena uses this sense of analysis but, as Sweeney argues, Aquinas likely acquires it from Albert the Great, who was aware of conflicting notions of resolution—that is, physiological analysis into matter and form (sense 1) versus the resolution to a first cause (sense 4) (Sweeney 1994, 218, 221–2).

Aertsen, who focuses more narrowly on Aquinas' use of resolution in metaphysics likewise attributes to Aquinas resolution as a discursive process of reasoning that gathers a simple common truth from many things and culminates in understanding (Aertsen 1996, 132). In Aquinas' metaphysics, resolution ultimately terminates in divine science, the highest of all the sciences. Aertsen distinguishes between the two kinds of resolution through which this is accomplished. Sweeney classifies both under sense 4), given that each moves upward to higher, simpler and more universal principles albeit in different ways (Aertsen 1996, 133). Aquinas pairs both resolutive processes with a corresponding composition.

The distinction drawn by Aquinas that Aertsen examines is between 4) a) resolution and composition *secundum rem* and 4) b) resolution and composition *secundum rationem.* In a resolution *secundum rem*, one demonstrates through extrinsic effects, arriving at the highest causes—namely, the immaterial substances—from their effects. The corresponding composition *secundum rem* demonstrates the effects through extrinsic causes. A resolution *secundum rationem* proceeds according to intrinsic causes and effects, arriving at the most universal forms from the more particular ones and culminating in "the consideration of being and that which belongs to being as such" (Aertsen 1996, 133). The corresponding composition *secundum rationem* proceeds in the reverse direction, beginning with more universal forms before attaining the most particular ones. As Aertsen demonstrated, since it culminates in separate substances, by resolution *secundum rem*, Aquinas cannot mean an analysis of natural things into their elements. Hence both he and Sweeney distinguish Aquinas' metaphysical use of analysis from sense 1). However, Aertsen attributes to Aquinas the view that composition is only possible for humans in mathematics, *contra* Sweeney, who views Neoplatonic analysis as a kind of judgment that follows composition (Sweeney 1994, 228; Aertsen 1996, 136).

Remnants of the Neoplatonic sense of resolution and Aquinas' *resolutio secundum rem* survive in Zabarella's account of a demonstrative proof known as the *regressus*. As stated, this proof starts from the effect, reasons syllogistically back to the cause, and then, following mental consideration, deduces the effect from the cause. Whereas the effect was known, albeit confusedly from the start, it is now understood scientifically through its cause. However, Zabarella, like other proponents of the *regressus*, calls the first downward movement of the proof a resolution and the upwards, resolutive movement toward the higher cause composition, possibly conflating this syllogistic working back to higher causes with the Aristotelian physiological and conceptual analyses (senses 1 and 2) by which one divides wholes into elements. In Sect. 5.2, I shall demonstrate that Zabarella does distinguish between resolution, in this sense, and Galen's analysis as order (sense 3). In Sect. 5.3, I shall demonstrate that Descartes invokes senses 2) and 3) of analysis.

In Sect. 5.4, I shall show that Hobbes does not distinguish between the various senses of analysis. This is unsurprising, given the potential for sense 3)—analysis as order—to become associated with sense 4)—Neoplatonic resolution—via Proclus' *Commentary on the First Book of Euclid's Elements.* As Sweeney highlights, for Proclus, the elements of geometry are not strictly the component parts of complex figures but are also the simpler and more general principles from which the figures proceed. For example, a line is simpler than a plane, a genus is simpler than a species, and common notions and general principles are simpler than determinate ones. Proclus conveys the sense that the simples are *causes* of the complex items associated with them. Hobbes likewise uses causal language to characterize this relationship. As Sweeney puts it, "For Proclus the movement of reason mirrors the order of being, i.e., its conclusions flow from and return to a single most simple principle, the One. This structure organizes Proclus' *Elements of Theology* and the *Liber de Causis* based on it" (Sweeney 1994, 219–220). A similar parallelism between the resolution to ever higher, simpler principles/causes and the order of the sciences in seventeenth-century logicians makes its way into Hobbes' method.[4]

5) Geometrical analysis, as found in Pappus of Alexandria, and synthesis in teaching. As the well-known study by Hintikka and Remes discusses, for Pappus,

> [A]nalysis is the way from what is sought—as if it were admitted—through its concomitants [*to akolouthon*] in order to attain something admitted in synthesis. For in analysis we suppose that which is sought to be entirely done, and we inquire from what it results, and again what is the antecedent of the latter, until we on our backward way light upon something already known and being first in order. (Hintikka and Remes 1974, 8)

This sense of analysis (5a) is often opposed to demonstration since the synthesis provides the proof. Sweeney traces Aquinas' uses of this sense of analysis back to a comparison in Aristotle's *Nichomachean Ethics* III, 1112b16–20. Someone who takes counsel resolves by assuming the goal and reasoning back to the means by which it will be accomplished to arrive at the action that will achieve the goal in a way that is similar to the analysis or resolution of a geometrical construction (Sweeney 1994, 229). In Sect. 5.2, I shall discuss the sixteenth- and seventeenth-century limitation of this sense of analysis to practical endeavors.

Via another route, however, a kind of analysis and the corresponding composition in the Neoplatonic sense becomes linked to mathematics, and in this sense (5b), it is part of a 'geometrical' approach to teaching theoretical matters. The link occurs in Boethius' *De hebdomadibus*, which aims to resolve certain puzzles about the

---

[4]Aertsen 1996 on p. 150 notes a similar parallelism between Aquinas' *resolutio secundum rationem* and *secundum rem*, claiming that in *De veritate*, Aquinas takes from Avicenna the idea that the firstness of being parallels the structure of demonstrative science. Conceptions of the human intellect are reducible to a first self-evident conception, just as propositions used in demonstrations are reducible to a first self-evident proposition. In his commentary on *Metaphysics* IV, Aquinas spells out how these parallel orders are linked more clearly than in his commentary on Boethius' *De hebdomadibus*. In his commentary on the *Metaphysics*, he claims that the first principle in the order of conceptions—in this case, the principle that something cannot simultaneously be and not be —is the foundation of what comes first in the order of demonstration. The principle of non-contradiction can itself be understood only if the mind understands the first conception of being.

good and its relationship to substance by deriving conclusions from a series of common conceptions laid out at the start as axioms, as mathematics does. Aquinas, in commenting on Boethius, reverses the Neoplatonic procedure and, like Zabarella, puts resolution first, followed by composition. Boethius writes, "Therefore, as customarily happens in mathematics and in other disciplines as well, I have set out first the terms and rules by which I shall develop all that follows" (Aquinas 2001, 3). Immediately following this claim, Boethius defines a common conception of the mind as "a statement that everyone approves on hearing" (Aquinas 2001, 3). In his commentary on the *De hebdomadibus*, Thomas Aquinas explicitly links Boethius' claims to Aristotle's method of resolution, on the one hand, and to the form of proof found in geometry on the other hand, thus creating a broader sense of geometrical analysis (5b) than that found in Aristotle's *Ethics* (5a). Aquinas elaborates that Boethius,

> ... states first that he intends to propose from the start certain kinds of principles, known through themselves, which he calls *terms* and *rules*: 'terms' because the resolution <back to prior principles> of all demonstrations stops at principles of this sort; 'rules', however, because through them one is directed to a knowledge of conclusions which follow. **{130}** From principles of this sort he intends to draw conclusions and to make known all that ought to be developed as following logically, as happens in geometry and in other demonstrative sciences. Therefore these are called 'disciplines,' because through them 'science' is generated in the 'disciples,' thanks to the demonstration which the master propounds. (Aquinas 2001, 10–11)

Here, resolution is linked to mathematics, but unlike in the *Nicomachean Ethics* sense, it is not merely an analysis back to the means of accomplishing a practical end, even though the corresponding synthesis implies the goal of imparting the scientific knowledge to students through a series of proofs. The first step rather resembles a Neoplatonic metaphysical resolution to higher principles that, once completed, will yield the fundamental terms and rules from which one can then compose the geometrical-like demonstrations of conclusions that constitute the scientific disciplines, enabling them to be taught. Sections 5.2 and 5.4 of this paper will reveal how late scholastic logicians and Hobbes link synthesis in senses 4) and 5)b) to the orderly method of teaching (sense 3).

5.2 Early Modern Scholastics

In this section, I shall demonstrate that the influential writings on method by the sixteenth-century Aristotelian logician, Jacopo Zabarella (1533–1589) acknowledge some of the distinctions between ancient methods of resolution appropriated and merged by medieval commentators. Section 5.4 will demonstrate that Hobbes' account of philosophical method blurs these lines, while Sect. 5.3 will argue that although Descartes' uses of the terms "analysis" and "synthesis" are closer to Zabarella's, analysis as a resolutive method that is central to the discovery of the immediately known principles of metaphysics is transformed into something different in Descartes' hands, given his rejection of scholastic metaphysics. An examination of two logic textbooks that prevailed in Descartes' and Hobbes' intellectual contexts illuminates several key differences between seventeenth-century views of analysis and synthesis and Zabarella's perspective.

Zabarella's *De Methodis* carefully distinguishes analysis and synthesis from what he calls "resolution" and "composition." For Zabarella, analysis and synthesis are methods used to order knowledge so that it is more easily grasped and taught, whereas resolution and composition are methods of demonstration. Like other Aristotelians, Zabarella holds that synthesis proceeds from the simpler—that is, from the universal or genus—to the species and from there to particular attributes of the object of knowledge, while analysis proceeds in the reverse direction. Synthesis is the preferred method of teaching, a point that Hobbes will echo. For example, in teaching physics, one would proceed from the genus of motion to the kinds of motions and finally to their properties. Analysis, which was often attributed to practical disciplines, begins with the end to be accomplished and, from there, works its way back to the means and finally to the starting points of the production process. Zabarella clearly uses analysis and synthesis in sense 3). Thus, they are not methods of demonstration and should not be confused with the resolutive and compositive phases associated of the *regressus* proof, which resemble sense 4)a).

In *De Methodis,* Zabarella defines method in the broad sense as: "an instrumental *habitus* of the intellect, which aids us in attaining knowledge of things" (Zabarella 1597, I.ii, 136).[5] He divides method, taken broadly, into order and method properly speaking. The task of method in the proper sense is to lead us from a known thing to knowledge of another, unknown thing, as when we are led from substantial change to knowledge of prime matter, or from eternal motion to knowledge of an eternal unmoved mover. This corresponds roughly to sense 4)a), Aquinas' Neoplatonic resolution *secundum rem.* Zabarella pairs this sense of resolution with the reverse process of demonstration by way of composition to develop a kind of scientific demonstration known as the *regressus*. This particular form of scientific proof falls under Zabarella's second sense of method (Zabarella 1597 ch. iv, 484–86).[6] Extensive study of Zabarella's theory of the *regressus* has resulted in both comparisons to and contrasts with the forms of demonstration used by Galileo Galilei, William Harvey, and René Descartes, but the relationship between

[5] All translations of this work are mine.

[6] In his *Liber De Regressu,* Zabarella gives a rather succinct example, taken from the first book of Aristotle's *Physics*, of the three parts of the demonstrative *regressus*. The first is the resolutive phase, by which we deduce confused knowledge of the cause from our confused knowledge of the effect. The second phase consists in the mental consideration of the cause known confusedly, so as to know it distinctly. The third phase consists in composition, by which the effect is deduced from the cause, now known distinctly. In the example Zabarella takes from Book I of Aristotle's *Physics*, we start from our confused knowledge that a certain effect occurs: the generation of a substance. We then reason back to the more fundamental principle (i.e., the cause of this generation). This is the *demonstratio quia* or τό ὅτι proof from what is more known to us to what is prior by nature.

early modern methods and the other branch of Zabarella's method remains understudied.[7]

My focus here is on the other sense of 'method' discussed in *De methodis*. This corresponds to 3) analysis as order and overlaps with 5)b) geometrical analysis in Aquinas' sense of a compositive method for teaching scientific knowledge. As Zabarella notes, method as order does not cause us to infer one thing from another but rather arranges [*disponere*] the things to be treated, as when the order of teaching demands that we first discuss the heavens and then the elements. In other words, it arranges the parts of a discipline (Zabarella, 1597, I.iii). Order takes precedence (over demonstration) because one must divide a discipline into parts before one can articulate the method that will lead us from the known to the unknown that is sought within each part (Zabarella, 1597, I.iii, 139).[8] For example, one must first treat of living things in general and then each individual species of living thing before finally seeking out methods to treat what is common to animals, to understand the nature of a particular animal and its accidental features (Zabarella 1597, I.iii, 139). Zabarella links 3) analysis as order to demonstrative proofs in a way that is reminiscent of Aquinas' appropriation of the method in Boethius' *De hebdomadibus* discussed under sense 5)b). For Zabarella, order seems to involve conceptual analysis into ever more universal genera in preparation for the acquisition of demonstrative knowledge of the less universal.

Zabarella adds that one must not state that the order is made randomly without any reason and internally by our choice: there has to be some certain reason or certain, necessary norm by which the correct arrangement and appropriate ordering is taken up (Zabarella 1597, I.iv, 140).[9] He also denies earlier views that ordering must always proceed from one thing, either a principle, medium, or end, and that correspondingly, there are three types of order: compositive, definitive and resolutive (Zabarella 1597, I.v, 140–141). Zabarella thus rejects this aspect of Galen's view, instead following Averroes in claiming that the procedure for ordering the sciences and all disciplines is found not in the essence of the objects sought, but in the manner of knowing things that is best and easiest for us. This is a major difference from Aquinas, according to Aertsen's reading of the latter. For Zabarella, when a science is ordered in one way rather than another, it is so ordered on the grounds that it will

---

[7] Studies of Galileo's and Descartes's methods include Wallace 1997 and Timmermans 1999. J.N. Watkins, relying on Randall and one passage from Harvey's *On Generation of Animals*, connects Harvey's method back to Zabarella's *regressus* (Watkins 1965, 64). This connection is contradicted, however, by more in-depth studies revealing that Fabricius, under whom Harvey studied medicine at Padua, drew heavily on Aristotle's biological writings in his method rather than Zabarella's writings on method; see Cunningham 1985, 211.

[8] Zabarella holds that one must treat of order first because it appears to be something more general, extending more widely than method, for it regards *scientia* as a whole and compares its parts. Method proper, by contrast, consists in the investigation of a single sought thing without any comparison to other parts of *scientia*.

[9] Following the distinction between order and method, Zabarella amends the common interpretation of the order of a discipline as an instrumental *habitus* or mental instrument, by means of which one is taught to appropriately arrange the parts of a doctrine.

be more easily and effectively learned in this way, rather than because of a natural order that exists outside the mind (Zabarella 1597, I.vi, 142–144).[10] Zabarella thus takes 3) analysis as order as producing a purely conceptual ordering that facilitates our ability to grasp the subject matter.

Zabarella nonetheless affirms Aristotle's view that the proper order is always from the universal to the particular on the grounds that we always investigate the essence or nature of a thing or its proper accidents. To know the nature of a thing, we must know its species, and this is only possible once we know the nature of the genus. Likewise, we know the accidents of the species when we know the accidents of the genus. Therefore, the easiest and most effective order of learning proceeds from knowledge of the genus, or the more universal, to the species and thence to accidents of the species (Zabarella 1597, I.iii, 139). Zabarella's sense of synthesis appears to blend sense 2), in that it presupposes a conceptual analysis that precedes the synthetic ordering; sense 3), in that it is a means of ordering rather than discovering knowledge; and sense 5)b), as found in Aquinas' geometrical method of teaching. However, most recent secondary literature on Zabarella mistakes what he means by "synthesis" for the composition that follows resolution in sense 4)a) Aquinas' *resolutio secundum rem*, whereby effects are demonstrated from causes in the second phase of Zabarella's *regressus*.

With these standard misconceptions corrected, we can see how Zabarella's writings on method informed the logic textbooks published by the influential Calvinist Scholastic, Bartholomaeus Keckermann (1572–1609) and Franco Burgersdijk, his Dutch follower who taught at the University of Leiden in the early seventeenth century, (1590–1635). Both textbooks underwent multiple editions and were commonly used in seventeenth-century England and the Netherlands, thus forming part of Descartes' and Hobbes' intellectual contexts.

Keckermann's definitions of 'synthesis' and 'analysis' in his *Systema Logica*, which combined elements of Ramism with scholastic logic, were commonplace. Like many contemporaries, he regards synthesis as a method for ordering the content of theoretical disciplines (sense 3), whereas analysis orders practical disciplines (sense 5a). Keckermann writes, "*The synthetic method is that by which the contemplative disciplines are thus disposed into parts so that it would have progressed from the universal subject of contemplation to the particulars and therefore from the simples to the composites*" (Keckermann 1613, 588). He then reiterates Zabarella's example of how this method orders physics,

[10] He points out that if the suitable order within each discipline were found in the natural order of its objects, then the compositive order would be the only valid order, since the simples and the principles of nature are prior by nature to the composites. However, the suitable order is the order by which we know more easily and more effectively, as seen by the fact that Aristotle often follows the resolutive order in his works (e.g., in Book VII of the *Metaphysics* and also the *De Anima* and *N.E.*). Nonetheless, a given order of learning will sometimes coincide with the natural order, which accounts for how some come to confuse the natural order with the order of knowing.

> where it is first treated of natural body in general, then of their affections and principles. Afterwards he descends to the species of natural bodies, namely simple body, heaven, the elements, afterwards the mixed, and this again either imperfectly mixed, or meteors, afterwards perfectly mixed: and this again either inanimate, as metals, minerals, or animate, and this either vegetating as plants, or sentient; and this again either irrational where all brute animals are treated: or rational, as man; and thus from the highest genus the lower species are reached. (Keckermann 1613, 589–590)

Keckermann notes that, like physics, both mathematics and metaphysics employ synthesis in sense 3). Following Zabarella, he writes, "*The analytic method is that by which the operative disciplines are disposed, thus so that from the notion of the end it will be progressed to knowledge of principles, or media, through which the end is introduced into its subject*" (Keckermann 1613, 589). Keckermann uses the standard Aristotelian example of the art of building a house, noting that analysis requires a prior notion of the end as the first principle of the operation from which one then progresses to the means of reaching the end. Here, he subordinates Aristotle's sense 5)a) to the Galenic sense 3).

Keckermann, like Zabarella, uses analysis and synthesis in sense 3) to designate the methods employed in the preliminary ordering of a discipline that precedes demonstration, not as methods of demonstration. Keckermann calls method as order the universal method, which he defines as "the director of the inferring discourse; it [method] serves as the director of ordering discourse, which is an act of the human mind or intellect proceeding from one part of doctrine to another, collecting and connecting them among themselves with the help of the method of teachers" (Keckermann 1613, 578). However, he criticizes Zabarella's definition of order, which he sums up as "the instrumental habit [*habitus*] of doctrine through which we are apt to arrange the parts of this discipline, in order that, in so far as it is possible, this doctrine be learned optimally and most easily" (Keckermann 1613, 581). He then approvingly quotes the definition offered by Zabarella's rival, Francesco Piccolomini: "Order" he [Piccolomini] says, "is the suitable [*congrua*] arrangement of several of the disciplines or parts of the discipline both among themselves and towards a first one produced by distinguishing from the nature of things by the diligent, so that they imitate in teaching [*disciplina*] the nature of things distinctly brought together and offer it to the souls of readers" (Keckermann 1613, 581). Keckermann later appears to espouse Piccolomini's view that the methodical order tracks the natural order, stating, "The process of method imitates the order of natural things, by progressing from things which are prior by nature and more known, to the posterior" (Keckermann 1613, 582). This aspect of Keckermann's view resembles Proclus' clarification of Neoplatonic analysis (sense 4). It is likely that Keckermann was influenced by Neoplatonism via Ramus. Keckermann adds, "And in this respect it is most rightly said that there is only one method, because the process from the prior and more known things by nature to the posterior is only one" (Keckermann, 1613, 583). Unlike Zabarella, who regarded methodical order as merely a cognitive order to facilitate learning, Keckermann, following the Neoplatonist view, takes the universal method used in teaching to reveal a natural order and hence as one. He thus anticipates later searches for a single method applicable to all sciences.

Burgersdijk is even more critical of Zabarella, rejecting his view that the genus of method is an "instrumental habit [*habitus*]" and claiming instead that it is a disposition or arrangement. Method concerns a "faculty of arranging which is imposed on things by artifice" (Burgersdijk, 1627, 376–377).[11] This artificial arrangement is directed toward the end of serving "the intellect and memory towards the better and easier perceiving of the proposed things and more faithful guarding [of them]" (Burgersdijk 1627, 377). It accomplishes this aim by placing that which is more known before the unknown—for example, placing principles before the things known from them. Burgersdijk distinguishes between natural and arbitrary methods, characterizing the natural method as "that which serves the order of nature and our distinct cognition" (Burgersdijk 1627, 378). He then equates the natural method with the didactic method (sense 3), rejecting the standard Aristotelian distinction between a method that starts from what is prior by nature and one that starts from what is better known to us. For Burgersdijk, "the same things are prior by nature which are more known to us as far as distinct cognition goes. For that cognition is distinct which corresponds to the things themselves and the order of nature" (Burgersdijk 1627, 378). Burgersdijk calls the universal method the "Total Method" and claims that it is "*that by which some entire discipline is arranged*" (Burgersdijk 1627, 380). Like Zabarella and Keckermann, he subdivides this total method into synthesis and analysis. "*The Synthetic method is that which progresses from the most simple principles towards those which are composed from those principles*" (Burgersdijk 1627, 380). Burgersdijk gives the same standard definition of analysis that Keckermann does.

In sum, influential early modern scholastic logicians, while appropriating Zabarella's identification of analysis with Galen's didactic method of ordering (sense 3), simultaneously reject his view that the ordering is merely cognitive and need not track nature's order. This move informs Descartes' and Hobbes' characterizations of philosophical analysis and synthesis.[12]

5.3 Descartes on Analysis and Synthesis

Descartes is often perceived as modeling his philosophical method after geometrical analysis. I shall first demonstrate that this view lacks support. Evidence for it stems from Descartes' suggestion in *Rules for the Direction of the Mind* that he was inspired by the secret art of analysis used by ancient mathematicians since he briefly mentions this lost art in Rule 4 (CSM I, 17). This is too vague to draw firm conclusions. In Part 2 of the *Discourse on the Method*, Descartes resolves to "take over all that is best in geometrical analysis and in algebra, using the one to correct all the

[11] All translations of this work are mine.

[12] As stated, I shelve the further question of whether their philosophical uses of 'analysis' and 'synthesis' inform and map onto methodical procedures that they employ in the sciences.

defects of the other" (CSM I, 121). Prior to arriving at this conclusion, he notes that despite their diversity of objects, all mathematical sciences only consider the relations and proportions between their objects. This leads him to the insight that he should "examine only such proportions in general, supposing them to hold only between such items as would help me know them more easily" (CSM I, 121). To facilitate his ability to consider such proportions independently, Descartes assumes that the proportions hold between lines "because I did not find anything simpler, nor anything that I could represent more distinctly to my imagination and senses" (CSM I, 121). From algebra, he then takes the practice of representing them by concise symbols. In the *Discourse*, Descartes appropriates the advantages of geometrical analysis and algebra in the means by which he will represent the cognitive elements of his chains of reasoning to better hold them in mind and not necessarily to guide the reasoning itself.

The four rules articulated in Part 2 of the *Discourse* include Rule 2), a broadly resolutive division of problems into the simplest, easiest parts to know, and Rule 3), a consequent gradual and orderly thought process by which one comes to know more and more composite things. This resembles Aristotelian analysis in sense 2) with Aquinas' addition of a corresponding synthesis. However, one cannot assume that this is the *philosophical* method that Descartes employs in the *Meditations.* As Descartes states, here, he is still engaged in applying his new method to mathematical problems: "as I practised the method I felt my mind gradually become accustomed to conceiving its objects [i.e., the proportions] more clearly and distinctly; and since I did not restrict the method to any particular subject-matter, I hoped to apply it as usefully to the problems of the other sciences as I had to those of algebra" (CSM I, 121). It is tempting to read into this a philosophical method for solving metaphysical problems, given that Part 4 of the *Discourse* prefigures the arguments of his later *Meditations on First Philosophy.* However, the Preface to the *Discourse* consists of a hodgepodge that summarizes various projects that Descartes had worked on at different times.[13] "Sciences" in the context of the passage in *Discourse* Part 2 either refers back to the mathematical sciences other than algebra (the mathematical sciences having been mentioned two paragraphs earlier) or, more broadly, to the other sciences that Descartes invokes in Part 1. There, he claims, "As for the other sciences, in so far as they borrow their principles from philosophy I decided that nothing solid could have been built up on such shaky foundations" (CSM I, 115). Either way, the "other sciences" do not include philosophy as in the metaphysical foundations of the *Meditations*. Furthermore, scholarly attempts to interpret Descartes' philosophical method, broadly speaking, as a kind of geometrical analysis have not clarified how the method adopted in the *Meditations* conforms to mathematical definitions and uses of these methods.[14] In the absence of any

[13] See Verbeek et al. 1996 on the circumstances in which Descartes composed it, at Reneri's urging, to accompany his *Geometry, Dioptrics, and Meteorology*.

[14] See, for instance, Raftopoulos 2003, 305 and Recker 1993. Tarek Dika's recent book confirms that Descartes' early problem-solving mathematical method is distinct from his later endeavors; see Dika 2023.

compelling textual evidence that the method Descartes claims to have used successfully to solve problems in mathematics and other sciences is the philosophical method of the *Meditations,* I turn my focus to his explicit characterization of analysis in the *Meditations* in light of the above five philosophical senses of analysis he inherited.

Descartes' fullest statements on the nature of analysis and synthesis occur in his reply to Mersenne's request, in the Second Set of Objections, that Descartes present the arguments of his *Meditations* in geometrical fashion. Descartes responds to Mersenne's request by first distinguishing between "two matters [*res*] in the geometrical manner of writing: namely, the order and the *ratio* of demonstrating [*rationem demonstrandi*]" (AT VIII, 155).[15] The 'ratio' of demonstrating is then further distinguished into a *ratio* of demonstrating through the way of analysis and a *ratio* of demonstrating through the way of synthesis. 'Ratio' here is standardly translated as '*method* of demonstrating', contrasted with the geometrical *order* of presentation. The Latin text reads as follows:

> Duas res in modo scribendi geometrico distinguo, ordinem scilicet, & rationem demonstrandi.
>
> Ordo in eo tantùm consistit, quòd ea., quae prima proponuntur, absque ullâ sequentium ope debeant cognosci, & reliqua deinde omnia ita disponi, ut ex praecedentibus solis demonstrentur. Atque profectò hunc ordinem quàm acuratissime in Meditationibus meis sequi conatus sum....
>
> Demonstrandi autem ratio duplex est, alia scilicet per analysim, alia per synthesim.
>
> Analysis veram viam ostendit per quam res methodice & tanquam a priori inventa est, adeo ut, si lector illam sequi velit atque ad omnia satis attendere, rem non minus perfecte intelliget suamque reddet, quàm si ipsemet illam invenisset.... (AT VII, 155)

The standard translation of the first sentence attributes to Descartes a distinction between a geometrical order [*ordinem*] of presentation and two methods [*rationes*] of demonstration. However, this reading makes little sense with respect to the next paragraph. That which Descartes calls 'order' there is not merely an order of presentation but resembles the ordering of Burgersdijk's natural method—that is, a conceptual ordering that simultaneously gets at what is ontologically prior by nature. In this order, "The items which are put forward first must be known entirely without the aid of what comes later; and the remaining items must be arranged in such a way that their demonstration depends solely on what has gone before" (CSM II, 110). The third sentence claims that Descartes tried carefully to follow this order in his *Meditations*. In his synopsis of the Second Meditation, Descartes writes, "the only order which I could follow was that normally employed by geometers, namely to set out all the premisses on which a desired conclusion depends" (CSM II, 9). He adds "A further requirement is that we should know that everything that we clearly and distinctly understand is true in a way that corresponds exactly to our understanding of it" (CSM II, 9). Given that the clear and distinct knowledge of God's existence provides the premise for concluding that things exist outside the mind, by "order," Descartes means a method of discovery that tracks the natural order, advancing

[15] Translations of this work are mine.

from that which is most basic, both conceptually and ontologically. This makes greater sense if "demonstrandi" in the first sentence of the above reply to the second objections is read as modifying "ordo" as well as "ratio" so that Descartes is distinguishing the one order of demonstrating displayed in his geometrical manner of writing [*modo scribendi*] the *Meditations* from the twofold "ratio" of demonstrating.

If the contrast drawn in the first sentence is not between a method of demonstrating and a mere order of presentation but rather between the proper order of demonstrating displayed in Descartes' manner of writing, and another variable aspect of the geometrical style of writing, then how should one translate the term 'ratio demonstrandi'? "*Ratio*" has a range of meanings in Latin, including "account," "relation," "procedure," "ratio," or "reason." A better translation might be, "I distinguish two matters in the manner of writing, namely, the *order* of demonstrating and the *reason* for demonstrating" (AT VIII, 155). Next, I shall explain why this translation better fits the text as a whole.

After discussing "order" in the final paragraph of the above-cited Latin text, Descartes employs the term "analysis" in discussing one of two approaches that fall under the "*ratio demonstrandi*." This paragraph's claim about analysis better fits my translation: "Analysis displays [*ostendit*] the way through which a thing is methodically and from the prior things [*a priori*] discovered as it were, thus so that that reader who is willing to follow and attend to all things sufficiently, would proceed and understand himself no less perfectly than if he had discovered it himself" (AT VIII, 155). Analysis thus tracks the order of demonstrating described in the second paragraph, but in this last paragraph, Descartes focuses on the reason or aim rather than the order of demonstrating. Hence, he describes analysis as a way of guiding the attentive reader methodically through one's order of discovery. Given that Descartes has already labeled the methodical procedure of the *Meditations* as "order" not "*ratio*," "*ratio demonstrandi*" here is most naturally read as "reason for demonstrating."

Another matter of confusion generated by the standard translation is hereby also resolved. In the first sentence of this last paragraph, "a priori" is typically assumed to refer to the *a priori* demonstrations from cause to effect in Aristotelian *propter quid* demonstrations: synthesis in sense 4)a), which constitutes the second phase in Zabarella's *regressus*. However, since such demonstrations employ the method of composition, whereas demonstrations from effect back to cause employ resolution/analysis in sense 4), Descartes' characterization of an "*a priori*" method of demonstration as analysis contradicts standard Aristotelian uses. However, he does not clarify why he uses "analysis" to describe what his contemporaries would call "synthesis." In my proposed translation, "*a priori*" refers to the things previously discovered. Descartes adds 'as it were' because technically the reader did not discover them. No clarification is required because his use of "analysis" follows common usage, referring to Zabarella's sense 3) of a conceptual ordering from what is better known to us for learning purposes. For Descartes, as for Keckermann and Burgersdijk, it tracks the true order back to metaphysical first principles.

My proposed reading of what Descartes means by "analysis" also fits better with his subsequent lead-in to his description of the "*ratio*" of demonstrating based on the way of synthesis, which concerns the fact that analysis fails to convince the inattentive or stubborn reader. Again, Descartes focuses not on the *methods* of demonstration but on the *reasons* why one might employ an analytic or synthetic approach to writing one's discoveries for the intended audience. Unlike analysis, synthesis wrests assent from hostile, inattentive readers, in that it.

> demonstrates [*demonstrat*] clearly that which is the conclusion and employs a long series of definitions, postulates, axioms, theorems and problems, so that if anyone denies one of the conclusions it can be shown at once that it is contained in what has gone before, and hence the reader, however argumentative or stubborn he may be, is compelled to give his assent. (AT VII, 156; CSM II, 111)[16]

However, this will not satisfy the other kind of reader, Descartes observes. Despite his use of the term "demonstrates," this passage in the Second Replies follows Zabarella's account of analysis and synthesis as methods of ordering knowledge already attained to facilitate learning or assent rather than describing demonstrative methods for discovering new knowledge. In this context, "demonstrates" is not used in its technical Aristotelian sense; rather, Descartes is simply saying that the way of synthesis "clearly designates the conclusion"—that is, the reader has minimal cognitive work to do. In this style of writing, used by geometers, the conclusion is spelled out and clearly separated from definitions, axioms, etc., thus precluding any confusion between it and a supporting premise (anyone teaching material that involves chains of reasoning knows that the two are often confused). Once the standard translation is corrected, Descartes' use of the terms "analysis" and "synthesis" is internally consistent and consistent with that of Zabarella. When engaging in metaphysics rather than solving problems in the mathematical sciences, Descartes treats analysis and synthesis as didactic means of ordering the subject matter suited to different kinds of audiences.[17]

Where Descartes diverges from Zabarella is in his assessment that it is "Analysis alone which is the true and optimal [manner] of teaching" (AT VII, 156). The way of analysis, unlike synthesis, is a style of writing that takes the reader through the methodical process (the order of demonstrating) by which the item of knowledge was discovered. In this regard, Descartes' account echoes Burgersdijk's natural method. For Descartes, analysis as order is not merely a cognitive ordering that facilitates learning but also tracks the natural order. He makes a point similar to Burgersdijk's claim that method serves "the intellect and memory towards the better and easier perceiving of the proposed things and more faithful guarding [of them]"

[16] The Latin reads, "Synthesis è contra per viam oppositam…. clare quidem id quod conclusum est demonstrat, utiturque longâ definitionum, petitionum, axiomatum, theorematum, & problematum serie…. sicque a lectore, quantumvis repugnante ac pertinaci, assensionem extorqueat…." (AT VII, 156). Translation modified by me.

[17] My reading fits a letter to Mersenne of November 13, 1639 in which Descartes refers to "Analysts" with opinions on the existence of God who are difficult to convince because they rely on the imagination, which aids them in mathematics but is of no use in metaphysical speculations (AT II, 622).

(Burgersdijk 1627, 377). Descartes highlights the following advantage to the synthesis used to teach geometry: "the breaking down of propositions to their smallest elements is specifically designed to enable them to be recited with ease so that the student recalls them whether he wants to or not" (CSM II, 111). However, according to Descartes, the analytic way is more suited to metaphysics than the synthetic way used in geometry, because the primary metaphysical notions, unlike geometrical notions, are not in accordance with our senses, and hence not everyone readily accepts them. Analysis is thus the way of writing that we must follow in metaphysics to overcome the obstacles posed by human cognition, since it facilitates the shedding of preconceived notions and renders "our perception of primary notions clear and distinct" (CSM II, 111).

Given Descartes' rejection of the Aristotelian hierarchy of species and genera, the natural order that Descartes' step-by-step procedure of the *Meditations* tracks is not a synthetic order from the most to the least universal. Rather, the *Meditations* begin with resolution in Sense 4) whereby the meditator arrives at the statement "I am, I exist," the simplest, most certain item of knowledge, which, though particular (unlike Aquinas' most universal conception of being), is immediately known. The way of analysis expresses this resolutive procedure in a manner that enables the attentive reader to optimally learn that which was discovered by following the natural order of discovery. Hence, Descartes also employs Galen's sense of analysis as a pedagogical order (sense 3). He thus appears influenced by Zabarella's use of the term and Burgersdijk's Neoplatonic conviction that natural method yields a cognitive order that both tracks the order of nature and facilitates learning.

Descartes exhibits another explicit philosophical use of "analysis" in his reply to an objection in the Sixth Set of Objections to the *Meditations*. It is unclear whether Descartes there speaks in terms that his interlocutor will understand or whether he accepts the interlocutor's labeling of the meditator's proof that s/he is a thinking thing as an "analysis." The objector holds that Descartes may be wrong in stating that the meditator is exclusively a thinking thing since the meditator might simply be a corporeal thing in motion. The objector challenges Descartes to clarify the proof that the meditator's thought precludes corporeal motion, asking, "Have you used your method of analysis to separate off all the motions of that rarefied matter of yours? Is this what makes you so certain? And can you therefore show us.... that it is self-contradictory that our thoughts should be reducible to these corporeal motions?" (CSM II, 278). The objector uses analysis in sense 1), demanding a physiological resolution into elements. Descartes resists the challenge, insisting instead that what was required (and presumably delivered) in this part of the *Meditations* was a conceptual resolution (sense 2). First, he restates the objector's point: "By 'reduced' I take it that they mean that our thought and corporeal motions are one and the same" (CSM II, 287). Descartes then adds,

> This mistake has obviously been made by those who have imagined that the distinction between thought and motion is to be understood by making divisions within some kind of rarefied matter. The only way of understanding the distinction is to realize that the notions of a thinking thing and an extended or mobile thing are completely different, and independent of each other; and it is self-contradictory to suppose that things that we clearly understand as different and independent could not be separated, at least by God. (CSM II, 287)

Descartes corrects his interlocutor, whom he takes to have confused the second Aristotelian sense of resolution, whereby one resolves the different notions or concepts linked in a proposition into more general elements, with a physical reduction. Regardless of whether Descartes here uses 'analysis' in this way as a concession to his interlocutor, his emphasis on the 'notions' of thinking and extended substance shows conclusively that the process of the second *Meditation* occurs at the conceptual level, though the accomplished reduction also tracks ontological divisions (CSM II, 287). Of course, this does not preclude other uses of analysis in his scientific works.

5.4 Hobbes on Analysis and Synthesis

Hobbes' pronouncements on philosophical analysis and synthesis are both more extensive and more aligned with prior senses than Descartes'. However, they are also subject to the confusions between different Scholastic senses of method that we encountered in interpreting Descartes' texts. Some conflations are due to Hobbes himself, others to our lack of context. As with Descartes, the scholastic legacy is palpable.

In *De Corpore*, Hobbes characterizes synthesis, which, unlike Zabarella, he equates with composition, as both the method of demonstration and the method of teaching (DC, 80). In this regard, his view of philosophical method aligns even more with Burgersdijk's take on sense 3) than Descartes'. Like Descartes, he differs from Burgersdijk on analysis. Analysis in the strict sense is not a method limited to practical philosophy for Hobbes. Rather, analysis is central to his theoretical philosophy, playing a similar role to Aquinas' resolution *secundum rationem* to the most universal forms and also paired with composition (sense 4b). The difference is that the most general concept arrived at through Hobbes' analysis, in this strict sense, is not being but motion (DC, 69). Confusion stems from Hobbes' conflation of a *quasi*-Neoplatonic sense of resolution to first principles with Aristotelian conceptual analysis in sense 2). Several examples illustrate this.

In Chapter VI "Of Method" in *De Corpore*, Hobbes gives two examples to illustrate the strictly analytical method that yields the universal notions that he claims we need to attain unqualified knowledge of things. Both begin with an idea, from experience, which is gradually resolved from the less to the more general:

i) The idea of this square is resolved or analyzed into 'plain, terminated with a certain number of equal and straight lines and right angles'. These concepts can then be resolved or analyzed further into the properties common to all material objects: "line," "plane," "angle," "straightness."
ii) The conception of gold is resolved into ideas of "solid," "visible," and "heavy." These ideas may then be further resolved or analyzed into successively more general ones, such as "extension" and "corporeity" until one arrives at the most general one: motion (DC, 68–69).

For Hobbes, once these ideas are analyzed down to the most general ones, which are also their simplest, conceptual elements, one has the causes of individual concepts of a square and gold. Immediately following these examples of analysis, he writes, "By the knowledge therefore of universals, and of their causes (which are the first principles by which we know the διότι of things) we have in the first place their definitions (which are nothing but the explication of our simple conceptions)" (DC,70). Hobbes' use of the term "cause" might suggest that the resolution of gold is a physiological reduction in sense 1) into actual elements of gold. For how can concepts *cause* our ideas? However, the next example, below, confirms that Hobbes uses "cause" in the broad sense of an explanatory factor. If we read 'cause' as akin to the intrinsic causes that Aquinas' resolution *secundum rationem* reveals, then Hobbes' claim makes perfect sense. This part of Hobbes' method thus echoes Neoplatonic division (sense 4) whereby one arrives at universal forms (or their equivalents), which are causal. Whereas Plotinus and Proclus do not call this 'resolution', Aquinas and other Scholastics do. Hobbes likewise labels this procedure 'analysis' conflating it with sense 2).

Sense 2 is also evident in a prior example which illustrates Hobbes' broad sense of 'cause'. He describes the experience of seeing something approaching. As it draws closer, the senses detect a certain shape and motion. Hobbes claims that we know by experiential cognition that this thing exists but not what it is and what its causes are, meaning that we know it in a confused way. *Scientia,* which he defines as *causal* knowledge, requires computation that begins with subtraction, the mental operation of resolution/analysis in sense 2. As when we subtract one number from another, analysis mentally separates out distinct features from the individual nature that encompasses them to arrive at the simple components of things. We first separate out body from our perception of the individual, then the property of being animated, and finally that of being rational (DC, 4–5). Hobbes explicitly argues that these elements are not physical parts of a thing but conceptual elements (DC, 67). Nor need the elements be linguistic entities, since we can ratiocinate without words (DC, 14). Analysis, strictly speaking, is a step-by-step conceptual separating out of general features, contained in an individual concept, from the concept as a whole. The conceptual elements form the basis for scientific ratiocination, which produces causal knowledge.

Although causal elements are non-physical, Hobbes does not regard analysis as merely an explanatory ordering to facilitate teaching and learning. Like Descartes, he regards analysis as a method of discovery that, although it does not resolve things into their material parts, gives insight into the universal natural order and faculties of bodies. Thus, his method also resembles the universal or total method of Keckermann and Burgersdijk. This is evident from Hobbes' account of accidents. The accidents of natural and artificial things are the object of scientific knowledge of causes, but these are not Aristotelian accidents.

> But most men will have it be said that *an accident is something,* namely some part of natural things, when, indeed, it is no part of them. To satisfy these men, as well as may be, they answer best that define an *accident* to be *the manner by which any body is conceived;* which is all one and the same as if they should say, *an accident is that faculty of any body, by which it works in us a conception of itself.* (OL I.91)

Hobbes holds that a method of analyzing/resolving our concepts into more general ones offers insight into fundamental bodily powers via the isomorphism of conceptual common accidents and bodily faculties. For Hobbes, as for Keckermann and Burgersdijk, analysis as the step-by-step resolution of our concepts into ever more basic, general concepts tracks the natural order. Thus, his view of analysis and synthesis, like Keckermann's and Burgersdijk's, incorporates the Neoplatonic sense.

Two details of Hobbes' method confirm that he likely drew on Keckermann's and Burgersdijk's logic textbooks, which circulated in England. First, as noted, like Burgersdijk, he regards synthesis as both a didactic method and a method of discovery. Second, Hobbes adopts the distinction between a universal/total method versus the particular method of demonstration. He writes that philosophers "… seek scientific knowledge [*Scientia*] simpliciter or indefinitely, that is, having posed no certain question, they [seek to] scientifically know as much as they can…" (OL I.60). In this case, they employ a method that is strictly resolutive and analytical (in sense 2) to arrive at universal notions that are compoundable by synthesis into definitions. However, when they seek to scientifically answer a particular question about the cause of certain phenomena, they employ the particular method using analysis and synthesis to construct syllogistic demonstrations. Here Hobbes confusingly uses the same terms for another sense of analysis/resolution—namely, Aquinas' resolution *secundum rem* and ensuing composition (sense 4a). Hobbes' particular method, like Zabarella's *regressus*, is premised on the view that scientifically solving problems requires the combination of resolution and composition.

Hobbes' view of the method that provides the principles or definitions to be used in problem-solving demonstrations is reminiscent of Burgersdijk's total method. Burgersdijk claims, "The natural method ought always to progress from universals to particulars; in that progression all the parts are to be connected by apt chains of transition" (Burgersdijk 1627, 380). He elaborates that universals are not merely better known than particulars, as far as distinct cognition goes, but that they also contribute to the acquisition of distinct cognition of the particulars since these universals are contained in their definition (Burgersdijk 1627, 380). This is consistent with Hobbes's use of synthesis, which builds on the results of the preliminary resolution of individual concepts, by syllogistically deducing scientific knowledge of individual natures and their differentiating accidents from universal definitions/principles revealed by analysis. Hobbes' linking of this preliminary resolution *secundum rationem* into principles to the subsequent scientific demonstrations echoes Aquinas' commentary on Boethius. On Aquinas' use of resolution and composition in Sense 5b), the resolution to what is immediately known, as in geometry, enables subsequent demonstrations of scientific conclusions in various disciplines for the purpose of teaching them.

For Hobbes, definitions are principles of demonstration and are themselves constructed by the synthetic combination of universal names. In addition to analysis as a resolution *secundum rationem* to principles followed by a compositive method used in scientific demonstrations, Hobbes invokes conceptual resolution in sense 2, whereby definitions are resolved into immediately known terms and a corresponding sense of composition that links terms into definitions:

> whensoever that thing has a name, the definition of it can be nothing but the explication of that name by speech; and if that name be given it for some compounded conception, the definition is nothing but a resolution of that name into its most universal parts. As when we define man, saying *man is a body animated, sentient, rational*, those names, *body, animated, & c.* are parts of that whole name *man*... (DC, 83)

The universal parts reached in such a resolution are not physical but somehow map onto common bodily faculties. For Hobbes, the gradual composition into ever more complex definitions yields a hierarchical order of disciplines that tracks nature's order. Hence, Hobbes also invokes synthesis in sense 3)—that of a Galenic ordering. Hobbes explains, "a line is made by the motion of a point, superficies by the motion of a line, and one motion by another motion, & c" (DC, 70–71). On the basis of such generative definitions, Hobbes envisions the gradual synthesis of a hierarchy of sciences resting on such definitions. *Scientia simpliciter*, which consists in attaining as much knowledge of the causes (in the broad sense) of things as possible, begins with the causes of the simple objects of geometry, such as lines or lengths generated from points in motion and surfaces generated from long bodies. Once these are demonstrated, we advance to the more complex phenomena of the science of motion, which are produced by the effects of one body's motion on others. The science of motion then provides the starting points for demonstrating the phenomena of physics, which are produced by the motions of the parts of bodies, including our sense organs. Hobbes asserts that we can progress in this manner all the way up to civil science and thus attain demonstrative causal knowledge in all sciences, including politics, that rests on the foundations of geometry. Scientific reasoning is thus built up in an orderly compositive procedure, starting from principles—namely, definitions—including generative definitions, giving us the ultimate conceptual/causal elements of bodies. From these, we syllogistically deduce more and more complex wholes which are the effects of these causes.

In civil philosophy, this lengthy procedure can be sidestepped as its principles can also be attained directly by analysis without prior knowledge of geometry and physics. In describing how, Hobbes combines conceptual resolution (2) and Neoplatonic resolution into first principles (4), claiming that to answer a question such as "*whether such an action be just or unjust*," one can break this proposition down into its terms (DC, 74). Then, one can resolve the term "unjust" into "fact against law" (its signification) and, in turn, resolve "law" into "command of the person(s) that have coercive power." Power is derived from the wills of the people who constitute this power for the purpose of living in peace. Finally, one arrives at the immediately known item of knowledge—namely, that human appetites and passions are such that people will always be at war if not restrained. From this, one can then compound to reach the answer to the question. Presumably, this method only works in civil science and not in theoretical sciences because we have direct knowledge of human appetites and passions through introspection but not the microstructures of bodies.

I have demonstrated that Hobbes's account of philosophical method contains vestiges of all senses of analysis/synthesis without distinguishing them. Today, much confusion stems from attempts to align his diverse claims with a single sense, found in Zabarella's *regressus*.[18] By situating Hobbes' claims within the wider tradition, it becomes clear that he incorporates medieval combinations of Aristotelian and Neoplatonic senses. The lines of transmission are unclear, but Thomist doctrines were available to Hobbes via Jesuit textbooks and other sources. Hobbes' main divergences from medieval uses are a) the incorporation of Keckermann's and Burgersdijk's view that the universal method yields a cognitive ordering that tracks the natural order, b) Burgersdijk's view that the synthetic order from more universal to less universal is both a method of discovery and teaching, and c) the replacement of the most general concept in which conceptual resolution culminates with motion. The latter, c), follows from Hobbes' elimination of the science of metaphysics. For him, there is no science of being in general, only of body, both natural and artificial. All other considerations belong to disciplines other than the scientific ones.

In conclusion, both Descartes' and Hobbes' pronouncements about philosophical analysis/synthesis contain substantial vestiges of medieval scholastic appropriations and merging of ancient methods of resolution/composition, in addition to innovations by the most prominent scholastic logicians in their environments. The influence of analytic/synthetic methods that canonical seventeenth-century philosophers inherited from scholastic philosophy was thus more substantial and far-reaching than was initially realized. Future research should consider whether/to what extent Descartes and Hobbes implemented these philosophical methods in their scientific endeavors.

References

Aertsen, Jan A. 1996. *Medieval philosophy and the transcendentals: The case of Thomas Aquinas*. Leiden/New York/Cologne: E.J. Brill.

Aquinas, St Thomas. 1961. *Commentary on Aristotle's* metaphysics. Trans. John P. Rowan. Notre Dame: Dumb Ox Books.

———. 2001. *Exposition on the hebdomads*. Washington, DC: Catholic University of America Press.

Aristotle. 1985. Metaphysics. In *The complete works of Aristotle*, ed. Jonathan Barnes, vol. II. Princeton: Princeton University Press.

Burgersdijk, Franco. 1627. *Institutionum Logicarum*. Leiden: Commelinus.

Cunningham, Andrew. 1985. Fabricius and the 'Aristotle Project'. In *Anatomical teaching and research at Padua: The medical renaissance of the sixteenth century*, ed. A. Wear, R.K. French, and I.M. Lonie, 195–222. Cambridge: Cambridge University Press.

[18] For a more detailed discussion of the nature and sources of these confusions, see Hattab 2014.

Descartes, René. 1984. Meditations on first philosophy. In *The philosophical writings of Descartes,* vol. II. Trans. John Cottingham, Robert Stoothoff, Dugald Murdoch, and Anthony Kenny. Cambridge: Cambridge University Press. [Cited as CSM II.]

———. 1985. Discourse on the method. In *The philosophical writings of descartes*, vol. I, Trans. John Cottingham, Robert Stoothoff, and Dugald Murdoch. Cambridge: Cambridge University Press. [Cited as CSM I.]

———. 1996. Meditationes de Primâ Philosophia. In *Oeuvres de Descartes*, ed. Charles Adam and Paul Tannery, vol. VII. Paris: Vrin. [Cited as AT VII.].

Dika, Tarek. 2023. *Descartes' method: The formation of the subject of science*. Oxford: Oxford University Press.

Hattab, Helen. 2014. Hobbes' and Zabarella's methods: A missing link. *Journal of the History of Philosophy* 52 (3): 461–485.

———. 2020. Methods of teaching or discovery? Analysis and synthesis from Burgersdijk to Spinoza. In *Teaching the new sciences in history of universities*, ed. Mordechai Feingold, vol. XXXIII/2, 86–112. Oxford: Oxford University Press.

Hintikka, Jaakko, and U. Remes. 1974. *Method of analysis: Its geometrical origin and its general significance*. Dordrecht: D. Reidel Publishing Company.

Hobbes, Thomas. 1839–1843b. Thomae Hobbes malmesburiensis opera philosophica, 5 vols, ed. Gulielmi Molesworth. London: John Bohn. [Cited as OL.]

———. 2005. De Corpore. In *The English works of Thomas Hobbes of Malmesbury*, vol. I. London: Elibron Classic. [cited as DC]

Keckermann, Bartholomaeus. 1613. *Systema Logicae*. Cologne: Apud Haeredis Guilielmi Antonii.

Morrison, Don. 1997. Philoponus and Simplicius on Tekmeriodic proof. In *Method and order in renaissance philosophy of nature*, ed. Daniel A. Di Liscia, Eckhard Kessler, and Charlotte Methuen, 1–22. Aldershot Brookfield VT: Ashgate.

Plotinus. 1969. *Enneads I*. Trans, A.H. Armstrong. Loeb Classical Library Vol. I. Cambridge, MA: Harvard University Press.

Raftopoulos, Athanassios. 2003. Cartesian analysis and synthesis. *Studies in History and Philosophy of Science* 34: 305–308.

Recker, Doren. 1993. Mathematical demonstration and deduction in Descartes' early methodological and scientific writings. *Journal of the History of Philosophy* 31 (2): 223–244.

Sacksteder, W. 1980. Hobbes: The art of the geometricians. *Journal of the History of Philosophy* 18: 131–146.

Sweeney, Eileen. 1994. Three notions of *Resolutio* and the structure of reasoning in Aquinas. *The Thomist* 58: 197–243.

Timmermans, Benoît. 1999. Descartes's conception of analysis. *Journal of the History of Ideas* 60: 433–447.

van Winden, J.C.M. 1959. *Calcidius on matter: His doctrine and his sources*. Leiden: E.J. Brill.

Verbeek, Theo, Jelle Kigma, and Philippe Noble. 1996. *De Nederlanders en Descartes*. Amsterdam: Maison Descartes [also translated into French].

Wallace, William. 1997. Galileo's regressive methodology: Its prelude and its sequel. In *Method and order in renaissance philosophy of nature*, ed. Daniel Di Liscia, Eckhard Kessler, and Charlotte Methuen, 229–222. Brookfield VT: Ashgate.

Watkins, J.N. 1965. *System of ideas*. London/Melbourne/Sydney: Hutchinson & Co.

Zabarella, Jacopo. 1597. *Opera Logica*. Cologne: Zetzner.

Dr Helen Hattab is the author of a monograph, *Descartes on Forms and Mechanisms* (Cambridge University Press, 2009) and numerous articles connecting Scholastic Aristotelian philosophical traditions with early modern natural philosophy, history and philosophy of science and metaphysics. She is currently a Full Professor of Philosophy at the University of Houston in the USA.

Chapter 6
A Fresh Look at Newton's Method of Analysis and Synthesis

Alan E. Shapiro

Abstract In Query 23, which Newton added to the Latin translation of the *Opticks*, he declared that investigations in natural philosophy should follow the mathematical methods of analysis, to discover principles and propositions, and synthesis, to demonstrate new propositions. He has proposed a method of reasoning and knowledge. In the next sentence, however, he states that analysis allows us to proceed from "compositions to ingredients," which appears to be chymists' method of decomposing material things. This study confirms that Newton did indeed follow mathematicians' concept of analysis and synthesis in his optical investigations. Through analysis he discovered principles or propositions—for example, that sunlight consists of rays of different colors—while the synthesis demonstrated new propositions, such as the formation of the rainbow and not that white light is composed of rays of different colors. With the exception of this single query, Newton never used the terms "analysis" and "synthesis" in his optical writings. Only in the second half of the eighteenth century, beginning with the French, were these terms applied to the decomposition and recombination of rays of light. The second part of this paper traces this transition to the modern usage of the terms "analysis" and "synthesis."

Keywords Query · Opticks · Hypothesis · Analysis · Synthesis · Decomposition

6.1 Introduction

Newton first introduced the terminology of his concept of the "method of analysis and synthesis" in 1706 in Query (Qu.) 23 in the *Optice*—the first Latin edition of the *Opticks*—which became Qu. 31 in the second English edition in 1717. The concept had a complex origin, for Newton was drawing on at least four intertwined traditions: the mathematical tradition of analysis and synthesis and its promise of

A. E. Shapiro (✉)
Program in History of Science, Technology and Medicine, Minneapolis, MN, USA
e-mail: shapi001@umn.edu

W. R. Newman, J. Schickore (eds.), *Traditions of Analysis and Synthesis*,
Archimedes 73, https://doi.org/10.1007/978-3-031-76398-4_6

certainty, the chymical tradition that he had been applying in the laboratory, and the philosophical and logical traditions. In his ongoing conflict with Cartesians and, later, Leibnizians, Newton wanted to stress that his method in both the *Opticks* and the *Principia* was based on experiment, not hypotheses, and he turned to the method of analysis and synthesis to make this point.[1] I shall begin with Qu. 23/31 and what became the canonical passage on the method of analysis and synthesis for Newtonians:

> As in Mathematicks, so in Natural Philosophy, the Investigation of difficult Things by the Method of Analysis, ought ever to precede the Method of Composition. This Analysis consists in making Experiments and Observations, and in <drawing general Conclusions from them by Induction, and admitting of no Objections against the Conclusions, but such as are taken from Experiments, or other certain Truths. For Hypotheses are not to be regarded in experimental Philosophy. … By this way of Analysis we may proceed from Compounds> to Ingredients, and from Motions to the Forces producing them; and in general, from Effects to their Causes, and from particular Causes to more general ones, till the Argument end in the most general. <This is the Method of Analysis:> And the Synthesis consists in assuming the Causes discover'd, and establish'd as Principles, and by them explaining the Phænomena proceeding from them, and proving the Explanations.[2]

In the following paragraph Newton explains how he applied this method in his discoveries on light in the *Opticks*. I shall return to this later.

Newton appears to be equating—or perhaps even confusing—several distinct meanings of "analysis" here. The first, which originated in chymistry, is straightforward: "analysis" is the decomposition or breaking down of a substance into simpler components, while "synthesis" is the formation of a substance by combining various material elements or components. A distinct concept of decompositional

---

[1] Shapiro 2004. When Bill Newman and Jutta Schickore first invited me to participate in a workshop on analysis and synthesis, I responded that I had nothing to contribute, because Newton's method of analysis was already understood. Little did I know then how correct Bill was when he urged me to look further into it and participate in the workshop. I also thank Bill, Niccolò Guicciardini, and Dmitri Levitin for their valuable comments and suggestions on my paper. Levitin's *The Kingdom of Darkness: Bayle, Newton, and the Emancipation of the European Mind from Philosophy* (2022) appeared after much of this paper was written. It is a masterful account of Newton's method—in particular, of his concept of mathematical certainty, and his two-stage model of scientific procedure (to be explained below) and their place in early modern science and philosophy. The book puts the material of this paper into the broader context of the evolving concept of mathematical certainty in the sixteenth and seventeenth centuries.

[2] Newton 1730/1952, 404–5; Newton 1984–2021, vol. 2, 412–13. This was the 26th and final paragraph of Qu. 23. When Newton expanded it in 1717 for Qu. 31, he divided it into two paragraphs. Angle brackets indicate an addition in the manuscript, which, in this case, is also an addition with respect to the 1706 version. I shall follow this notation throughout the paper and will also use strikethrough to indicate deletions; only changes that are of consequence for this study are indicated. The Latin version consistently translates "analysis" as *analytica* and "synthesis" as *synthetica*, but it also translated "composition" in the first sentence as *synthetica* (Newton 1706, 347). Newton composed the queries for the Latin edition in English, and Samuel Clarke translated them into Latin together with the text of the *Opticks*; see Newton 1984–2021, vol. 2, 21. All quotations from the queries in the Latin edition will be from Newton's English. The insertion indicated by the first pair of angle brackets replaced Newton's English: "arguing by them from compositions"; ibid., 378.

analysis was also applied to concepts and ideas in logic, philosophy, and mathematics. In this approach a concept or principle was decomposed into elementary or simple components, whereas "synthesis" involved combining the simple concepts into more complex or compound ones. When it is necessary to distinguish the two decompositional approaches, I shall refer to the former as the chymists' approach and the latter as the logicians.[3] For some reason, Newton and his followers did not adopt the logicians' decompositional meaning. However, as we shall see, it became quite widespread in the eighteenth century, particularly in France, and by the early nineteenth century was broadly adopted. Chymists' and logicians' decompositional concepts were initially distinct, but the sharp delineation between them had begun to dissolve by the end of the eighteenth century. Another wholly distinct meaning of "analysis" is more complex, and elements of it extend before Aristotle to early Greek geometers. In the broadest terms—in the long-interacting mathematical, philosophical, and logical traditions—"analysis" concerned the discovery of causes, principles, or propositions, while "synthesis" was concerned with demonstrating, proving, explaining, or teaching. This concept of analysis and synthesis has aptly been called "regressive," because it moves back stepwise until it arrives at a cause, principle, or proposition.[4] The regressive method of analysis and synthesis—like the logicians' decompositional approach—is about reasoning and knowledge rather than material things. The mathematical concept of analysis, as it was understood in the seventeenth century, involved the discovery of principles or that which is known, while synthesis was concerned with the proof or demonstration of problems and theorems.[5] It should be noted that the Latin terms "resolution" and "composition," the equivalents of the Greek "analysis" and "synthesis," were freely interchanged.

Components of the mathematical, philosophical, and logical traditions interacted on the concepts of analysis and synthesis over the millennia. However, the specifics need not detain us here, as my concern is with Newton's statement in Qu. 23/31 and the application of the methods of analysis and synthesis in optics in the seventeenth and eighteenth centuries. Newton was familiar with all four traditions. His long and deep involvement in the chymical and mathematical traditions requires no justification. I shall demonstrate that Newton was familiar with the contemporary logical tradition. He had also carefully read key early modern philosophers—in particular, Thomas Hobbes and René Descartes, who were deeply concerned with method.[6]

Among eighteenth-century Newtonians, particularly in England and the Netherlands, Newton's apparent—and "apparent" must be stressed—equating of

---

[3] On this terminology, see Beaney 2022, which provides an overview of analysis; see also Albury 1972. For the period preceding Newton, see Gilbert 1960 and Dear 1998.

[4] I have adopted the term from Beaney 2022.

[5] I am ignoring its application to geometrical constructions here. For that usage, see Guicciardini 2009, 35–7, 40–41.

[6] On Newton's familiarity with Hobbes's *De corpore*, particularly where he discusses analysis and synthesis, see McGuire and Tamny 1983, 219–21; see also Talaska 1988. Newton's life-long engagement with Descartes has long been a theme in Newton studies: see, for example, Westfall 1980.

the decompositional and regressive concepts of analysis was not accepted; rather, the two were strictly distinguished. The regressive concept was then applied to mathematics, mechanics, forces, and the *Principia*, and the decompositional concept to chymistry. Light and ingredients passed out of the picture, although they returned during the late eighteenth and early nineteenth centuries in the context of broader changes in philosophy and logic.[7]

In the expanded account of analysis and synthesis—at more than double its original length—in the second English edition of 1717, Newton added a passage on induction, much of which I have here omitted with ellipses. His aim was to argue that the methods of the *Principia* and of the *Opticks* are the same. In the General Scholium of the *Principia* in 1713, Newton had explained his method in terms of induction and in the queries of the Latin edition of the *Opticks* in 1706 in terms of analysis and synthesis. Now, in 1717, he appeared to espouse a single, unified method.

Previous accounts of Newton's method of analysis and synthesis have failed to take all four traditions into account.[8] It is essential to bear in mind that the decompositional and regressive concepts of analysis and synthesis were conceptually distinct; the same terms were being used with distinctly different meanings. The former, as I shall demonstrate shortly, was not applied to light until the late eighteenth century, at least among British and Dutch Newtonians, my principal focus for the Newtonian school. The situation in France differed considerably, as Newton's ideas and contributions were not widely adopted there until approximately mid-century. In particular, his claims regarding analysis and synthesis in Qu. 23/31 were simply not accepted as doctrine as they were in Britain and the Netherlands. The picture is rendered more complex by the fact that the chymists' decompositional concept transformed from an operational one on material substances to an epistemic one. By the late eighteenth and early nineteenth century, physicists and chemists began to apply the concept to light, and since then, this epistemic, decompositional meaning of "analysis" and "synthesis" has become standard in physics and other sciences. That is why today we freely speak of Newton's theory of light and color as an instance of analysis and synthesis, whereas Newton never applied those terms to light outside Qu. 23/31. My aim in this paper is first to clarify Newton's concept of the method of analysis and synthesis as it appears in Qu. 23/31 and then to trace the transformation of chymists' decompositional concept to the modern, epistemic

[7] On eighteenth-century Newtonians avoiding the use of the terms "analysis" and "synthesis" in optics, see note 40 below.

[8] The most thorough study of Newton's method of analysis and synthesis is Guicciardini 2009. He offers the insightful observation that, "These mathematical terms [analysis, *resolutio*, etc.] interacted in a complex way with the technical vocabulary pertaining to the philosophical, logical, chemical, and medical traditions," ibid., 2009, 34. See also his chapter in this volume. I hope to contribute to unravelling this interaction, although I ignore the fifth, medical, or Galenic tradition, because I believe that by the later seventeenth century, it had been incorporated into the philosophical–logical tradition. Levitin 2016 is valuable for its account of the relation of the method of analysis and synthesis to Zabarella and the philosophical tradition. See also Guerlac 1973, and Ihmig 2004 and 2005.

one. To accomplish this, I shall follow descriptions of the decomposition of white light and its compound nature from Newton's day to the early nineteenth century.

6.2 Newton on Analysis and Synthesis

Newton's public espousal of a two-stage method of natural philosophy may be traced back to the seventeenth century and the first edition of the *Principia.* He did not introduce the term "method of analysis and synthesis" for this process until Qu. 23 in the Latin translation of the *Opticks* in 1706. In the Preface he wrote that, "the basic problem of philosophy seems to be to discover the forces of nature from the phenomena of motions and then to demonstrate the other phenomena from these forces."[9] Here, he is clearly referring to analysis and synthesis without invoking those specific terms. Newtonians in the eighteenth century appear to have adopted this view of analysis and synthesis, one restricted to forces, as their general understanding of the concept. Ingredients and light largely dropped away. This was further reinforced by Roger Cotes's preface to the second edition of the *Principia,* in which he stressed the importance of the method of analysis and synthesis for natural philosophy but invoked it only for discovering the forces of nature. He stated that modern natural philosophers base their work on experiment, and that

> They do not contrive hypotheses, nor do they admit them into natural science otherwise than as questions whose truth may be discussed. Therefore they proceed by a twofold method, analytic and synthetic. From certain selected phenomena they deduce by analysis the forces of nature and the simpler laws of those forces, from which they then give the constitution of the rest of the phenomena by synthesis. This is that incomparably best way of philosophizing which our most celebrated author thought should be justly embraced in preference to all others.[10]

When I searched Newton's optical writings for language of the method of analysis and synthesis, I was surprised to find that they yielded no examples. With the exception of Qu. 23/31 itself, Newton did not use that language at all. I do not by any means wish to deny that Newton's demonstration of his new theory of light and color was a genuine instance of analysis and synthesis; indeed, it was, but in the modern, expanded decompositional meaning of the terms as well as in the regressive sense. Before I attempt to explain what happened, I shall present what I understand to be the essence of Newton's experimental use of analysis and synthesis—in the modern sense—in his optical work.

Newton began all presentations of his theory from the *Optical Lectures* through the *Opticks* with his basic prism experiment. He separated the different colored rays present in sunlight by refracting the light through a prism and projected them onto the opposite wall. When sunlight is passed through a prism (Fig. 6.1) so that it

[9] Newton 1999, 382.

[10] Ibid., 386.

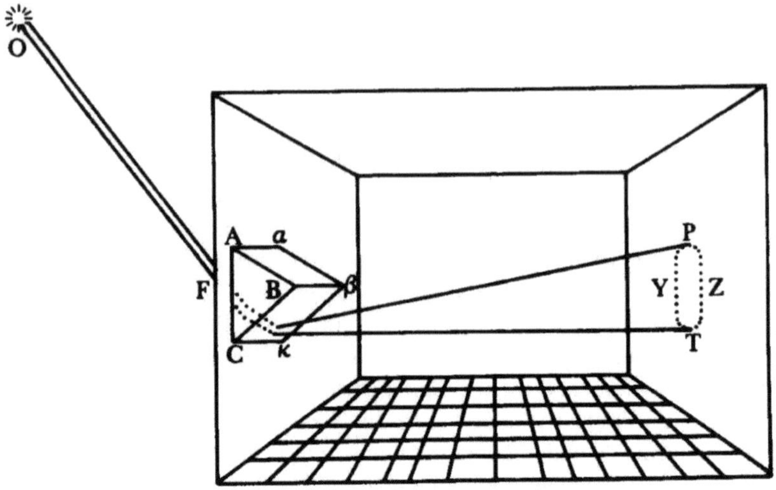

Fig. 6.1 Newton's basic prism experiment from his "Optical Lectures," 1670

undergoes equal refractions upon entering and leaving the prism, the rays of different color are refracted by different amounts and diverge and form an elongated colored image or spectrum *PT*. According to the then-received laws of optics, the image should be circular and yellowish-white like the sun itself. To allow the rays to separate and become elongated and distinguished into discrete colors, Newton found that he had to project the spectrum between 18 and 22 feet. This experiment and, more generally, prismatic refraction represented Newton's basic technique of resolution, decomposition, or analysis.

The strategy that Newton adopted to establish his theory may be simplified as follows: he aimed to establish, first, that sunlight consists of rays of different refrangibility and, second, that a one-to-one correspondence exists between degree of refrangibility and color. This decomposition of light by refraction may be seen as analogous to the chymical, decompositional concept of analysis and is the foundation of the entire theory. In the third stage, which is analogous to the chymical concept of synthesis, Newton performed experiments to demonstrate that when the colored lights of the spectrum are recombined by various means, they again form white light like sunlight. If these results are accepted, then sunlight consists of rays of different colors. This was the most fundamental and radical claim of his theory.

Examination of the successive drafts of Qu. 23 yields insights into what Newton meant by the ambiguous term "ingredients" in asserting that analysis proceeds "from compositions to Ingredients." It is important to recall that Newton wrote the queries for the Latin translation in English and that they were subsequently translated into Latin by Samuel Clarke.[11] To be sure, the passage on "ingredients" appears to refer to the decomposition of material bodies into the substances that compose

[11] See note 2 above.

them and reflects Newton's pursuit of chymists' decompositional approach. It was translated as "simplices" in the Latin translation of the *Opticks* and "simples" in the French, as the two translators apparently understood it to mean a chemical element or simple substance.[12] Thus, most readers on the continent would have had this material meaning in mind. I shall argue, however, that the term "ingredients" refers to the components of light and that Newton is pursuing the regressive rather than the decompositional form of analysis, although he developed a clever means of incorporating decomposition into a regressive framework.

Before proceeding to the drafts, I shall offer two, more general observations in support of this interpretation. First, since Newton's concern in this query is with method and the appropriate way of proceeding in natural philosophy, he would naturally have chosen the regressive concept of analysis and synthesis, which is concerned with causes and explanations. The chymical, decompositional concept is simply not relevant here. Second, Newton was above all a mathematician, and his concept of analysis and synthesis was grounded on the ancient mathematical concept, as expounded by Pappus of Alexandria.[13] The opening sentence of this paragraph of Qu. 23 with its "As in Mathematicks, so in Natural Philosophy…" was added only in the final draft in his attempt to argue that the *Principia* and *Opticks* follow the same method.[14] However, Newton had composed a draft preface for the *Opticks*—perhaps for the first English or, more likely, the first Latin edition—that contained a similar assertion regarding method in natural philosophy.[15] Here, Newton was arguing against Cartesian hypothetical physics and offering an alternative approach based on experiment and following mathematicians' method:

> As Mathematicians have two Methods of doing things wch they call Composition & Resolution & in all difficulties have recourse to their method of resolution <before they compound> so in explaining the Phaenomena of nature the like methods are to be used & he that expects success must resolve before he compounds. For the explications of Phaenomena are Problems much harder then those in Mathematicks. The method of Resolution consists in trying experiments & considering all the Phaenomena of nature relating to the subject in hand <& drawing conclusions from them> & examining the truth of those conclusions by new experiments & drawing new conclusions (if it may be) from those experiments & so proceeding alternately from experiments to conclusions & from conclusions to experiments untill you come to the general properties of things, [& by experiments & phaenomena have established the truth of those properties]. … But if wthout deriving the properties of things from Phaenomena you feign Hypotheses & think by them to explain all nature you may make a plausible systeme of Philosophy for getting your self a name, but your systeme will be little better then a Romance.[16]

---

[12] Newton 1706, 347; and Newton 1722/1955, 593.

[13] See Guicciardini 2009, 33–8.

[14] Cambridge University Library (henceforth CUL) Add. MS 3970, ff. 242r, 244v.

[15] McGuire 1970 first identified and published this draft. He dated it to the first English edition, and I supported that date, Newton 1984–2021, vol. 2, 19. More recently, Levitin 2022, 676, n. 81, dated it 2 years later to the first Latin edition, based on the reasonable argument that its contents agree with those of the queries added in the *Optice*, and I now lean towards his dating.

[16] CUL Add. MS 3970, f. 480v; the square brackets are Newton's; see McGuire 1970, 184–5; and Ducheyne and Dhondt 2021, 382.

Thus, from the beginning of his attempts to formulate a statement on his method for *Optice*, Newton insisted that experimental philosophers should follow a mathematical method and directed his method of analysis and synthesis against the hypothetical philosophy.

Newton first turned the two-stage method against hypotheses in a draft revision of the "Rules of Philosophizing" in the *Principia* in the early 1690s. At this early stage of his revision of the "Rules," Newton replaced the term "hypothesis" that he had used in the first edition with "axiom," before finally settling on "rule" in all later editions:

> Axiom 1: The most reliable method of philosophizing is that which, having set aside hypotheses, investigates the properties of things from phenomena and thereupon explains the operations and effects of those same things by means of the discovered properties. If the explanation of nature can, by this method, be reduced to a few properties of things, nothing further **would be left** than to investigate the causes of those properties.[17]

Although he does not yet use the terms "analysis" and "synthesis"—just as in the Preface to the first edition—it is the same two-stage process—namely, the discovery of properties from phenomena and explaining "those things" by these properties. In Qu. 23 "properties" would become "principles." A decade prior to Qu. 23 Newton turned his two-stage process—not yet called "analysis and synthesis"—to polemical purposes against hypothetical philosophy. It should be noted that in his drafts for Qu. 23 Newton not only introduced the "method of analysis and synthesis" in his campaign against hypothetical philosophy but also another key term—"experimental philosophy." Prior to that time, he had consciously avoided using the term, which was closely associated with the early Royal Society and Restoration England.[18]

6.2.1 Drafts of Qu. 23

I shall now return to the drafts of this query to establish the meaning of "ingredients" in Qu. 23. The first draft of this paragraph reads:

> I have hitherto proceeded <in this Book> by way of Analysis, arguing from effects to causes & from *compound bodies* to their *ingredients*. In the first Book I proceeded <first> by Analysis in searching into ye different refrangibility of the rays & the corresponding colours of light & then from those Principles compounded the explications of the colours of light refracted by Prisms those of the Rainbow & those of Natural bodies. In the second Book I proceeded by Analysis in searching out the fits of easy Reflexion & easy transmission of the rays of light, & then from this Principle compounded a further explication of the colours of natural bodies ~~& of the constitution of those bodies requisite for making those~~[19]

[17] Levitin 2021, 253, and 2022, 621; the phrase in bold is Levitin's restoration of the damaged manuscript.

[18] See Shapiro 2004.

[19] CUL Add. MS 3970, f. 244r, italics added. The drafts for Qu. 23/31, paragraphs 27 and 28 on analysis and synthesis are discussed and published in Ducheyne and Dhondt 2021, 360–6, 382–8.

In later drafts Newton replaced "compound bodies"—reflecting his belief in light corpuscles—with "compositions," presumably to remove an explicit reference to material bodies, though "compositions" still suggest chymists' concept of analysis. It is already apparent from this first draft that synthesis here involves the formulation of explanations and not the compounding of light from its components, as a decompositional approach would require. Moreover, in this first draft, Newton sets out his method of analysis and synthesis in the particular context of "this Book"—that is, of optics and his theory of light, not in the global context that the final version suggests.

The second draft helps to further clarify Newton's meaning:

> The business of Experimental Philosophy is only to find out by experience & Observation <not how things were created but> what is the present frame of nature. This inquiry must proceed first by Analysis in arguing from effects to causes & from compositions to <*ingredients*> ~~components~~. And when we have found <the principles> ~~the causes & <ingredients><components>~~ of things we may proceed by <Synthesis> ~~composition~~ from those Principles <to explain the things.> Of this method I gave instances in the two first books <proceeding first by Resolution & then by composition>[20]

Newton's initial intention in this draft was to replace "ingredients" with "components" prior to restoring "ingredients" through all later versions. These "components" or "ingredients," I hold, refer to the rays of different color and refrangibility that compose light. "Components" is a clear, neutral way of referring to the rays of different colors that evades the invocation of hypothetical light corpuscles. It is unclear why Newton rejected "components," for, as I shall demonstrate, he did use that term as well as "ingredients" in the *Opticks* for rays of different color in the context of color-mixing.

The final, published version shows more conclusively that by "ingredients" and "compounds" Newton was referring to light rays:

> In the two first Books of these Opticks I proceeded by Analysis to discover & prove the original differences of the rays of light in respect of refrangibility reflexibility & colour & their alternate fits of easy reflexion & easy transmission & the properties of bodies both opake & pellucid on which their reflexions & colours depend: & these discoveries being proved may be assumed as Principles in the method of Composition for explaining the phaenomena arising from them: an instance of wch Method I gave in the end of the first Book.[21]

---

[20] CUL Add. MS 3970, f. 243r, italics added. This is the first time that Newton introduced the term "experimental philosophy" in his writings. He also used it in the next draft (out of a total of four) of this paragraph, and then, for some reason that I cannot explain, in the final draft, he reverted to the common "natural philosophy"; see Shapiro 2004.

[21] Newton 1984–2021, vol. 2, 379; CUL Add. MS 3970, f. 286r. In 1716, this passage began a new paragraph. For the reader's convenience, I quote the opening sentences of this paragraph in Newton's English: "As in Mathematicks so in Natural Philosophy the investigation of difficult things by the method of Analysis ought ever to precede the method of Composition. This Analysis consists in making experiments & observations & in arguing by them from compositions to ingredients & from motions to the forces producing them & in general from effects to their causes & from particular causes to more general ones, till the Argument end in the most general: The Synthesis consists in assuming the causes discovered & established, as Principles; & by them explaining the Phaenomena proceeding from them, & proving the explanations."

At the end of the first Book, he explained the colors produced by prisms and the rainbow, which an earlier draft explicitly cited at this point.[22]

Newton states directly that the analysis in Book I established that rays of light differed in color and degrees of refrangibility, and the analysis in Book II established that the rays possess fits of easy reflection and refraction. Thus, according to his method of analysis and synthesis, Newton held that the analysis (in Book I) was the discovery of the principle (or series of propositions) that light consists of rays of different refrangibility and color and that the synthesis consists of explaining various phenomena, such as the colors of natural bodies and the rainbow by means of that principle. By formulating the analysis of light, which is decompositional, in proposition form—that is, as "principles"—Newton was able to transform that analysis into regressive form. Then, taking these as "principles," he explained and proved other phenomena, such as the colors of the rainbow and prisms. This is synthesis as explanation rather than synthesis as composition. Newton does not state that the composition of white light from the various colors is the synthesis, as the decompositional form requires.

In the text of the *Opticks* Newton used "components" twice and "ingredients" three times to designate rays of different color, all in descriptions involving the composition of light and color mixing. In his description of color mixing in the *Opticks*, Book I, Part II, Prop. 4, for instance, he describes mixing red and yellow spectral colors to produce an orange that looks just like the pure spectral orange. When the mixed orange is viewed through a prism, it "is changed and *resolved* into its *component* colours red and yellow," while the unmixed orange remains unchanged.[23] When describing the use of his color mixing circle he states that at a particular point in the circle "the main *ingredients* being the red and violet, the Colour compounded shall not be any of the prismatic Colours but a purple, inclining to red or violet."[24] Indeed, in the chemical portions of Qu. 23/31 he uses the term "ingredients" four times.[25] "Ingredients" was, of course, then a common word in the chemical literature. Boyle, for instance, used it one hundred times in *The Sceptical Chymist*. It was thus quite natural for Newton to apply that term to describe mixtures of different colors.

Newton's use of the terms "component" and "ingredients" in fact goes back to the initial publication of his theory of color in 1672. Proposition 4 in his "New theory about light and colors," states that "seeming transmutations of Colours may be made, where there is any mixture of divers sorts of Rays. For in such mixtures, the *component* colours appear not, but, by their mutual allaying each other,

[22] CUL Add. MS 3970, f. 242v.

[23] Newton 1704/1966, 97, italics added; and Newton 1952, 133. He also uses "components" in Newton 1704/1966, Bk. II, Pt. II, $_2$34; Newton 1730/1952, 229.

[24] Newton 1704/1966, Bk. I, Pt. II, Prop. 6, 116, 117; Newton 1730/1952, 156, 157, italics added. Newton also uses "ingredients" in the following paragraph, Newton 1704/1966, 117, Newton 1730/1952, 157, and in Newton 1704/1966, Bk. II, Pt. II, $_2$38; Newton 1730/1952, 232.

[25] Newton used "ingredients" once in paragraph 3, Newton 1984–2021, vol. 2, 362, and three times in paragraph 7, 366 in Qu. 23/31; Newton 1730/1952, 378, 384, 385.

constitute a midling colour."[26] Proposition 8 asserts that "Light is a Confused aggregate of Rays indued with all sorts of Colors…And of such a confused aggregate, as I said, is generated Whiteness, if there be a due proportion of the *Ingredients*."[27]

6.2.2 Newton and Chymists' Decomposition

Hitherto, I have been primarily concerned with analysis and synthesis in the later years of Newton's career, but I shall now turn to their role in his early years. In a carefully argued paper William R. Newman has shown that two types of arguments utilizing analysis and synthesis that Newton encountered in chymistry—principally in Boyle's works—would serve as a heuristic or guide in formulating his optical theory.[28] He is quite clear that he is concerned only with the period from 1664 through the publication and responses to Newton's "New theory about light and colors" in 1672. Newton's notes on Boyle are interspersed in the same notebook as his early optical essays. The first argument is "reduction to a pristine state." Here, a substance such as camphor is mixed with another, such as sulfuric acid, and forms a deep reddish solution and loses its perceptible properties, such as its odor and color. When water is added to the mixture, the camphor returns with all its original properties. Boyle's conclusion was that the camphor was present in the mixture but "hidden" among the corpuscles of the solution. This is an example of an apparently homogeneous substance actually being heterogeneous, just as Newton's claim regarding sunlight, and it involves synthesis followed by analysis. The second type of argument proceeds in the reverse order—that is, a decomposition or analysis followed by recomposition or synthesis that Boyle called "redintegration." Boyle decomposed saltpeter into its ingredients and then recombined those ingredients to arrive once more at saltpeter. The "structural similarity"—as Newman calls it—of Boyle's approach to that of Newton is striking. The first is similar to Newton's claim that light rays of different color are hidden in sunlight and then reduced to their simple or homogeneous state by refraction. The second recalls Newton's experiments in which sunlight that has been decomposed into rays of different color are subsequently recombined at the focus of a lens to again compose white. In Fig. 6.2 from the "New theory," a beam of sunlight *SF* is decomposed by the prism *ABC*, and the separated colors fall on the lens *MN*, which brings the rays to a focus at *e* where they again form white. By moving the screen *HI* along the beam one can see the colors gradually unite until at *e* they form white again, like the sun's original light,

[26] Newton 1672a, 3082, italics added; reprinted in Newton 1958, 54.

[27] Newton 1672a, 3083, italics added; Newton 1958, 55. In various drafts and responses to criticism of his theory, Newton used "component." I shall cite only Newton 1672b, 5005; reprinted in Newton 1958, 94: "7. Whether the component colours of each mixture be really changed; or be only separated when from that mixture various colours are produced again by Refraction?"

[28] Newman 2010, and lightly revised in Newman 2019, ch. 6.

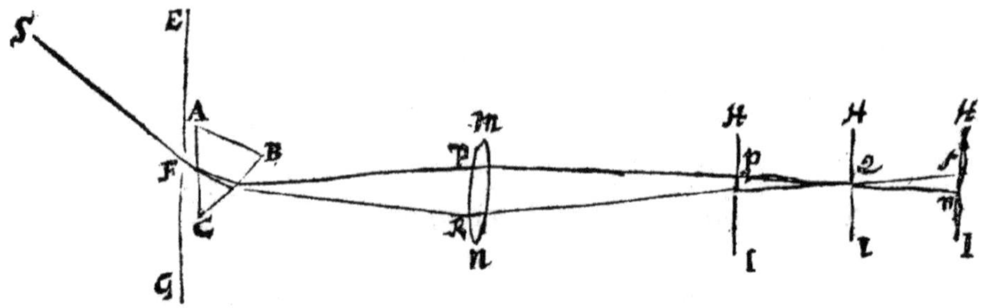

Fig. 6.2 Newton's experiment to decompose a light beam using a prism and then recompose it using a lens, from Newton 1672a

and then separate once more. This experiment with some variation is in all his presentations of the theory.[29]

While I am confident that Newton almost certainly recognized a structural similarity to the chemical cases, I also believe that he did not openly pursue the analogy between chemical and optical phenomena because of a fundamental physical dissimilarity. In the "reduction to a pristine state," Boyle explained that the sulfuric acid acted on the camphor by causing it to change its "texture"—that is, by causing a rearrangement of its particles. When light is refracted and separated into different colors, the refracting body—glass, for example—acts on the light rays. The light rays, Newton held, do not act on one another; rather, they are independent and immutable, a principle that, for him, was fundamental. In the chemical case, by contrast, the chemicals act on one another, but precisely what this involves for Boyle is obscure. This was of no help to Newton, who invoked the principle of immutability here. As a consequence of that principle, light rays do not act on one another, and as such, when they met at the focus of the lens in the preceding experiment, it followed that they could not mix there or act on one another to generate white. If the rays do not mix, why then do we see white there? Newton's response was that the sensations produced by the different colored light rays mix in the eye. No chemical equivalent to this concept exists.

Newton, however, did not recognize the principle of immutability until about mid-to-late 1671. The first time that he invoked that principle was in Lecture 6 of his *Lectiones opticae,* which was completed by that date.[30] It was a late addition because it was not yet a formal proposition as it would become in the revised version, the *Optica*.[31] Newton undoubtedly had a rough, tentative concept of color immutability before he recognized a formal principle for spectral colors. At this stage, before the formal principle, Newton would have seen a similarity with the chemical case. Once he recognized the principle, because of the physical difference between chemical

[29] See Newton 1984–2021, vol. 1, 115–25; and Newton 1704/1966, Bk. I, Pt. II, Prop. 5, Expt. 10.

[30] Newton 1984–2021, vol. 1, 18–20, 143–5.

[31] Color immutability was Prop. 2 in the *Optica* and fully demonstrated; Newton 1984–2021, vol. 1, 437, and 453–61.

and optical analysis and synthesis, while their common features had only heuristic value, there was no reason for Newton to call particular attention to analysis and synthesis. This is especially the case, because it would unduly emphasize a commitment to a corpuscular theory of light. Having publicly stated his suspicion in the *Principia* that all phenomena depend on the forces that the corpuscles of matter exert on one another, the physical difference between optical and chemical mixing was made explicit. Moreover, as stated at the beginning of the paper, I believe that Newton introduced the method of analysis and synthesis for polemical reasons against the Cartesians and Leibnizians and not particularly to elucidate his optical theory or his method. For this purpose, he required the regressive concept, which is concerned with causes and explanation and proof, rather than the chymists' decompositional one. To be sure, several decades after Newton the decompositional concept could be applied to light when chymists' physical, manipulative concept was replaced by an epistemic one.

6.2.3 Logic

It has been frequently observed that in the mathematical tradition that Newton invokes one searches for the known from what is unknown but assumed to be true, but he is actually applying the approach of the philosophical tradition whereby one searches in the opposite direction from an effect that is known to the unknown cause.[32] This problem may be resolved—at least from Newton's perspective—by recognizing the "new logic" that arose in the seventeenth century, which was significantly propagated by the Cartesian *Port-Royal Logic*, as *La logique, ou l'art de penser* (1662) by Antoine Arnauld and Pierre Nicole was known.[33] The "new logic" represented a shift from scholastic concern with syllogism and argument to a method of discovery. In the seventeenth century logic was still a standard part of the university curriculum. Newton had six books on logic in his library, including a 1687 Latin translation of *The Port-Royal Logic*.[34] *The Port-Royal Logic* incorporates both the philosophical and mathematical concepts of analysis. In a chapter on method, they explain,

[32] Perhaps the earliest to note of this was Stewart 1814, vol. 2, 365–70. More recent commentators on this point are Hintikka and Remes 1974, 106–07; Ducheyne 2005; and Guicciardini 2009, 324.

[33] *The Port-Royal Logic* was quite popular, and in England alone, "Between 1664 and 1700 it received eight London editions, one in its French text, four in Latin, and three in English"; Howell 1956, 351. On the term "new logic," see, for example, Howell 1956, ch. 6, "New horizons in logic and rhetoric."

[34] Harrison 1978, 59, and no. 980 on page 182. Ducheyne 2005, also turns to the logic tradition, but he claims that it is the Aristotelean tradition of logic that influenced Newton, and not the modern school of the *Port-Royal Logic*; see Levitin 2016, for a convincing refutation of Ducheyne's argument.

> there are two kinds of method, one for discovering the truth, which is known as *analysis*, or the *method of resolution*, and which can also be called the *method of discovery*. The other is for making the truth understood by others once it is found. This is known as *synthesis*, or the *method of composition*, and can also be called the *method of instruction*.[35]

They then enumerate four kinds of analysis, only the first of which need concern us here: "when we look for causes by effects. We know, for example, the different effects of a lodestone, so we look for its cause."[36] This is the regressive concept. Later in the chapter they proceed to explain "analysis as used by geometers":

> Suppose a question is presented to them, such as whether it is true or false that something is a theorem, or whether a problem is possible or impossible; they assume what is at issue and examine what follows from that assumption. If in this examination they arrive at some clear truth from which the assumption follows necessarily, they conclude that the assumption is true. Then starting over from the end point, they demonstrate it by the other method which is called *composition* [synthesis].[37]

Of course, if they fail, that which was proposed is false or impossible. Thus, in the account of analysis in this influential work, the mathematical concept of analysis is classified as just another form of analysis and not something of an entirely different nature. Thus, the problem of the direction of mathematical analysis in Qu. 23/31 should no longer be a historical problem. Newton clearly chose to call this "mathematical analysis" because he was so familiar and comfortable with that approach but also, I strongly suspect, because calling the method mathematical added to its polemical thrust in stressing that his approach was more certain than the hypothetical one.

As mentioned earlier, Newton did not use the terms "analysis" and "synthesis" in his optical writings, and British and Dutch Newtonians likewise did not apply those terms to light and optics. Newton's language for synthesis, in particular, for the compound nature of sunlight, is straightforward and unproblematic. He largely uses "composition," "compose," "compound" and their variants, along with "mix" and its variants. Recall that the "composition" family is the Latin equivalent of "synthesis." It is not at all clear whether Newton is using everyday language or the technical language of chymistry. For analysis, he used the Latin equivalent "resolved" once in the *Opticks* in Bk. I, Pt II, Prop. 4.[38] Otherwise he did not use "analysis" or "resolution" at all but rather a variety of expressions, stating, for instance, that the rays are "separated" or "diverge." He described the decomposition of sunlight in primarily geometric or spatial terms—that is, he tended to describe the rays as diverging from one another in space due to differences in their refraction. For example, in the

[35] Arnauld and Nicole 1996, Part 4, Ch. 2, 233.

[36] Ibid., 234. Arnauld and Nicole note here that "The greater part of what is said here about issues was taken from a manuscript by the late Descartes, which Clerselier was kind enough to lend us." Dugald Murdoch, the translator of the *Rules for the Direction of the Mind* in Descartes 1985–1991, vol. 1, 77, suggests that the four rules may be a paraphrase of Descartes' manuscript. The remaining three rules expound a decompositional concept. See note 47 below.

[37] Arnauld and Nicole 1996, Part 4, Ch. 2, 238.

[38] This passage is quoted at note 23, above.

Optical Lectures, he explains that the rays "will diverge from one another…insofar as any ray is disposed to undergo a greater or smaller refraction."[39] In all presentations of his theory Newton began by presenting the unexpected elongation of the sun's image when refracted by a prism as requiring an explanation—that is, as an instance of regressive analysis and synthesis. In the two extended accounts—the *Optical Lectures* and the *Opticks*—he explained the elongation by means of overlapping circular images of the sun. British and Dutch Newtonians largely followed Newton's spatial language and mode of presentation.

6.3 The French and Decompositional Analysis in the Eighteenth Century

I searched seventeen optical books (or relevant chapters of natural philosophy books) published by Newtonians in the eighteenth century—twelve British, four Dutch, and one Italian in English translation—for decompositional terminology. All but two of the books showed no instances of "decompose," "analysis," "synthesis," or "resolve" and their variants, except when recounting Newton's use of "resolve" or presenting his method of analysis and synthesis from Qu. 23/31.[40] In *The History and Present State of Discoveries Relating to Vision, Light, and Colours* (1772), Joseph Priestley, explaining the formation of the rainbow, states that after entering and leaving a raindrop, a light ray is "*decomposed* into as many small differently coloured pencils, as there are primitive colours in the light."[41] Priestley's usage reflects a shift to a decompositional concept of analysis and synthesis for light that was already under way in France. The other exception is a total outlier to me. Benjamin Martin was an instrument maker, itinerant lecturer–demonstrator, and popularizer of Newton. In his *Panegyrick on the Newtonian Philosophy* (1749)—which has only two paragraphs on light—he explains that the different magnitudes of the particles of light are "demonstrated from the *Analysis of light*, by Experiments of the *Prism*. …Hence the Doctrine of *Composition* and *Transmutation* of Colours…"[42] This is half a century earlier than the earliest use of "analysis" that I have found in mainstream British optics.[43] Several of these authors—namely, Henry Pemberton, Colin Maclaurin, Petrus van Musschenbroek, and Willem Jacob's

[39] Newton 1984–2021, vol. 1, 51.

[40] I searched the following works: Desaguliers 1719, 1744, vol. 2; Pemberton 1728; Smith 1738; Helsham 1739; Musschenbroek 1739; Algarotti 1742; 's Gravesande 1747; Maclaurin 1748; Martin 1749; Rowning 1753, vol. 2; Martin 1759; Priestley 1772; Harris 1775; Enfield 1785; Adams 1794, vol. 2; and Wood 1799.

[41] Priestley 1772, 274, italics added.

[42] Martin 1749, 23. In his later optical treatise (Martin 1759), however, he does not use "analysis" in that way.

[43] Young 1802, 395, refers to the "prismatic analysis of the colours of thin plates."

Gravesande—also presented an account of Newton's comments on analysis and synthesis in Qu. 23/31, and they all adopted his regressive concept.[44]

Newtonians continued to use "analysis" in its traditional but distinct chymical, decompositional sense. Musschenbroek, like others in the eighteenth century, used "analysis" in this chymical sense when he noted that, "One finds by chemical analysis all the principles that enter into the formation of a magnet."[45]

If Newtonians in Britain and the Netherlands essentially followed Newton's account of method with its regressive concept of analysis and synthesis in Qu. 23/31, the French followed a very different path. During the course of the eighteenth century the decompositional concept of analysis and synthesis replaced the regressive concept. For at least the first third of the eighteenth century, Cartesian natural philosophy remained dominant. Even when Newton's ideas, such as his theory of gravity and the composition of white light, were accepted, beginning in the second third of the century, the French were never such staunch, ideological Newtonians that they endorsed the method of Qu. 23/31.[46] In some of his writings Descartes appeared to endorse a decompositional approach, but in others he described the classical regressive method, as described by Pappus.[47]

In his groundbreaking development of analytical geometry Descartes applied algebra to geometrical curves in *La Géométrie*. Symbolic algebra initially derived its name "analytical" from the ancient concept of analysis, wherein one worked backward from an unknown to a known quantity—that is, in the regressive manner. However, in the course of the eighteenth century, particularly in France, analysis became identified with algebra, in both its symbolic form and in the new analysis or infinitesimal calculus as developed by Newton and Leibniz.[48]

---

[44] A large body of recent literature exists on the methodology of the Dutch Newtonians. See for instance: Schuurman 2004; Ducheyne 2014; van Besouw 2017; and Ducheyne 2017.

[45] "Cette pierre est un mixte, naturellement composé de fer, ou de la matiere du fer, de pierre, d'huile, & de sel; quelquefois d'autres principes concourent encore à sa composition: & ces ont, ou des métaux, des demi-métaux, &c. On trouve, par l'analyse chymique, tous les principes qui entrent dans la formation de l'aimant," Musschenbroek 1769, vol. 1, §DCCCCXLVII, 430; the original Latin (Musschenbroek 1762, vol. 1, 317) has "Analysi Chemica". To give one more instance of the chemical meaning of "analysis" among Newtonians, Desaguliers 1744, vol. 2, 367, quotes "Mr. *Lemery* the younger" as writing "in the Analysis, such inflammable Bodies produce Salt, Earth, Water, and a certain subtile Matter, which passes thro the closest Vessels."

[46] In the article "analytique" in the *Encyclopédie* D'Alembert quotes a French translation of paragraph 26 of Qu. 23/31 on analysis and synthesis with no comment whatsoever; Diderot and d'Alembert 1751–1772, vol. 1 (1751), 403–04.

[47] On the decompositional approach see, for example, *Discourse on Method*, Part Two, "The second [rule of logic], to divide each of the difficulties I examined into as many parts as possible and as may be required in order to resolve them better"; Descartes 1985–1991, vol. 1, 120. This is essentially a paraphrase of rule 13 in the *Rules for the Direction of the Mind*, ibid., 51. For the regressive approach, see *Objections and Replies* to Descartes' *Meditations*. In the Second Set of Replies, the objector suggested that Descartes put the *Meditations* into synthetic form "in geometrical fashion." In his reply, Descartes explained the relative virtues of the analytic and synthetic approaches and why he chose the analytic approach, ibid., vol. 2, 92, 110–11.

[48] Guicciardini, 2009, 39–40.

The transition in the concept of "analysis" becomes quite clear in mid-century in Denis Diderot and Jean le Rond d'Alembert's *Encyclopédie.* The sequence of articles on "analyse" presents the concept in a variety of areas, including grammar, literature, and chemistry, but it begins with d'Alembert's entry on mathematics:

> ANALYSIS … is properly the mathematical method to resolve problems while reducing them to equations. …
>
> Analysis, in order to solve problems, employs the aid of algebra, or generally the calculation of magnitude; also, these two words *analysis* and *algebra* are often regarded as synonyms. …
>
> *Analysis* is divided, with regard to its object, into *analysis of finite quantities*, and *analysis of infinite quantities.*
>
> *Analysis of finite quantities* is what is otherwise called *specious arithmetic* or *algebra....*
>
> *Analysis of infinite quantities,* or *of the infinite,* also called *the new analysis,* calculates the ratios of quantities which are taken as infinite, or infinitesimally small. One of its principal branches is *the method of fluxions,* or the *differential calculus....*
>
> The great advantage of modern mathematicians over the ancients comes principally from the uses they make of analysis.[49]

The classical, mathematical concept of analysis—that is, in Greek geometry and Newton's mathematics—was regressive. Here, we can see how in the eighteenth century, the French turned away from Newton's synthetic, geometric approach to calculus and replaced it with analysis, which they considered to be equivalent to algebra.

The article on analysis in logic likewise introduces a decompositional concept of analysis:

> *Analysis* consists in going back to the origin of our ideas to elucidate their generation and to make different compositions and decompositions in order to compare them from all aspects which can show their relations. … It is not with the assistance of general propositions that one searches for the truth, but always with the help of a kind of calculation, i.e., in composing and decomposing notions in order to compare them in the most favorable manner to the discoveries one has in sight.[50]

---

[49] "ANALYSE … est proprement la méthode de résoudre les problèmes mathématiques, en les réduisant à des équations. …

"L'*Analyse*, pour résoudre les problèmes, employe le secours de l'Algebre, ou calcul des grandeurs en général: aussi ces deux mots, *Analyse*, *Algebre*, sont souvent regardés comme synonymes. …

"L'*Analyse* est divisée, par rapport à son objet, en *Analyse des quantités finies*, & *Analyse des quantités infinies*.

"*Analyse des quantités finies*, est ce que nous appellons autrement *Arithmétique spécieuse* ou *Algebre. …*

"*Analyse des quantités infinies*, ou *des infinis*, appellée aussi *la nouvelle Analyse*, est celle qui calcule les rapports des quantités qu'on prend pour infinies, ou infiniment petites. Une de ses principales branches est *la méthode des fluxions*, ou *le calcul différenciel. …*

"Le grand avantage des Mathématiciens modernes sur les anciens, vient principalement de l'usage qu'ils font de l'*Analyse*"; Diderot and d'Alembert 1751–1772, vol. 1(1751), 400–01.

[50] "L'*analyse* consiste à remonter à l'origine de nos idées, à en développer la génération & à en faire différentes compositions ou décompositions pour les comparer par tous les côtés qui peuvent en montrer les rapports. … Ce n'est point avec le secours des propositions générales qu'elle cherche la vérité: mais toûjours par une espece de calcul, c'est-à-dire, en composant & décomposant les

6.3.1 Condillac and Analysis

Etienne Bonnot de Condillac was the most influential advocate of the concept of decompositional analysis through his *Essay on the Origin of Human Knowledge*, which appeared in 1746, five years before the *Encyclopédie*. He was friends with d'Alembert and was widely read by the encyclopedists. Condillac rejected the regressive concept of synthesis as a fruitful method of discovery in favor of decompositional analysis:

> The uselessness and abuse of principles is especially apparent in synthesis, a method that appears to prohibit the truth from appearing unless it has been preceded by many axioms, definitions, and other supposedly fertile propositions. ...If mathematicians' ideas are exact, it is because they are the product of algebra and analysis.[51]

He explained that analysis "consists only in composing and decomposing our ideas in order to make different comparisons, and to discover in this way the relations between them and the new ideas they can produce."[52]

Language and algebra play a prominent role in Condillac's *La logique, ou les premiers développemens de l'art de penser*, which appeared shortly after his death in 1780. He argued that we think only by means of names (or words, signs, or symbols) that represent ideas and that we analyze only by means of language.[53] Algebra—"the language of mathematics, is the simplest of all languages"[54]—is presented as the ideal analytic tool:

> I shall not say with the mathematicians that algebra is a kind of language: I say that it is a language and that it can be nothing else....
>
> Algebra is, in fact, an analytic method: but it is no less a language for that, if all languages are themselves analytic methods. ... But algebra is very striking proof that the progress of the sciences depends solely upon the progress of their languages; and that well-made languages alone could give to analysis the degree of simplicity and precision of which it is capable ...

---

notions pour les comparer, de la maniere la plus favorable, aux découvertes qu'on a en vûe." ibid., 401, italics added. The article is by Claude Yvon (1714–1789), a priest.

[51] "L'inutilité & l'abus des principes paroît surtout dans la synthèse: méthode où il semble qu'il soit défendu à la vérité de paroître qu'elle n'ait été précédée d'un grand nombre d'axiomes, de définitions & d'autres propositions prétendues fécondes..... Si les idées des mathématiciens sont exactes, c'est qu'elles sont l'ouvrage de l'algèbre & de l'analyse"; Condillac 1746, Sect. II, Ch. 7, §63, vol. 1, 96–7.

[52] "Elle ne consiste qu'à composer & décomposer nos idées pour en faire différentes comparaisons, & pour découvrir, par ce moyen, les rapports qu'elles ont entre elles, & les nouvelles idées qu'elles peu vent produire," ibid., §66, 102.

[53] "We can analyse only by means of language. ...we think only with the aid of words"; Condillac 1980, 211. "Languages are so many analytical methods...Analysis is made and can only be made, with signs"; ibid., 225. This book is a facing-page translation of Condillac 1780, *La logique*, and the first phrase in each quotation here is a postil.

[54] Condillac 1980, 285.

> Well-made languages could do this, I say: for in the art of reasoning as in the art of calculating, everything is reduced to compositions and decompositions.[55]

By the third quarter of the eighteenth century in France, the decompositional concept of analysis, which was represented by symbolic algebra, had attained dominance.

Condillac's works, with their concept of analysis and views on language and algebra, were already well known in France when Antoine Lavoisier drew attention to them in the preface to his *Traité élémentaire de chimie* (*Elements of Chemistry*, 1789) and emphasized their influence on his thinking. He quoted from Condillac's *Logic* both at the beginning and end of the preface and stated that when he began the book, his "only object was to extend and explain more fully the Memoir" that he had read to the Academy of Sciences 2 years earlier "on the necessity of reforming and completing the Nomenclature of Chemistry."[56] In that memoir he explained Condillac's views in greater detail: "Languages … are also analytical methods, by the means of which, we advance from the known to the unknown, and to a certain degree in the manner of mathematicians…Algebra is the analytical method [par] excellence … Even, a moment's reflection is sufficient to convince us that algebra, is in fact a language." He immediately noted that these concepts have "been explained with infinite exactness and perspicuity in the *Logic* of the abbé de Condillac, a work which can never be too much studied by the youth that dedicate themselves to the sciences."[57] Noting the spread of the decompositional concept of analysis, Georg Christoph Lichtenberg in Göttingen wrote in a notebook some time between 1789 and 1793 that, "Whichever way you look at it, philosophy is always analytical chemistry."[58]

6.4 A New Concept of Analysis in Optics

After this brief interlude on Condillac's influence, I shall now return to the gradual adoption of the decompositional concept of analysis and synthesis in optics. By the time of the *Encyclopédie*'s publication, French writers on optics were applying the term "décomposer" and its variants to light, which, as I have shown, the Newtonians had avoided.[59] In his *Leçons de physique experimentale* in 1758, Jean Antoine

---

[55] Ibid., 303–05.

[56] Lavoisier 1790/1965, xii.

[57] Morveau 1788, 4–5, from Lavoisier's paper in the collection, "Sur la nécessité de réformer & de perfectionner la nomenclature de la chimie." On Lavoisier and Condillac see Albury 1972; Levere 1990; Beretta 1993, 187–206; and Bensaude-Vincent 2010, 473–89.

[58] Lichtenberg 1990, 162. Beaney 2022 led me to this quotation.

[59] The British may have avoided "decompose" because of possible confusion with the related word "decompound," which means the contrary. The *Oxford English Dictionary Online,* 2000– defines it as to "repeatedly compound; compounded of parts which are themselves compound." Newton himself used "decompound" this way in Newton 1704/1966, Bk. I, Pt. II, Prop. 5, Expt. 10, 101; Newton 1730/1952, 138.

Nollet freely used "décomposée," and Section III of Leçon XVII of the *Leçons,* on the properties and nature of colors, is entitled "On decomposed (*décomposée*) light, or on the nature of colors."[60] In the *Encyclopédie* itself, in 1765, the article on light (*lumière*) by d'Alembert uses "composé" and "décomposer" in discussing the colors of light.[61] However, Gabriel-François Venel's article "décomposition" (chemistry) does explain that "chemical decomposition is better known in the art under the name of *analysis*."[62] Nicolas Louis de La Caille in 1764 described the problems that arise in designing optical instruments that are caused by the decomposition (*décomposition*) of the light rays after refraction.[63] By the last quarter of the century the use of "décomposé" became common in French. The chemist Claude-Louis Berthollet, of course, used "décomposer" and "analyser" for chemical descriptions, but he also applied "décomposer" to light in his *Elements of the Art of Dyeing* (1791).[64] One year later the chemist Antoine-Francois Fourcroy also used "décompose" for light as well as "analyse," explaining that, "In refraction light is *decomposed* into seven rays, the red, the orange, the yellow, the green, the blue, the indigo, and the violet…. This *decomposition* by the prism is a kind of *analysis* of light."[65]

The writings of the leading optical scientists at the beginning of the nineteenth century show that "analysis" was becoming widely applied to light on both sides of the Channel. In 1802, Thomas Young referred to the "prismatic analysis of the colours of thin plates."[66] Four years later René Just Haüy in France in his textbook entitled the chapter on color "On decomposed (*décomposée*) light, or of colors."[67] In recounting an experiment from Newton's *Opticks*, Bk I, Pt I, Prop. 2, expt. 9, in which light is totally reflected from the base of a right-angled isosceles triangle, Haüy observed that this experiment "serves therefore to confirm, in some way, by way of synthesis (*synthèse*) that which the preceding experiments had established by a contrary operation that can be compared to analysis (*analyse*.)"[68] In his *Traité*

[60] Nollet 1758, vol. 5, 336. He writes that "Before Newton no one had imagined that light could be decomposed (*décomposer*)…."

[61] Diderot and d'Alembert 1751–1772, vol. 9 (1765), 721.

[62] "La *décomposition chimique* est plus connue dans l'art sous le nom d'*analyse*"; ibid., vol. 4 (1754), 699.

[63] La Caille 1764, 102, 104. Dutour 1773, freely used "décomposition" and its variants in his paper beginning with its title, "Considérations optiques. IVe mémoire sur la décomposition de la lumiere dans le phénomene des anneaux colorés, produit avec un miroir concave."

[64] Berthollet 1791, vol. 1, 12.

[65] "En se refrangeant, la lumière se *décompose* en sept rayons, le rouge, l'orangé, le jaune, le vert, le bleu, l'indigo, & le violet. … Cette *décomposition* par le prisme est une espèce d'*analyse* de la lumière"; Fourcroy 1792, 7, italics added.

[66] For Young's usage, see note 43 above. In 1817, Young also used "analysis" in his article "Chromatics," for the *Encyclopaedia Britannica;* reprinted in Young 1855, vol. 1, 282.

[67] "Part VIII. De la lumière, Ch. 3. De la Lumière décomposée, ou des Couleurs"; Haüy 1806, vol. 2, 192.

[68] "L'expérience … servait donc à confirmer, en quelque sorte, par la voie de synthèse, ce que les précédentes avaient établi par une opération contraire que l'on pourrait comparer à l'analyse"; ibid., vol. 2, 207–08. This passage is not in the first edition of 1803.

in 1816, Jean-Baptiste Biot entitled the section on color "The analysis (*analyse*) of light," while the running head for this section is "The decomposition (*décomposition*) of light."[69] By the 1830s, the decompositional concept of analysis and synthesis had been firmly established.[70]

The Scottish philosopher Dugald Stewart's *Elements of the Philosophy of the Human Mind* in 1814 marks the shift in the concept of analysis and synthesis from regressive to decompositional. In what he called the method of "experimental or inductive logic," we must discover the laws of nature, or general facts, from observations "by a sort of *analysis* or decomposition." He noted that,

> In fact, the meaning of the words *analysis* and *synthesis*, when applied to the two opposite modes of investigation in physics, is extremely analogous to their use in the practice of chemistry. The chief difference lies in this, that, in the former case, they refer to the logical processes of the understanding in the study of *physical laws*; in the latter, to the operative processes of the laboratory in the examination of material substances.[71]

Here, he is clearly expounding the decompositional concept.

Stewart devotes an entire part of the chapter on inductive logic to "the Import of the Words Analysis and Synthesis, in the Language of Modern Philosophy," to prevent "readers from falling into the common error of confounding the analysis and synthesis of the Greek Geometry, with the analysis and synthesis of the Inductive Philosophy."[72] His objection turns on the direction of analysis and synthesis in the mathematical and physical cases, which I discussed earlier in this paper. "Sir Isaac Newton himself has," Stewart wrote, "in one of his Queries, fairly brought into comparison the Mathematical and the Physical *Analysis*, as if the word, in both cases, conveyed the same idea." He goes on to quote paragraph 26 of Qu. 23 on analysis and synthesis in its entirety and notes—quite properly—that the first sentence "has been repeated over and over by subsequent writers."[73] He observes,

> The meaning conveyed by the word Analysis, in Physics, in Chemistry, and in the Philosophy of the Human Mind, is radically different from that which was annexed to it by the Greek Geometers, or which ever has been annexed to it by any class of modern Mathematicians. In all the former sciences, it naturally suggests the idea of a decomposition of what is complex into its constituent elements.[74]

Stewart continues in this vein, but this suffices to show that he does not acknowledge either the legitimacy of the regressive concept of analysis and synthesis that

[69] Biot 1816, vol. 3, 383.

[70] Herschel 1830, vol. 2, §406, 406, widely used analysis and synthesis in his article "Light." Writing, for instance, that "In order to justify the term *analysis*, or *decomposition*, as applied to the separation of a beam of white light into coloured rays, we must show by experiment that white light may again be produced by the *synthesis* of these elementary rays." This family of terms was used by Brewster 1831; and Powell 1833. The terms "decomposition" and "recomposition" were now being used along with "analysis" and "synthesis."

[71] Stewart 1814, 308, 333, 334. Niccolò Guicciardini brought Stewart's book to my attention.

[72] Ibid., 353, 354–5.

[73] Ibid., 367.

[74] Ibid., 368.

Newton invoked, or that Newton was invoking the method of mathematicians to buttress the certainty of his works.

Stewart's book reveals that our modern decompositional concept of analysis and synthesis in the physical sciences had become firmly established—beyond chemistry—by the first decades of the nineteenth century. This transition of the concept of analysis from one that was operational on material substances to an epistemic one had already occurred in chemistry under the influence of Condillac and Lavoisier.[75] The similar transition of the concept of analysis has not been previously demonstrated for a physical science and, in particular, for optics. It is clear that the French played a major role in effecting the shift in the analysis and synthesis concept's meaning. I strongly suspect that changes in logic also contributed to it, and until further investigation of these concepts is undertaken in other contexts—Germany, in particular—the story will remain incomplete.

References[76]

Adams, George. 1794. *Lectures on natural and experimental philosophy: Considered in it's present state of improvement. Describing, in a familiar and easy manner, the principal phenomena of nature; and shewing, that they all co-operate in displaying the goodness, wisdom, and power of God.* 5 vols. London: R. Hindmarsh.

Albury, William Randall. 1972. *The logic of Condillac and the structure of French chemical and biological theory, 1780–1801*. PhD dissertation. Johns Hopkins University.

Algarotti, Francesco. 1742 *Sir Isaac Newton's theory of light and colours, and his principle of attraction, made familiar to the ladies in several entertainments*. Trans. Elizabeth Carter. 2nd ed., 2 vols. London: G. Hawkins.

Arnauld, Antoine, and Pierre Nicole. 1996. *Logic or the art of thinking: Containing, besides common rules, several new observations appropriate for forming judgment*. Trans. Jill Vance Buroker. 5th ed. Cambridge: Cambridge University Press.

Beaney, Michael. 2022. Analysis. In *The Stanford encyclopedia of philosophy*. https://plato.stanford.edu/archives/sum2022/entries/analysis. Accessed 23 March 2023.

Bensaude-Vincent, Bernadette. 2010. Lavoisier lecteur de Condillac. *Dix-Huitième Siècle* 42: 473–489.

Beretta, Marco. 1993. *The enlightenment of matter: The definition of chemistry from Agricola to Lavoisier.* PhD Dissertation. Uppsala University, Uppsala Studies in History of Science, 15. Canton: Science History.

Berthollet, Claude-Louis. 1791. *Éléments de l'art de la teinture*. 2 vols. Paris: Firmin Didot.

Biot, Jean-Baptiste. 1816. *Traité de physique expérimentale et mathématique*. 4 vols. Paris: Deterville.

Brewster, David. 1831. *Treatise on optics*. The cabinet cyclopaedia. London: Longman, Rees, Orme, Brown & Green, and John Taylor.

de Condillac, Étienne Bonnot. 1746. *Essai sur l'origine des connaissances humaines*. 2 vols. Amsterdam: Pierre Mortier.

---

[75] Beretta 1993, 201, refers to Lavoisier's "epistemological generalization" of the notion of chemical analysis; see also Sect. 14.2 of Bensaude-Vincent's paper in this volume.

[76] When the date of publication is given as two dates separated by a backslash, these represent the original date of publication and that of a reprint.

———. 1780. *La logique, ou les premiers développemens de l'art de penser; ouvrage élémentaire, que le conseil préposé aux Écoles Palatines avoit demandé, & qu'il a honoré de son approbation*. Paris: L'Esprit, Debure l'ainé.
———. 1980. *La logique/Logic*. Trans. W. R. Albury. Janus Series, 6. New York: Abaris Book.
de La Caille, Nicolas Louis. 1764. *Leçons élémentaires d'optique*. new ed. Paris, H. L. Guerin & L. F. Delatours.
de Morveau, Louis Bernard Guyton. 1788. *Method of chymical nomenclature, proposed by Messrs. De Morveau, Lavoisier, Bertholet, and De Fourcroy.* Trans. James St. John. London: G. Kearsley.
Dear, Peter. 1998. Method and the study of nature. In *The Cambridge History of Seventeenth-Century Philosophy*, ed. Daniel Garber and Michael Ayers. 2 vols. vol. 1, 147–177. Cambridge: Cambridge University Press.
Desagulier, John Theophilus. 1719. *Lectures of experimental philosophy: Wherein the principles of mechanicks, hydrostaticks, and opticks, are demonstrated and explained at large, by a great number of curious experiments*. 2nd ed. London: W. Mears, B. Creake, and J. Sackfield.
Desaguliers, John Theophilus. 1744. *A course of experimental philosophy*. Vol. 2. London: W. Innys.
Descartes, René. 1985–1991. *The philosophical writings of Descartes*. Trans. John Cottingham, Robert Stoothoff, and Dugald Murdoch. 3 vols. Cambridge: Cambridge University Press.
Diderot, Denis, and Jean Le Rond d'Alembert. 1751–1772. *Encyclopédie ou Dictionnaire raisonné des sciences, des arts et des métiers par une Société de Gens de lettres*. 17 plus 11 vols. Paris: David l'aîné, Briasson, Le Breton, Durand.
Ducheyne, Steffen. 2005. Newton's training in the Aristotelian textbook tradition: From effects to causes and back. *History of Science* 43: 217–237.
———. 2014. 's Gravesande's appropriation of Newton's natural philosophy, Part I: Epistemological and theological issues. Part II: Methodological issues. *Centaurus* 56 (31–55): 97–120.
———. 2017. 's Gravesande's and van Musschenbroek's appropriation of Newton's methodological ideas. In *Reading Newton in Early Modern Europe*, ed. Elizabethanne Boran and Mordechai Feingold, 192–243. Scientific and Learned Cultures and Their Institutions, 19. Leiden/Boston: Brill.
Ducheyne, Steffen, and Frederik Dhondt. 2021. Isaac Newton explicating his natural philosophical method: A study of the development of the methodological statements in the queries to the *Opticks*. *Revue belge de Philologie et d'Histoire/Belgisch Tijdschrift voor Filologie en Geschiedenis* 99: 343–390.
Dutour, Etienne François. 1773. Considérations optiques. IVe mémoire sur la décomposition de la lumiere dans le phénomene des anneaux colorés, produit avec un miroir concave. *Observations et Mémoires sur la Physique, sur l'Histoire Naturelle et sur les Arts et Métiers* 2: 349–373.
Enfield, William. 1785. *Institutes of natural philosophy, theoretical and experimental*. London: J. Johnson.
Fourcroy, Antoine Francois. 1792. *Philosophie chimique, ou vérités fondamentales de la chimie moderne, disposées dans un nouvel ordre*. Paris: Cl. Simon.
Gilbert, Neal W. 1960. *Renaissance concepts of method*. New York: Columbia University Press.
's Gravesande, Willem Jacob. 1747. *Mathematical elements of natural philosophy, confirm'd by experiments: Or, an introduction to Sir Isaac Newton's philosophy.* Trans. John Theophilus Desaguliers. 6th ed. 2 vols. London: W. Innys, T. Longman and T. Shewell, C. Hitch, and M. Senex.
Guerlac, Henry. 1973. Newton and the method of analysis. In *Dictionary of the history of ideas*, ed. Philip Wiener, 4 vols, 3, 378–391. New York: Charles Scribner's Sons.
Guicciardini, Niccolò. 2009. *Isaac Newton on mathematical certainty and method*. Cambridge, MA: MIT Press.
Harris, Joseph. 1775. *A treatise of optics: Containing the elements of the science; in two books*. London: Sold by B. White.
Harrison, John. 1978. *The library of Isaac Newton*. Cambridge: Cambridge University Press.

Haüy, René Just. 1806. *Traité élémentaire de physique,* 2nd ed. 2 vols. Paris: Courcier.
Helsham, Richard. 1739. *A course of lectures in natural philosophy.* Published by Bryan Robinson. London: John Nourse.
Herschel, John. 1830. Light. In *Encyclopædia metropolitana*, ed. Edward Smedley. Second Division, Mixed Sciences, vol. 2, 341–586. London: Baldwin and Cradock.
Hintikka, Jaakko, and Unto Remes. 1974. *The method of analysis: Its geometrical origin and its general significance.* Boston Studies in the Philosophy of Science, 25. Dordrecht: D. Reidel Publishing Co.
Howell, Wilbur Samuel. 1956. *Logic and rhetoric in England, 1500–1700*. New York: Russell & Russell.
Ihmig, Karl-Norbert. 2004. Die Bedeutung der Methoden der Analyse und Synthese für Newtons Programm der Mathematisierung der Natur. *History of Philosophy & Logical Analysis* 7: 91–119.
———. 2005. Newton's program of mathematizing nature. In *Activity and sign: grounding mathematics education*, ed. Michael H.G. Hoffmann, Johannes Lenhard, and Falk Seeger, 241–261. New York: Springer.
Lavoisier, Antoine Laurent. 1790/1965. *Elements of chemistry, in a new systematic order, containing all the modern discoveries*. Trans. Robert Kerr. Edinburgh: William Creech; facsimile reprint New York: Dover Publications.
Le Grand, Antoine. 1672. *Institutio philosophiae secundum principia domini Renati Descartes nova methodo adornata & explicata, in usum juventutis academicæ*. London: J. Martyn.
Levere, Trevor H. 1990. Language, instruments, and the chemical revolution. In *Nature, experiment and the sciences: Essays on Galileo and the history of science,* ed. Trevor H. Levere and William R. Shea, 207–233. Boston Studies in the Philosophy of Science, 120. Dordrecht/Boston: Kluwer Academic.
Levitin, Dmitri. 2016. Newton and scholastic philosophy. *British Journal for the History of Science* 49: 53–77.
———. 2021. Newton on the rules of philosophizing and hypotheses: New evidence, new conclusions. *Isis* 112: 242–265.
———. 2022. *The kingdom of darkness: Bayle, Newton, and the emancipation of the European mind from philosophy*. Cambridge: Cambridge University Press.
Lichtenberg, Georg Christoph. 1990. *Aphorism.* Trans. R.J. Hollingdale. Harmondsworth: Penguin.
Maclaurin, Colin. 1748. *An account of Sir Isaac Newton's philosophical discoveries*. London: Printed for the Author's Children.
Martin, Benjamin. 1749. *A panegyrick on the Newtonian Philosophy. Shewing the nature and dignity of the science, and its absolute necessity to the perfection of human nature; the improvement of arts and sciences, the promotion of true religion, the increase of wealth and honour, and the completion of human felicity.* London: W. Owen, J. Leake, and J. Frederic.
———. 1759. *New elements of optics; or, the theory of aberrations, dissipation, and colours of light*. London: Printed for the Author.
McGuire, J.E. 1970. Newton's "principles of philosophy": An intended preface for the 1704 *Opticks* and a related draft fragment. *British Journal for the History of Science* 5: 178–186.
McGuire, J.E., and Martin Tamny. 1983. *Certain philosophical questions: Newton's Trinity Notebook*. Cambridge: Cambridge University Press.
Newman, William. 2010. Newton's early optical theory and its debt to chymistry. In *Lumière et vision dans les sciences et dans les arts*, ed. Michel Hochmann and Danielle Jacquart, 283–307. Geneva: Droz.
———. 2019. *Newton the alchemist: Science, enigma, and the quest for nature's "secret fire"*. Princeton: Princeton University Press.
Newton, Isaac. 1672a. A letter of Mr. Isaac Newton… containing his new theory about light and colors. *Philosophical Transactions* 6 (80): 3075–3087.

———. 1672b. A serie's of quere's propounded by Mr. Isaac Newton, to be determin'd by experiments, positively and directly concluding his new theory of light and colours.... *Philosophical Transactions* 7 (85): 5004–5007.

———. 1704/1966. *Opticks: Or, a treatise of the reflexions, refractions, inflexions and colours of light.* London: Sam Smith and Benj. Walford; facsimile rpt: Brussels: Culture et Civilisation.

———. 1706. *Optice: Sive de reflexionibus, refractionibus, inflexionibus coloribus lucis libri tres.* Trans. Samuel Clarke. London: Sam. Smith & Benj. Walford.

———. 1717. *Opticks: Or, a treatise of the reflections, refractions, inflexions and colours of light.* 2nd ed., with additions. London.

———. 1722/1955. *Traité d'optique sur les reflexions, refractions, inflexions, et les couleurs de la lumiere.* Trans. Pierre Coste. 2nd ed. Paris: Montalant; facsimile reprint Paris: Gauthier-Villars.

———. 1730/1952. *Opticks: Or, a treatise of the reflections, refractions, inflexions and colours of light*, based on the 4th edition London, 1730. London: G. Bell & Sons: 1931; reprint: New York: Dover Publications.

———. 1958. In *Isaac Newton's papers and letters on natural philosophy and related documents*, ed. I. Bernard Cohen. Cambridge, MA: Harvard University Press.

———. 1984–2021. *The optical papers of Isaac Newton: Vol. 1. The optical lectures, 1670–1672. Vol. 2. The 'Opticks' (1704) and related papers ca. 1688–1717*, ed. Alan E. Shapiro. Cambridge: Cambridge University Press.

———. 1999. *The Principia: Mathematical principles of natural philosophy.* Trans. I. Bernard Cohen, Anne Whitman, and Julia Budenz. Berkeley: University of California Press.

Nollet, Jean Antoine. 1758. *Leçons de physique experimentale.* Vol. 5. 2nd ed. Paris: Hippolyte-Louis Guerin & Louis-François Delatour.

Oxford English Dictionary Online. 2000–. Oxford University Press. https://www-oed-com.ezp2.lib.umn.edu/view/Entry/48357?rskey=puK7zG&result=1#eid. Accessed 20 Mar 2023.

Pemberton, Henry. 1728. *A view of Sir Isaac Newton's philosophy.* London: S. Palmer.

Powell, Baden. 1833. *A short elementary treatise on experimental and mathematical optics designed for the use of students in the university.* Oxford: D. A. Talboys.

Priestley, Joseph. 1772. *The history and present state of discoveries relating to vision, light, and colours.* London: J. Johnson.

Rowning, John. 1753. *A compendious system of natural philosophy: With notes containing the mathematical demonstrations, and some occasional remarks. In four parts.* 2 vols. London: Sam Harding.

Schuurman, Paul. 2004. *Ideas, mental faculties and method: The logic of ideas of Descartes and Locke and its reception in the Dutch Republic, 1630–1750.* Brill's Studies in Intellectual History, 125. Leiden/Boston: Brill.

Shapiro, Alan E. 2004. Newton's 'experimental philosophy'. *Early Science and Medicine* 9: 185–217.

Smith, Robert. 1738. *A compleat system of opticks in four books, viz. a popular, a mathematical, a mechanical, and a philosophical treatise. To which are added remarks upon the whole.* 2 vols. Cambridge: Printed for the Author.

Stewart, Dugald. 1814. *Elements of the philosophy of the human mind.* Vol. 2. Edinburgh: George Ramsay.

Talaska, Richard A. 1988. Analytic and synthetic method according to Hobbes. *Journal of the History of Philosophy* 26: 207–237.

van Besouw, Jip. 2017. *Out of Newton's shadow: An examination of Willem Jacob's Gravesande's scientific methodology.* PhD dissertation. Vrije Universiteit Brussel.

van Musschenbroek, Petrus. 1739. *Essai de physique.* Trans. Pierre Massuet. 2 vols. Leyden: S. Luchtmans.

———. 1762. *Introductio ad philosophiam naturalem.* 2 vols. Leyden: Sam. et Joh. Luchtmans.

———. 1769. *Cours de physique experimentale et mathematique.* Trans. Joseph-Aignan Sigaud de la Fond. 3 vols. Paris: P. Fr. Didot le jeune.

Westfall, Richard S. 1980. *Never at rest: A biography of Isaac Newton*. Cambridge: Cambridge University Press.
Wood, James. 1799. *The elements of optics: Designed for the use of students in the university*. Cambridge: J. Burges, Printer to the University.
Young, Thomas. 1802. An account of some cases of the production of colours, not hitherto described. *Philosophical Transactions* 92: 387–397.
———. 1855. *Miscellaneous works of the late Thomas Young,* ed. George Peacock and John Leitch. 3 vols. London: J. Murray.

Alan E. Shapiro is Professor Emeritus of History of Science and Technology, University of Minnesota. He is the editor of *The Optical Papers of Isaac Newton.*

Chapter 7
Descartes, Leibniz, and Newton on Analysis and Synthesis

Niccolò Guicciardini

Abstract Early modern European mathematicians understood the terms "analysis" and "synthesis" according to the definitions provided by Pappus in the *Collectiones mathematicae*, which had been available in Latin translation since 1588. This chapter surveys the meanings that the two Pappusian methods acquired in the works of Descartes, Leibniz, and Newton to appreciate the different approaches that these authors adopted to the new symbolical analytical methods of algebra and calculus and to the synthetic tracing of curves. Tracing curves was important not only for the construction of geometrical solutions but also for several practical applications.

Keywords Pappus · Descartes · Leibniz · Newton · Mathematics · Algebra · Calculus

7.1 The Early Reception of Book VII of Pappus's Collectiones Mathematicae

As is well known, in the last, twenty-third *quæstio* of the Latin *Optice* (1706), Newton emphasizes a similarity between the methods to be followed in mathematics and those to be followed in physics (*physica*, translated as *natural philosophy* in the English *Opticks* of 1718).[1] It is not my purpose here to discuss this famous, often studied, and difficult-to-interpret passage to which Alan Shapiro (this volume)

[1] "Quemadmodum in Mathematica, ita etiam in Physica, investigatio rerum difficilium ea Methodo, quæ vocatur *Analytica*, semper antecedere debet eam quæ appelatur *Synthetica*" (Newton 1706, 347). This *quæstio* became Query 31 in the second English edition of the *Opticks* (1718), where we read: "As in Mathematicks, so in Natural Philosophy, the Investigation of difficult Things by the Method of Analysis, ought ever to precede the Method of Composition." (Newton 1718, 380).

N. Guicciardini (✉)
Dipartimento di Filosofia "Piero Martinetti", Università degli Studi di Milano, Milano, Italy
e-mail: niccolo.guicciardini@unimi.it

W. R. Newman, J. Schickore (eds.), *Traditions of Analysis and Synthesis*, Archimedes 73, https://doi.org/10.1007/978-3-031-76398-4_7

devotes a fine essay. In fact, I shall have little to say about it here. Nonetheless, opening my chapter with reference to Newton's famous words allows me to address the following question: what mathematical methods was he referring to with the contemporary reader in mind, who apparently needed few explanations of the meanings of the terms "methodus analytica" and "methodus synthetica"? As we shall see, the answer to this question is far from straightforward. References to the distinction and complementariness between the methods of analysis and synthesis are pervasive in early modern mathematics. As with all tropes, the two terms were not explicitly defined but rather were proposed in different contexts that qualified their meanings in a variety of (sometimes contrasting) ways.

The starting point for the historian of early modern mathematics is the 1588 publication of Federico Commandino's Latin translation of Pappus's *Collectiones mathenaticae*.[2] The seven surviving books of Pappus's Συναγωγή, composed in Alexandria in the fourth century AD, were already in circulation among European humanists and mathematicians, but it was the printing press that saw the work attain its more widespread circulation and fame.[3] Most notably, the incipit of Book VII, in which Pappus intriguingly described the ancient "Domain of Analysis," posed a challenge to the "geometers" active in the Latinate world. In his famous address to Hermodorus, Pappus alluded to a "resource" that the ancient Greeks had possessed that supposedly allowed them to "solve problems." Pappus claimed that this promising resource had been expounded on in a series of works that were regrettably lost for early modern mathematicians, particularly the three books of Euclid's *Porisms*. The seventh book of the *Collectiones* consisted in an incomplete presentation of these lost works. Pappus assumed that his readers had access to them, his aim being to introduce and comment on these texts, filling any gaps. For early modern mathematicians, it was an arduous and challenging task to "divine"—as they used to say—the lost ancient works on the "Domain of Analysis."

Pappus made a distinction between "analysis" and "synthesis" (Pappus 1986, 82–5). Analysis was often conceived of as a method of discovery or problem solving that, working backward step-by-step from what is sought as though it had already been achieved, eventually arrives at what is known. Synthesis proceeds in the opposite direction: it starts from what is known and, working through the consequences, arrives at what is sought. On the basis of Pappus's authority, it was often stated that synthesis reverses the steps of analysis[4] and that it was synthesis that provided rigorous proofs. This gave rise to the widespread belief that the ancients had kept the method of analysis hidden and had published only the rigorous synthetical method either because they considered the former to be not wholly demonstrative or

[2] Pappus 1588: this work was published posthumously thanks to the editorial work carried out by Guidobaldo Del Monte. Other editions appeared in Venice, 1589, Pesaro, 1602, and Bologna, 1660. The last was revised by Carlo Manolessi: Pappus 1660. For a critical edition of Book VII, see Pappus 1986.

[3] See Pappus 1986, 62–3 and Rose 1976, 222–79.

[4] On ancient analysis and synthesis, see Fabio Acerbi's commentary in Euclid 2007, 439–523.

because—in imitation of the practice of Pythagoras' sect—they wished to conceal the method of discovery. Such ideas were shared by many people, including François Viète, perhaps the most creative mathematician active in the late sixteenth century. It should be noted that the Greek terms analysis and synthesis were interchangeable with the Latin *resolutio* and *compositio* or *constructio*. These mathematical terms interacted in a complex way with the technical vocabulary pertaining to the philosophical, logical, chemical, and medical traditions.[5]

Pappus made another distinction that was of momentous importance for early modern mathematicians: that between problems and theorems. A problem calls for a construction achieved via permitted means. It starts from certain elements considered to have been already constructed either by assumed axioms and postulates or by previous constructions. A problem ends with "what was to be done" (e.g., *quod facere/fecisse oportebat* or *quod erat faciendum*). A theorem, by contrast, requires a deductive proof, a sequence of propositions, each following from the previous one by permitted inference rules. The starting point in the deductive chain may be either axioms and postulates or previously proved theorems. A theorem ends with "what was to be demonstrated" (e.g., *quod ostendere/demonstrare oportebat* or *quod erat demonstrandum*). According to Pappus, therefore, analysis has two types: "problematic" and "theorematic," the former referring to problems, the latter to theorems.[6] In problematic analysis, one starts from a sought construction as given and deduces from it constructions that are either already found or given by postulates. In theorematic analysis, one starts with a sought proposition and deduces from it either already-proven theorems or axioms and postulates. However, it is clear that early modern mathematicians were mainly concerned with the analysis of geometric problems, and their much greater emphasis on problems than on theorems is itself an interesting feature of their mathematical agenda.

In short, early modern European mathematicians had to engage with a text that alluded to a mathematical procedure that promised the resolution of problems but eluded a clear definition. Many began to identify Pappusian analysis with algebra, which underwent significant developments in the second half of the sixteenth century. François Viète was explicit in defining his new "art of discovery" as an analytic art, as the title of his masterpiece, *In artem analyticem isagoge* (1591), reveals. The "discovery" of Pappus's *Collectiones* was, of course, part of a broad humanist movement involving the editing and printing of classical mathematical works by scholars such as Apollonius, Archimedes, and Vitruvius.[7]

---

[5] On the reception of the methods of analysis and synthesis in the modern period, see Otte and Panza 1997.

[6] We follow Alexander Jones's translation from the Greek, which renders θεωρητικόν and προβληματικόν with "theorematic" and "problematic" (Pappus 1986, 82–3). In Commandino's Latin, these are *contemplativum* and *problematicum*, respectively (Pappus 1588, 157v).

[7] This research field was pioneered in Rose 1976. A recent monograph focused on sixteenth-century Paris is Oosterhoff 2018. On Italian mathematical humanistic culture, see Marr 2011.

The attitude of European mathematicians in the period under consideration in this chapter ranged from philological interest in the restoration of the original texts (in Latin, Greek, and Arabic) to a more theoretical active engagement in "filling the gaps" or even "going beyond" ancient mathematical achievements, often drawing inspiration from Pappus's intriguing Book VII. This humanistic movement interacted in a complex way with the agendas of engineers, architects, painters, opticians, musicians, map-makers, and those generally engaged in the application of mathematics for practical purposes: a plethora of early modern mathematicians who, in the literature, are grouped under the label "mathematical practitioners." Given that this is not a period term, however, I shall refrain from using it here. I shall also avoid polarizing my narrative into two categories: the humanists on one side, intent on editing classical works, and the technicians on the other, dirtying their hands with instruments. Indeed, the Renaissance editors of Apollonius', Vitruvius', and Archimedes' works frequently emphasized that the rediscovery of the ancient mathematical treasures had an import for applications. The likes of Francesco Maurolico and Federico Commandino typically emphasized that the knowledge of conics was essential in the making of sundials.[8]

Book VIII of the *Collectiones* played a major role in this respect, since in it Pappus provided valuable information about the mechanical contrivances of the ancients—most notably, Heron and Archimedes. Bernardino Baldi, a disciple of Commandino, was among the first to commend Heron's achievements, and he did so by underlining the importance of the method of resolution and composition. In the introductory "Discorso di chi traduce" of his Italian translation of Heron's *Automata* (1589)—a work that exhibited the power of the subordinated mathematical sciences in producing theatrical wonders, whereby short, but complex, mythological plays were performed mechanically without human intervention—he praised Heron's method for its beauty ("il bell'ordine e metodo"). According to Baldi, the author of the *Automata* allows the reader to understand the functioning of the machines of his own invention through his adoption of a "resolutive method" ("il suo metodo è risolutivo"), since he begins by indicating the aim he intends to achieve ("egli ci dà quanto intende fare, cioè il fine") and then proceeds backward with "order" until he encounters the "principles" of mechanics. Heron then reverses the steps in the composition until he guides his reasoning from the principles to the intended aim ("quei principi che adoperati con ordine contrario da chi desidera di comporre guidano al fine intento") (Heron 1589, 13r). It is interesting to note that Baldi's praise of the pedagogical merits of an exposition structured according to the double methods of resolution and composition is not found in Heron's text, but

---

[8] See, e.g., Maurolico, in the Preface to "De lineis horariis" states that knowledge of conics was preliminary for those who wrote about gnomonics ("qui de gnomonica ratione conscripserunt"). Marurolico 1575, 162 and 263. Similar statements may be found in Commandino's commentary to his edition of Ptolemy's, *De analemmate liber* (1562, 58–59). I thank Elio Nenci for his suggestions.

rather plays a rhetorical role in his "Discorso."[9] As we shall see in the sections that follow, this is a characteristic of most of the early modern mathematical texts that we shall comment upon: the methods of analysis and synthesis were invoked in statements concerning the nature of mathematics aimed at emphasizing both continuities and discontinuities with the Greek tradition.

The reception of Pappusian analysis and synthesis in early modern Europe is thus to be viewed as situated at the intersection of different activities aimed both at the restoration of texts and practical purposes. Early modern mathematicians were thus operating at the crossroads between the mutilated heritage of the classical tradition, which they strove to recover, and the yet unfulfilled promises of the new science, which they began to apply with practical goals in mind.[10] The historian often perceives a sense of anxiety in the early modern mathematicians' understanding of what the purpose of mathematics should be. The recovery of ancient texts, such as those attributed to Archimedes and Heron, seemed to indicate that the results and methods employed by the ancient Greeks may be of interest not only for the sake of mathematical generality and beauty but also for practical purposes (say, for the functioning of pulleys and levers). However, not all of the mathematical techniques invented or inspired by the ancients could be implemented in the workshop or the arsenal. Meanwhile, the practitioners, such as the numerical table-makers and the mariners who sought to chart the *curva nautarum*, the sea route that intersects the meridians at a constant angle, often proposed methods—such as the logarithms—that clearly lay beyond the purview of the ancient worthies. In this chapter, I shall offer a study of how this intersection and conflict of interests shaped the approach to analysis and synthesis adopted by three giants of seventeenth-century mathematics—René Descartes, Gottfried Wilhelm Leibniz, and Isaac Newton—for whom Pappus' compilation meant a great deal.

7.2 Descartes on Analysis and Synthesis

7.2.1 Ancient and New Analysis and Synthesis

Descartes conceived of his mathematical study of plane curves in terms of algebraic equations, as expounded in the celebrated geometrical essay appended to the *Discours de la méthode* (1637), which marked the fulfillment of his project in the *Regulae ad directionem ingenii* (1628ca)—namely, to construct a *mathesis universalis*, an algebraic reasoning concerning "order and measure irrespective of the

[9] I thank Claudia Cristalli for pointing this out. See also Hattab (this volume) for medieval and early modern ideas on the preferred method of exposition for teaching purposes.

[10] A pioneering paper is Bennett 1986. For a recent assessment, see Cormack et al. 2017.

subject-matter."[11] Famously, and somewhat scandalously, in that essay, entitled *La géométrie* (1637), he departed from the widespread admiration for the ancients, claiming that they lacked a method of discovery as powerful as that which he had found,

> for otherwise they would not have put so much labour into writing so many books in which the very sequence of the propositions shows that they did not have a sure method of finding all, but rather gathered together those propositions on which they had happened by accident (Descartes 1637/1954, 17).

Descartes' bold conviction that the new algebraic methods represented a break with, and a decisive improvement of, ancient geometry is somewhat exceptional. The commonly held position was to regard the new algebra, analytic geometry and—later, in the seventeenth century—calculus as developments, or even rediscoveries, of findings belonging, at least *in nuce*, to the Greek tradition. In the *Géométrie*, though, Descartes profiled himself as an innovator. In the *Discours* and in the *Principia philosophiae*, writing as a philosopher, Descartes claimed to be rebuilding metaphysics from scratch. His ambitions were equally grandiose in the *Géométrie.*

Why did Descartes conceive his algebra as a new "method of analysis"? Briefly put, because translating a geometrical problem into a system of equations is possible by assuming what is sought—algebraically represented by *indeterminées*—as given. The algebraists—apparently following Pappus' prescriptions to Hermodorus—deduce conclusions from a system of equations until that which is sought, the *indeterminées*, are expressed in terms of what is known—namely, the coefficients.[12] However, the geometer could not conclude with this analytic procedure, according to Descartes. A synthesis, a composition, had to be provided—namely, a geometric construction of the roots of the equation. This may be best explained through two examples that are afforded pride of place in the *Géométrie*: the so-called "Pappus problem" and the trisection of an angle.

7.2.2 The Pappus Problem

This problem calls for the construction of a plane curve that satisfies certain conditions (see Fig. 7.1). When translated into algebra, as Descartes found, it yields a second-degree algebraic equation in two unknowns. This is the end result of the analytical part of Descartes' problem-solving procedure applied to this problem. Descartes was able to show—in what was a considerable result for his time—that

[11] See Rule IV of the *Regulae* in Descartes 1984–1991, I, 15–20, esp. 19 and, for a general survey, Rabouin 2009. See also Hattab (this volume).

[12] See Descartes 1637/1954, 6–9.

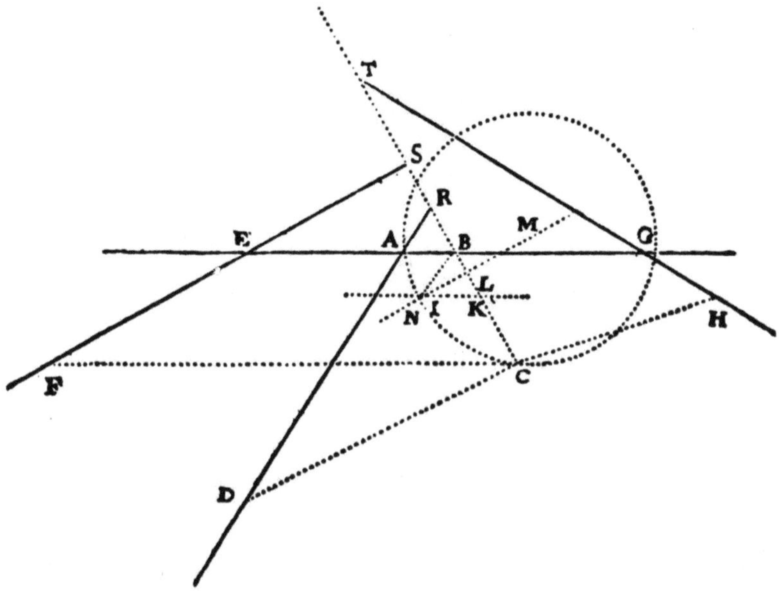

Fig. 7.1 Diagram of the Pappus problem. The Pappus problem of four lines was typically worded as follows: having four lines given in position (indicated by solid lines), it is required to find the locus of points *C* from which drawing four lines (indicated by dotted lines) to the four lines given in position and making given angles with each one of the given lines the following condition holds: the rectangle of two of the lines so drawn shall bear a given ratio to the rectangle of the other two. In this case, the locus is a circle. (Source: Descartes 1637/1954, 61)

because the equation is a second-degree one, the locus sought is a conic section.[13] In the synthetic part, the conic was traced (in Fig. 7.1 above, it is a circle).

The resolution and composition of the Pappus problem played an important rhetorical role in the *Géométrie* because Descartes—on the basis of Pappus' account in the seventh book of the *Collectiones*, which he quoted—quite rightly claimed that the ancients could not tackle its generalization to *n* lines. As we shall see, this boastful statement was challenged by Newton, who was able to provide a geometrical solution of the four-line locus "as required by the Ancients," and claimed that his solution was simpler and more elegant than Descartes'. The generalization to *n* lines remained beyond the scope of Newton's geometrical methods, however.

[13] Descartes' solution to the Pappus problem has received considerable attention in the literature: see Bos 2001, 271–83, 313–34.

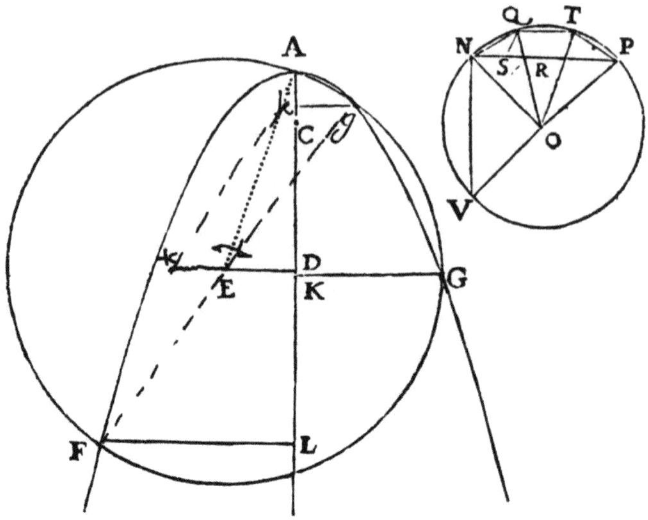

Fig. 7.2 Diagram for the problem of angle trisection in Descartes' *Géométrie*. (Source: Descartes 1637/1954, 206)

7.2.3 Angle Trisection

We can appreciate the Cartesian methods of analysis and synthesis at work in greater detail by considering a simple problem taken from the *Géométrie*: the trisection of an angle. Here, one seeks to divide a given angle *NP* into three equal parts—that is, the length of the chord *NQ* (see Fig. 7.2) must be found given the length of the chord *NP*. To find the solution, Descartes first resolves the problem into a third-degree algebraic equation in one unknown. He names $NO = 1$ (that is, as unit segment he chooses the radius of the circle), $NP = q$ (the chord of the given angle), and $NQ = z$ (the chord to be found). From the similarity between triangles *NQO*, *QRN*, and *RSQ*, he obtains that *NO* is to *NQ* as *NQ* is to *QR*, as *QR* is to RS (*NO*:*NQ*::*NQ*:*QR* ::*QR*:*RS*). From this proportionality follows this equation:

$$z^3 = 3z - q \tag{7.1}$$

Having found Eq. (7.1), Descartes does not seek to calculate its roots. Rather, he provides a geometric construction as a solution of the trisection problem: a construction that resembles similar techniques developed long before in Islamic mathematics. Descartes "constructs the equation" by intersecting the circle and parabola. Given a fixed parabola of *latus rectum* equal to 1, Descartes determines (in function of the coefficients of the equation) the position of the center *E* (in Fig. 7.2) of a circle that intersects the parabola in four points: *A* (the vertex), g, *G*, and *F*. The ordinates *kg*, *KG*, and *FL* have lengths equal to the two positive roots and single negative root of the equation. The segment *kg* is the sought chord *NQ*, which is thus constructed mechanically by tracing the circle and the parabola.

It should be borne in mind that the problems in Euclid's *Elements* are solved by constructions obtained via intersections of circles and straight lines. The circle and straight lines are, of course, generated by the use of compass and straightedge regulated by the opening postulates. Here, Descartes is extending the tools allowed in the *Elements*: indeed, to construct the sought chord, he uses both a compass tracing a circle and an instrument tracing a parabola. It is the use of these two tracing devices that makes the solution of the angle trisection problem possible.

The Cartesian technique may be contrasted with the approach that would be adopted today. Nowadays, we would seek the solution to the trisection problem by calculating the roots of Eq. (7.1). If a geometric representation of the roots were required, we would seek the intersection of the graph of $y = z^3 - 3z + q$ (a cubic curve) with the z-axis (a straight line). However, Descartes is envisaging a solution not in numerical terms but in strictly geometrical ones. Rather than deploying the intersection between a cubic and a straight line, Descartes prescribes the use of two conics (a circle and a parabola), which are simpler to trace than a cubic. Indeed, a mechanism for tracing a cubic is going to be more complex than the parabolograph and the compass deployed in Descartes' example. The point to be emphasized here is that Descartes is seeking solutions to geometrical problems in terms of the constructions allowed by an extension of the Euclidean postulates. His analytical method prescribes how geometrical problems may be translated into algebraic equations, such as Eq. (7.1), but the solutions are not provided in algebraic language: rather, they are geometrical constructions in the spirit of Greek geometry, notwithstanding Descartes' claim to be making a break with the ancient mathematical tradition.

7.2.4 Descartes and the Lens-Grinders

Descartes' interest in providing constructions of problems in terms of the intersection of plane curves led him to devote many pages of the *Géométrie* to curve-tracing devices of his own invention. This topic brought him into contact with technicians who deployed such mechanical tools not in a world of paper but in the laboratory. The mechanical generation of curves was clearly already part of the classical geometrical canon: mechanical generations of conics, conchoids, *quadratrices* and spirals occurred in geometrical constructions detailed in Greek and Islamic treatises. For the mathematicians active in the early modern period, a curve-tracing device was often not so much a theoretical construct as it was an instrument to be applied in one's workshop. The curve traced by an instrument could serve as the conic surface of a lens, the hyperboloid surface of the fuzee of a clock, or the cycloidal shape of the teeth of a wheel. The tension between theoretical and practical methods to which I alluded above (Sect. 7.1) is often present in this field: it is far from obvious that the curve-tracing devices depicted in the engravings adorning mathematical books could actually be implemented for practical purposes.

In *La géométrie*, Descartes studied a class of curves, the Cartesian ovals, that were of paramount importance for his optical work. He sought to avoid spherical aberration (whereby the parallel rays of incoming light do not converge on the same point after passing through a spherical lens). He proved that a lens shaped by the revolution of a Cartesian oval (the hyperbola being an example) is not subject to such aberration. In *La dioptrique*, which is one of the essays that, together with the *Géométrie* and *Les météores*, was published in appendix to the *Discours de la méthode*, Descartes provided a detailed description of curve-tracing devices and a lens-grinding machinery designed to produce hyperbolic lenses. He discussed these topics extensively with high-ranking men of letters, including Constantijn Huygens, mathematicians such as Florimond de Beaune and Claude Mydorge, and well-known artisans, such as Jean Ferrier.[14]

Catherine Wilson has called for "an examination of the interaction between the history of science and the history of technology that takes into account the problems that arise in connection with the idea that science based on the use of machines and instruments gives a truer, better, or deeper account of the world" (1995, 70). In his pioneering paper devoted to the "mechanics' philosophy," Jim Bennett (1986) has called into question the distinction between natural philosophy and the "mechanical arts." Domenico Bertoloni Meli (2006) has demonstrated that practical machines functioned not only as engineering tools but also as tools of knowledge, since nature itself was portrayed as being ultimately grounded in mechanical elements.

Recent studies by Jean-François Gauvin, D. Graham Burnett, and Anita McConnell, have shed light on the role of machines, most notably curve-tracing devices, in the Cartesian intellectual enterprise. Gauvin notes that "machines, according to Descartes, ought to resemble natural philosophical ideas; their design, consequently, needed to be generated by the method" (2006, 188). Similarly, Burnett notes that Descartes' aim in his project for a lens-grinding machine was to mechanize it in such a way that the skill of the artisan, a variable and uncertain contribution, would become redundant, precisely by making the instrument automatic (Burnett 2005, 19–20). Following Henk Bos's insightful study, historians of mathematics can confirm that also the curve-tracing devices that Descartes proposed in the *Géométrie*—and which he called the "new compasses" extending the Euclidean tools allowed in the postulates of the *Elements*—were intended to function in an exact and controllable way, as the various elements composing them moved automatically in function of "one single motion" of one component (see Fig. 7.3) (Bos 2001, 237).

However, one should be careful not to make too direct a connection between advanced mathematics, such as the methods for drawing ovals outlined in the

[14] John A. Schuster discusses Descartes' correspondence with Ferrier, both in the 1620s when they worked on refraction in Paris with Claude Mydorge, and in the 1630s when Descartes, now in the United Provinces, conceived of a machine to grind lenses. See Schuster's chapter in Cormack et al. 2017, 63–4.

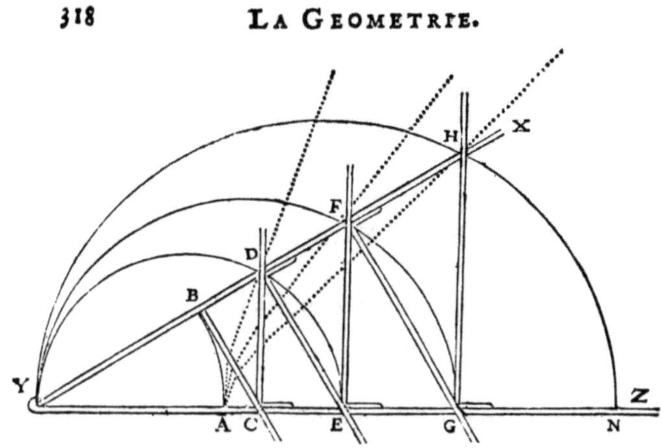

Fig. 7.3 Descartes' mesolabum. The arrangement of the sliding rulers is such that when the angle *XYZ* increases from 0, points *D*, *F*, and *H* describe the dotted curves. (Source: Descartes, *Géométrie* 1637/1954, 46)

Géométrie, and the lens-grinding machine described in the *Dioptrique*. As we have suggested above, the relationship between geometry and algebra on the one hand, and the practical applications of mathematics, on the other, was not easily defined and therefore often had to be renegotiated. In many instances, it was the "craftsman" who contributed most substantially to mathematical innovation. The hierarchical subordination of "applied" mathematics relative to "pure" mathematics occurred much later in the nineteenth century. Furthermore, Vera Keller, with her studies (2010, 2022) devoted to Cornelis Drebbel, an "inventor [who] showcased the fusion of disciplines in an era of new hybrids" (2010, 64), cautions against a too-easy classification between high-ranking natural philosophers on the one hand and "humble" craftsmen on the other, a classification—and again a subordination—that many of the so-called practitioners might have objected to.[15] The production of non-spherical lenses was pursued by seventeenth-century lens-grinders, such as Jean Ferrier, independently of advanced mathematical theorizing (McConnell 2016, 76–105). As Burnett (2005, 41–2) has pointed out, Descartes' young friend and mentor, Isaac Beeckman, had already attempted to produce an "astigmatique" hyperbolic lens in 1622 to solve the problem of color fringes. It appears that the Dutch polymath was inspired by Johannes Kepler's *Ad Vitellionem paralipomena,* in which the German astronomer attempted to eliminate spherical aberration with a hyperbolic lens using a refraction table borrowed from the medieval optical tradition of Witelo and relying on theoretical arguments that frustrated him (Kepler 1604, 106–09). Kepler even promoted the hyperbolic shape based on his observation that a cross section of a

[15] I thank Vera Keller for her comments and her email of July 13, 2023.

cow's eye looked like a hyperbola (Burnett 2005, 15–16).[16] The practice of lens-making, the mathematics of algebraic curves, the mechanical operation of curve-tracing instruments, and the anatomical study of the eye overlapped in ways whose complexity would come as little surprise to early modern scholars.

7.3 Leibniz on Analysis and Synthesis

7.3.1 The Characteristica Universalis *and the New Mathematical Analysis*

Leibniz became a mathematician at a somewhat late stage in his intellectual career. It was between 1672 and 1676, while visiting Paris on a diplomatic mission, that he developed a keen interest in the new methods that had emerged from the work of Descartes, Blaise Pascal, John Wallis, Isaac Barrow, and Newton, among others. In his youth, he had already achieved significant results in the field of logic, having developed the idea of a *characteristica universalis*, a universal symbolic language (the Latin *character* means "symbol" or "mark") that would allow the mechanization of all reasoning via an automatically regulated manipulation of symbols. Such an algebraized language would constitute an "art of discovery" by means of which knowledge could be expanded for the benefit of mankind. In this field, "analysis" meant—as in the Ramist tradition—the breaking down of "composite" concepts into their "formal" constituents, until one reaches the "simple parts, or undefinable terms," so as to reduce reasoning to algebraic equivalences between symbols representing concepts.[17] "Synthesis," by contrast, meant the "combination" of "more composite" concepts either from "less composite ones," the formal constituents mentioned above, or even from the most basic undefinable terms.[18] The symmetrical

[16] I thank Tawrin Baker for sharing the typescript of his essay now published as (2023), which sheds much new light on the influence that medical–anatomical investigations of the eye exerted on Descartes's *Dioptrique*, which investigated the deterioration of vision due to injury, disease, and old age, and how to extend and perfect humanity's visual powers; see also Baker (this volume).

[17] As Osvaldo Ottaviani puts it, "Given Leibniz's account of truth in terms of conceptual inherence, a demonstration is nothing else than a 'chain of definitions', starting from a set of propositions or axioms which may be (provisionally) taken as primitive. Resorting to a well-ordered series of definitions of the main philosophical concepts, then, would constitute the first step toward a rigorously demonstrative approach to metaphysics." Email (December 24, 2022). See Ottaviani 2022.

[18] In the *Dissertatio de arte combinatoria* (Leipzig, 1666) one reads, "Analysis haec est: I. Datus quicunque Terminus resolvatur in partes formales, seu ponatur ejus definitio; partes autem hae iterum in partes, seu terminorum definitionis definitio, usque ad partes simplices, seu terminus indefinibiles" (A VI, 1, 194–195) = "The analysis is this: (1) To resolve any given Term into its formal parts, that is, to lay down a definition of it; and to resolve those parts again into parts, that is, to lay down a definition of the definition of the terms, right down to simple parts, or undefinable

process of analysis and synthesis in Leibniz's logic mirrors the double process of "theorematic" analysis and synthesis described by Pappus (see Sect. 7.1), where analysis begins with a proposition to be proved and deduces from it either already-proven theorems, or the most fundamental propositions, the axioms and postulates. In synthesis, meanwhile, one begins either from already-proven theorems or even starts from axioms and postulates to deduce the sought proposition.[19] Thus, in mathematical "theorematic" analysis and synthesis, axioms and postulates stand for the simple parts, or undefinable terms, of Leibniz's logic. In this context, Leibniz placed a high value on the power of algebraic reasoning to free the mind from the "burden of the imagination," thereby allowing the mechanization of reasoning. This idea, the high value attributed to a blind use of reasoning, also emerges in the mathematical work that led to the discovery of the differential and integral calculus during the mathematician's Parisian sojourn. As is well known, Leibniz began publishing the algorithms of the calculus in 1684 in papers that appeared in a journal he had been instrumental in founding, the *Acta eruditorum.* It should be noted, however, that Leibniz's views on the "blind use of reasoning" are not entirely consistent. In some of his writings, particularly those on "analysis situs," he places a high value on geometrical interpretation and intuition.[20]

The philosophical and mathematical definitions of analysis and synthesis played a key role in Leibniz's long intellectual career in ways that have been the object of innumerable studies.[21] The exploration of such a dauntingly complicated issue lies beyond the purview of this chapter. Indeed, Leibniz dealt with analysis and synthesis not only in his logical and metaphysical writings but also in his works devoted to metaphysics, jurisprudence, and medicine. Suffice it to say here that Leibniz perceived an analogy between the ways in which analysis is practiced in the different fields mentioned above: as we have just seen, the analytical and synthetical methods he was interested in were all symbolical, or algebraic, thus allowing reasoning to unfold in a mechanized way, in a way freed from the vagaries of subjective imagination. On the other hand, Leibniz was keenly aware of the differences between logical/philosophical and mathematical analyses: while logicians distinguish, name, and order concepts (as the Ramists do with their dichotomic *tabulae*),

terms" (DCA, 139). For the synthesis, see, e.g., "omnes Notiones derivatae oriuntur ex combinatione primitivarum, et decompositae ex combinationes compositarum" (GPS VII, 293) = "All derivative concepts, moreover, arise from a combination of primitive ones, and the more composite concepts from the combination of less composite ones" (L, 230).

[19] The symmetry of the two processes of analysis and synthesis, with synthesis reversing the process of analysis, has been much discussed by mathematicians, as it is often unclear how the steps might be reversed (see Otte and Panza 1997). On this problem in another context, see Bensaude-Vincent (this volume).

[20] I am grateful to Alessia Salierno for bringing this to my attention.

[21] Of particular interest are Schneider 1974 and Picon 2021. The classic Couturat 1901 remains a valid reference.

mathematicians. as he wrote in the middle of the 1680s, "order propositions according to their dependence upon each other."[22]

As Leibniz wrote in a letter probably addressed to Jean Chapelain (?) in the first half of the 1670s,[23] it is the scientific nature of the mathematical method that makes it superior to that of the philosophers, which is instead purely verbal (*verbifica*). Therefore, philosophers and jurists—according to Leibniz—should emulate mathematicians, following Euclid's example rather than those by Ramus or by Lull.[24] As a matter of fact, as Mugnai has demonstrated, Leibniz was both a critic and a beneficiary of the Renaissance tradition of those who, like, Ramon Lull, Ramus (Pierre de la Ramée), Johannes Wirth, and later Thomas Hobbes, Seth Ward, Jakob Thomasius, and Johann Heinrich Alsted, had distanced themselves from Aristotelian syllogistics and looked with interest at a symbolization of reasoning in imitation of the mathematicians.[25] From Leibniz's point of view, however, the logic of philosophical (and juridical) reasoning should not be merely a combinatorics restricted to simple notions but should apply to complex symbolical expressions (which can bear a truth value) and order them according to algebraic rules. After setting himself this ambitious task, Leibniz was able, in the years from 1676 to 1690, to develop an algebraized logic that, when rediscovered by Louis Couturat in the early twentieth century, was interpreted as an anticipation of mathematical logic (Couturat 1901, 1903). As far as the method of analysis is concerned, however, mathematicians were divided, according to Leibniz, into those who limited themselves to the Cartesian analysis and those who ventured into the new analysis of the infinite and infinitesimal.

[22] "As a boy I learned logic, and having already developed the habit of digging more deeply into the reasons for what I was taught, I raised the following question with my teachers. Seeing that there are categories for the simple terms by which concepts are ordered, why should there not also be categories for complex terms, by which truths may be ordered? I was then unaware that geometricians do this very thing when they demonstrate and order propositions according to their dependence upon each other." De synthesi et analysi universali seu arte inveniendi et judicandi, Summer 1683 to early 1685 (?). A VI, 4 n. 129 (at p. 538). English transl. in (L, 229).

[23] Leibniz to Jean Chapelain (?), early 1670s, (A II 1, 88). Similar statements occur frequently in Leibniz's manuscripts and correspondence. For example, in Leibniz to Hermann Conring, January 11/12, 1670, (A II 1, 48–49), and Leibniz to Jakob Spener, December 11/12, 1670 (A II 1, 115). On Leibniz's criticisms of the Ramist tradition and his defense of the method of the mathematicians, see Marine Picon's paragraph entitled "Méthode 'divisive' et méthode 'scientifique,'" in Picon 2021, 63–67. See also, Schneider 1974.

[24] Leibniz, in several instances, after stating that his method of reasoning in philosophical matters is modeled on Viète's and Descartes' method of analysis, refers to the combinatorics of Lull and Athanasius Kircher. It appears that what he meant is that his philosophical analysis was akin to, but also transcended, the combinatorics of the Lullian and Ramist traditions. For example, see his letter to the Duke of Hanover, Johann Friederich, dating October 1671 (A II I, 261).

[25] A highly informative overview of Leibniz's thought on the *characteristica* is provided in the "Introduction" to *Leibniz's Dissertation on Combinatorial Art* by Massimo Mugnai (DCA, 1–56).

7.3.2 Transcendental Curves

One of the motivations for the invention of the new calculus was the attempt to overcome the limitations of the Cartesian method expounded in the *Géométrie*. In that celebrated essay Descartes had excluded "mechanical" curves—which Leibniz termed "transcendental"—on the basis that they lacked "exactness" (Breger 1986; Knobloch 2006). Put simply, Descartes accepted only curves that are loci of polynomial equations, such as

$$x^3 - 3axy + y^3 = 0 \tag{7.2}$$

However, the new science required mathematicians to go beyond—indeed, to "transcend"—the limitations of the Cartesian canon. In Christiaan Huygens's *Horologium oscillatorium* (Huygens 1673), a mechanical curve, the cycloid, proved its importance: only a pendulum that is forced to swing along a cycloidal arc is exactly isochronous. This was just one example—the most celebrated at the time—of how the "inexact" curves that Descartes excluded were necessary for the new mathematized natural philosophy, a discipline that required the calculation of curvilinear areas and volumes and the rectification of curves.

The new Leibnizian calculus made it possible to achieve these results, but the price was equations that were not always finite polynomials in which all the symbols represented finite magnitudes, as in the example (7.2) just provided. "Infinitesimal" magnitudes might occur in the equations of the Leibnizian calculus. Indeed, the new calculus implied the use of infinitesimal magnitudes and infinite series and products. This new mathematical theory came to be known, after L'Hospital's textbook (L'Hospital 1696), as the *analysis of the infinitely small.*

The equations made possible by Leibniz's calculus were "differential equations"—namely, equations in which not only finite but also "differential" quantities occurred, such as

$$\frac{dy}{dx} = \frac{s}{a} \tag{7.3}$$

where x and y are the Cartesian coordinates, a is a constant, and s is the arc-length of plane curves (to be determined by solving the equation). The ratio of the differential quantities dy and dx, the infinitesimal increments of the abscissa and ordinate, gives the slope of the tangent of the sought curves. Finding the curves was therefore often termed "the inverse-tangent problem." In this case, the solution curves are the *catenariae*, having the shape of a free-hanging inextensible chain suspended from two points.

Leibniz's new analysis, particularly differential equations, proved its usefulness in applications across a variety of fields, including ballistics, the calculation of the volumes of barrels, horology, and navigation. However, it aroused the suspicions of many because it implied the introduction into mathematics of the uncertainties related to the concept of infinite and infinitely small magnitudes. Indeed, the

Cartesian analysis based on polynomials came to be known as "common analysis," while the *nova methodus* that Leibniz was proposing was often referred to as "new analysis."

7.3.3 Constructions

Today, the solution to a differential equation, such as (Eq. 7.3), would be symbolic. It would be a formula representing a class of "functions":

$$y = \frac{a}{2}\left(e^{x/a} + e^{-x/a}\right) \tag{7.4}$$

Instead, for Leibniz and the early practitioners of calculus—most notably, Jacob and Johann Bernoulli, Jacob Hermann, or Pierre Varignon—a symbolic solution such as (Eq. 7.4) was not appropriate. They aimed at a geometric construction of the solution curves, in this case of *catenariae*. Such a requirement—a survival of the Pappusian prescription according to which a synthesis, a construction, must follow the analysis—slowly faded away. Indeed, geometric or mechanic constructions of the solution curves of differential equations soon became too complicated to achieve. Thus, in the course of the eighteenth century, the Leibnizian "new analysis" acquired complete autonomy from geometrical representation.[26]

7.3.4 Leibniz and the Engineers

While one might expect that Leibniz, a towering diplomat and metaphysician, would have little interest in the practical applications of mathematics, we should not forget that this *homo universalis* spent many years of his life in the Harz Mountains designing and perfecting machines such as wind-mills, in the hope that they might improve the extraction of silver. It would, of course, be absurd to define Leibniz as an "engineer," and indeed he did not define himself as such. However, he did interact with engineers, and though the relationships were not always smooth, he spoke their language and was able to engage with them in a relatively technical dialogue. References to the analytical and synthetical methods surface sporadically in Leibniz's technical writings—for example, those related to the cohesion of matter.[27]

Leibniz aimed to avoid "unforeseeable disturbance factors, such as human failure" in the operation of mills: the purpose of his engineering work was to develop automatic control mechanisms to avoid "excessive damaging strain on the machine"

[26] On this topic, see Bos 2001, 420–8 and Blåsjö 2017, 98–9; 134–40.

[27] See the contraposition between a *methodus inveniendi analytica* and a *methodus synthetica sive combinatoria* in *De firmitate corporum,* January–March 1683 (A VII, 3, 202).

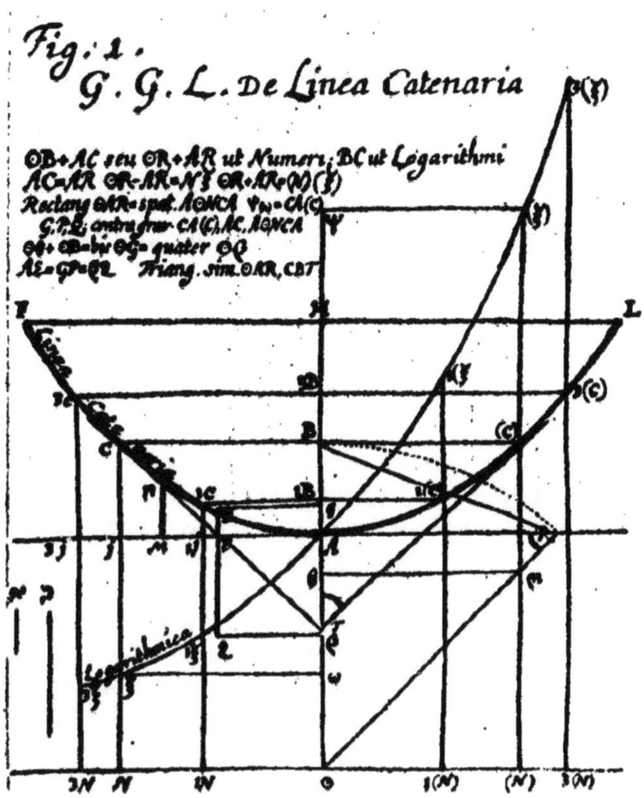

Fig. 7.4 The shape of the *catenaria* and its associated logarithmic curve. (Source: Leibniz 1691, 278, Table VII, Fig. 7.1)

(Hecht and Gottschalk 2018, 12). Automation also guided Leibniz's calculating machine. Famously, he developed a stepped reckoner that embodied his mechanization of reasoning project in the most visible way.

Even while busily competing in the solution of problems associated with analytic mechanics, such as that of finding the shape of a free-hanging (inextensible) chain suspended from two points—the so-called *catenaria* considered above (see Eq. (7.3))—Leibniz conceived the bold idea that the practice of keeping a chain in one's pocket and allowing it to hang when required could be used to calculate logarithms (see Fig. 7.4) (Blåsjö 2017, 139–140). Rather than using logarithms (see formula (7.4)), to calculate the shape of this transcendental curve, the shape of the chain could be used as a calculation aid. Furthermore, it was known that an inverted *catenaria* is the shape of a self-supporting bridge. Thus, this curve was indeed important for engineering purposes.

Leibniz conceived the differential and integral calculus as part of his youthful project to develop a *characteristica*, a dream that he shared with many contemporaries intent on creating a perfect, universal language. Its mechanization—that is, its functioning according to rules codified in an algorithm—was envisaged by Leibniz as a tool for the fulfillment of a grandiose project to reform the Republic of Letters. As Maria Rosa Antognazza notes, Leibniz's plan to create a logical calculus "was meant to play a pivotal role in [his] efforts toward reconciliation at a time of enormous religious, political, and intellectual upheaval" (Antognazza 2016, 34)· For the "benefit of public happiness" (A VI, 4, 525), Leibniz aimed to developing a *scientia generalis*, a collaborative international and interconfessional enterprise carried out by academies, such as the Berlin academy that he helped to found (Antognazza 2009, 1–14).

7.3.5 Ancient and New Analysis and Synthesis

How did Leibniz envisage his work on calculus vis-à-vis the ancient Greek tradition? While he never fulfilled his repeated promises to publish a history of the discovery of infinitesimal analysis, a work announced as the *Scientia infiniti*, he left manuscript evidence of his views on the historical development of calculus. First, Leibniz underlined both the continuity and break with the ancient past: modern analysis was superior to that of the ancients because it was formulated using symbolic algorithms. Second, he distinguished two traditions: an Apollonian tradition perfected by the symbolisms of Viète and Descartes and an Archimedean tradition perfected by his own calculus symbolism.

A typical statement, occurring in a manuscript datable to 1698 and likely to be identified as a draft of an introduction to the *Scientia infiniti*, runs as follows:

> Indeed, as we have often advised, Geometry has two parts, wholly different in kind from one another, one treated more by Apollonius, the other more by Archimedes. The former treats the magnitude only of straight lines, whereas of curves it treats only their position, as determined by the magnitude of straight lines; the latter measures the curved quantities themselves or determines those which depend on them. So you can say that the former is more determinative, the latter more dimensional. Those who deal with the first one, Apollonius and company, improve only those [parts], in which there is nothing which could not follow from imagination. Archimedes, however, [...] seems to have conceived in his mind certain infinitely small lines, by the aid of which he discovered many outstanding theorems. [...] Certainly, he hid the art of discovery so well that it seems no one had matched it until our century.[28]

Thus, according to Leibniz, the Apollonian "part" of geometry is "determinative"—that is, it allows the determination of the *loci* (e.g., the conic sections) by consideration of the relationship between straight lines, e.g., the relationship between the abscissa x and the ordinate y of a parabola expressed by the equation:

---

[28] Niedersächsische Landesbibliothek, Hannover, shelfmark LH 35, 7, 10, Bl. 1–4 (on 1v–2r), in (Gerhardt 1875, 595). I am quoting from the forthcoming edition and translation by Richard Arthur and Osvaldo Ottaviani, provisionally entitled *Leibniz on the Metaphysics of the Infinite* and accepted for publication by Oxford University Press.

$$ay = x^2 \tag{7.5}$$

Furthermore, in the Apollonian "part" nothing transcends our imagination, since the lines to which it refers are all finite. Not so the Archimedean "part," since in this case, "infinitely small lines" that cannot be "imagined" are assumed. The Archimedean "part" of geometry is devoted to the determination not so much of the position of loci as of their "dimension," e.g. of the areas and volumes, of curvilinear figures.

After identifying these two traditions, Leibniz proceeds to distinguish two analyses of the Moderns, the analysis of Viète and Descartes, and the "new analysis of the infinite," due to Leibniz himself. These two recent analytical methods have an advantage over the ancient ones: they provide a *filum meditandi*, a thread of reasoning based upon a symbolic algorithm. Thus, Leibniz writes,

> After so many advances in knowledge, however, there was still lacking what seemed most of all to be desired, the very thing, namely, that was missing to make the common Geometry rich with outstanding discoveries, until this was supplied by the works of Viète and Descartes. Of course, in long chains of reasoning and a multiplicity of figures the mind is disturbed and the imagination confounded, unless there is as it were a thread in the labyrinth which governs our paths; [...] This *filum meditandi* produced characters appropriate for thought, whose use in Mathematics we call the Calculus [...] So, already more than twenty years ago I undertook to supply that need by giving specimens from time to time, until I had managed to publish the very foundations of this new Analysis of the Infinite in the Leipzig *Acta eruditorum*.[29]

Leibniz was thus praising the calculus as a *cogitatio cæca* and promoting the "blind use of reasoning" among his disciples. Nobody, according to Leibniz, could follow a long reasoning without freeing the mind from the "effort of the imagination" (Pasini 1993, 205). When we turn to Newton, we find a remarkably different approach to the Greek mathematical tradition, and particularly to the methods of analysis and synthesis.

7.4 Newton on Analysis and Synthesis

7.4.1 Ancient and New Analysis

Newton does not contrast Archimedes to Apollonius, as Leibniz repeatedly does in his surviving writings datable to the 1690s. Both are praised and put on a par with one another, not so much for the power and generality of their methods as for the conciseness and beauty of their geometrical constructions. Furthermore, in the mathematical writings penned by Newton after the early 1680s, the ancient Greeks'

[29] Ibid., Bl. 4r (see Gerhardt 1875, 598–599). On Leibniz's infinitesimal calculus, see Bos 1974 and Arthur and Rabouin 2020.

methods are invoked as an alternative to the symbolic approach to geometry championed by Descartes and Leibniz.[30]

As Tom Whiteside has detailed in his magnificent edition of Newton's mathematical papers, the English mathematician only began studying ancient Greek geometry in the late 1670s or early 1680s. Up to then, he had worked primarily on modern mathematics, "common" and "new" analysis: he had busied himself with algebra, the classification of cubics, series, interpolations, and fluxions (e.g. in the study of tangents, curvatures, and quadratures). The young Newton proudly self-fashioned himself as a mathematician who was contributing to the advancement of the moderns' "analysis," which through the use of infinite series ("infinite equations," as he called them) could solve "almost all problems." In a famous letter for Leibniz addressed to Henry Oldenburg in 1676, Newton wrote (translation by H. W. Turnbull),

> From all this it is to be seen how much the limits of analysis are enlarged by such infinite equations: in fact by their help analysis reaches, I might almost say, to all problems. (NC, II, 39)

The extant manuscripts reveal that during the period just before the composition of the *Principia* (1687) Newton studied Pappus' *Collectiones* in depth (MP, IV, 274–335). His interest shifted from the moderns' new analysis to the "domain of analysis" of the ancients. His aim, shared by many of his contemporaries, was to restore the ancients' lost method of discovery, the *Analysis Veterum*, which Newton conceived of as a geometrical method based on the projective invariant properties of plane curves.

This feature of mature Newtonian analysis should be emphasized. The analysis in which he was now interested was a geometrical rather than symbolical method. It was during the eighteenth century, for reasons too complex for discussion here, that a shift occurred in the meaning of the term "analysis," which came to denote an algebraic method, while "synthesis" indicated a geometric one. For eighteenth-century mathematicians, writing in an analytic/synthetical style meant using an algebraic/geometrical method.

7.4.2 *The Elegance of Ancient Geometry*

Newton often peppered his geometrical research on Pappusian analysis with some rather heated anti-Cartesian statements and more generally with pronouncements addressed against the moderns' algebraic methods, championed by Descartes. Newton expounded on the superiority of the geometry of the ancients over the algebra of the moderns even in his lectures on (modern) algebra, which he deposited, in

[30] See the references to Archimedes and the method of exhaustion—understood as a rigorous foundation of the methods based on indivisibles/infinitesimals—in Cavalieri, Wallis, Isaac Barrow, Ismael Boulliau, Huygens, and James Gregory in Malet 1996, 15–17, 37, 40–1, 50, 52–3, 56, 80.

keeping with the statutes of the Lucasian Chair, in the University Library and which were published in 1707 under the title *Arithmetica universalis*. It is in the latter parts of these lectures that we find a praise of geometry over algebra. In the final section of the lectures (whose editor, William Whiston, titled "Appendix"), Newton praises the geometrical constructions of problems by the "first geometers" above those of the "recentes":

> for anyone who examines the constructions of problems by the straight line and circle devised by the first geometers [a primis Geometris] will readily perceive that geometry was contrived as a means of escaping the tediousness of calculation by the ready drawing of lines.[31]

The "recentes," rather, by introducing arithmetical terms into geometry, have lost "the simplicity in which all elegance of geometry consists" (MP, V, 429). Indeed, in his writings on geometry from the 1680s and 1690s, Newton is particularly concerned with praising the greater elegance and simplicity of the ancients' geometrical constructions.

7.4.3 *Postulates as Mechanical Constructions*

The criticism that Newton advanced against Descartes and the modern mathematicians who follow the Cartesian method, is best explained through two examples that Newton himself often proposed: the problem of angle trisection and the Pappus problem. It should be emphasized that Newton's insistence on these two problems when comparing the ancients' methods with those of the moderns is intended as a criticism of Descartes, since in the *Géométrie*, it is precisely these problems that are given pride of place as proof of the Cartesian method's superiority over its old geometrical counterpart.

Newton compared Descartes' solution to the problem of the trisection of an angle with that attributed to Archimedes in the *Book of Lemmas* to bring the greater simplicity and elegance of the latter into sharper relief. According to Newton, Descartes' method (see Sect. 7.2.3) leads to a construction that has little to do with the purpose a geometer might have in mind when trisecting an angle. Certainly, by intersecting circle and parabola, Descartes constructs the required segment, the sought chord. However, Descartes introduces two auxiliary figures—a circle and a parabola—that are external to the angle that one is asked to trisect. Instead, Archimedes' method allows one to construct the sought angle using a simple geometrical procedure (known as *neusis*) performed on the figure at hand. Furthermore, the Archimedean procedure is such that the synthesis is merely, as Pappus prescribed, the inversion of the analysis.

Let us briefly examine the trisection of the angle proposed in Proposition 8 of the *Book of Lemmas*, at the time attributed to Archimedes, that held such great

[31] Translation from Latin by D.T. Whiteside in (MP, V, 429).

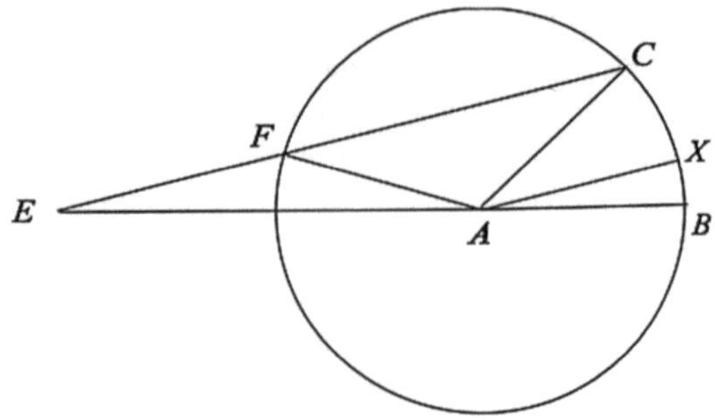

Fig. 7.5 Diagram for the trisection of the angle according to the pseudo-Archimedean *Book of Lemmas*. (Source: author's drawing)

fascination for Newton (see Fig. 7.5). One is required to trisect a given angle *BAC*. In the analysis, it is assumed that the trisecting angle *BAX* is given (one assumes that the problem has been solved). Next, draw the circle with radius AB. Then, draw the parallel *CFE* to *XA* passing through point *C*, meeting the circle in *F*, and prolong the diameter until *BAE* meets *CFE* in *E*. It can be proved that the segment *FE* is equal to the radius *AB* of the circle. Here the analysis ends, because if a *neusis* construction is allowed (technically, is "postulated"), the problem has been reduced to a construction that can be performed. In the synthesis, the steps just considered are reversed. One is required to trisect a given angle *BAC*. First, draw the circle with radius AB and a line passing through *C*. Second, rotate this line until the segment *FE* placed between the circle and the prolongation of the diameter is equal to the radius *AB*. The second step is the *neusis* construction. If *FE* = *AB*, the angle *BEC* trisects *BAC*.

The complication of Descartes' method, Newton claimed, derives from his use of algebra, which introduces something external to geometry and therefore leads to an unnatural construction:

> For almost all problems have a natural way of being solved [...] Whence happens it, I think, that the ancients, whose aim was composition, frequently arrived at simpler conclusions than the moderns, who are more devoted to algebra.[32]

The Archimedean trisection is possible, by the "drawing of a single line," when the Euclidean postulates are extended in such a way that *neusis* constructions are allowed. These constructions consist in fitting a segment of given length, such as *AB*, between two given plane curves, such as the circle and the prolongation of the diameter, in such a way that the segment or its extension passes through a given point *C*. A *neusis* construction might be performed using a ruler—with two marks as the end-points of the given segment—that can be rotated around the point *C*.

Thus, Newton sought to extend the Euclidean postulates by allowing constructions other than the circle and the straight line, as prescribed at the outset of the *Elements*. Before practicing geometry, one must learn how to operate using straightedge and compass. So, Newton often states, geometry is based on "mechanical practice," since the postulates prescribe the manner in which mechanical operations may be legitimately performed using straightedge and compass. For this reason, the

[32] Translation from Latin by D.T. Whiteside in (MP, VII, 251).

postulates premised to geometrical practice belong to mechanics, according to Newton.

Furthermore, there are easy postulates, accessible to a *tiro*—the Euclidean ones—and more advanced postulates, accessible to a *peritus*—such as *neusis*. Indeed, *neusis* constructions are possible once mechanical instruments that can fit a given segment in between two given curves are accepted. Such an instrument, a *mesolabum*, an instrument that can trace a curve known as "conchoid," must be admitted alongside the straightedge and compass to construct the angle that trisects a given angle according to the Archimedean procedure (see Fig. 7.5).

Similarly, Newton solved the Pappus problem of three or four lines by deploying a tracing instrument of his own invention (see Fig. 7.7). Rather than using algebra, as Descartes had done (see Sect. 7.2.2), Newton resorted to a construction of conic sections that he had developed, possibly based on inspiration drawn from Apollonius's *Conics* and Jan de Witt's *Elementa curvarum linearum*, printed as an appendix to the second Latin edition (1659–1661, II, 163–164) of Descartes's *Géométrie*. Newton showed that the problem can be solved if one mechanically traces a conic passing through five given points. In the early 1680s, Newton wrote,

> Descartes in regard to his accomplishment of this problem makes a great show as if he had achieved something so earnestly sought after by the Ancients and for whose sake he considers that Apollonius wrote his books on conics. With all respect for so great a man I should have believed that this topic remained not at all a mystery to the Ancients. For Pappus informs us of a method for drawing an ellipse through five given points and the reasoning is the same in the case of the other conics. And if the Ancients knew how to draw a conic through five given points, does any one not see that they found out the composition of the solid locus? […] To reveal that this topic was no mystery to them, I shall attempt to restore their discovery by following in the steps of Pappus' problem.[33]

Newton's geometrical solution to the Pappus problem, which was hardly of any use for gravitation theory, was given pride of place in Section 5, Book 1, of the *Principia*. In Lemma 19, Newton solved the vexing "problem of four lines" not by a "calculus but by a geometrical composition, such as the ancient required."[34]

After the late 1670s, Newton regularly expressed his admiration for ancient geometry. He often repeated the idea that Archimedes and Apollonius had achieved better (that is more elegant, simpler and more appropriate to geometry) solutions because they had conceived of curves not as defined by equations but as generated by tracing mechanisms (the straightedge for the straight line, the compass for the circle, a system of linked rulers for the conics, the *mesolabum* for the conchoid, etc.), thus founding geometrical constructions upon a mechanical practice that extended the Euclidean postulates. This is also evident in the incipit of the *Principia*: in the "Praefatio ad lectorem," after citing Pappus, Newton claims that geometry is based on mechanical practice, which is to say that it is based upon postulates (Newton 1687, A3r).

---

[33] Translation from Latin by D.T. Whiteside in (MP, IV, 275, 277).

[34] "Atque ita Problematis veterum de quatuor lineis ab Euclide incaepti & ab *Apollonio* continuati non calculus, sed compositio Geometrica, qualem Veteres quaerebant, in hoc Corollario exhibetur" (Newton 1687, 75).

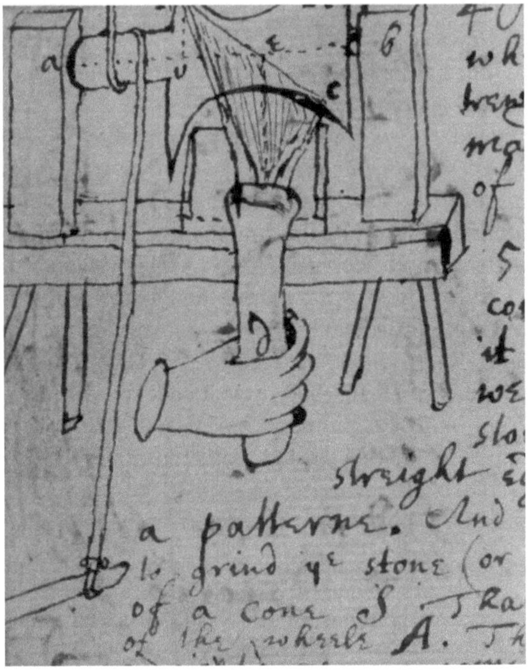

Fig. 7.6 Lens-grinding machine (1665–1666). Source: Cambridge University Library, MS Add. 4000, fol. 26v. (Reproduced by kind permission of the Syndics of Cambridge University Library)

7.4.4 Newton and the Gaugers

The mechanical tracing of curves other than the straight line and circle did not serve merely theoretical purposes—namely, to extend Euclidean geometry to problems that cannot be constructed using straightedge and compass. Newton often conceived of curve-tracing instruments as real tools for use in the laboratory—for example, in the workshops of "glass-grinders" and "spectaclemakers," such as Christopher Cock or that "Mr. Cooper" with whom he collaborated.[35] Justifiably famous is the machine that Newton devised for shaping a wheel according to a hyperbolical profile, so that it could later be used for grinding lenses, most likely inspired by the Cartesian program expounded in the *Dioptrique* (see Fig. 7.6).

Historians of mathematics have perhaps afforded insufficient attention to the very "physical" form that Newton's construction of conics takes in a letter dated 1672 and addressed to the mathematical accountant and publisher John Collins (Fig. 7.7). That construction, which in the *Principia* appears as a depiction of the abstract world of pure geometrical objects, acquires a very real form in Newton's letter and is accompanied by a detailed description of how one might be obtained in a laboratory. The tracing of conics found several applications at the time, as evidenced by the subtitle of the work, the *Geometria organica*, that Frans van Schooten devoted to the topic in (van Schooten 1646). The Dutch editor of Descartes' Latin

[35] See Newton to Hooke, November 28, 1678 (NC, II, 303), and Levitin and Mandelbrote 2023.

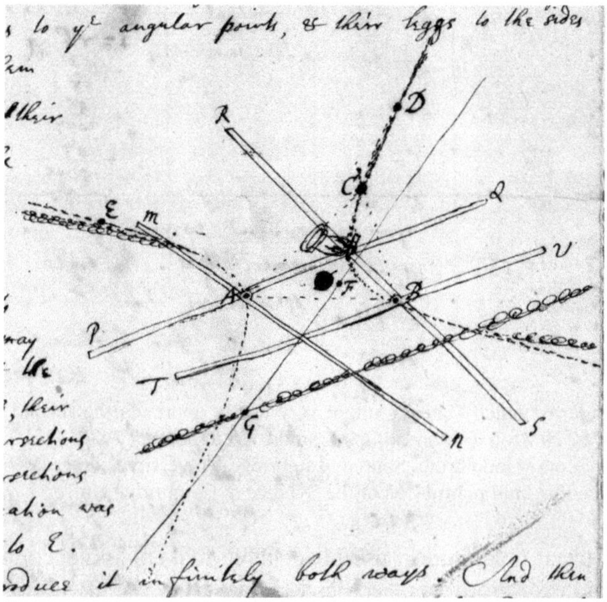

Fig. 7.7 Construction of a conic through five given points. The conic—in this case, a hyperbola—passes through the five given points *A, B, C, D,* and *E*. Newton to John Collins (20 August 1672). Source: Cambridge University Library, MS Add. 3977.10, fol 1v. (Reproduced by kind permission of the Syndics of Cambridge University Library)

Geometria (1649, 1659–1661) made it clear that he was not simply interested in pure geometry, for his work—as he emphasized in the preface—was useful not only to geometers but also to opticians, designers of sundials, and mechanicians (*geometris, opticis, praesertim vero gnomonicis & mechanicis utilis*).

It is particularly interesting to note that the drawing accompanying the letter to Collins is quite realistic and suggests that Newton made use of real instruments to trace curves. Indeed, in a manuscript belonging to the Macclesfield Collection (MS Add.9597/2/3) Newton describes his "organon" in rather practical terms:

> Two rules... are to be manufactured so that their legs... can be inclined to each other, at will, in any given angle.... And at the junctures … there should be a steel pin-point around which the rules may be rotated while the pin is fixed on some given point … as its centre. To be sure, the steel nail by which the legs of a sector are joined might be finely sharpened at one end, and on the other threaded to take a nut more or less tightly (as the need arises) which will clamp the legs of the sector in the given angle. (MP, II, 135)

One should not forget that the young Newton was practicing mathematics by following the tradition of "mixed mathematics." His small pocketbook (MS Add. 4000), from which I have drawn the above illustration of a lens-grinding machine (Fig. 7.6), deals with algebra, methods for calculating tangents, areas and volumes, alongside optics, music, and navigation. Indeed, the first applications of Newton's great mathematical discoveries, such as the binomial theorem and quadrature techniques (integration in Leibnizian terms), were the calculation of logarithms

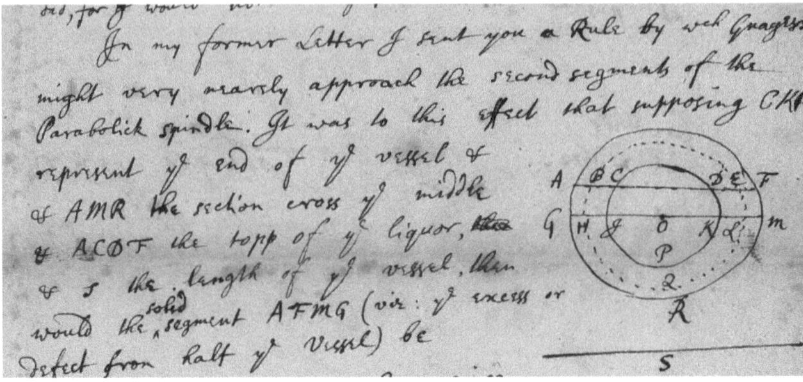
In my former Letter I sent you a Rule by wch Gaugers might very nearly approach the second segments of the Parabolick spindle. It was to this effect that supposing CKP represent ye end of ye vessel & AMR the section cross ye middle & ACOF the topp of ye liquor, & S the length of ye vessel, then would the solid segment AFMG (viz: ye excess or defect from half ye vessel) be

Fig. 7.8 "[A] Rule by which Gaugers might very nearly approach the second segments of the Parabolick spindle." Newton to John Collins, Cambridge, 2 October 1672. This unpublished letter was identified by Scott Mandelbrote. Source: Cambridge University Library, MS Add. 9597/2/12, fol. 1r. (Reproduced by kind permission of the Syndics of Cambridge University Library)

and trigonometrical magnitudes, useful for table-makers and accountants, such as John Smith, and excise officers, such as Michael Dary, with whom Newton corresponded (Beeley 2019).

To consider just a single example, the letter to Collins reproduced as Fig. 7.8 offers evidence of Newton's interest in using the binomial theorem for developing techniques useful to "gaugers"—that is, accountants involved in the measurement of the contents of barrels (for customs or other purposes). The gaugers were in dire need of methods for calculating the volumes of solids of revolution. Newton discussed such methods with Collins and Dary, who had to establish the volumes of "the variously shaped vats, hogsheads, and barrels" for the Excise Office (Beeley 2024; Morel 2024, 194–8; Wess 2024, 209–10). Newton's involvement in such a practically oriented task is an interesting feature of his early correspondence with the London mathematicians.

It is worth noting the different attitude toward mathematical practice that Newton exhibited relative to those of Descartes (Sect. 7.2.4) and Leibniz (Sect. 7.3.4). The two great Continental mathematicians and metaphysicians sought to eliminate the vagaries and uncertainties associated with technicians' dexterity. The Cartesian and Leibnizian machines embodied an ideal of automation that ultimately outstripped the technicians' competence. Such an ideal was in line with the high value that they attributed to the automation of mathematical language through the *mathesis* or the *characteristica universalis*. Newton, by contrast, praised geometrical constructions over the blind manipulation of symbols, not least because geometrical methods required prowess on the mathematician's part, whereas algebra—as he allegedly used to claim—was the method used by the "bunglers of mathematics" (Hiscock 1937, 42). Newton admired the inventive creativeness of the ancient geometers, whom he emulated. Their procedures could not easily be codified. Somewhat similarly, one may surmise, he had a high regard for the competence of the London "gaugers" and table- and spectacle-makers with whom he corresponded, co-operated, and competed. While Descartes ultimately addressed university professors, proposing a new metaphysics and a new mathematics as a substitute for Aristotelian philosophy and syllogistics, and while Leibniz addressed the Republic of Letters, proposing the *characteristica* as a tool for a *scientia generalis* to be adopted in the academies, Newton appears less ambitious. His audience—his intended readers as a mathematician—were the characters of Eva Taylor's prosopography (1954) of the "mathematical practitioners of Tudor and Stuart England," which—not without reason, it seems to me—includes an entry for a practitioner named "Isaac Newton."[36]

7.5 Concluding Remarks

In this chapter, I have dealt with the influence of the Greek methods of analysis and synthesis in early modern mathematics, with particular attention to Descartes, Leibniz, and Newton. Inevitably, I have omitted much. However, I hope that this chapter will help better contextualize the statements on analysis and synthesis made by three influential early modern mathematicians. Their statements should be seen as part of the rhetorical strategies that early modern mathematicians pursued to place their own work within a historical narrative that took ancient Greece as its point of departure. They reflect the variety of ways in which the mathematical methods, so incompletely and somewhat mysteriously described in the *Collectiones*, were received. Pappus' text served different purposes for different authors. Thus, examining the changing meanings of the terms analysis and synthesis can help us to understand the different, sometimes divergent, agendas of early modern mathematicians and the ways in which they fitted these agendas within different narratives about the development of mathematics.

A vast corpus of literature has been devoted to analysis and synthesis in the history of mathematics. This literature, often of exceptionally high quality (Otte and Panza 1997), aims to interpret in a logically cogent way, historically and philosophically, how the actors understood and used the two methods. In this chapter, I have attempted to extend the investigation to date by examining texts that might appear peripheral or even unrelated to analysis and synthesis. Thus, I have considered the notion of construction in its various declinations. From this broader perspective, the mathematical notions of analysis and synthesis lose some of their logical precision and acquire a rhetorical dimension. While it may seem grandiose to quote Paolo Rossi in this context, I agree with his suggestion that the history of science

> is an activity to some extent distinct from a "philosophical" or "epistemological" history of science, [since] it has more to do with the ambiguous and elusive realm of ideas, metaphors,

[36] For a collection of recent studies on the topic, see Hantsche 1996.

> worldviews, preferences, and choices than with the logical structure of scientific theories. (Rossi 1999, 36)[37]

Mathematics might be portrayed as a discipline characterized by its precision, yet when viewed in its historical development, its concepts and methods are better described as "ambiguous" and "elusive," or perhaps "fluid," as the mathematical historian Henk Bos once wrote (2004, 65). In the mathematical texts produced in the period considered in this chapter, there is often a sense of tension, a complex dialectic, between tradition and innovation, between the humanistic recovery of past texts and the practical goals of engineers, between the library and the court on the one hand and the arsenal, the battlefield, the construction yard on the other. These two levels were not separate but were related in a complex way: it is telling that from the outset, in Urbino, the mathematicians active in sixteenth- and seventeenth-century Europe often did both things at the same time. Cartographers drew new maps while reading Ptolemy, architects working on the Duomo of Milan were also editing Vitruvius. Commandino, Del Monte, Baldi, and Muzio Oddi in Urbino were humanists at court translating from Greek and in the field advising engineers on fortification and artillery. Early modern mathematicians related their contested ideas of analysis and synthesis to the past and the present in different ways, using different rhetorical strategies. They espoused different attitudes toward past texts and the questions of how and why they should be recovered, but they also adopted different approaches to what constituted the modernity of mathematics, to how best to move forward in the interest of not only imitating or "divining" but also surpassing the ancient methods.

Acknowledgments I am grateful to the organizers and participants of the Sawyer Seminar Conferences held from September 29 to October 2, 2021, and on February 10–11, 2023 for their comments, most notably Tawrin Baker, Claudia Cristalli, Vera Keller, Julia Kursell, Bill Newman, Alan Shapiro, and Jutta Schickore. Osvaldo Ottaviani (Radboud University, Nijmegen), Stefano Gulizia and Elio Nenci (Università degli Studi di Milano), and Paolo Rubini (Leibniz-Edition—Arbeitsstelle Berlin) provided invaluable suggestions. I am grateful to Scott Mandelbrote for having shared his knowledge of Newton's manuscripts and unpublished letters with me. This research was funded under the Project PRIN 2022 PNRR (Project Code 2022CCKT93) awarded by the Italian Ministry of University and Research (MUR).

Abbreviations

A: *G. W. Leibniz: Sämtliche Schriften und Briefe*. Berlin-Brandenburgische Akademie der Wissenschaften und Akademie der Wissenschaften zu Göttingen. Berlin: Akademie Verlag, 1923—

[37] The translation from Italian is mine.

DCA: G.W. Leibniz. *Dissertation on Combinatorial Art* (translated with introduction and commentary by Massimo Mugnai, Han van Ruler, and Martin Wilson). Oxford: Oxford University Press, 2020.

GMS: *Leibnizens mathematische Schriften* (7 vol.), ed. C.I. Gerhardt. Berlin: A. Asher & Co./Halle: H. W. Schmidt, 1849–1863 (reprinted Hildesheim: G. Olms, 1962).

GPS: *Die philosophischen Schriften von Gottfried Wilhelm Leibniz* (7 vol.), ed. C.I. Gerhardt. Berlin: Weidmann, 1875–1890 (reprinted Hildesheim: G. Olms, 1965).

L: *G. W. Leibniz: Philosophical Papers and Letters: A Selection Translated and Edited, with an Introduction* (2nd ed.), ed. L.E. Loemker. Dordrecht: Reidel, 1969.

MP: *The Mathematical Papers of Isaac Newton* (8 vol.), ed. D.T. Whiteside, with the assistance of Michael A. Hoskin and Adolf Prag. Cambridge: Cambridge University Press, 1967–1981.

NC: *The Correspondence of Isaac Newton* (7 vol.), eds. Herbert W. Turnbull, John F. Scott, A. Rupert Hall, and Laura Tilling. Cambridge: Cambridge University Press, 1959–1977.

References

Antognazza, Maria Rosa. 2009. *Leibniz: An intellectual biography*. New York: Cambridge University Press.

———. 2016. *Leibniz: A very short introduction*. Oxford: Oxford University Press.

Arthur, Richard T.W., and David Rabouin. 2020. Leibniz's syncategorematic infinitesimals II: Their existence, their use and their role in the justification of the differential calculus. *Archive for History of Exact Sciences* 74: 401–443.

Baker, Tawrin. 2023. The medical context of Descartes's *Dioptrique*. In *Descartes and medicine: Problems, responses and survival of a Cartesian discipline*, ed. Fabrizio Baldassarri, 121–140. Turnhout: Brepols.

Beeley, Philip. 2019. Practical mathematicians and mathematical practice in later seventeenth-century London. *British Journal for the History of Science* 52: 225–248.

———. 2024. Mathematical businesses: Seventeenth-century practitioners and their academic friends. In *Beyond the learned academy: The practice of mathematics, 1600–1850*, ed. Philip Beeley and Christopher Hollings, 272–311. Oxford: Oxford University Press.

Bennett, J.A. 1986. The "mechanics' philosophy" and the "mechanical philosophy". *History of Science* 24: 1–28.

Bertoloni Meli, Domenico. 2006. *Thinking with objects: The transformation of mechanics in the seventeenth century*. Baltimore: Johns Hopkins University Press.

Blåsjö, Viktor. 2017. *Transcendental curves in the Leibnizian calculus*. Amsterdam: Academic.

Bos, Henk J.M. 1974. Differentials, higher-order differentials and the derivative in the Leibnizian calculus. *Archive for History of Exact Sciences* 14: 1–90.

———. 2001. *Redefining geometrical exactness: Descartes' transformation of the early modern concept of construction*. New York: Springer.

———. 2004. Philosophical challenges from history of mathematics. In *New trends in the history and philosophy of mathematics*, ed. Tinne Hoff Kjeldsen, Stig Arthur Pedersen, and Lise Mariane Sonne-Hansen, 51–66. Odense: University Press of Southern Denmark.

Breger, Helmut. 1986. Leibniz Einführung des Transzendenten. In *300 Jahre 'Nova Methodus' von G. W. Leibniz: (1684–1984), Studia Leibnitiana, Sonderheft 14*, ed. A. Heinekamp, 119–132. Stuttgart: Franz Steiner Verlag.

Burnett, D. Graham. 2005. Descartes and the hyperbolic quest: Lens making and their significance in the seventeenth century. *Transactions of the American Philosophical Society* 95 (3): 1–152.

Cormack, Lesley B., Steven A. Walton, and John A. Schuster, eds. 2017. *Mathematical practitioners and the transformation of natural knowledge in early modern Europe*. Cham: Springer.

Couturat, Louis. 1901. *La logique de Leibniz, d'après des documents inédits*. Paris: Félix Alcan (reprinted Hildesheim: G. Olms, 1969).

———. 1903. *Opuscules et fragments inédits de Leibniz : extraits des manuscrits de la Bibliothèque Royale de Hanovre*. Paris: Félix Alcan.

Descartes, René. 1637/1954. *La géométrie*. In *Discours de la méthode, pour bien conduire sa raison et chercher la vérité dans les sciences : Plus la Dioptrique, les Météores, et la Géométrie, qui sont des essais de cete méthode*, 297–413. Leiden: Maire, 1637. (Edition cited: *The Geometry of René Descartes with a Facsimile of the First Edition*, edited and translated by D. E. Smith and M. L. Latham. New York: Dover, 1954).

———. 1649. *Geometria, à Renato des Cartes anno 1637 gallicè edita*. Leiden: Ex Officina J. Maire.

———. 1659–1661. *Geometria, à Renato des Cartes anno 1637 gallicè* edita, 2 vols. Amsterdam: apud Ludovicum et Danielem Elzevirios.

———. 1984–1991. *The Philosophical Writings of Descartes*, vol. 1–3. Trans. John Cottingham, Robert Stoothoff, Dugald Murdoch, and Anthony Kenny. Cambridge: Cambridge University Press.

Euclid. 2007. *Tutte le opere: Testo greco a fronte*. Introduction, translation and notes by Fabio Acerbi. Milano: Bompiani.

Gauvin, Jean-François. 2006. Artisans, machines and Descartes' organon. *History of Science* 64: 187–216.

Gerhardt, C.I. 1875. Zum zweihundertjährigen Jubiläum der Entdeckung des Algorithmus der höheren Analysis durch Leibniz. *Monatsbericht d. Königlichen Preussische Akademie der Wissenschaften zu Berlin*: 595–599.

Hantsche, Irmgard, ed. 1996. *Der 'mathematicus': Zur Entwicklung und Bedeutung einer neunen Berufsgruppe in der Zeit Gerhard Mercators, Referate des 4. Mercator Symposiums 30.–31. Oktober 1995.* Duisburger Mercator-Studien, Bd. 4. Bochum: Brockmeyer.

Hecht, Hartmut, and Jürgen Gottschalk. 2018. The technology of mining and other technical innovations. In *The Oxford Handbook of Leibniz*, ed. Maria Rosa Antognazza, 526–540. Oxford: Oxford University Press.

Heron. 1589. *Di Herone Alessandrino de gli automati, overo machine se moventi: Libri due tradotti dal greco da Bernardino Baldi.* Venice: Appresso G. Porro.

Hiscock, W. 1937. *David Gregory, Isaac Newton and their circle: Extracts from David Gregory's memoranda 1677–1708*. G. ed. Oxford: printed for the editor.

Huygens, Christiaan. 1673. *Horologium oscillatorium: sive, de motu pendulorum ad horologia aptato demonstrationes geometricae*. Paris: F. Muguet.

Keller, Vera. 2010. Drebbel's living instruments, Hartmann's microcosm, and Libavius's Thelesmos: Epistemic machines before Descartes. *History of Science* 48: 39–74.

———. 2022. "Communicated only to good friends and philosophers": Isaac Beeckman, Cornelis Drebbel, and the circulation of artisanal philosophy. In *Knowledge and culture in the early Dutch Republic: Isaac Beeckman in context*, ed. Klaas van Berkel, Albert Clement, and Arjan van Dixhoorn, 393–412. Amsterdam: Amsterdam University Press.

Kepler, Johannes. 1604. *Ad Vitellionem paralipomena, quibus astronomiae pars optica traditur: potissimùm de artificiosa observatione et aestimatione diametrorum deliquiorumq[ue] solis & lunae, cum exemplis insignium eclipsium, habes hoc libro, lector, inter alia multa nova, tractatum luculentum de modo visionis, & humorum oculi usu, contra opticos & anatomicos*. Frankfurt am Main: apud Claudium Marnium & haeredes Ioannis Aubrii.

Knobloch, Eberhard. 2006. Beyond Cartesian limits: Leibniz's passage from algebraic to "transcendental" mathematics. *Historia Mathematica* 33: 113–131.

L'Hospital, Guillaume-François-Antoine de. 1696. *Analyse des infiniment petits : pour l'intelligence des lignes courbes*. Paris: Imprimerie Royale.

Leibniz, Gottfried Wilhelm. 1691. De linea in quam flexile se pondere proprio curvat, ejusque usu insigni ad inveniendas quotcunque medias proportionales & logarithmos. *Acta eruditorum* June: 277–281.

Levitin, Dmitri, and Scott Mandelbrote. 2023. Newton as theologian, artisan, and chamberfellow: Some new documents. In *Collected wisdom of the early modern scholar: Essays in honor of Mordechai Feingold*, ed. Anna Marie Roos and Gideon Manning, 251–275. Cham: Springer.

Malet, Antoni. 1996. *From indivisibles to infinitesimals: Studies on seventeenth-century mathematizations of infinitely small quantities*. Bellaterra: Universitat Autònoma de Barcelona.

Marr, Alexander. 2011. *Between Raphael and Galileo: Mutio Oddi and the mathematical culture of late Renaissance Italy*. Chicago: University of Chicago Press.

Maurolico, Francesco. 1575. *Opuscula mathematica: nunc primum in lucem aedita, cum rerum omnium notatu dignarum. Indice locupletissimo. Pagella huic proxime contigua, eorum catalogus est*. Venice: apud Franciscum Franciscium Senensem.

McConnell, Anita. 2016. *A survey of the networks bringing a knowledge of optical glass-working to the London trade, 1500–1800,* ed. Jenny Bulstrode. Cambridge: Published by the Whipple Museum of the History of Science.

Morel, Thomas. 2024. Mathematics and technological change: The silent rise of practical mathematics. In *Volume 3: A cultural history of mathematics in the early modern age*, ed. Jeanne Peiffer and Volker Remmert, 179–206. Bloomsbury Press.

Newton, Isaac. 1687. *Philosophiae naturalis principia mathematica*. London: J. Streater.

———. 1706. *Optice: sive de reflexionibus, refractionibus, inflexionibus & coloribus lucis libri tres: authore Isaaco Newton,... latine reddidit Samuel Clarke,... accedunt tractatus duo....* London: impensis Sam. Smith & Benj. Walford.

———. 1718. *Opticks: Or, a treatise of the reflections, refractions, inflections and Colours of light (the Second Edition, with Additions)*. London: printed for W. and J. Innys, printers to the Royal Society.

Oosterhoff, Richard. 2018. *Making mathematical culture: University and print in the circle of Lefèvre d'Étaples*. Oxford: Oxford University Press.

Ottaviani, Osvaldo. 2022. Inside Leibniz's metaphysical laboratory: Two draft texts from 1710. *The Leibniz Review* 32: 55–105.

Otte, Michael, and Marco Panza, eds. 1997. *Analysis and synthesis in mathematics: History and philosophy*. Dordrecht/London: Kluwer Academic Publishers.

Pappus. 1588. *Mathematicae collectiones, à Federico Commandino urbinate in latinum conversae, & commentariis illustratae*. Pesaro: apud Hieronymum Concordiam.

———. 1660. *Mathematicae collectiones a Federico Commandino urbinate in latinum conversæ, & commentarijs illustratæ: in hac nostra editione ab innumeris, quibus scatebant mendis, & præcipuè in græco contextu diligenter vindicatæ*. Bologna: ex typographia H.H. de Duccijs.

———. 1986. *Pappus of Alexandria Book 7 of the Collection. Part 1. introduction, Text and Translation. Part 2. Commentary, Index and Figures*, ed. with translation and commentary by Alexander Jones. New York: Springer.

Pasini, Enrico. 1993. *Il reale e l'immaginario: La fondazione del calcolo infinitesimale nel pensiero di Leibniz*. Torino: Sonda.

Picon, Marine. 2021. *Normes et objets du savoir dans les premiers essais leibniziens*. Paris: Classiques Garnier.

Ptolemy. 1562. (and Federico Commandino). *Liber de analemmate a Federico Commandino urbinate instauratus, & commentariis illustratus, qui nunc primum eius opera e tenebris in lucem prodit: eiusdem Federici Commandini liber de horologiorum descriptione*. Rome: apud Paulum Manutium Aldi f.

Rabouin, David. 2009. *Mathesis universalis : l'idée de "mathématique universelle" d'Aristote à Descartes*. Paris: Presses Universitaires de France.
Rose, Paul Lawrence. 1976. *The Italian renaissance of mathematics: Studies on humanists and mathematicians from Petrarch to Galileo*. Geneva: Droz.
Rossi, Paolo. 1999. *Un altro presente: Saggi sulla storia della filosofia*. Bologna: Il Mulino.
Schneider, Martin. 1974. *Analysis und Synthesis bei Leibniz*. Bonn: Inaugural Dissertation.
Taylor, Eva G.R. 1954. *The mathematical practitioners of Tudor & Stuart England*. Cambridge: for the Institute of Navigation at the University Press.
van Schooten, Frans. 1646. *De organica conicarum sectionum in plano descriptione, tractatus: geometris, opticis, praesertim verò gnomonicis & mechanicis utilis, cui subnexa est appendix, de cubicarum aequationum resolutione*. Leiden: ex officinâ Elzeviriorum.
Viète, François. 1591. *In artem analyticem isagoge: Seorsim excussa ab opere restitutæ mathematicæ analyseos, seu, algebrâ novâ*. Tour: apud Iametium Mettayer Typographum Regium.
Wess, Jane. 2024. Mathematics and technological change: Sublime mathematics and practical applications. In *Volume 4: A cultural history of mathematics in the eighteenth century*, ed. Maarten Bullynck, 193–221. Bloomsbury Press.
Wilson, Catherine. 1995. *The invisible world: Early modern philosophy and the invention of the microscope*. Princeton; Chichester: Princeton University Press.

Niccolò Guicciardini holds two degrees (in physics and philosophy) awarded by the Università degli Studi di Milano. He teaches history of science at the Department of Philosophy of the Università degli Studi di Milano. He has written about the mathematics and natural philosophy of Isaac Newton and its reception in the eighteenth century.

Chapter 8
Knowing Diseases and Medicines Forward and Backward: Analysis and Synthesis from Galen to Early Modern Academic Medicine

Evan R. Ragland

Abstract Early modern academic physicians followed Galen's models on many points, including reasoning from signs of diseases back to hidden causes and from causes to signs. Galen's *Art of Medicine* (*Ars medica*, *Ars parva*, *Tegni*) famously spurred reflections on the methods of teaching and discovery, which included his categories of analysis, synthesis, and dissolution of the definition. Medieval and early modern physicians at times described analysis as reasoning from effects to causes, and synthesis from causes to effects. Diagnosis often depended on reasoning backward from signs to hidden seats of diseases and causes (analysis), and from causes and organs to signs and symptoms (synthesis). At patients' bedsides, physicians also moved forward in time, observing disease progression and the effects of therapeutic interventions. From the early 1600s on, physicians at Leiden University followed the Paduan model and used postmortem dissections to confirm or go beyond the diagnoses and prognoses made during daily clinical teaching in the local hospital. These Galenic physicians did not develop radically new medical frameworks, but they did gradually add to the store of pathological and therapeutic knowledge. They based these gradual innovations on their humanist scholarship, as well as on their knowledge of diseases and therapies going forward with living patients, and their inferences from postmortem evidence back to the historical causes of patients' diseases and deaths. By the mid-1600s, this established pedagogical practice allowed physicians and students pushing new medical theories to generate important new pathological knowledge, notably of consumption (phthisis). By the later 1600s, and into the 1700s, it appears that physicians still reasoned back from signs to hidden anatomical states, using causal principles of natural philosophy, but reserved the terms "analysis" and "synthesis" for mathematics and, especially, chemistry. Although the new Paris medicine, around 1800, explicitly eschewed theoretical systems and causal principles in favor of correlating and counting close descriptions of anatomical and clinical phenomena, the practices of reasoning back

E. R. Ragland (✉)
Department of History, University of Notre Dame, Notre Dame, IN, USA
e-mail: eragland@nd.edu

W. R. Newman, J. Schickore (eds.), *Traditions of Analysis and Synthesis*,
Archimedes 73, https://doi.org/10.1007/978-3-031-76398-4_8

from symptoms to impaired organ functions and anatomical states revealed in post-mortem dissections appear significantly continuous across the early modern period.

Keywords Analysis · Synthesis · Medicine · Disease · Pathology · Galen · Leoniceno · Da Monte · Fernel · Consumption

> There are three types of teaching in all, each with its place in the order. First is that which derives from the notion of an end, by analysis. Second is that from the putting together of the findings of an analysis. Third is that from the *dialysis* of a definition: and it is this which we shall now embark on. (Galen, *The Art of Medicine* Ki.305, trans. P. N. Singer, *Galen: Selected Works*)

> Therefore affected seats [*sedes*] are discerned from the symptoms, from the wounded action, from the quality of the excrements, from swelling beyond nature, from pains, from a fault in the color of something that follows, either in the whole body or in one part, or in two, and especially in the eyes and the tongue. (Galen, *De Affectorum Locorum Notitia* [*De locis affectis*] (1520), 22r–22v)

> Diseases, as we said, dwell in the parts. However many ways therefore those become faulty, there are just so many diseases. (Johannes Heurnius, *Institutiones Medicinae* (1638), 516)

8.1 Introduction

Analysis and synthesis, and especially analysis, informed and inspired seemingly ceaseless discussions and even practices in premodern medicine. While medieval and early modern physicians almost always agreed that analysis and synthesis were important methods, orders, or ways, they disagreed on the details. Analysis and synthesis appear in canonical places and in many different forms in Galen's influential writings. As was common, his writing on this theme was not always consistent or systematic. In hindsight, and with the help of later commentators, we can see that Galen used roughly five different meanings for analysis and synthesis, and concentrated, like the physicians after him, on analysis. In general, analysis moved from effects, symptoms, and signs—or body parts and compounds—back to causes or principles, and synthesis moved the other way. Ancient, medieval, and early modern commentators wrestled with Galen's meanings, and added their own distinctions, such as the difference between logical and material analysis and synthesis. Analysis garnered the most attention and was not always paired with synthesis. In this long medical tradition, analysis was often presented as a process for identifying a cause by making inferences from phenomena to the hidden causes most consistent with those phenomena. Possible causes, and the kinds of causes, very often came packaged from the principles of natural philosophy, notably the hot, the cold, the dry, and the wet. Even early modern critics of Galen endorsed the structure or order of reasoning from effects or symptoms in patients' bodies back along supposed causal lines to hidden organ impairments and the causes of impairment. After all, such conditions of impairment constituted most definitions of disease, from Galen on.

This chapter comprises four parts: it moves from a survey of Galen's mentions of analysis and synthesis to select ancient and medieval commentaries, to sixteenth-century debates over the terms and practices, and then into the seventeenth-century practices of medical diagnosis and postmortem dissection. With frequent pathological dissections from the later sixteenth century into the seventeenth, physicians could finally sense directly the morbid states that rational analysis had long predicted. We will travel mostly through ancient Rome, medieval and Renaissance Italy (and France), and late-sixteenth- and early-seventeenth-century Leiden in the Low Countries. Even into the end of the seventeenth century, physicians continued to express multiple meanings of analysis and synthesis, from logical decomposition to breaking down substances into components by chymical means, or the composition of drugs.

Throughout, the structure of reasoning from symptomatic effects to hidden causes remained central to discussions and practices, and even embraced significant innovations in pathology. For example, through Galenic methods and practices of analysis, physicians gradually produced the vital finding that consumption (phthisis, or, roughly, tuberculosis) came about from the development of ulcers and tubercles in the lungs.

In sum, this chapter demonstrates the existence and main features of a long and robust tradition of analysis and synthesis—and especially analysis—in academic medicine through a sampling of Greek and Latin texts. Especially given the evidence of important discussions and practices in ancient medical synthesis (Totelin, this volume) and early modern anatomy (Baker, this volume), I suggest that we ought to add medicine to four traditions identified by Alan Shapiro (this volume): the chemical, mathematical, logical, and natural philosophical traditions. As we will see, medical writers engaged with all these learned traditions, and developed their own theories and practices.

8.2 Galen's Works as Sources for "Analysis" and "Synthesis"

The expansive and ramifying works of the ancient Roman physician and philosopher Galen (129–c. 200) displayed multiple presentations and meanings for analysis and synthesis for the future generations who read his texts avidly across the medieval and early modern periods. Galen presented analysis and synthesis in at least five different ways: (1) ways or methods of ordered teaching; (2) geometrical and architectural methods, which included testing by making and using; (3) philosophical methods for moving from effects to causes or principles; (4) for analysis, as anatomical methods for isolating parts and testing claims; and (5) rational practices for testing and making drugs. We will briefly survey each in turn.

8.2.1 Teaching Methods

Two of Galen's most important works will set the stage for our longer tour of medical "analysis" and "synthesis" in premodern academic medicine. First, his mature summary, *The Art of Medicine*, briefly articulated three modes of teaching, offering a key *topos* for commentary and debate in later centuries. Next, we will examine his *Method of Medicine*, whose early books, especially, show the application of these methods in therapeutic practice, namely through the identification of hidden causes in patients' bodies by tracing back from surface symptoms to fundamental organ states and other morbid dispositions.

For medieval and early modern physicians, as well as philosophers, talk of "analysis" and "synthesis" would have brought to mind the beginning of Galen's *Art of Medicine*. Here is the canonical passage, courtesy of P. N. Singer's translation, with parenthetical Latin terms from a 1544 Latin translation (Galen 1997e; Acakia 1544):

> There are three types of teaching [*doctrina*] in all, each with its place in the order. First is that which derives from the notion of an end, by analysis [*resolutio*]. Second is that from the putting together [*compositio*] of the findings of an analysis. Third is that from the *dialysis* [*dissolutio*] of a definition: and it is this which we shall now embark on.

Although Galen used the method of *dialysis* or dissolution of a definition in *The Art of Medicine*, later commentators concentrated on the modes of analysis and synthesis. As John Herman Randall showed now over 80 years ago, this passage was a *topos* for commentary on method and medicine among medieval and early modern physicians, commentary that initially appeared as a small trickle among medieval physicians and built up to a steady stream in the 1500s (Randall 1940, 1961). We need not accept Randall's thesis (and, in my view, we should not) that Aristotelian discussions of method over the centuries, especially those by professors at the University of Padua, created *the* method of analysis and synthesis that Galileo and, hence, modern science adopted. After all, as his critics quickly pointed out, although the Paduan Aristotelians modeled their language on the methods of geometry, there is no significant talk of mathematics in their discussions (Wightman 1964; Edwards 1967). Galileo's science was thoroughly mathematical, and especially Archimedean (Bertoloni Meli 2006, 50–79). Clearly, as this volume demonstrates much further, the ways and orders of analysis and synthesis were many and changing.

The Art of Medicine, the mature summary of Galen's medical system, or the *Techne iatrike*, was also later called the *Tegni*, the *Ars parva*, and the *Ars medica*. After Randall, the studies of Per-Gunnar Ottoson and Nancy Siraisi have drawn particular attention to this work (Ottoson 1984; Siraisi 1981). It was a major medical teaching text from the Late Antique period, throughout Arabic medieval texts from the ninth century on, and formed an essential part of Latin medical teaching and learning from the later twelfth-century versions of the *Articella* (Kaye 2014; Ottoson 1984).[1] Most recently, Joel Kaye has demonstrated that Galen's discussions

[1] For acceptance of this work as genuine today, see Hankinson (2008a, b), 237, n. 28.

of "balance" or equilibrium in *The Art of Medicine* stimulated a revolutionary shift in medieval intellectuals' conception of "balance" from a static, simple form—as in a balanced scale or lever—to a dynamic, self-ordering, complex, interdependent conception (Kaye 2014). This notion is important for understanding Galenic pathology, diagnosis, and therapy, as we will see, since an imbalanced mixture of the four qualities (hot, cold, wet, and dry) in a part of the body often generated disease. Galen's *Art of Medicine* shaped medical teaching widely from the Late Antique period in the Hellenized Mediterranean, from the ninth century in Arabic sources, and from at least the late 1100s in Latin medical teaching (Kaye 2014, 137–138; Bouras-Vallianatos and Zipser 2019). Galen intended this text as the key to his extensive other writings, and later readers and teachers tended to treat it this way.

8.2.2 Geometry, Architecture, and Testing

A striking discussion of the pair "analysis and synthesis" appears prominently in an extended discussion of the conception, designing, and construction of accurate timekeeping devices—notably sundials and water clocks—in Galen's *On the Diagnosis and Cure of Errors of the Soul*. There, he describes how people such as "architects" (like Galen's own father) must use "analysis" to first reduce the problem to shapes suitable for the design, then use "analysis and synthesis" to see how each design should be done and to determine the best instruments to draw such a timekeeping device, and then actually construct it and test it in use (Galen 1997d, 138–142). Galen takes care to describe the proper construction of sundials and water clocks. As R. J. Hankinson has observed, Galen has in mind many stages of analysis and synthesis, with each stage the methodical search for an answer to a different problem: designing the structure in question, achieving a flat surface, picking out instruments for engraving, building and testing the device in various ways, etc. (Hankinson 2009, 227–228).

Galen's model is geometry, in which analysis moves from conceptions of a specific problem back to well-conceived criteria or principles, and then back along "the same path in the opposite direction in order to put the solution together" (Galen 1997d, 138). As in geometry, in the construction of timekeeping devices and other architectural things, "the theory is confirmed by the solution itself when it is discovered." He also sharply contrasts this constructive method of "analysis and synthesis" with the "shameless, ill-considered rubbish" often spoken in philosophy. "One who constructs a sundial or water-clock wrongly is refuted by the clear evidence of the facts; but the refutation of philosophical positions is not so immediately clear" (Galen 1997d, 142). As Teun Tieleman notes, Galen claims originality in identifying the procedure of testing with "synthesis," and may well have been the first to do so (Tieleman 2002). Galen's enthusiasm for such procedures of geometric and engineering analysis and synthesis likely grew from his own rescue from Pyrrhonian skepticism by the "incontrovertible truth" demonstrated in mathematical arts (Galen 1997b, 18). Galen had first learned mathematics from his father, an architect, and had witnessed the truth of mathematical calculations as tested in accurate predictions of eclipses, sundials, water clocks, and other architectural geometric and mathematical designs synthesized into tangible constructs.

8.2.3 Philosophical Demonstration and Reasoning from Effects to Causes or Principles

With geometry as a model, Galen used a broadly Aristotelian method of moving from effects to causes or principles and back to effects. He drew from Stoic logic, but mainly followed Aristotelians, who appreciated geometric demonstration (*apodeixis*) (Tieleman 2008, 52; Hankinson 2008a, 165–9).[2] In this approach, the physician reasons from the symptoms and signs available to the senses back through causal chains to the fundamental principles of disease and health, and then back to the effects. Therapeutic interventions allow the physician to confirm the identification of causes. Just as with the prediction of eclipses in astronomy, if the intervention accurately predicts the cure of the disease, the physician can have confidence that he has identified and affected the cause(s) or disease state(s) as well as possible (e.g., at least "for the most part") (Tieleman 2008, 62; Hankinson 1991, 120–121).

In his works on method, Galen references these meanings, and a notion of analysis as demonstration. His massive and influential *Method of Medicine* includes a discussion of proper demonstration (*apodeixis*) as a centerpiece of the methodological discussions in the first book. In Galen's terms, his opponents do not understand demonstration since they have never studied "geometry, arithmetic, formal logic, analytics, or indeed logical theory of any kind" (Galen 1991, I 3.15, p. 16.) As Hankinson notes, it is unclear what, exactly, Galen meant by "analytics" here—whether the technical procedures of the geometers (*analusis*) or the Aristotelian analytics of philosophical demonstration and inference—but it was certainly "a method for arriving at the first principles of any science; and as such, it cannot presuppose them" (Hankinson 1991, 112–113, 124). Later, he adds a demand for proper logical training in method for physicians, similar to the remarks from *Errors of the Soul* mentioned above.

But how could one apply this sort of mathematical method to another art or *technē* beyond geometry or architecture, namely, to medicine? Galen clearly demanded that the true method for finding and treating diseases ought to be a logical one, and one parallel to the methods of geometry, architecture, and other successful arts based on and tested by reason and experience. Other passages in Galen's *Method of Medicine* make his therapeutic methodology clearer. Ultimately, all knowledge and practice must be discovered by a logical (*logos*-based) method and confirmed by reason and experience:

---

[2] For geometry as exemplary of "analysis," see Aristotle, *Nichomachean Ethics* III.3 1112.b12–24, in Aristotle (1984).

> Logical methods have the power to discover what is sought, while there are two criteria for the confirmation of things that have been discovered, namely reason [*logos*] and experience [*peira*]. (Galen 1991 I 3.3, 11–12)

The leader of the so-called Methodists, Thessalus, ought to have tested and demonstrated his claims by either reason or experience, and should not have shoe-horned all diseases into the fluid and the constipated (*rhoōdes* and *stegnon*), and the mixed (Galen 1991, I 3.3, 11–12). These practitioners, like Galen's other opponents, did not follow "Hippocrates" (at least, Galen's self-legitimizing vision of Hippocrates) in first establishing proper definitions and divisions that define and then distinguish diseases according to the species or kinds of things (Galen 1991, I 2.2–2.4 and 3.3–3.13).

In this emphasis on definitions and true divisions, developed through the method of *dialysis* of the definition, Galen drew explicitly from Aristotelian natural philosophy and method. Thus "rational," "footed," and "bipedal" picks out "man," but the first book of Aristotle's *Parts of Animals* demonstrated how difficult the task of finding proper divisions and definitions is in practice (Galen 1991 I 3.8–3.10).[3] So, for physicians the method was first to

> accurately to define what disease, symptom, and affection are, and to distinguish in what ways each of the aforementioned things resemble one another and in what ways they differ, then to try and cut them into their proper *differentiae* according to the method which the philosophers have taught us. (Galen 1991 I 3.8–3.10)

Galen then began, as Aristotle did in his works on animals, with common notions (Aristotle 1984 *Parts of Animals* I.3, 643b9–12). People commonly say they are healthy when they have no impediment in the activities (*energeiai*) of all their bodily parts. But when they

> become aware that some one of their natural functions (*dunameis*) is beginning to perform either badly or not at all, they consider themselves to be sick in that section (*meros*) of the body whose activity they see to be impaired. (Galen 1991 I 5.4, p. 22)

Activities of parts are active movements, with sight the activity of the eye (Galen 1991 I 6.1). This activity depends on the structure of the eye, with its crystalline humor, which in turn has its properties of perfect purity and transparency due to its particular mixture of the fundamental qualities of hot, cold, wet, and dry. "For it has been shown that each thing is such as it is on account of its blend of Hot, Cold, Wet, and Dry" (Galen 1991 I 6.5). Fundamental mixtures and structures, such as different contractile fibers, gave the different parts of bodies their powers or faculties (*dunameis*) relative to different substances (Hankinson 2014).

Across his works, though, Galen had in mind moving from what is perceptible to the senses back to first principles established by the intellect. Thus, the physician reasons from the effects of patients' symptoms, or the effects of drugs' powers (*dunameis*), to the first principles already established by the "best philosophers":

---

[3] As Hankinson points out, Galen departed from Aristotle in his model *differentiae*. Hankinson 1991, 102–03.

the hot, the cold, the wet, and the dry (Galen 1997a, 202). Galen insisted throughout his works that the theory of the hot, cold, wet, and dry qualities as the fundamental constituents of mixtures was the true philosophy of Hippocrates (Hankinson 2008b, 217–22).[4]

Here he has in mind the Hippocratic treatise *Nature of Man*, which Galen elevated to the true foundational work of Hippocrates. Unusually among the other Hippocratic texts, *Nature of Man* argued at length for a strict philosophical system based on the fundamental principles of hot, cold, wet, and dry as the constituents of all things (King 2013; Nutton 2013, ch. 5). This text also emphasized the basic bodily humors formed from these qualities: blood, phlegm, and bile (both yellow and black). Other Hippocratic works directly criticized the hot, cold, wet, and dry principles, notably *On Ancient Medicine*, while others defended different principles, such as the fatty and the glutinous; air; bile and phlegm; or fire and water. As is well known, Aristotelian philosophers later developed more extensive accounts of how the hot, cold, wet, and dry mixed to form simple or homoiomerous substances, which formed composite or anhomoiomerous substances. Aristotle, the son of a physician, was a philosopher who engaged deeply with medical works of his time, begins book two of his *Parts of Animals* with a short summary:

> For wet and dry, hot and cold, form the material of all composite bodies; and all other differences are secondary to these, such differences, that is, as heaviness or lightness, density or rarity, roughness or smoothness, and any other such properties of bodies as there may be. The second degree of composition is that by which the homogeneous parts of animals, such as bone, flesh, and the like, are constituted out of the primary substances. The third and last stage is the composition which forms the heterogeneous parts, such as face, hand, and the rest. (Aristotle 1984, *Parts of Animals*, 646a18–24)[5]

Similarly, Galen took the hot, cold, wet, and dry mixtures of the simple (homogeneous or homoiomerous) parts, as well as the fibers and structures of composite parts, as the grounds for their natural functions or "faculties" (*dunameis*) (Ragland 2022, ch. 3; Hankinson 2014). As he put it in his work on pulse prognosis, "the substance of a faculty of the individual parts is attributed to the fitting temperament of the individual parts" (Galen 1821–1833, vol. XI, 244). Faculties are then relational properties revealed by the regular effects of two interacting substances—for instance, aloe regularly strengthens stomachs, binds wounds, and cleanses eyes, and so we say that aloe has faculties or powers of strengthening, binding, and cleansing these objects (Galen 1997c, 151).

[4] For a thoughtful discussion of the limits of material explanation in Galen's thought, see Singer, "Levels of Explanation in Galen." Galen insisted that only a divine craftsman-like power could account for the complex structures and integration of the thousands of powers or faculties of a living body.

[5] For Aristotle and medicine, see Van der Eijk 2009, 8–15.

8.2.4 Analysis in Anatomical Method and Practice

Galen applied "analysis" in his discussion of anatomical method and anatomical practice. First, Galen explicitly described anatomy as a mode of "analysis." In his *Constitution of the Art of Medicine*, Galen argues that physicians must know well all the parts of the human body, just as architects and builders need to know all the parts of a house. The architect knows what a house is and how it functions "through analysis and dialysis" (Galen 2016, 23–5; K I 230–1). He continues: "In the same way, we know the human body through anatomy" (Galen 2016, 25; K I 231). Here, Galen paired analysis with "dialysis," which later writers glossed as division or dissolution, as we will see. But Galen's emphasis on analysis in anatomical practice and reasoning is consistent and clear.

Galen also used analysis and synthesis in his anatomical practice, with analysis playing key roles in discovery and in disconfirming rival hypotheses. In his studies of Galen's experimental and logical method in the early books of *On the Doctrines of Hippocrates of Plato*, Tieleman has argued that Galen used analysis for anatomical research and discovery (Tieleman 2002). In short, Galen sought to resolve the philosophical debates of his time over the seat of the soul—either in the heart, as Aristotle and the Stoics had it, or in the brain, as "Hippocrates" and the Platonists argued—by systematic argumentation using first principles and the careful deployment of anatomical experimentation as well as common experience. As Tieleman points out, he likely followed the method of Aristotle who argued in *Parts of Animals* that the student of nature ought to follow something like "the plan adopted by the mathematicians in their astronomical demonstrations, and after considering the phenomena presented by animals, and their several parts, proceed subsequently to treat of the causes and the reasons why" (Aristotle 1984, *Parts of Animals* 639b7–10).[6] Famously, Aristotle followed a two-stage method in his studies of living things, beginning with *historia* or description of "the that" (*to hoti*) and then philosophizing to find causes and principles or "the reason why" (*to dioti*) (Lennox 2021).

In Galen's anatomical practice, analysis involved the isolation of anatomical structures and actions while, as we saw in his remarks on the construction of geometric devices, he identified synthesis with trialing the device in use. Through anatomical procedures, Galen proceeded to confirm a modified Platonic theory of the tripartite soul, and rule out rival theories (usually by a form of *modus tollens*), by performing anatomical experiments that systematically damaged organs' and vessels' structures and actions to demonstrate their functions (Tieleman 2002). For instance, he ligatured arteries systematically to show that the pulse originates in the heart, and he attempted to demonstrate the functions of the brain by systematically wounding the different ventricles of the brain, and by cutting or ligaturing the nerves, veins, and arteries going to the brain. As we will see, as in his anatomical

[6]The following discussion in the rest of part 1 shows that Aristotle endorses this position, although he frames it here as a question. Tieleman 2002, 267 also points out that Herophilus introduced into medicine the move from perceptible phenomena to unseen causes.

method, in his discussion of teaching and therapeutic or clinical methods Galen argued for the importance of reasoning from phenomenal effects back to first principles, and forward again to the effects. Later anatomists would refine and elaborate these methods and concepts, for which see especially Baker (this volume).

8.2.5 Testing and Composition of Medicines or Drugs

Galen often wrote about the "synthesis" of medicines, especially in his *Method of Medicine*. In this work he used "synthesis" (*sunthesis*) to describe the composition of medicines from multiple ingredients (Galen 2011, 255–9; K X 165–8). His approach is a small slice of the larger practice of Greek medicinal synthesis discussed by Totelin (this volume). Similarly, in the shorter *Method of Medicine to Glaucon*, he described how combining ingredients with known powers or faculties allowed physicians to "synthesize" (*suntithenai*) compound medicines to generate the drawing, repelling, drying, flesh-growing, or other effects needed to cure diseases—especially to remove corrupt matter or blockages, expel morbid matter, restore organ temperaments, and restore continuity to bodily parts (Galen 2011, 460–1; K XI 81).

In Galen's works, we also find rules for rigorously isolating and testing the faculties or powers of medicinal substances, as Philip van der Eijk has shown (Van der Eijk 2009, 279–98). These faculties worked in virtue of their qualitative mixture, relative to the mixture of the part of the body they operated upon (as well as their "total substance" in some cases, such as poisons and purgatives). As elsewhere in his writings, Galen presents experience (*peira*) as the "teacher" (*didaskalos*) of reason. But the complexity of drug–body interactions, and variations of drugs and bodies, makes it clear that not just any experience will suffice (Van der Eijk 2009, 280). Instead, the physician must follow Galen's method and use the right "qualifications" or conditions (*diorismoi*) to isolate causal relations, avoid confounding causes, and generate "qualified experience" (*diorismene peira*). A few of the qualifications will give a sense of the method: one should first test small amounts of a substance on a healthy, temperate body; the physician should consider whether the drug's particles were thick or thin (which would affect how far they could penetrate into solid parts); the physician should know the patient's natural temperament, especially how hot and cold they were, and whether the disease made the patient's body preternaturally hot or cold; the drug should be tested on a simple disease caused by a bad temperament or a mixture of a part or parts, etc. (Van der Eijk 2009). Galen applied these rules in practice at times. In his influential work on medicinal simples, Galen recounted how he smeared the burning plant *thapsia* on his own thighs, and then after 4 or 5 h of allowing the inflammation to grow tested different antidotes—of which vinegar proved the most effective (Ragland 2022, 209). In sum, Galen's drug testing, like his anatomical experimental method and his therapeutic method, relied on exploring and isolating causal factors in relevant phenomena, moving from effects back to putative causes.

So far, we can see that Galen articulated several different meanings for "analysis" and "synthesis." In *The Art of Medicine* he presented them as forms of teaching or presenting knowledge. Analysis begins from the end goal and works backward. Synthesis puts together the steps found through analysis. On the other hand, in his *Errors of the Soul*, Galen presented "analysis" and "synthesis" as hands-on methods for constructing and testing devices to solve specific problems. In this view, they are methods of discovery and proof, even proof-of-concept in the operation of a device. To construct a reliable timekeeping device, one would have to act like a good architect or geometer and first move from the problem to the principles, then actually make the device and test whether it worked as needed. In *Method of Medicine* and other works on method, Galen described a parallel therapeutic method. Finally, we have the analysis of drug powers by conditions that attempt to isolate causal pathways and avoid confounding accidental causal interference, and then synthesis in the making of drugs.

All these methods exhibit, at least to some degree of resemblance, similar organization or order: parallel structures of taking a particular phenomenon or effect, working back to prior principles (often causal principles), and then assessing this analysis by mentally retracing the steps from the principles to the effects, or, in the case of constructing devices or making drugs, noting the utility and effects of the things constructed in operation.

8.3 Some Ancient and Medieval Elaborations and Debates

Medieval physicians writing in Arabic famously systematized, elaborated, and critiqued Galen's extensive and sometimes contradictory writings (Bouras-Vallianatos and Zipser 2019). Galen had argued, often polemically, that physicians ought to follow the logical methods of the mathematical practitioners and the Aristotelians. But it was not at all clear how one could do so, and if and when Galen's prescriptions for proper method applied to teaching, the discovery of the principles of an art or science, or the discovery of causes in *this* particular patient here and now.

Of course, Galen's extensive writings were not the only game in town. Around the same time as Galen, other philosophers drew on Platonic and Aristotelian traditions to articulate several different senses of "analysis." For instance, Alexander of Aphrodisias (late second and early third century AD) distinguished logical or methodical analysis from material analysis:

> They are called *Analytics* because the reduction of any compound to the things from which it is compounded is called analysis [*analusis*]. Analysing is the converse of compounding, for compounding [*sunthesis*] is a route from the principles to what depends on them, whereas analysing is a return route from the end up to the principles. Geometers are said to analyse when they begin from the conclusion and proceed in order through the assumptions made for the proof of the conclusion until they bring the problem back to its principles. Again, if you reduce compound bodies to simple bodies, you use analysis; and if you reduce each of the simple bodies to the things on which their being depends—that is to say, to matter and form—you are analysing. If you reduce compound syllogisms to simple ones you

> are said to analyse in a special sense of the word, and so too if you reduce simple syllogisms to premises on which their being depends. (Alexander of Aphrodisias 2013, 49–50)

Here, Alexander summarizes different forms of analysis—geometric, material, metaphysical, and syllogistic—as instances of reducing compounds into their components. Synthesis moves the other way, compounding components or principles together. Alexander's comments, and his distinction between logical or conceptual analysis and material analysis, would become part of the scholarly conversation into the early modern period.

Looking back to ancient medical writings, it is clear that systematic texts on diseases, causes, and treatments used at least implicit forms of analysis and synthesis when they reasoned from phenomenal effects back to first principles and causes, and then from causes to effects. Authors of manuals on the methodical practice of medicine—in the *practica* tradition, to use the later Latin term—had strong precedents for arranging their texts by the anatomical sites of disease states. Ancient Mesopotamian medical tablets from the eleventh century BC, for example, adopted a standard and easy-to-use format of listing diseases from head to foot. With the aim of passing down practical instruction for the art of medical practice, writers described the specific symptoms, the prognosis, and perhaps favored therapies (Scurlock 2018).

We can perhaps see a similar structure in later texts in book four of Celsus' first-century AD Latin *De medicina*, which discusses diseases and some treatments from the head down through to the extremities. Similarly, Paul of Aegina of the seventh century dedicates the third book of his Greek *Medical Compendium in Seven Books* to discussions of diseases and treatments from roughly the top of the head (baldness) to the feet (corns). As Adrian Wilson has pointed out well in his study of concepts of pleurisy and pleuritis from Hippocratic sources into the eighteenth century, concepts of disease framed in terms of clusters of symptoms existed in tension and productive interaction with concepts framed in terms of anatomical localization throughout premodern medicine (Wilson 2000).

8.3.1 Medieval Arabic Medical Sources

In medieval Arabic sources, we see a continued emphasis on diseases conceived as conditions impairing the functions of parts of the body and sustained attempts to think through, and think with, Galen's notions of analysis and synthesis. Al-Razi (ca. 864–865 to 925–935) or Rhazes, produced influential guides to the practice of medicine, notably his treatises on smallpox and measles, and recorded hundreds of histories of patient interactions, which we might call case histories. As Emilie Savage-Smith argues, in over 900 case histories, a few diseases are caused by humors (e.g., headaches caused by yellow bile vapor), but humors are not at all primary for pathology, and "therapy is never couched in humoral terms" but, rather, in terms of the four qualities (Savage-Smith 2013, 92). The "balance to be reinstated during the restitution of health had to be centred not on the balance of the four humours but on the balancing of the four primary qualities… and the six 'non-naturals'" (Savage-Smith 2013, 101–02).

Another influential physician, the Egyptian Abu'l Hassan Ali ibn Ridwan Al-Misri (ca. 988–1061) wrote a commentary on Galen's *Art of Medicine* that was later translated into Latin by Gerard of Cremona (Pormann and Savage-Smith 2007, 44).[7] Using a 1557 Latin version, William Edwards has argued for the importance of Ibn Ridwan's commentary to medieval and early modern Latin debates over analysis and synthesis in *The Art of Medicine*, as well as to medicine and philosophy more generally (Edwards 1967).[8] Rather than seeing Galileo as the first to merge mathematical and philosophical traditions through a method of analysis and synthesis, from effects to causes and back, Edwards argues that historians should look to Ibn Ridwan's influential commentary. For later Latin medieval physicians, Ibn Ridwan marked the beginning of a tradition of commentary and set several points of the debate. First, he directly identified the "analysis" at the beginning of *The Art of Medicine* with geometrical analysis *and* with Aristotle's demonstration of "the that" (*to hoti* or *quia*) in the *Posterior Analytics*. Second, and rather naturally, Ibn Ridwan also identified Galen's "synthesis" or *compositio* with Aristotelian demonstration of "the reason why" (*to dioti*, or *propter quid*), and, presumably, with geometrical synthesis. As Edwards points out, the methods of geometers and of physicians seem quite different, since geometers begin with something they want to prove as if it is known, then work back to principles already known and established. In contrast, physicians reason from symptoms and signs that are known from the senses back to hidden causal states that are possible or probable, and the inference only gains greater probability if the administration of a remedy fit for the proposed cause produces the expected effect (Edwards 1967). As we have seen, though, in his other works, Galen understands "synthesis" in terms of the constructive or predictive practices of architects, astronomers, and instrument-makers, who also demonstrate the accuracy of their causal claims by mastering the materials and producing the expected effects. He likely saw significant similarity between geometrical and medical forms of analysis and synthesis if the physicians could rely on true philosophical principles to intervene reliably to affect patients and diseases.

Ibn Ridwan seems to have had in mind not the practice of diagnosis and therapy of physicians, but medicine as a body of knowledge. This knowledge is divided into two parts—the acquisition and conservation of health—and ultimately descends to its foundations in "known bodies of knowledge [*scientias notas*], all the way to those things which must be posited as the principles of medicine" (Ibn Ridwan in Turisianis 1557, 175v). Increasingly, physicians presented those principles as established by natural philosophy—especially the mixtures of hot, cold, wet, and dry that

[7] Note that Ibn Ridwan, sometimes called "Hali" or "Haly," should be distinguished from Ali ibn al-Abbas al-Majusi (fl. ca. 983) or Haly Abbas.

[8] See Turisianus (1557), 175r–175v. Edwards calls Ibn Ridwan "Haly," following the 1557 text, in which he is named "Hali son of Rodbon."

form the temperaments or complexions of simple parts, which make up the composite parts, and which significantly determine organs' functions (Chandelier 2018).

Ibn Sina (ca. 970–1027) or Avicenna, systematized and distilled Galen's sprawling works and added important twists of his own, especially in his hugely influential *Canon* (Fancy 2020). In this work, analysis and synthesis appear clearly in his discussion of the second internal perceptive faculty, the thinking or cogitative faculty. This faculty "disposes sense impressions stored in the imagination" and "rearranges them through synthesis and analysis" (Avicenna 1993, 1:116). With this activity, the thinking faculty produces forms such as are received from the common sense, and variant forms no one has sensed, "such as a flying man." Given the relatively terse nature of the *Canon*, Avicenna does not seem to have commented at length on the beginning of the *Ars medica*, as later authors did. But his remarks on following signs and symptoms to understand the locations, dispositions, and causes of disease track Galen's methods. In this way, the "analysis and synthesis" in his remarks on the thinking faculty render understanding of healthy and diseased organ function reliable.

To see his method at work, we will turn to the third book of the *Canon*, which discusses diseases from head to heel by the part of the body whose function is impaired. Like Galen, Ibn Sina understood health primarily in terms of organ function, such that "an organ with perfect functions is a healthy organ" (Avicenna 1993 1:185). Earlier, in his definition of disease, we read the following:

> Disease is the abnormal condition of the human body which, by itself, produces functional disorder as a primary consequence, and that is either an intemperament or an abnormal composition…. An example of cause is the hot catarrh; that of disease is the ulcer in the lungs; and that of symptom is redness of the cheeks and the curving of the nails. (Avicenna 1993, 1:119)

In this definition, Ibn Sina is clearly describing phthisis or consumption, since he moves from the traditional cause (hot catarrh dripping down into the lungs), to the defining lesion favored by Galen and other writers (ulcer in the lungs), up to the characteristic symptoms of red cheeks and curved nails (Meinecke 1927; Ragland 2022, 364–77). Signs and symptoms of other diseases follow a similar inferential pattern, such that "weakness of the liver is indicated by stools and urine resembling the washing of fresh meat" (Avicenna 1993, I:185). A wavy pulse and pain in the chest indicate swelling and disease in the substance of the lungs (rather than the pleura). Ibn Sina also makes distinctions among diseases, summarizing much of Galen's work on the differentia of diseases:

> Diseases are named in various ways: after the organs carrying them, as *dhat al-janb* or pain in the sides (pleurisy) and *dhat al-riya* or pain in the lungs (pneumonia)[;] after their symptoms, *sar'* or falling down (epilepsy); after their causes, for example we say melancholia, after their resemblance, as we say leontiasis or elephantiasis… (Avicenna 1993 I:128)

Latin authors, drawing on Avicenna's text for centuries, would express similar ideas, but add elaborations, debates, and anatomical dissections.

8.3.2 Medieval Latin Sources

Arabic medieval sources were vital for discussions in medieval Latin texts. The twelfth-century anatomical text, *The Second Salernitan Demonstration*, for instance, uses Arabic terms for anatomical parts and repeatedly attributes the causes of diseases to the "gathering" of humors in certain organs (Wallis 2010, 164–166). When fluid gathers in the place between the trachea and esophagus, it causes one type of quinsy; when it gathers in the lung, it causes difficulty of inspiration (versus when it gathers outside the lung and causes difficulty of respiration); and when humors gather in the membranous capsule around the heart, it causes syncope (Wallis 2010, 163–5). Thus, the causes sought in reasoning back to symptoms—and in this case illuminated by the first-person descriptions of the dissections of pigs—were often the anatomically localized causes emphasized in the ancient and medieval traditions surveyed so far.

Medieval physicians writing in Latin often translated the Greek terms "analysis" and "synthesis" as "resolutio" (or "dissolutio") and "compositio," or resolution and composition. Taddeo Alderotti (d. 1303), a professor of medicine at Bologna was the first known writer to describe the movement of "composition" or "synthesis" from causes back to effects as a *regressus* (Edwards 1967, 63; Siraisi 1981, 239). This terminology continued, with changes of use and meaning, through the end of the 1500s, as is well known. In the next scholarly generation, though, at nearby Padua, Pietro d'Abano (ca. 1257–1315) echoed Galen's description of the threefold path of analysis (or *dissolutio*), synthesis (or *compositio*), and dissolution of the definition in his *Conciliator*, as Randall noted long ago.[9]

But Pietro also described how the physician can use these general "paths" to structure knowledge of causal relations and symptoms in specific cases (rather than giving a general method of diagnosis). He gave this example: "Take, therefore, something sought in the mind according to the analytical teaching, for example humoral fever, and break it up into subject and predicate, namely a putrid fume or vapor immediately seeking the heart…" (Ottoson 1984, 115). This fume or vapor, he continued, is caused by local putrefaction, which is caused by impaired transpiration, which is caused by an excess of fat, which is caused by an excess of food, which is caused by excessive appetite, which is caused by a cold stomach, which is

[9] Latin from Randall 1961, 31–2, n. 5. "In all the teachings which run according to an order of progression there are three orders. One of these is that which follows according to the way of conversion and dissolution [*dissolutio*]. And according to it you set up the thing you are aiming toward and the knowledge of which you are seeking in your mind according to the end of its completion; then you must examine closely what is next to it, and the next from that, without which the thing cannot stand; nor is it at all completed until you come to the principle of its completion. And the second is according to the way of composition [*compositio*], and follows the opposite of the first path. And in it you begin from the thing at which you arrived according to the way of analysis, and then return to those things and then put them together, one to another, until you come through to the last of them. And the third is according to the way of the dissolution of the definition." My translation.

caused by a cold temperament, humors, or environment, such as cold air. Thus, in analysis, the physician conceptually distinguished the phenomena of interest, then reasoned back along a causal pathway to prior causes and principles.

The method of synthesis begins from a qualitative cause of excess cold, or the principle found by analysis, and works downward to the cold stomach and, hence, its impaired function of concocting food, to the excess appetite, the excess of food, the excess fat and humors, the obstruction of transpiration, the local putrefaction, and finally to the resulting putrid fume that causes the fever. This is clearly an example of something like Galen's method of therapeutic analysis, found in his *Method of Medicine* and other works on method, at work in the academic medicine of the thirteenth and fourteenth centuries. Though Pietro was later declared a heretic, in part due to his strong advocacy for astrology and naturalistic denial of some miracles, and his remains disinterred and burned, his works had wide influence in the 1400s and after.

By the 1400s, we find Jacopo (or Giacomo) da Forlì (d. 1414), teaching at Bologna in the 1380s and at Padua by 1400, and writing commentaries on Galen's *Art of Medicine*, the Hippocratic *Aphorisms*, and Avicenna's *Canon* (Ottoson 1984, 53). His commentaries were probably frequently used as textbooks (Siraisi 2001, 116, n. 7). In his commentary on Galen's *Art of Medicine* (then most often called the *Tegni*, from its Greek title, *Techne iatrike*), Jacopo emphasized analysis or "resolutio":

> First note that resolution is twofold, namely natural or real, and logical. Real resolution, although taken as if multiple and improperly, is yet properly the separation and division of a thing into the parts composing the thing itself. Logical resolution, however, is called so metaphorically, and the metaphor is taken in this way: for just as the composed [thing] while it is resolved the parts are separated from one another, thus whatever by itself persists in its own simple being, when a logical resolution happens, the thing first comprehended confusedly is comprehended distinctly so that the parts and causes touching its essence are comprehended distinctly. As when you first comprehend a fever, and you have the concept of fever general and confused; *resolve*, then, the fever into its causes; since any one exists either from the heating of a humor or of the spiritus or of the parts; and again [if] it exists from the heating of a humor, either of the blood, or of phlegm, etc.; finally, you will come through to the specific and distinct cause and knowledge. (Da Forlì 1487, a2)

These are clear examples of prominent medieval physicians elaborating on Galen's texts, terms, and ideas, but with more refined and concrete directions. Importantly, Jacopo distinguished clearly between analysis or resolution as the physical separation of natural things into their parts, and logical resolution. His categories and examples of "real resolution" and "logical resolution" extend the earlier distinction made by Alexander of Aphrodisias, and with concrete examples. Of course, as historians such as Michael McVaugh have emphasized at length, it is almost impossible to determine what such physicians did in practice (McVaugh 1971, 2009). As Ottoson reminds us, these medieval physicians writing in Latin mainly concerned themselves with understanding Galen and with coming up with different schemes for ordering established knowledge, although at times they did give specific examples that suggest they did use methods of "analysis" and "synthesis" in diagnosis (Ottoson 1984, 124–6).

By the 1500s, physicians assumed a distinction between methods of teaching and discovery, which might suggest that such methods of "analysis" and "synthesis" might not have been used explicitly in diagnosis or to determine treatment (Maclean 2007, 201). However, the general structure of reasoning from perceptible symptoms back to hidden causes, notably causal principles provided by natural philosophy, characterized much of the medical thinking from the ancient and the medieval periods.

8.4 Leoniceno, Da Monte, Fernel, and Argenterio: Analysis in Commentary, Teaching, and Some Practice from the Early to Mid-Sixteenth Century

In the 1500s in particular, physicians and scholars built textual bridges across the ancient, medieval, and early modern sources. They produced new translations of Galen's works and engaged with the long tradition of medieval Arabic and Latin scholarly texts. Early modern discussions of analysis and synthesis in academic medical and philosophical texts generally commented on passages from Aristotle and Galen. As with discussions of method in general, scholars attempted to explicate the original passages by the ancient authors, and engaged with centuries of commentary to do so (Gilbert 1960, xxiii–xxvi). Galen usually appeared as the most reliable commentator on Aristotle, and writers usually sought to find internal consistency in their works and harmony across the texts of the two ancient authors (Gilbert 1960, xxvi, 13–19). Critics, however, did not find it difficult to point to tensions and even contradictions in Galen's writings, and between Galen and Aristotle.

We will briefly examine the views of four influential physicians: Niccolò Leoniceno, Giambattista da Monte, Jean Fernel, and Giovanni Argenterio. Leoniceno (1428–1524) was a leading humanist physician and professor of the late 1400s and early 1500s and wrote an extensive discussion of Galen's notions of analysis and synthesis, one that inspired or provoked responses throughout the sixteenth century. Da Monte (1498–1551), a disciple of Leoniceno, preferred a life dedicated to the pedagogical formation of physicians to one of publishing his own works or even a courtly appointment (Bylebyl 1979, 346). Instead, he oversaw the definitive Giunta editions of the Latin texts of Galen's works, and dedicated himself to Paduan students, lecturing, and bedside teaching in private homes and the nearby San Francesco hospital, leaving it up to his loyal disciples to publish all the teachings they hoarded from lectures and clinical instruction notes. In contrast, Jean Fernel (1497–1558) established his reputation as a leading scholar and synthesizer, writing his own textbooks of the whole of medicine that grafted astrological and Neoplatonic ideas into a largely Galenic trunk, earning him a place as a physician of the French royal court (Henry and Forrester 2003). Argenterio (1513–1572), taught at the University of Pisa from 1543 to 1555, roughly the same time that Da

Monte taught at the rival Padua, and deployed his own humanist erudition and clinical experience to mount a skeptical critique of Galenic medicine, pointing out its inconsistencies, lack of true demonstrations, and questionable status as *scientia* (Siraisi 1990). Taken together, these sections give further support to historians since Randall's thesis, who have looked beyond Padua—for instance, to Leoniceno's Ferrara or Fernel's Paris (Gilbert 1960; Nutton 2022).

Here, I show that Galen's discussions of analysis and synthesis, combined with the revival of his works on the method of medicine and anatomical practice, inspired influential academic physicians to frame their teaching and practice in terms of analysis and synthesis, but particularly analysis. While they continued to discuss synthesis, and the third mode of ordered teaching, *dialysis* of a definition, analysis took center stage as a method useful in teaching and medical practice. All four physicians here, even the anti-Galenist critic Argenterio, accepted both the utility of analysis for identifying the seats and causes of diseases and the Galenic emphasis on the anatomical localization of diseases. For at least Fernel and Argenterio, postmortem dissections could reveal to the senses through hands-on anatomical analysis what reasoning, according to the method of analysis, could discover in diagnosis: the anatomical seat of the disease and its morbid conditions. In this period, analysis and synthesis in academic medicine appear vividly and frequently as modes of ordered teaching (*doctrinae*), and increasingly as modes or methods or orders of diagnosis and discovery in practice.

8.4.1 Leoniceno: Humanist Recovery of Galenic Analysis and Synthesis

First, let us look to Leoniceno's expansive commentary on the three ordered ways of teaching or *doctrinae* Galen described at the beginning of his *Art of Medicine*. Like his humanist colleagues, Leoniceno sought to recover the true meanings of Galen's text through better translations, ongoing critical engagement with ancient and medieval commentaries on Galen's works, and his own interpretations and arguments. Some scholars compiled previous commentators' discussions into single volumes, textually unifying the centuries-long debates (Champier 1516?). Leoniceno had followed the example of humanist historian Angelo Poliziano (1454–1494) and wielded his excellent linguistic skills and memory to expose and cut down the errors in the Latin versions of traditional medical texts. His 1492 *On the Errors of Pliny and Others in Medicine* made his reputation and exemplified his program for refining scholarship to get at the original Greek texts and their meanings, all in the service of better knowing the *things* of medicine: medicinal herbs and minerals, organs and simple body parts, and patients in sickness, convalescence, and health (Nutton 2022, 100–102).

Leoniceno sought to engage with other scholarly traditions of analysis and synthesis—from philosophy or mathematics, for example—define terms and methods, and mark out the medical meanings of analysis and synthesis. His 1508 discussion of the preface to Galen's *Art of Medicine* extended over some 34 folio pages, drawing together texts from ancient and medieval writers and highlighting important distinctions and contradictions (Leoniceno 1509). In the preface to the work, Leoniceno had trumpeted his scorn for the "infantile and barbaric speech" of Pietro d'Abano, who wrote "rubbish" by including the cause of the disease in its definition (Leoniceno 1509, Aiiir). Yet, as Siraisi has shown elegantly, Galen's own texts displayed inconsistencies on just this point, in some places drawing neat lines between the cause of the disease, the disease, and the symptoms, while at other places allowing for causes and diseases to be called symptoms. Similarly, Galen emphasized disease as a condition of the body that primarily impaired a function, but he was unclear or waffled on what counted as a condition or affect of the body (*diathesis*) stable enough to be called such, unlike, say, mere convulsions (Siraisi 2002, 224–229). Leoniceno, like nearly all his colleagues, also emphasized disease as a disposition or condition (*affectus*) impairing bodily functions, against nature.

Much like his medieval sources (and targets), Leoniceno translated Galen's "*analysis*," "*synthesis*," and "*dialysis*" in the beginning of *The Art of Medicine* as "*resolutio*," "*compositio*," and "*dissolutio*," respectively. He explicitly engaged with a long list of authors, including Plato, Aristotle, Galen, Porphyry, Ammonius, John Philoponus, Alexander of Aphrodisias, Simplicius, Proclus, Ibn Sina (Avicenna), Ibn Rushd (Averroes), Ibn Ridwan, Pietro d'Abano (the Conciliator), and Turisanis (Pietro Torrigiano, the Plusquam Commentator).

Faced with this scholarly mob, Leoniceno sought to make some clarifications and rebuttals. Leoniceno agreed with the general structures of analysis and synthesis established in the tradition so far, namely that analysis ascends from things caused or from the effects to the causes, or from composite things to simple things, while synthesis moves from causes to effects, or from simples to composites. But he wanted to distinguish the *syllogistic method* of Aristotle as a mode of demonstration—one of the *four* ancient modes or methods of dialectical teaching or demonstration—from Galen's *three* modes of ordered teaching (*doctrinae*), as expressed in the beginning of *The Art of Medicine* (Leoniceno 1509, 32r–32v; cf. Gilbert 1960, 103). As a skilled humanist, Leoniceno also worked his way through dozens of texts, drawing quotations and paraphrases to make distinctions and follow (or construct) traditions.

First, Leoniceno discussed logic and modes of dialectic. Platonic philosophy, he wrote, had established four modes of teaching or dialectic: the divisive, the definitive, the demonstrative, and the resolutive or analytical (Leoniceno 1509, 21v, 24r). According to Leoniceno, the Platonic and Aristotelian sects are not so far apart on these four modes of teaching and the multiple meanings of resolution or analysis in the demonstrative mode—although Aristotle gets the glory for teaching the resolution and composition of syllogisms. Properly, analysis or *resolutio* is the disentangling of composite syllogisms into simple ones. In general, *compositio* proceeds from the simple to the compound, *resolutio* from the compound to the simple (Leoniceno 1509, 22v). As Philoponus and Alexander had it, the resolution of the whole into its parts, of course, appears in various ways, from the composite to the simple, as in geometry, or natural bodies into the four elements, or from particulars to universals, and from what is (knowledge *quia*) to the cause or reason why (knowledge *propter quid*), as well as the reduction of syllogisms into propositions (Leoniceno 1509, 23r, 25v). Resolution from effects to causes seems to count as *demonstratio quia*, even though Aristotle emphasized demonstration *from* the principles believed per se, rather than from the effects.

Next, Leoniceno considered the mathematicians' texts. The resolution or analysis of the mathematicians seems to be different, Leoniceno argues, since it does not usually end with finding principles but consists in assuming what one wants to prove and then proceeding according to this until one arrives at something true, as in the thirteenth book of Euclid's *Elements*. Proclus, though, "the most excellent mathematician according to the judgment of all the Greeks," attested that ancient mathematicians used analysis no less than the other three dialectical modes. In his report of Proclus' testimony, and his gloss of the books of the ancient mathematicians, Leoniceno brought his humanist erudition to bear against Pietro d'Abano's denial that mathematicians used analysis to find a demonstration *propter quid* (Leoniceno 1509, 26r). The Conciliator himself was led astray by the Arab commentators, notably Ibn Ridwan and Averroes, whose bad translations and arguments distorted Aristotle's texts with thousands of mistranslations and shadowed "an earth darkened with the darkness of ignorance" (Leoniceno 1509, 26v).

Leoniceno repeatedly commented on Galen's texts at length, arguing that his three ordered modes of teaching could also be modes of demonstration, and that Galen used the three ordered modes of teaching to structure his works. As we have seen, Galen also prized geometry as an example of coming to know by analysis and synthesis, and Leoniceno acknowledged Galen's comparison but also focused on medical method. Diagnostic and therapeutic methods followed the resolutive way, as in the eleventh book of the *Method of Medicine*. There, Galen argued that to cure a putrid fever one had to follow two signs indicating action, one from the fever as excess heat to a cooling therapy and the other from putridity somewhere in the body as the cause of the excess heat (Leoniceno 1509, 27r–27v; Galen 2011, 121–5). In this case, Leoniceno portrayed Galen's text as using analysis to find the cause of the fever, and then to intervene to remove this cause. He made a clear distinction between analysis as discovery (of either knowledge *quia*, or of causes or principles) and analysis as teaching (Leoniceno 1509, 30v). Clearly, Galen used different modes in his different works: *The Art of Medicine* explicitly used the definitive ordered way of teaching, *On the Elements According to Hippocrates* employed the synthetic way, and the *Constitution of Medicine* used the analytical way (Leoniceno 1509, 35r–35v). Leoniceno's characterizations fit the structures of Galen's texts reasonably well. The *Elements* book progressed from the elemental qualities of hot, cold, dry, and wet, to mixtures, humors, and other simple parts as a prolegomenon to *On the Natural Faculties*. This structure moved from principles or simple causes to effects and composites. *Constitution*, a short, mature summary like *The Art of Medicine*, began with the natural and healthy functions of the body—which together constituted health, the end goal of medicine—and then resolved back to defects in the functions, then the components of matter, mixtures of the body parts, simple and compound parts, and then diagnosis, drugs, prognosis, prophylaxis, and the restoration of health.

Strikingly, Leoniceno demoted Galen from the position of the first to use the analytical method, since other physicians had done so before him, yet insisted that Galen had used it, and the synthetic mode, excellently in his actual practice. Obviously, Galen used the synthetic method in the composition of medicines, "which is according to the judgment of the physician, taking into account the various dispositions of the sick patients, and the diversities of times, and rationally following the indications, and books on the curative art" (Leoniceno 1509, 35r). On the analytical method, Leoniceno declared that Torrigiano was wrong to claim that Galen was the first physician to use it, and all the other ancients lacked the requisite subtlety of reasoning.

For Leoniceno, the method of moving from signs and symptoms or other accidents back to their causes was the method of the Hippocratic writers and other physicians. For instance, following Galen's practice we would perceive the signs of a fever, a thick pulse, and pain in the head and reason back to putridity causing the fever:

> But insofar as it pertains to the diagnosis of fever, Galen himself did not judge that we are able [to] do it by any other method more skillful than that by which we thoroughly examine the hidden diseases and their causes, that is, from the accidents which are posterior. (Leoniceno 1509, 38r)

Through his teaching, Leoniceno passed on the rational search for the causes of diseases through analysis and, so, true diagnosis and treatment, to his students. As one student, Giovanni Manardi, argued, "true physicians" did not care as much for names of diseases, but followed a method to "inquire into the substances and causes of diseases according to division and resolution, from which they elicit curative indications, and acquire intentions, with which they may find instruments for driving out diseases from human bodies" (Manardi in Wightman 1964, 371). As we will see further in the next sections, similar expressions and practices spread widely through university teaching in the sixteenth century.

8.4.2 Da Monte and the Analytical Order of Galenic Medicine

In the 1540s, especially, Giambattista da Monte, Leoniceno's student, sought to recover Galen's rational method of diagnosing and treating diseases by isolating their true causes. In this, he elevated analysis, making the analytical "order" (*ordo*) a way of knowing that involved reasoning from effects to causes, and even included division, definition, and composition (synthesis). A leading professor of medical practice in the mid-sixteenth century, Da Monte taught this methodology to many hundreds of students and followed it in his practice.

Da Monte objected to practicing medicine without analysis of the causes. The ninth book of al-Razi's *Ad Almansorem* was the standard basis for the head-to-toe sections of medical *practica* courses, but al-Razi tended to limit his discussions of diseases from head to toe to correlating clusters of diagnostic signs and remedies (Bylebyl 1991, 167). This was the Empiric or Empiricist approach, and not the rational or methodical one. Da Monte used his lectures and commentaries on al-Razi's text as occasions for going deep into the rational, and analytical, search for causes. For example, al-Razi "proceeds empirically," and "empirics omit the indications" or the signs and symptoms that point to the cause and nature of the disease and, so, what the physician must remove or change (Da Monte 1554b, 104r, 6r). But Da Monte followed Galen in insisting that the physician will know nothing about practice unless at the same time we distinguish all the causes and kinds of diseases (Da Monte 1554b, 14v). The order of division is necessary for finding the definitions, especially as part of analysis or resolution. Delirium, for example, must have its seat in the impaired action of the principal part of the soul that has its seat in the brain or spinal cord. As in his other works, Da Monte discussed Galen's three orders (analysis, synthesis, and dissolution of the definition), but across his texts the analytical order does most of the work isolating diseases and causal pathways (Da Monte 1554b, 4r–6r).

In his lectures and bedside teaching sessions, Da Monte constructed and practiced a sophisticated Galenic method using division, analysis, and synthesis to discover and then intervene on the causes of diseases. As Jerome Bylebyl and Craig Martin have shown, Da Monte used a rational method of reasoning carefully from signs and symptoms back to the impaired functions and conditions of impairment (the diseases), and the causes of diseases (Bylebyl 1993; Martin 2022). Da Monte prized the method of division for isolating the nature and causes of disease, the rational inference to the disease and its causes from the consideration of signs, and the discovery of proper treatment according to curative indications (Bylebyl 1991, 175, 185).

In his commentary on Galen's *Method of Medicine to Glaucon*, Da Monte insisted with pride that he had taken his method from Galen, but set it down clearly and discussed method in general (Bylebyl 1991, 178). He surveyed the four ancient philosophical methods—demonstration, division, resolution (or analysis), and composition (or synthesis)—and noted that division and especially resolution (analysis) were essential. Division distinguished functions, parts, and treatments. Resolution or analysis involved the sequential division of disease in general down to its species and then causes, which gave the physician knowledge of its essence and indicated the proper treatments. Further, Bylebyl argues that Da Monte ordered his medical doctrine with an analytical or resolutive order, beginning with health as the goal of medicine, but ordered his teaching according to division (Bylebyl 1991, 178–9). But no single method would allow the physician to jump from even the lowest levels of generality to the particular concerns of medicine: curing *this* individual patient. Da Monte trusted the universal explanatory principles of Galenic medicine, which allowed him to reason from signs perceptible to the senses, especially the functions and operations of each patient, to universal judgments "as under a cloud" (Da Monte in Bylebyl 1991, 178).

In the face of the ongoing controversy over what Galen meant by invoking analysis, synthesis, definition, and division, Da Monte attempted to find consistency by making distinctions. Da Monte's commentary on Galen's *Art of Medicine* made further distinctions between analysis or resolution as a "way" (*via*) and as an "order" (*ordo*). The analytical or resolutive order, for Da Monte, was very much what we have seen so far: one begins with the end one wants to teach or discover in an art or science and then moves from the things more known toward the principles, hunting and seeking for causes (Da Monte 1554a, 36r–36v). Health depends on a healthy body, which consists of healthy parts of the body and all of their operations working together in harmony according to nature (Da Monte 1554a, 37r). The parts ultimately depend on their elemental qualitative mixtures or *kraseis* (sg. *krasis*) for their operations or faculties. Ultimately, the resolutive *order* embraces division, definition, and composition, since it begins from the end, subdivides the parts making up the end, and then resolves the parts. Thus, the analytical or resolutive order holds "the whole *scientia*" (Da Monte 1554a, 42v). The synthetic or compositive order goes the other way, taking up the parts and principles found by analysis and running along causal pathways to the symptoms (or from ingredients to compound drugs) (Da Monte 1554a, 37v–39r). The resolutive "way" (*via*), though, does not begin from the notion of an end or goal, but appears rather like the divisive method of taking what is manifold in nature and making logical distinctions. Thus, we can define a human as substantially rational and accidentally mortal, and then consider what is mortal and rational in itself and how those inhere in one substance (Da Monte 1554a, 41r).[10]

Anatomy was "the alphabet of medicine," without which one could never have medicine (Da Monte 1554a, 179r). Indeed, proper medical analysis always involved anatomical analysis. Just as each direction of motion for the eyes depended on one of six different muscles, so all the different actions and faculties of the body depended on different anatomical parts. Galen famously was "most skilled in anatomy," and his *Art of Medicine* considered the parts of the body and their faculties and diseases from head to foot (Ragland 2022, 251–2). Like Galen, Da Monte taught that "the majority of diseases is discovered in the interior organic parts, which cannot be known without correct incisions, and anatomy" (Da Monte 1554a, 193r, 300r).

The operations or faculties of a body are impaired according to the varied nature of its parts, and to know "the disease existing in the part, which is known from the operations" one must know the anatomical composition and temperament, or basic qualitative mixture, of the part (Da Monte 1554a, 301r). "Thus you see that anatomy is necessary not only in diseases from temperament, but it is highly necessary in those of composition, and break of continuity. With these things supposed we come down from anatomy into knowledge of the disease existing there in the organic members" (Da Monte 1554a, 301r-301v).

[10] It is not clear why this should be called resolutive or analytical.

So much for Da Monte's programmatic remarks, drawing on the long traditions of medical commentaries. Did any of this matter for his practice or the practice of his students? We should return first to disease theory and therapy. Like other Galenic physicians, Da Monte defined disease as "a preternatural disposition that clearly disturbs the functions" (Da Monte in Bylebyl 1991, 186). Thus, discovering the impaired functions became the central goal of his Galenic medicine (Bylebyl 1993, 50, 53). In therapy, too, the physician had to make divisions, moving from the symptoms of putrid fever to its cause in a putrefied humor, and then to the division of humors into their anatomical locations and types (Bylebyl 1991, 187). From an impaired pulse, the Galenic theory of the origins of the pulse in the vital faculty of the heart and arteries allowed the physician to reason back to a condition in the heart. From the increased heat (the fever) and the impaired pulse, he reasoned back to a putrid fever. Then the physician had to localize the putrid matter in an organ, vessel, or other part of the body and expel it via purgative drugs or another therapeutic means.

As Bylebyl, Martin, and Stolberg have shown, Da Monte employed his methods of analytical division in the hunt for causes in his actual practice, as recorded in his hundreds of *consilia* (Bylebyl 1991, 184–8; Martin 2022, 47–51; Stolberg 2014, 645–9). Da Monte taught his students to construct a comprehensive *historia* of a patient, including the symptoms and signs of disease, the environment, ingesta, and anything else that might cause conditions of disease, and then he pressed students to methodically reason through chains of causes from symptoms to disturbed functions and anatomical seats and conditions to internal and external efficient causes of disease states (Martin 2022, 47–8).[11] A patient suffering from a paralyzed middle finger prompted the analysis into a fault in either the finger (the organic part) or the spirit sent from the brain along the nerve. But the patient appeared warm and full of blood and spirit, so it had to be the finger's temperament itself, since a bad temperament would impair its function. This temperament likely came from cold fluxes descending from the head, which in turn resulted from both exposure to cold winds and excessive cold phlegm produced in the stomach from a rich diet and a lack of work. Therapy targeted the ongoing proximate or "conjoined" cause, namely the fluxes, and involved warming the finger and evacuation of the fluxes by means of diuretics, laxatives, and bloodletting (Martin 2022, 48). This style of medical analysis in practice may not be compelling to present-day readers or patients, but it matched Da Monte's Galenic science.

[11] Martin points out that Da Monte used Galen's categories of conjoining, antecedent, and primitive causes.

8.4.3 Fernel's Diagnostic Analysis from Symptoms to the Seats of Diseases and Postmortem Dissections

As part of this scholarly movement, Jean Fernel (1497–1558) crafted his immense and influential *Medicina* (1554), revised and expanded as *Universa Medicina* (1567), with many editions into the seventeenth century. Although he is probably best known for his advocacy of Neoplatonic and other additions and refinements of Galenic medicine, Fernel's work presents a thorough systematization of Galenic medicine. The first part of his compendia, *Physiologia*, has been called "the fullest and most clearly organized exposition of Renaissance Galenism that was ever written," and it earned a great deal of commentary and influence (Henry and Forrester 2003, 5). His reputation as a scholar and practitioner won him a place as physician to King Henry II from 1556. In his *Universa Medicina*, Fernel began the section on "*physiologia*" as a distinct category of knowledge with some reflections on the order of teaching inspired by Galen. After averring that our mind, when free of the body, "clearly perceives the uncovered and unveiled essences of things," Fernel insisted that embodied minds use the senses and reason to grasp hidden matters and establish first principles (Fernel 2003, 15). He then wrote:

> The supreme faculty of investigating is what all the most respected Philosophers called Analysis [*Analusin*], that is, dissolution. Which of course proceeds either from the whole and universal to the parts and singulars, or from the composite to the simples, or from the effect to the cause, or from the posterior things to the prior things serially, it investigates those more hidden causes, from which individual things proceeded originally. In the opposite direction to this is the other, the method of putting things together, which nature especially follows and sometimes art itself, it links together the parts into the whole, the simples into the composite, the causes to the effects, the prior things to the posterior things, sets up first of all that which was to be investigated last by dissolution. (Fernel 2003, 14, my translation).[12]

Fernel claimed that philosophers who wanted to "link everything together clearly and by a reliable chain of demonstration" began with analysis. His chief examples were Euclid in geometry, Ptolemy in astronomy, and Aristotle in philosophy (Fernel 2003, 15).[13] Following their lead, Fernel makes the human body the start of medicine, since it is the subject of the art and because it is most knowable by the senses. The rest of *Physiologia* textually broke down the human body into its parts, beginning with the bones, ligaments, and muscles (as did the main anatomical sections of Avicenna's *Canon*) and continuing with the vital and nutritive internal parts, then the head, nerves, veins, arteries, etc., through to hair and nails. In the second book, Fernel explicitly turned to the "analysis" of the parts of the body into simple, or homoiomerous, and composite parts. True philosophers, he admonished his readers, ought to follow established examples and begin with the senses, and then follow

[12] Fernel has very similar remarks in the preface to the second book of the first version of what would become his *Physiologia*, in Fernel 1542, 47r–47v.

[13] Forrester translation amended slightly.

those effects perceived by the senses back to their causes. With the human body laid open to the senses, with its parts separated out, Fernel proceeded to investigate each part's elemental mixture, temperament, power, faculties, spirit, and heat.

> When these things have been found and perceived by analysis [*analysi*], then will become clear by the order of composition [*compositionis*] what are the efficient causes of all, which humors are born from these, what are the functions of the individual parts, and what is the natural management of them all. (Fernel 2003, 180, my translation)

Like other Galenic physicians, Fernel clearly framed the philosophical investigation of the body through anatomy as a process of analysis and synthesis, or dissolution and composition. First, the anatomist separated and distinguished the different parts one from another. Second, with the parts revealed to the senses, the anatomist–philosopher (as a learned physician ought to be) followed the way of analysis to find the causes of the parts and their properties. Finally, the way of synthesis or composition showed the common efficient causes, humors, functions, and integrated workings of the parts.

Famously, Fernel repeatedly embedded practical concerns and even postmortem dissections into his new, systematic textbooks (Siraisi 2002, 230–1, 240; Stolberg 2018, 73, 76–7; Ragland 2022, 262, 370). Anatomy was the landscape of the processes of disease and health:

> Thus, just as Geography should be thoroughly learned to give credibility to history, so should a description of the human body be well learned for the medical subject. (Fernel 2003, 178, my translation)

For Fernel, this extended not only to practices of diagnosis and therapy but to the increasingly important practice of postmortem dissection for generating new pathological knowledge. Throughout his major work on diseases, *Pathologia* (1555), Fernel aimed to teach his readers how to identify and treat "the affect seat" or the "affected part" (*sedes affecta* or *pars affecta*).[14] Following Galen's *Method of Medicine*, *Constitution of Medicine*, and *On the Affected Places*, as well as the rise of anatomical dissection across the sixteenth century, Fernel emphasized the anatomical localization of diseases (Siraisi 2002, 234; Ragland 2022, 263). In a chapter from his *Pathologia* entitled "By what method the affected seat is to be investigated from the signs," Fernel laid out his analytical method of diagnosis to the anatomical seat of disease:

> Every investigation leads from that which is perceptible to sense to abstruse and hidden causes, and that which is last in its origin and the order of causes, occurs first in the investigation.... damaged function, or abnormal excrement, or pain... first leads us to the suspicion of an affected part.... almost no symptom exists alone, but in one and the same disease many always go together. Therefore when one has heard everything and thought over everything, if the symptoms agree in signifying one and the same affected part, the affected site should be investigated.... For example, suppose we decide to inquire into certain things about difficulty of breathing. Because when that function is injured, it is established that one of the organs that serve breathing is affected, see if there is any defect peculiar either to

[14] For references, see Fernel 1555, 10r, 33r, 34r, 35r, 36r, 50v, 54r, 89r, 94r, 100v, 110v, 120v, 121v, 126r, 127v, 164r, 172v, and 176r.

> the throat or the trachea. If nothing appears wrong with these, the cause must be in the lungs or thorax or diaphragm. When stertor is heard, and there is a bothersome cough without a sense of pain, the fault is in the lungs. When pain presses in the chest, around the false ribs of the right side, it should be picked out and discerned from other signs. For if there is a sharp pain with a continual fever, with a cough that expels very bloody spittle, this is an index of pleurisy. If, on the other hand, the pain is serious and with a continual fever, with a dry cough, this brings a conjecture of the inflammation of the liver. (Fernel 1555, 34r–34v)[15]

While today we would describe this as a process of differential diagnosis, Fernel framed it in terms of analysis: moving from what is perceptible to the hidden causes, positioning the symptoms as effects from which one could reason back to the impaired functions of organs or simple parts, and making inferences to the chains of causation linking hidden anatomical conditions to perceptible symptoms.

Diseases had to be known in terms of their anatomical seats, and postmortem dissections could reveal diseased organs and parts to the senses. Fernel certainly put a high value on this sort of knowledge of diseased anatomical seats:

> I considered that no disease was known and examined deeply enough, unless it was held verified and as if discerned with the eyes which seat in the human body primarily suffered, which was the affect in it beyond nature, whence this proceeded ... or finally what interior cause fostered it. (Fernel 1555, 93v)

As Michael Stolberg has argued, Fernel used pathological phenomena from post-mortem dissections to make sense of the causes and conditions of diseases. In one case, Fernel argued against common opinion by reporting that the brains of epileptic patients showed abscesses rather than obstructions (Stolberg 2018, 76–66). Fernel's dissections of patients' lungs sometimes revealed "true stones," some "extremely hard and solid, others of the consistency of old cheese, others just beginning, with the hardness of gypsum" (Fernel 1555, 113r). Since Fernel followed Hippocratic tradition in thinking that phlegm dripping down or extravasated blood caused such morbid deposits in the lungs, he seems to be hinting toward a process of development through accretion and hardening—from liquid to cheese-like substance to stones. He also asserted, though, that even a "soft, tender substance of the lungs" could become corrupt, without any morbid humor, leading to abscesses and consumption (Fernel 1555, 115r).

By at least the 1590s, as we will see, physicians used more extensive postmortem dissections to begin sketching a more detailed history of the gradual development of tubercles in patients killed by consumption. But this section has provided consistent evidence for Fernel's use of the order of analysis both for teaching in his textbooks, and for the discovery of the anatomical seats and causes of diseases in his medical practice. By at least the mid-1500s, postmortem dissections appear as making perceptible the key stages of medical analysis.[16]

---

[15] I have added to the translation in Siraisi 2002, 233.

[16] There was a much longer tradition of postmortem dissections, beginning in the late 1200s with many private postmortems among Italian families and in cases of purported sanctity, as Katharine Park has described in Park, 2010. See also Siraisi 2001, 226–52, and Stolberg 2019.

8.4.4 Argenterio Against Galenic Pretensions, But in Favor of Analysis, Anatomical Localization, and Postmortems

Giovanni Argenterio, working and writing at the same time or just after Da Monte and Fernel, attacked Galen's claims in print nearly 200 times, even as he adopted Galenic concepts and modes of thought (Siraisi 1990). Duke Cosimo I of Tuscany recruited Argenterio to a chair in medical theory at the University of Pisa based on Argenterio's roughly 9 years of medical practice in Lyons and Antwerp (Siraisi 1990, 165). In the mid-1540s, he wrote a commentary on the Hippocratic *Aphorisms* and began a two-decade project writing a commentary on Galen's *Art of Medicine*. While Da Monte had used the standard works of Ibn Sina or al-Razi (Avicenna or Rhazes) in his lectures as starting points and targets of critique, Argenterio dismissed Ibn Sina from the curriculum. Although Argenterio lectured on Galen's texts, he did not spare the "prince of medicine" or his modern champions, writing sharp critiques of Leoniceno, Da Monte, and others he thought followed Galen into inconsistency and error.

Argenterio's massive commentaries on *The Art of Medicine* run to over 750 double-column folio pages and brim with learned critiques, skepticism, and counsel for medical theory and practice. His text displayed his humanist erudition as a sharp blade, setting off the Greek lines from Galen's text and pulling out terms for investigation. Leoniceno was wrong, he wrote, to claim that the Greek term *ennoia*, usually translated as "notion," could include the desire for an end as distinct from the idea of an end. After all, you could desire the end of medicine, which is health, without essential knowledge and so be unable to perform the analysis necessary to achieve health. What is more, mental states in general are not desiderative (Argenterio 1610, col. 21). Da Monte did worse to argue that the analytical way was distinct from the analytical order. He "disputes *ad nauseam*, saying nothing certain or consonant with reason or ancient authority, as anyone can see" (Argenterio 1610, col. 22). For Argenterio, what Da Monte wants to call the analytical way or the analytical order both fell under "analytical method" very aptly. The analytical method embraces many structurally similar processes: moving from the inferior to the superior, from the *principiata* to the principles, from the particulars to the universals, from effects to causes, from the conclusion to the premises, and from the imperfect to the perfect (Argenterio 1610, col. 21). After resolution finds these origin points, synthesis explains the effects or particulars by them (Argenterio 1610, col. 24). Through both, the philosopher or physician finds a single "way" (*via*), just as the geographic line from Athens to Thebes is the same as that from Thebes to Athens. "Method" in general is the *via et ratio* in which something is first, something else second, and so on, in a process of investigating and finding (Argenterio 1610, col. 32).

When Argenterio gave an example of analysis, he followed Galen in taking health as the end of medicine, then health as a condition of the body according to nature, then the different parts of the body performing natural actions, and so on down to elemental mixtures of the parts. Like his targets of criticism, he defined health in anatomical terms:

> The health of an organic part consists from the commoderation of those from which it is composed; they are composed, moreover, from the due conformation, magnitude, number, conjunction, and union of the similar parts; it is necessary for the Physician to know Anatomy, since from this the structure of the organic parts is most powerfully discerned. (Argenterio 1610, col. 31)

From the elemental mixtures and structures of the parts come the faculties, and so the actions and usefulness of the parts for the healthy body. Similarly, to really know bodies in health and disease one would have to analyze all signs and symptoms, as well as the nature, genera, differentia, and powers of diseases, and causes of diseases (Argenterio 1610, col. 32).

Galen had failed to do all this and more. Siraisi has shown that Argenterio simultaneously critiqued Galen for his seeming contradictions and adopted a pathology that remained largely Galenic (Siraisi 1990, 169–71). Galen was inconsistent on what counted as a "condition" or "affect" of the body (*diathesis*), and he waffled on whether disease was a condition that impaired function or anything at all that impaired function (Siraisi 2002, 229). Worse, Galen never gave a reliable way for distinguishing diseases from symptoms or causes. After all, Galen confessed that since a symptom was something that happened in the body beyond the natural, diseases and the causes of diseases could be called symptoms (Siraisi 2002, 225).

Worse still, Galen's system misled Argenterio's own medical practice, encouraging him to blithely and mistakenly watch the gradual death of his wife from consumption. One fall in Pisa, Francesca Damiana, "my wife, a most beautiful, praiseworthy woman, was twenty years old," when she began to spit blood (Argenterio in Siraisi 1990, 171). She feared the signs of almost certain death, but Argenterio followed Galen's confused advice that spitting blood without a cough or phlegm indicated a diseased seat *above* the lungs. If true, Francesca faced only a wound in her throat or head, not a fatal ulcer of the lung. But then came the phlegm, pus, cough, and fever; by the end of the month of May, Francesca was dead. Strikingly, while Argenterio did not suffer his own wife to be dissected, he drew pathological conclusions from other postmortems. In 1 year, two other young women of Pisa died from the same disease. The wife of a fellow physician lived for 2 years with the symptoms, and after she died her body was dissected. There, "we found in the highest part of the lung, which is underneath the first ribs, an ulcer, in which one could scarcely place a chestnut, and she never spit up blood, because the ulcer was located beyond the vessel of the lung" (Argenterio 1610, col., 687–8). From this, Argenterio concluded that not all who suffer from consumption spit up blood, nor suffer coughing, but only do so when the ulcer is in the deeper parts of the lung where there are greater blood vessels.

With Argenterio, we see that even a critic of Galen and Galenic physicians, who mounted sharp skeptical critiques of pretensions to knowledge, embraced key elements of the longer story of medical analysis and synthesis. Argenterio accepted the anatomical localization of disease, the rational processes of diagnosis, and even postmortem pathology. By the end of the sixteenth century, medical professors increasingly developed these methods to embrace innovations in pathology.

8.5 Analysis in the Lecture Hall and at the Bedside in Early Modern Leiden

Analysis as a way of teaching and a way of discovering the causal pathways of diseases by rational method suffused the academic discourse of sixteenth-century medical schools. This continued into the seventeenth century, as pedagogical practices, texts, and people trained in Italian universities moved north. Here, we move to the Low Countries and find Galenic rational diagnosis presented again as analysis. This provides a pedagogical link into medical practice and innovations, discussed in the following sections.

In his expansive, and popular, textbook of medical theory, the Leiden professor of medical theory and practice Johannes Heurnius (or van Heurne, 1543–1601) sought to make the vast range and power of the best of Galenic medicine accessible to students. He set out a combination of pathology and therapeutic method that was aimed primarily at the accurate identification of the impaired parts in patients' bodies:

> Diseases, as we said, dwell in the parts. However many ways therefore those become faulty, there are just so many diseases. (Heurnius 1638, 516)

In this section, we will turn to Leiden University, which boasted one of the most popular and innovative medical schools of the early modern period, and which modeled its pedagogy on that of Padua where many of the early professors had studied (Ragland 2022, 74–82). From its founding in 1575 until around 1640, the teaching there was solidly Galenic. Professors also performed their own anatomical research, finding new bones in the ear, for example, and critiquing past anatomists such as Galen and Vesalius.

In early 1587, Heurnius published a version of his lectures on practical medicine as his *New Method of the Practice of Medicine* (*Praxis Medicinae Nova Ratio*, 1587, 1590, etc., 1650). In over 500 double-columned pages, he detailed his "new" method for precisely knowing the qualitative variation of all the parts, variations of the hot, cold, dry, and wet qualities that impaired functions and, so, constituted diseases. Like Fernel, he wrote frequently of the "sick part" or parts involved in each disease. He freely acknowledged that his approach followed Galen's pathology, diagnostic method, and therapeutics, and frequently cited Galen's *Art of Medicine*, *Method of Medicine*, and *On the Affected Places*. He claimed his new method of targeting the qualitative variations of the parts and then precisely weighing out the right amounts of ingredients to make medicines for correcting their qualities would allow the physician to make safe, effective, and agreeable remedies with just a few grains of each ingredient (Ragland 2022, chs. 4 and 5).

Like Galen, he described the natural qualitative variations of the principal parts: an individual's naturally hot brain, cold brain, wet brain, dry brain, then a hot, cold, wet, and dry heart, etc. He then cataloged the morbid qualitative variations of the parts away from their natural functional states. To do this, he worked exhaustively through how students should perceive and make judgments or conjectures about a long list of signs:

> 1. From the condition itself, that is, from the disease, the cause of the disease, or the pressing symptom. 2. From the temperament of the whole body of the sick patient. 3. From the part occupied by the disease. 4. From the powers of the sick patient. 5. From the ambient air. 6. From age. 7. From daily custom. 8. From the particular nature of the one who is occupied. 9. From exercise. 10. From the length or brevity of the disease. 12. From the four times of the disease, namely, beginning, increase, height, and decline. 13. From the particular paroxysms of the diseases. 14. From the ordained functions of nature. 15. From the powers of medicines. 16. From the influx of the stars. (Heurnius 1650, 465)

He always taught his students to begin with the qualitative variations of hot, cold, dry, and wet, in the mixtures of the parts which impaired functions and often constituted "the condition itself"—that is, the disease. Like Galen, Fernel, Da Monte, and other physicians, he performed differential diagnosis, looking for consistent connections between clusters of symptoms and causal chains inside his patients' bodies. He also explicitly connected his method to "analysis." First, as elsewhere, he discussed differential diagnosis and the various causes of different signs or symptoms.

> The particular symptoms of the parts very manifestly reveal the sick part [*pars aegra*]: so that disgust toward food speaks of weaknesses of the stomach: blood excreted through the bottom which resembles water in which was washed recently butchered meat, indicates a weak liver…if someone is morbidly drawing *spiritus*, he will be tinged with a ruddy jaw, and will expectorate foam, and then we will affirm that he has inflammation of the lungs. Therefore, we are not confident of the kinds [of diseases] from the redness alone, for that man might be florid, as with others, from nature, or it could be a forerunner of a coming crisis. Thus, in the hepatic excretion already mentioned, we must explore whether it is flowing out from hemorrhoids; or from swelling and heaviness of the liver, given the color of the face, and then out from the subordinate gland. (Heurnius 1650, 477)

So far, Heurnius has presented the classic diagnostic practice of differentiating among diseases by the presence or absence of their characteristic symptoms. He uses very striking visual imagery, such as the liver disease indicated by bloody excreta also mentioned by Avicenna, and notes the importance of joining symptoms to establish different organs as seats of diseases. A palpably swollen liver, color in the face, and bloody excreta *together* establish that the liver is impaired; a ruddy jaw and coughing up foam *together* indicate inflamed lungs; other jaws could be naturally and healthily ruddy, or be made red by the oncoming crisis of a fever. Strikingly, this example comes directly from Galen's *On the Affected Places*, although Heurnius did not give the usual reference to Galen's works in the margin or in the main text (Galen 1520, 60v, 49r–49v). Apparently, he knew Galen's book so well it stealthily and naturally populated his thinking.

Heurnius goes on, explicitly leading this kind of differential diagnosis back to hidden causes in terms of "analysis":

> Sometimes from one symptom we recognize something, thus one disease revealed another: as when the body is wasting, this thinness will be an indication that the way of nourishment has been blocked up; closed by an infarction of the bulging parts of the liver; we will diagnose this infarction from the labor and heaviness of the liver, which it undergoes in the second coction, when the aliment is transferred through the little veins of the liver. Thus, analysis [*analysis*] makes the affect clear.
>
> In addition, sometimes the nature of the affected place reveals the disease: because there are diseases particular to each part, since neither the lung nor the ligament feel pain; thirst is very hard in scurvy; the intestines are wearied by worms; the kidneys and bladder by the stone. Thus, it is easy to know that these signs narrated by us do not have equal power everywhere. For the excrements and, along with these, but even more, pains, surely show the afflicted part. (Heurnius 1650, 477–8)

Here, "analysis" is clearly the way of reasoning back from external signs to interior hidden causes. In this case, visible whole-body wasting indicates a blockage in passages of the liver, which impairs the liver's primary function of concocting nutrifying venous blood.

So Heurnius and other Galenic physicians used the term *analysis* to describe the *practice* of diagnosis and reasoning to causes. What was more common, especially at Leiden, was the practice of combining bedside diagnosis and treatment with postmortem dissections. With the unfortunate patient dead, physicians, surgeons, and medical students could see, touch, and smell the evidence of the diseased state of a patient's organs. In their own words, postmortem dissections revealed the "causes" of disease and death to those present.

8.5.1 Analysis and Synthesis in Clinical Teaching and Postmortem Dissections at Leiden, 1636–1658

Now, I will give a sketch of clinical teaching and postmortem dissections in Leiden's Galenic medical teaching. The practices of bedside teaching clearly followed the Galenic methods of differential diagnosis and the path of analysis from perceptible effects back to causes. This reveals another strand of medical analysis. With regular postmortem dissections, physicians, professors, and students could finally directly *see* (and touch and smell) the hidden anatomical and humoral states predicted using the differential diagnosis and philosophical principles discussed in the previous sections.

In the postmortem dissections, physicians and surgeons *displayed* these causes to the senses. This gave evidence from dissected cadavers greater epistemic weight than the conjectures of analysis in diagnosis, however well the physician practiced his method of reasoning from signs to hidden causes in living patients. Records from the physicians, students, and university officials all freely describe the process of observing postmortem dissections as resulting in the physicians and surgeons *demonstrating* or *confirming* the causes of disease and death. Reports often used the verbs *demonstrare* or *confirmare* to describe how the professors revealed the causes of disease and death to the students in their pathological dissections (Heurnius 1656, 4, 10; Paaw 1657, 25, 30).

These causes included the material changes of organs: lungs corrupted and eroded away by pus and filled with abscesses or ulcers; kidneys and livers blackened and shriveled, or, like the lungs, also obstructed by abscesses; digestive tracts eaten away by caustic poisons (Paaw 1657, 18–19, 23–6). Once they knew these causes clearly by direct sensory perception, physicians had to construct chains of reasoning that moved from these causes back to the prior clinical symptoms. Mentally moving *backward and forward in time*, physicians created chains from these causes to the correlating symptoms and signs observed and reported while the patient still lived.

Throughout these "conjectures" or chains of inferences, though, the direct sensation of material states revealed through dissections had the greatest epistemic weight. Notably, physicians often reached the fundamental causal *principles* of the hot, cold, wet, and dry primary qualities only *indirectly*. They could feel and see dried up parts, but assumed that blackened, dry, and shriveled organs came about through excessive heat, which correlated with patients' prior fevers or reports of internal hot sensations.

As early as 1636, and probably earlier, Leiden followed the model of Padua from a century earlier and instituted the regular pedagogical practice of hospital bedside instruction combined with postmortem dissection. Here is the official announcement from the university:

> The Professor, with both the ordinary city doctors ... with the students, together with a good surgeon, will visit the sick persons in the public hospitals, and examine the nature of their internal diseases, as well as all their external accidents, and debate their cures and surgical operations, prescribing medicines according to the order of the hospital, and, also, will open all the dead bodies of the foreign or unbefriended persons there and show the causes of death to the students. (Molhuysen 1913–1923, 3:312)

In the program established in 1636 by Otto Heurnius (or van Heurne, 1577–1652), the son of Johannes, professors, surgeons, and students performed the clinical observations and treatments among twelve beds—six for men, six for women—in the hospital wards, and performed the dissections in the anatomy room of the hospital (Beukers 1988; Ragland 2022, ch. 7). From published versions of the professors' anatomical diaries, we get a rich sense of their anatomical practice and pathological thinking.

Even before the founding of this program, professors performed bedside instruction and postmortem dissections in private homes, or, it appears, a room near the ox and pig market in the northwest corner of the city (Paaw 1657, 18, 31; Blaeu 1652). In these private anatomies, they dissected cadavers beyond those of "foreign or unbefriended persons" to include many local townsfolk, members of aristocratic families, several students, and even Professor Johannes Heurnius himself (Ragland 2022, ch. 6). They also used the two or three public dissections every winter in the anatomy theater to demonstrate pathological states of the cadavers' organs and parts, and the anatomy professor Pieter Paaw (or Pauw or Petrus Pavius, 1564–1617), announced in a 1615 work that the whole point of such public anatomies was to prepare "practical physicians… who would thoroughly understand the affects and diseases of the individual parts of the whole body, and of those what is required for their curing" (Molhuysen 1913–1923, 1:58; Paaw 1615, preface)

Putting pathological organs and parts into series across cadavers helped the anatomists to find stable causes and conditions of diseases. In his postmortem dissections, Paaw looked for similarities among the morbid states. Thus, two women's bodies dissected years apart revealed a strange "yolk-like" fluid in their hearts, which Paaw noted and described as the cause of their deaths (Paaw 1657, 11; Paaw 1615, 145).

Paaw also assisted the local magistrates in cases of suspected poisoning. In one case in 1594, his dissections revealed the corroded and blackened upper parts of the stomach, undoubtedly caused by "some corrosive, caustic drug" (Paaw 1657, 19). Two years later, in another case of suspected poisoning, Paaw did *not* find the distinctive phenomena of erosion and blackening. Also in 1596, two other cases of suspected poisoning reached similarly opposing conclusions based on the presence or absence of clearly visible and tangible lesions along the upper digestive tract. In sum, Paaw and his colleagues routinely relied on these dissections to "confirm" (*confirmare*) or disconfirm earlier diagnoses and suspicions.

The cadavers of patients were *not* blank canvases for receiving any and all imagined portraits of disease history. To identify the material histories of diseases, and even their causes, physicians, surgeons, and students needed changes perceptible to sight, touch, or smell. As usual, the solid organs and simple parts, especially the chief organs of the brain, heart, and liver, were the primary sites of inspection. Only when they found no lesions on the organs and solid parts did they consider morbid states of the humors and spirits.

The Leiden student Thomas Bartholin (1616–1680) reported on the usual procedures of clinical teaching and dissection in 1638:

> With the recent sabbath day elapsed, in the public Hospital in individual weeks the proteges of Aesculapius [Asclepius] and the Practical Professor of Medicine come together, so that they may inquire into the nature of diseases, their causes and remedies, and I was present for the dissection of a human cadaver. We were occupied in the investigation of the hidden cause of death, but truly from the more principal parts, which were preternatural, we found none, for which reason the cause was referred to the spirits or humors. (Bartholin to Worm, 3 October 1638, in Worm 1751, 653)

Bartholin, notably, would go on to become one of the most eminent anatomists in Europe, and he attempted to compile a comprehensive pathological anatomy from the hundreds or thousands of reports of postmortems done in early modern Europe, as did Théophile Bonet in his 1679 *Sepulchretum* (Rinaldi 2018). Similarly, Professor Otto Heurnius recorded the 1638 dissection of the body of Johannes Hax, in which they could not locate any clear cause of morbid developments. Hax had sustained a contusion of his ribs, which then began to putrefy. No organ of the abdomen displayed sure signs of preternatural affects; not even the nearby lungs appeared corrupted. A fair amount of pus had collected in the abdomen, but "when we had accurately investigated the cause of this, we were not able to grasp any cause by ocular confidence, nor comprehend any reason" (Heurnius 1656, 16). In contrast, when visible and tangible phenomena clearly showed damaged organs, especially damaged organs whose functions had been impaired during the life of the patient, the physicians concluded that such lesions were the cause of disease and death.

Just as in the order or method of analysis discussed for so long and so widely among medical professors, through postmortem dissections, Leiden professors sought to make perceptible the anatomical causes of diseases and link them to the symptoms observed while the patients had lived. As Otto put it in 1639, in the case of Sara Mente, "with the Chest and Abdomen opened, we detected every cause of the malady" (Heurnius 1657, 19). In Mente's case, the visibly putrefied lungs, which were also filled with abscesses, were the clear seat of disease and death. Like so many people, she had died of a consumptive lung disease, something the many gathered students could see clearly in her opened cadaver. For another patient's body, in 1636, that of the clothmaker Joannes de Neeff, Otto "demonstrated, explaining causes of death; dissection performed by Mr. Joannis Camphusius, ordinary surgeon of the Republic of Leiden" (Heurnius 1657, 4). In de Neeff's case, his trouble breathing, chest pains, and wasting away also followed from the visibly corrupted lungs; the failure of the lungs to assist in the generation of vital spirits resulted in an impaired liver, which failed to nourish the body.

Postmortem evidence—phenomena revealed to the senses of sight and touch—had the final say, but reports from patients could direct the initial stages of the search for the causes and paths of diseases, especially reports of the types and location of pains. As Otto taught his own students in a 1638 case, it was of great importance in medical practice "to observe and distinguish the various kinds, size, and duration of pains, and from these we know the part affected, the cause of the affect, and the outcome" (Heurnius 1657, 16). A jabbing pain indicates an affected membrane, and a spread-out pain a membrane distended with copious matter. An oppressive or heavy pain acts in a fleshy part, and a pulsing pain acts in the part that suffers, with the arteries woven in it. Different parts had characteristic pains with their common diseases: a heavy pain in the kidney indicated a stone in the flesh or sometimes an abscess, while a pulsing pain meant that the stone cut into the pelvis. From such site-specific pains, together with the "nature of the injured part," and from the constitution of the disease, the physician should derive the foundations of all curative indications, or sure signs for treatment (Heurnius 1657, 23).

8.5.2 Making New Knowledge About Consumption (Phthisis)

Consumption killed everywhere, and nearly always, in early modern Europe. Francesca Damiana, Argenterio's young wife, was not the only one terrified of spitting blood. As Da Monte put it, "Where one is cured, fully a hundred are dead" (Da Monte in Heurnius 1602, 143). In the later seventeenth century, the English physician Thomas Sydenham lamented that deaths from consumption made up nearly two-thirds of patients felled by chronic diseases (Sydenham 1848, 332). This deadly disease attracted the attention of physicians from the ancient through the early modern periods. The concept of consumption, in its symptoms and pathology, remained fairly stable until postmortem dissections from the late sixteenth century allowed anatomists to analyze, isolate, and reveal the growth of pulmonary ulcers and tubercles that caused the symptoms and eventual deaths of sufferers.

By the time Galen wrote about consumption or phthisis in the first century AD, Hippocratic writers and other medical authors had already established its clear profile (Meinecke 1927; Ragland 2022, 364–87). As the name indicated (from the Greek *phthon* for "wasting") it was a wasting disease, in which the whole body seemingly liquefied and sloughed away. Patients also exhibited a mild whole-body fever and showed clear difficulties in breathing, as well as the more distinctive signs of coughing up pus and blood. Galen followed the Hippocratic writers in locating the seat of consumption in the lungs (though they also described other kinds, such as consumption of the back), and defined it in terms of a specific anatomical lesion: ulcer of the lungs. The Hippocratics had probably used evidence from butchered animals to conclude that consumption involved "swellings" (or *phumata*) in the lungs, but also claimed that one could hear the pus sloshing in the patients' lungs, pus that later appeared visibly—and stank horribly—when patients coughed it up (Meinecke 1927, 383).

For Galen, consumption was clearly contagious, and likely passed from person to person through poisonous exhalations. But he also ascribed its causes to blood from ruptured veins due to trauma, excess cold, or pleurisy; from the buildup and corruption of phlegm in the lungs; and from blockage of blood flow in the lungs. In each case, stagnating fluid became corrupt and ate away at the lungs, producing ulcers and pus (Meinecke 1927, 389–90).

Early modern Galenists, freed by their medieval forebears from the ancient taboo against human dissection, increasingly sought to find the causes of consumption by cutting into patients' lungs. In the 1550s, Fernel, in Paris, published his account of consumption, defined as "the ulceration of the lung by which the whole body is gradually liquefied" and indicated by the symptoms of fever, wasting, and spitting blood and pus (Fernel 1555, 115r). As we have seen, he found a range of ulcers, stones, and cheese-like concretions in the lungs of patients who had suffered from respiratory diseases. Yet he did not clearly link the symptoms of consumption to the gradual development of ulcers and tubercles in the lungs.

Around 1600, roughly the same time he dissected the cadaver of his former colleague, Johannes Heurnius, at Leiden University, Paaw also dissected several bodies of consumptive patients. In 1593, he "publicly dissected the body of a Young man in the Hospital" who had died of consumption. He revealed to the senses pus-producing abscesses and hardened "stones":

> Here and in each part of the lung [there were] various purulent abscesses, because in various places phlegm was noticed so hardened in the lung that they [the abscesses] seemed to be stones to the touch, and indeed in not a few places it reached a perfect stony hardness. (Paaw 1657, 15)

Dissections of the bodies of two other consumptive patients revealed similar lesions. At the same time, the learned German surgeon Fabricius Hildanus found similar evidence of ulcers and abscesses in the lungs of consumptives, including some the size of goose eggs (Hildanus 1682, 53).

At Leiden a few decades later, around 1640, Otto Heurnius dissected several bodies of consumptive patients. In one, he revealed the "cause of death seated in the lungs, namely that the whole parenchyma was filled with tiny tubercles from a crude viscous matter" (Heurnius 1656, 22). In another, he found pus-filled parts of lungs and abscesses, and noticed other, smaller lesions, describing the lungs as "marbled with dark and pale droplets" (Heurnius 1656, 10). Finally, in the body of a Scot, David Jarvis, Otto found "many tiny tubercles [*tubercula*] from a crude, viscous matter" (Heurnius 1656, 21). Otto did not *explicitly* connect the smaller and larger lesions across time along a developmental pathway in which small droplets or tubercles grew into larger abscesses and ulcers. But the notion of the gradual erosion of the lungs, and the range of tubercle sizes, may have urged this connection.

By the 1660s, Leiden physicians such as Franciscus Sylvius (Frans Dele Boë, 1614–1672) and Johannes van Horne (1621–1670) collaborated with leading students such as Nicolaus Steno, Reinier de Graaf, and Jan Swammerdam to create a fully experimental and experimental*ist* program of research in anatomy, chymistry, and practical medicine (Ragland 2017a). They explicitly declared experiments to be the foundation and test of knowledge claims and tried to make good on their promises in their practices.

Like his Galenic predecessors, Sylvius adopted a definition of disease in terms of impaired or injured function, *and* he described analysis as a process of identifying causes. For him, a disease was "a faulty Constitution of a Man impairing some Functions" (Sylvius 1695, 56). Since he taught a new, chymical matter theory, he cast aside the older fundamental principles of hot, cold, wet, and dry primary qualities in favor of chymical principles, especially acids and alkalis. In anatomy, he described the canonical method or order of analysis. In this case, ongoing debates about the anatomical process of respiration urged him to investigate the matter through careful anatomical experiments.

> So that for this reason in a matter doubtful and full of quarrels we may resolve more happily the proposed difficulty, we ought to follow the Analytic method in the deduction of this, namely beginning from the Effects running into the external senses, and to this point more known to us, indeed prior and per se noticed, and proceeding step by step to the Causes.... (Sylvius 1695, 30)

This is clearly the same language and structure of thought we have seen for some time, here applied to anatomical experiments. His Cartesian opponents had argued for a circular movement in which the chest rose and so pushed air into the lungs, inflating them. But Sylvius showed by anatomical analysis and experiments that the muscles of the diaphragm and chest increased the volume of the pulmonary cavity, thus inflating the lungs, so that they expanded and drew in air (Ragland 2017a, 346). The structure and actions of the muscles of the chest and diaphragm contracted to enlarge the lungs, which passively inflated. Experiments on live animals showed similar phenomena, with air blown into or sucked out through a tube inserted into a dog's trachea the sole perceptible cause of the inflation and deflation of the lungs.

Experiments allowed him to isolate possible causes. Wounding the heart or nerves did not change the expansion and contraction of the lungs, but wounding the chest did, even though air entered the nose and trachea. Sylvius also used "analysis" and "synthesis" to discuss the decomposition and recomposition of chymical substances—for instance, in his discussion of Johann Rudolph Glauber's "analysis and synthesis of nitre," which clearly demonstrates what the ingredients of nitre are by breaking it down into its components and then putting these components together to reconstitute the nitre (Sylvius 1695, 211, 775).

Sylvius and his colleagues and students explicitly aimed to make new discoveries by experimentation in anatomy, chymistry, and the practice of medicine. As elsewhere in early modern academic medicine, the language and practices of experimentation, testing, or "making a trial" came to almost eclipse the language of analysis and synthesis (Ragland 2017b). In "practical experiments" in the practice of medicine, they tried to discover new pathological and therapeutic knowledge. Three sets of student notes from the daily hospital teaching and frequent postmortem dissections reveal their day-to-day practices.[17] A student would observe the patients and write, in the words of one, "with a flying stylus" while the professor asked questions about the symptoms, diagnosis, prognosis, and therapies of each patient (Merian in Sylvius 1695, 70). They divided their small notebooks into sections by patients' names over time, and then later combined these daily records into longer case histories for each patient (Hepburne 1660–1666). The professors tweaked and changed their remedies and treatments in response to patients' progress—for those who weakened, for example, they reduce the dosages or stopped using stronger drugs; for increased pain, they prescribed more opium medicines; and if the patients died, students and professors used their bodies to discover new anatomical structures. For example, they tested for connections between the brain and the sinuses, which had long been claimed in prior medical traditions, and professors left hospital cadavers for the students to pursue their ownresearch on sweat glands or to perform more detailed pathological analyses. As with previous professors, revealing the hidden causes to the senses ought to have settled debates. As Sylvius put it, "in the accurate opening and demonstration of their bodies it was revealed to all whether they had judged rightly or wrongly about that Disease, and from that the rest of the things considered" (Sylvius 1695, 907). Visible lesions on organs and parts, especially the heart, liver, and other primary parts, indicated the seats and causes of diseases.

The close observation of dozens of consumptive patients and the dissection of the cadavers of those who died allowed Sylvius to build a new theory of consumption, one organized around the development of tubercles in the lungs. As with his predecessors, from Argenterio to Otto Heurnius, putting postmortem evidence into series generated stronger knowledge claims for the pathological causes and processes of consumption:

[17] Merian in Sylvius (1695), Hepburne (1660–1666), Sylvius (1681).

> I saw not just once *Glandulous Tubercles in the Lungs*, smaller and larger, in which there was ever varied Pus contained, as shown by dissection. These Tubercles gradually dissipated into Pus, and I consider that the things contained in their thin membranes should be considered *Abscesses*, and from these I recognized that frequently *Phthisis had its origin*. (Sylvius 1695, 692)

Earlier student notes record Sylvius pointing to the cause of consumption in lungs filled with "little hard particles, which usually developed gradually into abscesses" (Sylvius 1681, 731). Unlike prior physicians, Sylvius united consumption and scrofula, primarily a disease of swollen and corrupt glands outside the lungs, with pulmonary consumption. Although their symptoms differed greatly, they shared the same characteristic lesions developing along the same pathway or life history: from small, millet-like spots to droplets to small tubercles and abscesses, then larger ones filled with a cheese-like substance, and finally open ulcers and pus.

Over the next century and a half, Sylvius's work on consumption was well-accepted, and passed into standard pathological references, notably Morgagni's study of the *Seats and Causes of Diseases* in the mid-1700s (Morgagni 1980, 1:656–8). Looking even farther ahead, we can see that Sylvius's anatomical-clinical method, built on the pedagogical and pathological practices of his Galenic predecessors, also compares strikingly well with that of the hero of the "new" French hospital medicine, R. T. H. Laennec. Writing in the early nineteenth century, Laennec defined pathological anatomy in terms of correlating the altered functions and symptoms of diseases with the visible alterations found in patients' bodily organs through postmortem dissections.

> Pathological anatomy is a science which has the goal of the knowledge of the visible alterations which the state of disease produced in the organs of the human body. The opening of cadavers is the means of acquiring this knowledge; but, in order for it to become of direct use and an immediate application to practical medicine, it must be joined to the observation of symptoms or the alterations of functions [*fonctions*] which coincide with each species of alteration of the organs. (Laennec 1812–1822, 2:46–47)

Famously, Laennec had his new invention, the stethoscope, with which he claimed to hear the "music" of patients' chests, which supposedly allowed him to distinguish different diseases of the chest while patients lived, and then to correlate these distinctions with the detailed evidence of postmortem dissections. Our own present-day research on auscultation strongly suggests that he could not actually do this reliably, as even modern methods have a low success rate in detecting respiratory diseases.[18] But even granted Laennec's new tool and method for hearing directly the changes in the lungs, his method is similar to the way of analysis and synthesis developed in and out of early modern Galenic academic medicine. Moreover, Laennec's account of the disease of consumption matched that of Sylvius from a century and a half earlier, and in strikingly similar ways. Both unified their theories around the growth of tubercles in the body, as revealed by postmortem dissections of several dozen cadavers. Both rejected extravasated blood as a proximate cause,

[18] A 2020 meta-analysis argues that the pooled sensitivity of lung auscultation is 37% and the specificity is 89%. Arts et al. 2020.

given its absence in dissections. Both argued that strong grief was an antecedent cause, with Laennec highlighting the deep grief of women in a religious order whose community had been laid low by consumption. Interestingly, while Sylvius followed tradition in accepting that some or most forms of pulmonary consumption were contagious, Laennec rejected contagion based on his own experience with the seemingly successful cauterization of a tubercle on his finger (Ragland 2022, 382–7).

There were many continuities. As we have seen, though, over this long period of time there were many modifications and refinements. Notably, as is often the case in medicine, physicians made finer and clearer distinctions. Moreover, the importance, quantity, and scope of evidence from postmortem dissections—evidence long prized for knowing the seats and causes of diseases—reached much greater heights in the later 1700s and early 1800s. Laennec's friend, Pierre Bayle, based his account of the six species of consumption on dissections of some 900 cadavers (Rey 1993). Like so many before and since, Bayle and Laennec died of tuberculosis.

By the time Laennec was writing, dictionary definitions of "analysis" and "synthesis" often concentrated on the chemical meanings: roughly dissolution into components and recombination into compounds, respectively. Following up on mathematical developments in the previous two centuries, they also included discussions of mathematical "analysis" using algebra. But Diderot's entry, "Detailed Explanation of the System of Human Knowledge," still contained echoes from the longer history of discussions of methods of analysis and synthesis. "But in demonstration, either one goes back from the thing to be demonstrated to the first principles, or one descends from the first principles to the thing to be demonstrated; whence are born analysis and synthesis" (Diderot 1751, xlviij).

8.6 Conclusions

Analysis and synthesis in premodern academic medicine were many things to many people, and the relationship between them appears in different ways. Notably, most commentators studied here emphasized analysis—and the meanings and practices they attached to it often shared a similar structure—and participated in an ongoing conversation. Discussions of analysis and synthesis often engaged with Galen's many meanings and many textual *loci*. Famously, Galen presented *analusis* and *sunthesis* in at least five related forms: modes of ordered teaching; geometrical and architectural methods of testing by making and using constructions; philosophical methods for moving from effects to causes or principles; anatomical methods for isolating parts and testing ideas about them; and rational practices for testing and making drugs. Commentators, from medieval Arabic and Latin authors and early modern European academic medicine, argued over the diversity of meanings but nearly always agreed on the shared structure: lines of causes, or reasoning from effects to causes, and back.

There was a great deal of continuity in physicians seeking to move from symptoms and other phenomenal signs back to causes, notably causes understood in terms of anatomically localized material states and functions. Even when "analysis" and "synthesis" ceased to be used by physicians to describe teaching from more particular knowledge to more general principles, or the discovery of causes in a particular patient from the symptoms, physicians, even of the new Paris medicine, still thought along similar lines. These forms of medical reasoning were not formal philosophical demonstrations. They concerned particular patients and relied for the most part on universal principles supposedly established prior to medical practice. But even critics such as Argenterio shared similar views of the structure of the inferences, even as they took a more skeptical stance about the epistemological status of the medical processes of analysis and synthesis in diagnosis, pathology, and therapeutics.

Galenic medicine had long concentrated its pathological thinking on impaired or injured organs and other anatomical parts. The rise of human dissection in Europe since the late 1200s very gradually allowed physicians to sense, finally, the morbid states of these organs and some of the causes of disease. By the middle of the sixteenth century, at least, physicians increasingly joined this form of reasoning from surface symptoms back to hidden causes to new analytical methods of postmortem dissections. These postmortem dissections finally revealed to the senses the diseased organs and parts that had long been inferred from rational analysis. Explicit discussions of diagnosis and anatomical and pathological discovery by means of "analysis" or its cognates continued well into the seventeenth century. There, the Galenic methods of reasoning, from stable clusters of surface symptoms back to impaired organs and the proximate efficient causes of these impairments, joined with widespread and frequent postmortem dissections. As physicians from the late 1500s and early 1600s increasingly dissected down to pulmonary ulcers, tubercles, cheese-like substances, and stony concretions in the lungs of consumption patients, they put these findings into series and developmental sequences to explain the nature, symptoms, and progression of consumption.

We do not want to be led back into superficial claims for grand stories of progress. But rather than seeing a "great break" or "rupture" in the history of medicine around 1800, as many histories of medicine claim, I would argue for much more continuity (*contra* Foucault 1975, 179; Bynum 2008, 15, 55). Knowing diseases backward, from signs in the clinic to hidden causes, and forward, from the causes revealed to the correlating symptoms, has a long history. Galenic physicians added regular postmortem dissections to close clinical observations from the mid-1500s through the 1600s. In so doing, they cleared Galen's paths of "analysis" and "synthesis" for more direct sensory perception of the causes of disease and death. Which is, of course, what Galen himself always wanted.

References

Alexander of Aphrodisias. 2013. In *Alexander of Aphrodisias: On Aristotle prior analytics 1.1–7, trans*, ed. Jonathan Barnes, Susanne Bobzien, Flannery S. Kevin, and J., and Katerina Ierodiakonou. London/New York: Bloomsbury Academic.

Argenterio, Giovanni. 1610 [1566]. Ioannis Argenterii in Claudii Galeni Pergameni, Artem Medicinalem Commentarii Praefatio. In *Opera Iohannes Argenterii*. Hanau: Typis Wechelianis apud haeredes Claudii Marnii.

Aristotle. 1984. *The complete works of Aristotle: The revised Oxford Translation*, ed. Jonathan Barnes. 2 vols. Princeton: Princeton University Press.

Arts, Luca, Endry Hartono Taslim Lim, Peter Marinus, Leo van de Ven, and Heunks, and Pieter R. Tuinman. 2020. The diagnostic accuracy of lung auscultation in adult patients with acute pulmonary pathologies: A meta-analysis. *Nature: Scientific Reports* 10: 7347.

Avicenna (Ibn Sina). 1993. *Canon of medicine*. 2 vols. English translation of the critical Arabic text. New Delhi: Dept. of Islamic Studies Jamia Hamdard.

Bertoloni Meli, Domenico. 2006. *Thinking with objects: The transformation of mechanics in the seventeenth century*. Baltimore: Johns Hopkins University Press.

Beukers, Harm. 1988. Clinical teaching in Leiden from its beginning until the end of the eighteenth century. *Clio Medica* 21: 139–152.

Blaeu, Joan. 1652. *Toonneel der steden van de Vereenighde Nederlanden*. Amsterdam.

Bouras-Vallianatos, P., and Barbara Zipser, eds. 2019. *Brill's companion to the reception of Galen*. Leiden: Brill.

Bylebyl, Jerome. 1979. The School of Padua: Humanistic medicine in the sixteenth century. In *Health, medicine and mortality in the sixteenth century*, ed. Charles Webster, 335–370. Cambridge: Cambridge University Press.

———. 1991. Teaching *Methodus Medendi* in the renaissance. In *Galen's method of healing*, ed. Fridholf Kudlien and Richard J. Durling, 157–189. Leiden: Brill.

———. 1993. The manifest and the hidden in the renaissance clinic. In *Medicine and the five senses*, ed. W.F. Bynum and Roy Porter, 40–60. Cambridge: Cambridge University Press.

Bynum, W.F. 2008. *The history of medicine: A very short introduction*. Oxford: Oxford University Press.

Champier, Symphorien. 1516. *Index eorum omnium quae in hac arte parva Galeni pertractantur*. Lyon: Jean Marion.

Chandelier, Joël. 2018. Averroes on medicine. In *Interpreting Averroes: Critical essays*, ed. Peter Adamson and Matteo Di Giovanni, 158–176. Cambridge: Cambridge University Press.

Da Forlì, Jacopo. 1487. *Expositio super tres libros Tegni Galeni*. Pavia.

Da Monte, Giambattista. 1554a. *In artem paruam Galeni Explanationes*, ed. Valentinus Lublinus. Venice: Apud Balthassarem Constantinum ad Signum Divi Georgii.

———. 1554b. *In nonum librum Rhasis ad Monsorem Regem Arabum Expositio*, ed. Valentinus Lublinus. Venice: Apud Balthassarem Constantinum ad Signum Divi Georgii.

Diderot, Denis. 1751. Detailed explanation of the system of human knowledge. In *The encyclopedia of Diderot & d'Alembert collaborative translation project*. Trans. Richard N. Schwab and Walter E. Rex. Ann Arbor: Michigan Publishing, University of Michigan Library, 2009. http://hdl.handle.net/2027/spo.did2222.0001.084 (Accessed 19 Mar 2024) Originally published as "Explication détaillée du système des connoissances humaines," *Encyclopédie ou Dictionnaire raisonné des sciences, des arts et des métiers*, 1: xlvii–li (Paris, 1751).

Edwards, William F. 1967. Randall on the development of scientific method in the School of Padua—A continuing reappraisal. In *Naturalism and historical understanding: Essays on the philosophy of John Herman Randall, Jr*, ed. John P. Anton. Albany: State University of New York Press.

Fancy, Nahyan. 2020. Verification and utility in the Arabic commentaries on the *canon of medicine*: Examples from the works of Fakhr al-Dīn al-Rāzī (d. 1210) and Ibn al-Nafīs (d. 1288). *Journal of the History of Medicine and Allied Sciences* 75: 361–382.

Fernel, Jean. 1542. *De naturali parte medicinae Libri septem*. Paris: Apud Simonem Colonaeum.
———. 1555. *Pathologiae Libri septem*. Venice: Petrus Bosellus excudendum curabat.
———. 2003. *The* Physiologia *of Jean Fernel (1567): Translated and Annotated by John M. Forrester*. Philadelphia: American Philosophical Society.
Foucault, Michel. 1975. *The birth of the clinic: An archaeology of medical perception*. Trans. A. M. Sheridan Smith. New York: Vintage Books.
Galen. 1520. *De Affectorum Locorum Notitia, libri sex, Guilielmo Copo Basiliensi interprete*. Paris: officina Simonis Colinaei.
———. 1821–1833. *Opera Omnia*, ed. C. G. Kühn. 20 vols. Leipzig.
———. 1991. *On therapeutic method*. In *Galen on the therapeutic method, books I and II*. Translated with an Introduction and Commentary by R. J. Hankinson. Clarendon Later Ancient Philosophers. Oxford: Clarendon Press.
———. 1997a. Mixtures. In *Galen: Selected works*. Translated with an introduction and notes by P. N. Singer, 202–289. Oxford/New York: Oxford University Press.
———. 1997b. My own books. In Galen: Selected works. Translated with an introduction and notes by P. N. Singer, 3–22. Oxford/New York: Oxford University Press.
———. 1997c. The soul's dependence on the body. In *Galen: Selected works*. Translated with an introduction and notes by P. N. Singer, 150–176. Oxford/New York: Oxford University Press.
———. 1997d. The affections and errors of the soul. In *Galen: Selected works*. Translated with an introduction and notes by P. N. Singer, 100–149. Oxford/New York: Oxford University Press.
———. 1997e. The art of medicine. In *Galen: Selected works*. Translated with an introduction and notes by P. N. Singer, 345–96. Oxford/New York: Oxford University Press.
———. 2011. *Method of medicine*, vol. I: Books 1–4. Edited and Translated by Ian Johnston, G. H. R. Horsley. Loeb Classical Library 516. Cambridge, MA: Harvard University Press.
———. 2016. *On the constitution of the art of medicine*. In *The art of medicine. A method of medicine to Glaucon*. Edited and Translated by Ian Johnston. Loeb Classical Library 523. Cambridge, MA: Harvard University Press.
Galen and Martin Acakia [Akakia]. 1544. *Claudii Galeni Pergameni Ars Medica*. Venice.
Gilbert, Neal. 1960. *Renaissance concepts of method*. New York/London: Cornell University Press.
Hankinson, R. J. 1991. Galen on the therapeutic method, books I and II. Translated with an Introduction and Commentary by R. J. Hankinson. Clarendon Later Ancient Philosophers. Oxford: Clarendon Press.
———. 2008a. Epistemology. In *the Cambridge companion to Galen*, ed. R. J. Hankinson, 157–83. Cambridge Companions to Philosophy. Cambridge: Cambridge University Press.
———. 2008b. Philosophy of nature. In *The Cambridge companion to Galen*, ed. R. J. Hankinson, 210–241. Cambridge companions to philosophy. Cambridge: Cambridge University Press.
Hankinson, R.J. 2009. Galen on the limitations of knowledge. In *Galen and the world of knowledge*, ed. Christopher Gill, Tim Whitmarsh, and John Wilkins, 206–242. New York: Cambridge University Press.
———. 2014. Galen and the ontology of powers. *British Journal for the History of Philosophy*. 22 (5): 951–973.
Henry, John, and John Forrester. 2003. Introduction. In *The Physiologia of Jean Fernel (1567)*. Translated and Annotated by John M. Forrester, 1–12. Philadelphia: American Philosophical Society.
Hepburne, Robert. 1660–1666. *Sloane MS 201*. British Library.
Heurnius (van Heurne), Johannes. 1602. *De Morbis pectoris liber*. Leiden: Ex Officina Plantiniana, Raphelingii.
———. 1638. *Institutiones medicinae*. Leiden: Ex Officina Ioannis Maire.
———. 1650. *Praxis Medicinae Nova ratio*. Rotterdam: Ex Officinâ Arnoldi Leers.
Heurnius, Otto (Otto van Heurne). 1656. *Historiae et Observationes quaedam Rariores ex Praxi et Diario in Jean Fernel, Universa Medicina Primùm quidem studio & diligentiâ Guiljelmi Plantii, Cennomanni elimata, Nunc autem notis, observationibus, & remediis secretis Ioann. & Othonis HeurnI, Ultraject. et Aliorum Praestantissimorum Medicorum scholiis illustrata. Cui*

accedunt Casus & observationes rariores, Quas Cl. D.D. Otho Heurnius in Academia Leydensi Primarius Medicinae practicae, Anatomiae & Chirurgiae Professor, in diario practico annotavit. Utrecht: Gisbertus à Zijll & Theodorus ab Ackersdijck.

Hildanus, Guilhelmus Fabricius. 1682. *Opera Observationum et Curationum Medico-Chirurgicarum*. Frankfurt: Sumptibus Ioan Ludov Dufour.

Kaye, Joel. 2014. *A history of balance, 1250–1375: The emergence of a new model of equilibrium and its impact on thought*. Cambridge: Cambridge University Press.

King, Helen. 2013. Female fluids in the Hippocratic corpus: How solid was the humoral body? In *The body in balance: Humoral medicines in practice*, ed. Peregrine Horden and Elisabeth Hsu, 25–52. New York/Oxford: Berghahn Books.

Laennec, R. T. H. 1812–1822. Anatomie Pathologique. In *Dictionnaire des Sciences Médicales*, ed. Nicolas Adelon (1812–1822).

Lennox, James. 2021. Aristotle's biology. In *The Stanford Encyclopedia of Philosophy* (Fall 2021 edition), ed. Edward N. Zalta, https://plato.stanford.edu/archives/fall2021/entries/aristotle-biology/. Accessed 2 June 2024.

Leoniceno, Niccolò. 1509. *Nicolai Leoniceno Vicentini in Libros Galeni e Graeca in Latinam Linguam a se Translatos Praefatio Communis*. Ferrara: per Joannem Macciochium.

Maclean, Ian. 2007. Logic, signs and nature in the renaissance: The case of learned medicine. In *Ideas in context*. Cambridge: Cambridge University Press.

Martin, Craig. 2022. Galenic causation in the theoretical and practical medicine of Giambattista Da Monte. In *Galen and the early moderns*, ed. M. Favaretti Camposampiero and E. Scribano, 37–53. Switzerland: Springer Nature.

McVaugh, Michael. 1971. The *Experimenta* of Arnau of Vilanova. *Journal of Medieval and Renaissance Studies* 9: 107–118.

———. 2009. The 'experience-based medicine' of the thirteenth century. *Early Science and Medicine* 14: 105–130.

Meinecke, Bruno. 1927. Consumption (tuberculosis) in classical antiquity. *Annals of Medical History* 9: 379–402.

Molhuysen, P. C., Ed. 1913–1923. *Bronnen Tot de Geschiedenis Der Leidsche Universiteit*. 7 vols. Rijks Geschiedkundige Publicatiën, nos. 20, 29, 38, 45, 48, 53, and 56. The Hague: Martinus Nijhoff.

Morgagni, Giovanni Battista. 1980. *The seats and causes of diseases investigated by anatomy*. Trans. Benjamin Alexander. 2 vols. Mount Kisco, NY: Futura Publishing Company, Inc.

Nutton, Vivian. 2013. *Ancient medicine*. Abingdon/New York: Routledge.

———. 2022. *Renaissance medicine: A short history of European medicine in the sixteenth century*. London/New York: Routledge.

Ottoson, Per-Gunnar. 1984. *Scholastic medicine and philosophy: A study of commentaries on Galen's Tegni*, 1300–1450. Naples: Bibliopolis.

Paaw, Petrus. 1615. *Primitiae Anatomicae de Humani Corporis Ossibus*. Leiden: Ex Officina Iusi à Colster.

——— (Pieter Pauw or Pavius). 1657. Observationes Anatomicae. In *Thomas Bartholin, Historiarum Anatomicarum Rariorum, Centuriae III & IV*. Copenhagen: Ex Typographia Adriani Vlacq.

Park, Katharine. 2010. *Secrets of women: Gender, generation, and the origins of human dissection*. New York: Zone Books.

Pormann, Peter E., and Emilie Savage-Smith. 2007. *Medieval Islamic medicine*. Washington, DC: Georgetown University Press.

Ragland, Evan R. 2017a. Experimental clinical medicine and drug action in mid-seventeenth-century Leiden. *Bulletin of the History of Medicine* 91: 331–361.

———. 2017b. 'Making trials' in sixteenth- and early seventeenth-century European academic medicine. *Isis* 108: 503–528.

———. 2022. *Making physicians: Tradition, teaching, and trials at Leiden University, 1575–1639*. Leiden: Brill.

Randall, J. 1940. The development of scientific method in the School of Padua. *Journal of the History of Ideas*. 1 (2): 177–206.

———. 1961. *The school of Padua and the emergence of modern science*. Padua: Editrice Antenore.

Rey, Roselyne. 1993. Diagnostic différentiel et espèces nosologiques: le cas de la phtisie pulmonaire de Morgagni à Bayle. In *Maladies, médicines et sociéties: approches historiques pour le présent, Actes du Vie colloque d'histoire au present*, ed. François-Olivier Touati, 185–200. Paris: L'Harmattan.

Rinaldi, Massimo. 2018. Organising pathological knowledge: Théophile Bonet's *Sepulchretum* and the making of a tradition. In *Pathology in practice*, ed. Silvia De Renzi, Marco Bresadola, and Maria Conforti, 39–55. Abingdon/New York: Routledge.

Savage-Smith, Emilie. 2013. Were the four humors fundamental to medical practice? In *The body in balance: Humoral medicines in practice*, ed. Peregrine Horden and Elisabeth Hsu, 89–106. New York and Oxford: Berghahn Books.

Scurlock, Joann. 2018. Mesopotamian beginnings for Greek science? In *Oxford handbook of science and medicine in the classical world*, ed. Paul T. Keyser and John Scarborough, 35–46. Oxford/New York: Oxford University Press.

Siraisi, Nancy. 1981. *Taddeo Alderotti and his pupils: Two generations of Italian medical learning*. Princeton: Princeton University Press.

———. 1990. Giovanni Argenterio and sixteenth-century medical innovation: Between princely patronage and academic controversy. *Osiris* 6: 161–180.

———. 2001. *Medicine and the Italian universities 1250–1600*. Leiden: Brill.

———. 2002. Disease and symptom as problematic concepts in renaissance medicine. In *Res et Verba in der Renaissance*, ed. Eckhard Kessler and Ian Maclean, 217–240. Wiesbaden: Harrassowitz.

Stolberg, Michael. 2014. Bedside teaching and the acquisition of practical skills in mid-sixteenth-century Padua. *Journal of the History of Medicine and Allied Sciences* 69: 633–664.

———. 2018. Post-mortems, anatomical dissections and humoral pathology in the sixteenth and early seventeenth centuries. In *Pathology in practice*, ed. Silvia De Renzi, Marco Bresadola, and Maria Conforti, 79–95. Abingdon and New York: Routledge.

Sydenham, Thomas. 1848. *The works of Thomas Sydenham, M.D*, ed. William Alexander Greenhill and R. G. Latham. London.

Sylvius, Franciscus Dele Boë. 1681. *Opera Medica...Accessit huic Editioni hactenus ineditum collegium Nosocomicum ab Authore habitum*. Geneva: Apud Samvelem de Tovrnes.

———. 1695. *Opera Medica...Editio Nova, Cui accedunt Casus Medicinales*. Utrecht/Amsterdam: Apud Guillelmum van de Water and Apud Antonium Schelte.

Tieleman, Teun. 2002. Galen on the seat of the intellect: Anatomical experiment and philosophical tradition. In *Science and mathematics in ancient Greek culture*, ed. C.J. Tuplin and T.E. Rihil, 257–273. Oxford: Oxford University Press.

———. 2008. Methodology. In *The Cambridge companion to Galen*, ed. R. J. Hankinson, 49–65. Cambridge Companions to Philosophy. Cambridge: Cambridge University Press.

Turisanus [Pietro Torrigiano de Torrigiani]. 1557. *Plusquam Commentum in parvam Galeni Artem*. Venice.

Van der Eijk, Philip. 2009. *Medicine and philosophy in classical antiquity: Doctors and philosophers on nature, soul, health and disease*. Cambridge: Cambridge University Press.

Wallis, Faith. 2010. *Medieval medicine: A reader. Readings in medieval civilizations and cultures*. Toronto: University of Toronto Press.

Wightman, William P.D. 1964. Quid Sit Methodus?' 'Method' in sixteenth century medical teaching and 'discovery. *Journal of the History of Medicine and Allied Sciences* 19: 360–376.

Wilson, Adrian. 2000. On the history of disease-concepts: The case of pleurisy. *History of Science* 38: 271–319.

Worm, Ole, et al. 1751. *Olai Wormii et ad Eum Doctorum Virorum Epistolae*. Copenhagen.

Evan R. Ragland is Associate Professor of History at the University of Notre Dame, and Co-Director of the Program in History and Philosophy of Science. His research examines the long history of scientific and medical experimentation and method, medical pedagogy, and pathology and disease concepts. He is currently working on a study of the emergence of experimental medicine (chymistry, anatomy, and clinical experiments) in the seventeenth century, and a history of the changes in ethical visions or the goods of knowing nature in the early modern period.

Chapter 9
Cutting Through the Epistemic Circle: Analysis, Synthesis, and Method in Late Sixteenth- and Early Seventeenth-Century Anatomy

Tawrin Baker

Abstract This chapter examines several late sixteenth-century anatomical works that adopted either the terms and/or methods of analysis and synthesis. Following a general discussion of the various roles that images could play in anatomy, I compare methodological discussions of anatomy as analysis and synthesis with the actual dissection procedures and results of anatomical investigation of the eye found in works by Costanzo Varolio (1543–1575), André du Laurens (1558–1609), Girolamo Fabrici d'Acquapendente (1533–1619), and Giulio Casserio (1552–1616). The "method of anatomy" is discussed from several perspectives: as a dissection procedure leading to discovery, with physiological/theoretical implications only implied; as a pedagogical ordering of a text or course on anatomy, whereby one chooses either analysis or synthesis as one's method; as an art (analysis) or a science (synthesis); as human artifice (analysis) vs. nature's processes (synthesis/genesis); and finally as an ordered procedure, wherein one first dissects animal bodies, discovers temperaments and structures, and then conceptually reassembles the newly discovered and understood parts, with the aim of understanding the animal soul.

Keywords Sixteenth-century anatomy · Eye · Dissection · Order · Art · Costanzo Varolio · André du Laurens · Girolamo Fabrici d'Acquapendente · Giulio Casserio

9.1 Introduction

What was anatomy (literally, "cutting up" or "cutting apart") in the late sixteenth century, and how was it related to analysis and synthesis? The initiation of a new age of anatomical discovery in the sixteenth century that challenged long-held ideas about the structures, activities, and purposes of the parts of human and animal bodies is well attested. Many sixteenth-century anatomists invoked analysis and

T. Baker (✉)
Independent Scholar, South Bend, IN, USA

W. R. Newman, J. Schickore (eds.), *Traditions of Analysis and Synthesis*, Archimedes 73, https://doi.org/10.1007/978-3-031-76398-4_9

synthesis—using either analogous Latin terms (*resolutio* and *compositio*) or, in some cases, the Greek terms themselves—as a *topos* relevant to ideas about anatomical practice and pedagogy and to establish an ordering principle for their works. We can see this particularly clearly in discussions of "anatomical method." Beyond mere discussions, however, we can investigate dissection techniques and anatomical histories, and analyze how authors incorporated these into the substance of their anatomical treatises. That is, we can also examine conceptual and material *practices* of analysis and synthesis, whether or not authors framed their anatomies using those Greek terms or their Latin equivalents.

Understanding anatomical analysis and synthesis as empirical, material practices raises both practical and philosophical issues. Similar to arguments against alchemists' claims about the products of their analysis—for example, the argument that alchemists were generating new substances via fire analysis rather than separating out the original substance's ever-present components—doubts were often raised as to whether the inanimate parts seen, felt, or otherwise sensed in a dissection were identical to the substances found in a living animal body. One well-known dispute, for example, concerned the supposed pores in the septum of the heart. This structure was essential to Galenic physiology: prior to the discovery of the pulmonary transit and the circulation of the blood, the pores provided a passageway between the veinous and arterial systems, and Galen also argued that a crucial transformation from veinous blood into arterial blood (substances with distinct powers) occurred during this passage. Although some claimed to find them (or at least traces of them) during dissection, most simply argued that the pores closed upon death; the theoretical—and, arguably, logical—need to connect the veins and arteries overrode empirical anatomy.

Furthermore, in the sixteenth century (again, often following Galen), it was generally believed that spirit or pneuma—a subtle substance in an animal body responsible for its most elevated natural powers—rapidly dissipated from the animal body after death. Finally, and perhaps most fundamentally, as Aristotle says throughout his corpus (and Galen following him), a dead eye (or any dead body part) is such in name only. The part's nature, intrinsic activity, or fundamental purpose within the animal oeconomy—seeing, for the eye; grasping, for the hand; and whatever disputed capacities were attributed to the heart—is absent to the dissector. This was often framed in terms of the loss of the part's form or *esse*, its what-it-is-to-be.

As a consequence of these issues anatomists and natural philosophers were faced with an epistemic circle: it seems one can only determine the true living structure of the dead body part or organ under the anatomist's knife by appealing to some physiological theory, but according to both Galen and Aristotle (and thus virtually every early modern anatomist) this physiological theory should be rooted in sensible knowledge of the parts and organs, particularly that obtained via dissection. How did early modern anatomists break this circle? Dissection is an artificial procedure that necessarily destroys the integrity of the whole to reveal its components. Did anatomists address how to be sure that their anatomical resolution revealed the body's natural divisions—especially without appealing to theory *a priori*?

Synthesis perhaps raised even greater problems. Dead body parts could not be reassembled, in the method of Dr. Frankenstein, to test or confirm notions about how body parts function individually, much less how the activities of parts, organs, or several organs work together for the sake of the animal. Anatomists could observe living bodies, but the knowledge that such observation yielded was limited. In the sixteenth century, animal vivisection offered one workaround, but again with limited results: the relevance to humans was debatable, and only certain parts of animal bodies were revealed via the knife, often imperfectly (Shotwell 2013). Indeed, many vivisections were simply those described by Galen. Perhaps the best known of these procedures, adopted from Galen and performed publicly by many early modern anatomists, involved ligating the recurrent laryngeal nerves of a vivisected dog to take away its voice; the anatomist would then demonstrate the larynx's connection to the animal spirits in the brain by removing the ligation, as evidenced by the dog's voice returning (Shotwell 2013, 173–5). Teun Tieleman has argued that this, and many other arguments from experiments and experiences with cut or ligated nerves and their effects on the functions of the parts of animals, had been used by Galen in *De placitis Hippocratis et Platonis* (*On the Doctrines of Hippocrates and Plato*) as an argument, explicitly in the form of analysis and synthesis, that the brain is the seat of the intellect (Tieleman 2002).[1] However, I have not identified any sixteenth-century anatomists who used analysis and synthesis to describe such experiments derived from Galen, perhaps because, on this topic, they drew on either *De usu partium* or *De Locis affectis*, rather than the more philosophical *De placitis*.

Experience from medical practice offered another route to knowledge via synthesis. One of the surgeon's jobs was to unite body parts that had become separated (Paré 1575, 716), and thus various surgical successes could yield a kind of evidence via nature's power of synthesis. However, this was not part of an organized method, and the uniqueness of each case made it difficult to generalize results. Observing diseases in humans that impaired the capacities of parts and organs (perhaps combined with autopsy after death) was another means of probing the actions and ends of living parts, organs, and systems but only indirectly and, again, only partially. As we shall see, some considered synthesis (understood as genesis) to be the exclusive domain of nature. In other cases, anatomists resorted to a conceptual synthesis, but how to best do so was debated.

To date, there exists no comprehensive and reliable historical investigation of early modern anatomy understood as analysis and synthesis. To the extent that the issue has been examined, scholars have largely reached conclusions similar to that of Roger French in his account of the Galenic tradition understood broadly (French 1984, 144):

> In general anatomy was taught theoretically, by the 'synthetic' method beginning with elements, qualities and similar parts to organic parts, which was said to follow nature in her construction of the body. Practical anatomy was 'analytic' and followed the sequence of dissections, in which the similar parts were (in theory) the final product.

---

[1] I thank Evan Ragland for pointing this out and should note that I have benefited greatly from Ragland's account, in this volume, of Galen's discussions of analysis and synthesis.

We also read accounts similar to that given by Domenico Laurenza (2003, VIII–IX), who makes a strong connection between anatomy and scholastic discussions of resolution and composition; Laurenza also links the "synthetic" anatomical method with deduction, rendering it theoretical, and the "analytic" method with induction, linking it with empirical investigation (32). As we shall see, both characterizations are misleading, particularly the latter. Most recently, Fabrizio Bigotti has addressed analysis and synthesis in anatomy and medicine, investigating new sources and offering perspicacious insights into several key issues. Bigotti's work includes an account of student notes taken during lectures given by Hieronymus Fabricius ab Aquapendente (Bigotti 2021), as well as an investigation of Bassanio Landi (1525–1585); the latter is a largely overlooked but was, in his time, significant figure (Bigotti 2023). Bigotti notes that Landi championed the Platonic tradition in which—also following Vesalius—a rational order of division was followed rather than that of physical dissection. Bigotti also addresses the issue of analysis and synthesis in this period generally (2019, 36–8) but does not offer anything approaching a complete survey. A comprehensive account is still lacking. Such an account is needed, however, given the diversity of positions adopted during the period in question together with scholars' growing recognition of the impact that anatomy had on seventeenth-century transformations to science generally. The present chapter is intended as a preliminary and partial sketch of a more systematic history.

As we shall see, in composing their treatises on anatomy, early modern medical authors did not necessarily consider anatomical analysis/resolution to be empirical or indeed to involve actual dissection at all. Much less did they understand the "anatomical method," when taken as analytical, to be an inductive method, and nor did they typically understand the synthetic anatomical method as explicitly deductive—analogous to Euclidean geometry, say. Rather, these topics were generally taken up in the context of teaching, as a way to best order the growing body of anatomical knowledge for the sake of student learning. Manual anatomical practice was not systematically aligned with these teaching objectives. Authors largely appear to have taken as their point of departure Galen's *Ars medica*, in which he argues that the order of *presentation* of the medical art should be either synthetic, analytic, or definitional. Thus, authors chose either analysis or synthesis as a way of structuring their publications and organizing their university lectures: that is, as a way of teaching anatomy.

During this period major shift occurred in Padua, where dissection (and vivisection) practices became systematically integrated into a "method of anatomy." Following a synthesis of Galen's anatomical works (*On Anatomical Procedures*, *On the Doctrines of Hippocrates and Plato*, and *On the Usefulness of the Parts*) in dialogue with Aristotle's treatises on animals (*History of Animals*, *De anima* and the *Parva naturalia*, and *On the Parts of Animals*), the so-called "School of Padua," especially in the works of Aquapendente and Casseri, argued in favor of an anatomical method that is closer to what we might call an empirical, investigative method. In this Paduan synthesis of Galen and Aristotle, "method" as the organization of information (also called *ordo*) and "method" as a means to uncover and grasp universal causes became aligned, arguably as a consequence of Jacopo Zabarella's criticism of accounts of "method" in medical authors (Edwards 1960; Mikkeli 1992). Moreover, in turn-of-the-century Padua, "method" as autoptic knowledge-gathering included the *discovery* of structure, action, and teleological function (the latter referred to as *usus, utilitas,* or *officium*). All of this also occurred in the context of teaching medical students. In short, in the "School of Padua" toward the end of the sixteenth century, anatomical analysis and synthesis were united into an ordered sequence, aimed simultaneously at the investigation of nature and at teaching medical students. This ordered sequence aimed to discover structures and to give new, better anatomical *historia;* to give a causal, teleological account of those parts and their functions; and, perhaps most importantly, it was designed to teach not just anatomical facts but also the method of anatomical discovery itself. Nevertheless, calling this practice "analysis and synthesis" is problematic, given that the Paduan anatomists did not themselves use those terms or their Latin equivalents.

This chapter examines how practitioners understood anatomy in terms of resolution/analysis or composition/synthesis. Comparison of methodological discussions of anatomy as analysis and/or synthesis with the actual dissection procedures and anatomical findings contained in those works is crucial. To render the topic more manageable and because of my interest and expertise, I investigate these questions as they pertain to a single organ: the eye. I examine how it was anatomized during the second half of the sixteenth century and how an understanding of the primary action of a living eye—namely, seeing—was related to anatomical investigation. This survey is not systematic; I have focused on just a few important or interesting texts. More work is called for.

This paper begins by examining some of the ways in which the term "anatomy" was used around the middle of the sixteenth century in relation to the organ of vision, particularly in connection with printed anatomical images. Lower-status surgical specialists dealt with most eye conditions during this period, and physicians and learned surgeons thus had little professional interest in the eye during the first half of the century. Examination of vernacular works thus gives perspective on the broad sense of what was intended by the term "anatomy" in treatises addressing the eye, and also on how visual and textual depictions interacted generally.

Following this the writings of two anatomists—Costanzo Varolio (1543–1575) and André du Laurens (1558–1609)—are discussed. I examine their statements on the "methods of anatomy" as well as their arguments for the proper definition of anatomy. Varolio's *De nervis opticis* (1573) was influential within its narrow domain, while the impact of his posthumous *Anatomia, sive de resolutione corporis humani* (1591) has received little attention. By contrast, the works of du Laurens impacted medical education enormously from the turn of the seventeenth century into its first half.

I then turn to the Paduan anatomists Hieronymus Fabricius ab Aquapendente (Fabrici d'Acquapendente, 1533–1619) and Julius Casserius (Giulio Casseri, 1552–1616). Both created ambitious anatomical projects in which the stages of physical dissection and conceptual composition are integrated, with the second (conceptual) stage building from the empirical foundation established in the first stage. This explicit methodological framework is given in the introductions to their works. I also discuss how this framework relates to their anatomical practice, as revealed in the main texts and images themselves, and I argue that their discussions of method are closely aligned with their practices.

I conclude with a summary of how these authors explicitly or implicitly cut the epistemic circle inherent to anatomical investigation. In addition to the traditions identified by Alan Shapiro (this volume)—namely, the chemical, mathematical, logical, and natural-philosophical—the anatomical tradition may constitute a separate one. Perhaps it is better to say that, in the anatomical tradition, the terms "analysis" and "synthesis" were *topoi* for addressing issues related to the anatomical method and to pedagogy generally. Such discussions in some respects overlapped with—but in others were independent of—analysis and synthesis in medicine proper (Ragland, this volume). I suggest that the Paduan "anatomical method," which could be cast as a kind of analysis and synthesis, was broadly influential on seventeenth-century science. However, I do not contend that this constitutes a tradition of "analysis and synthesis" *per se*, for reasons that will be made clear.

9.2 Meanings of Anatomy in the Sixteenth Century: Image and Text

Images played an important role in sixteenth-century anatomy. Their presence in anatomy is perhaps unsurprising: as far back as Aristotle we find diagrams are referred to when a detailed account of spatial properties, including shapes and relative positions, of various anatomical parts was needed (Lennox 2018). But their role is complicated. Modern assumptions about how and why anatomical images (and images generally) should be created and used impede our understanding of the visual culture of this period.[2] Moreover, the term "anatomy" has always, it seems, been used both literally and figuratively. For example, Aristotle uses ἀνατομή in the sense of logical as well as literal dissection.[3]

In the sixteenth century, particularly in vernacular works, a diagram alone might be described as an anatomy. An example of this is a pamphlet titled *A New, Highly Useful Booklet for Recognizing Sicknesses of the Eye, and also an Image [Figur] or Anatomy of the Eye....* Published in 1538 and republished in 1539 in Strasbourg by Heinrich Vogtherr (likely the author of both the text and the image), the "anatomy" itself is just a labeled figure, without any accompanying textual description, any account of how to dissect the eye, or any sense that the figure was generated from direct experience. The second edition places the figure on the title page itself

[2] See Murdoch (1984), Lüthy and Smets (2009), Kusukawa (2012), and Fay and Jardine (2014). For sixteenth-century images of the eye specifically, see Raynaud 2020 and Baker 2023.

[3] For a logical use, see 98a1–98a12; for a literal use, see 497a30–497a35. See also the OED, which holds that Aristotle uses ἀνατομή "in the sense of logical dissection or analysis." "anatomy, n.". OED Online. Oxford University Press. https://www-oed-com/view/Entry/7179 (accessed December 24, 2022).

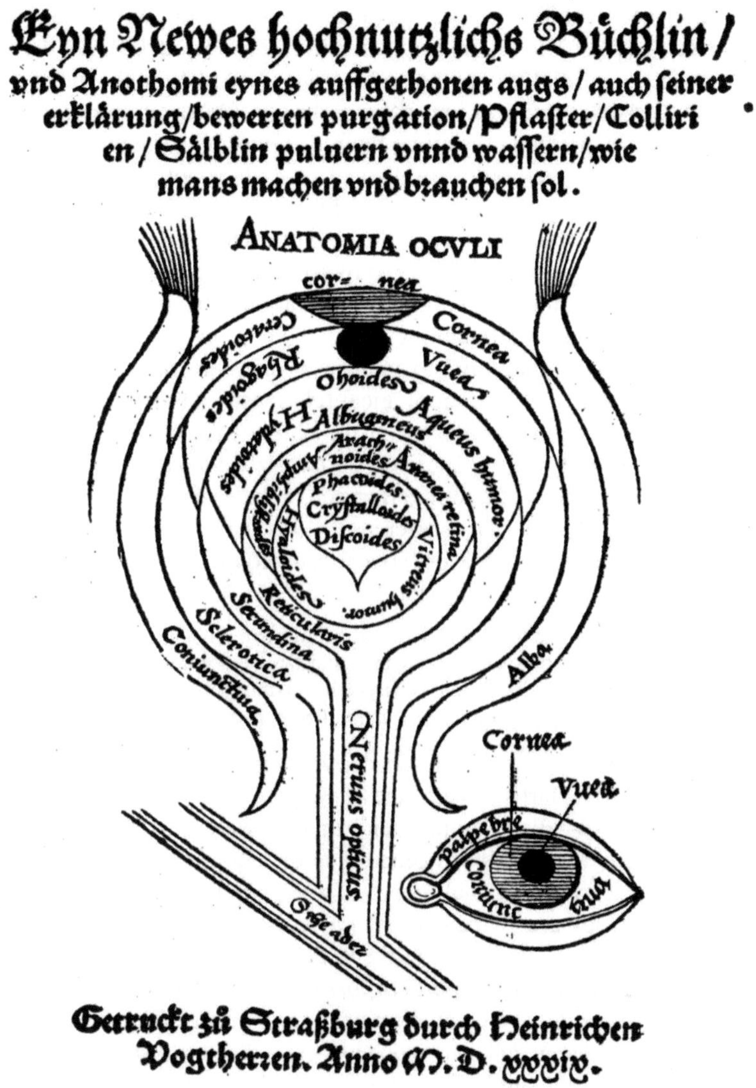

Fig. 9.1 An ocular anatomy as simply a labeled picture without accompanying textual description or account of dissection procedure (Eyn Newes hochnützliches Büchlin 1539, 1)

(Fig. 9.1), while the first provides it on a separate page (Eyn newes hochnützliches Büchlin 1538).[4]

[4] For more on Vogtherr, see Muller (1987, 1997).

This image is ultimately derived from Gregor Reisch's *Margarita philosophica*, originally published in 1503 but reprinted (and plagiarized) throughout the sixteenth century (Reisch 1503, *219r) (Cunningham and Kusukawa, 2003). By the time of Vogtherr's image in 1538, this copying and recopying resulted in the inclusion of as many terms as possible for the various parts of the eye. Other differences have crept in, but the distinctive, pointed crystalline humor in Reisch has been preserved. This image is not intended to be read naturalistically—it does not give us any sense of how an eye would appear when dissected and is not intended to accurately convey information about the sizes and shapes of the parts. Thus, one important sense of "anatomy" during this period is simply knowledge of the technical terms for the body's parts and a basic understanding of the relationships among the parts, most importantly that of contained/containing.

The use of the term "anatomy" to indicate a diagram that primarily conveys information about the names, the topological relationships (connection, enclosing/being enclosed), and the axial ordering of the parts of the eye is also seen in the famous ophthalmologist Georg Bartisch's *Ophthalmodulea* (1583), written in German but peppered with Latin terms. Bartisch includes some textual description, but he uses "anatomy" primarily in reference to his two famous sequences of flap images of the brain and eye. These are woodcuts based on his own drawings. On the anatomy of the head, he writes (Bartisch 1996, 5r),

> the head is the chief, top, and highest member, also a true natural mother, site, and dwelling spot of the eyes. The eyes have and receive their substance (*substanz*), foundation (*fundament*), and origin from the head. Then it is deserved and of importance that one sets out the counterfeit (*Contrafactur*), anatomy (*Anatomiam*), figure, and form.[5]

I have supplied Bartisch's Latin or Latin-derived terms here, which are also highlighted in the original. Bartisch calls these images "Contrafactur." According to Sachiko Kusukawa, so-called "counterfeit" images were not necessarily, if ever, drawn from life; they were meant to generate in the reader the impression of a singular event, not to serve as a visual record of any particular dissection (Kusukawa 2012, 15). As was the case with the pamphlet just discussed, Bartisch's images were not derived from an actual dissection of the brain or the eye but took as their model Andreas Vesalius' famous *De humani corporis fabrica* (1543). Bartisch generated his images of the brain and eye by combining several images from the *Fabrica*, and at times he merged several *Fabrica* woodcuts representing distinct, incompatible stages of dissection into a single flap image. Immediately after presenting his flap sequence on the brain, he writes (Bartisch 1996, 6r),

> One has an actual, clear, short, complete, and certain report and demonstration of the appearance, nature and all conditions of the head. This is to prepare one. Each man may see and consider the origin of the eyes.

---

[5] Note that this translation preserves the formatting and pagination of the original; all translations are from this edition, but I have inserted the original German/Latin terms that are significant for my analysis.

Thus, his diagrams are meant to delineate key parts, provide technical names, convey the generally accepted account of the substance of those parts (in terms of their tangible qualities, especially with respect to hot, cold, wet, and dry), and to show how key parts of the brain are connected to the optic nerve and thereby to the parts of the eye.

On his anatomical diagrams of the eye (Fig. 9.2), he writes (Bartisch 1996, 8v),

> The clear representation and figure of the eye follows. From this the correct and actual anatomy, state, nature, and condition of the eyes is to be seen and found. One lifts up a part or leaf one after another of the largest central figure.

Here, while the images in and around this flap sequence were derived from Vesalius's *Fabrica*, certain features were modified to conform with his extensive manual experience in the operation known as cataract couching (Baker 2017, 54–60). At the end of Bartisch's section on the anatomy of the eye, we read (Bartisch 1996, 10r),

> Therefore one has the entirely complete, clear, and actual report of the entire eye as it is constructed, created, and positioned. Henceforth it must be a very simple, rough, and ignorant person who does not want to prepare himself with these figures and descriptions or be able to follow them.

Turning to Bartisch's visual source material, we find a far more complex relationship between anatomical texts and images. Vesalius' *Fabrica* contains accounts of ocular anatomy as a dissection process—the *ratio administrandi organi*—as well as the results of that process, namely the *fabrica* of the eye. Note that *fabrica* here conveys the senses of both structure and craftsmanship (and thus teleology). Vesalius dissected numerous eyes to determine the shapes, sizes, colors, and temperaments of its simple parts and then reconstructed the eye based on his prior understanding of what today we would call physiology. In this case, Vesalius largely follows Galen's physiology of vision, according to which copious amounts of visual spirit fill out the eye; these spirts are sent to the eye via a foramen in the optic nerve and are present particularly in the foremost aqueous chamber.

Vesalius' image (Fig. 9.3) may be read simultaneously as a composition and an analysis of the eye, although the synthesis of the eye is prioritized. I have used a 1568 unauthorized edition of the *Fabrica*, partly to emphasize the diffusion of Vesalius' images and because this edition combines both of his depictions of the eye, the top cross-section and the bottom composition/dissection sequence, on a single page. Differences from the original are negligible for the purposes of the present investigation. Looking at the bottom series, reading from the top-left to bottom-right, we have what I call the "divine craftsman" sequence, in which the eye is constructed from the innermost, homeomerous part—the crystalline lens—outwards (Baker 2023, 222). Reading in reverse we have an account of Vesalius' ocular dissection practice as given in both the *Fabrica* and the notes of a medical student, Baldasar Heseler, who attended Vesalius' famous Bologna dissection in 1540 (Heseler 1959). This image does *not* convey what one might expect to actually see in any dissection: the parts (all either liquid or flexible) are drawn according to Vesalius' conception of their idealized, living configuration. It is clear that, throughout, Vesalius is incorporating Galenic ocular physiology, according to which a great

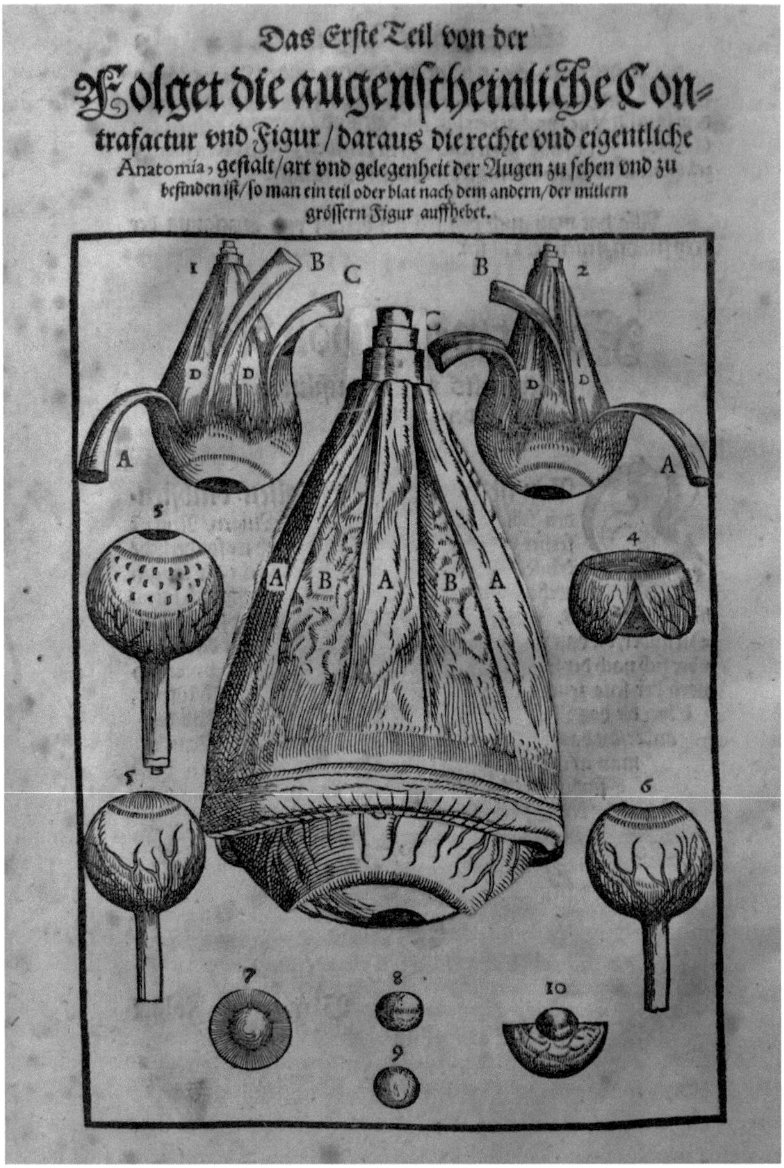

Fig. 9.2 Georg Bartisch's "anatomy" as a diagram with textual description (not shown). Bartisch copied the outer figures from Vesalius's *Fabrica*, but some details of his central flap-image (containing six leaves attached at the top of the optic nerve) were modified to fit with his experience treating eyes and couching cataracts (Bartisch 1583, 8v.) (Courtesy Lilly Library)

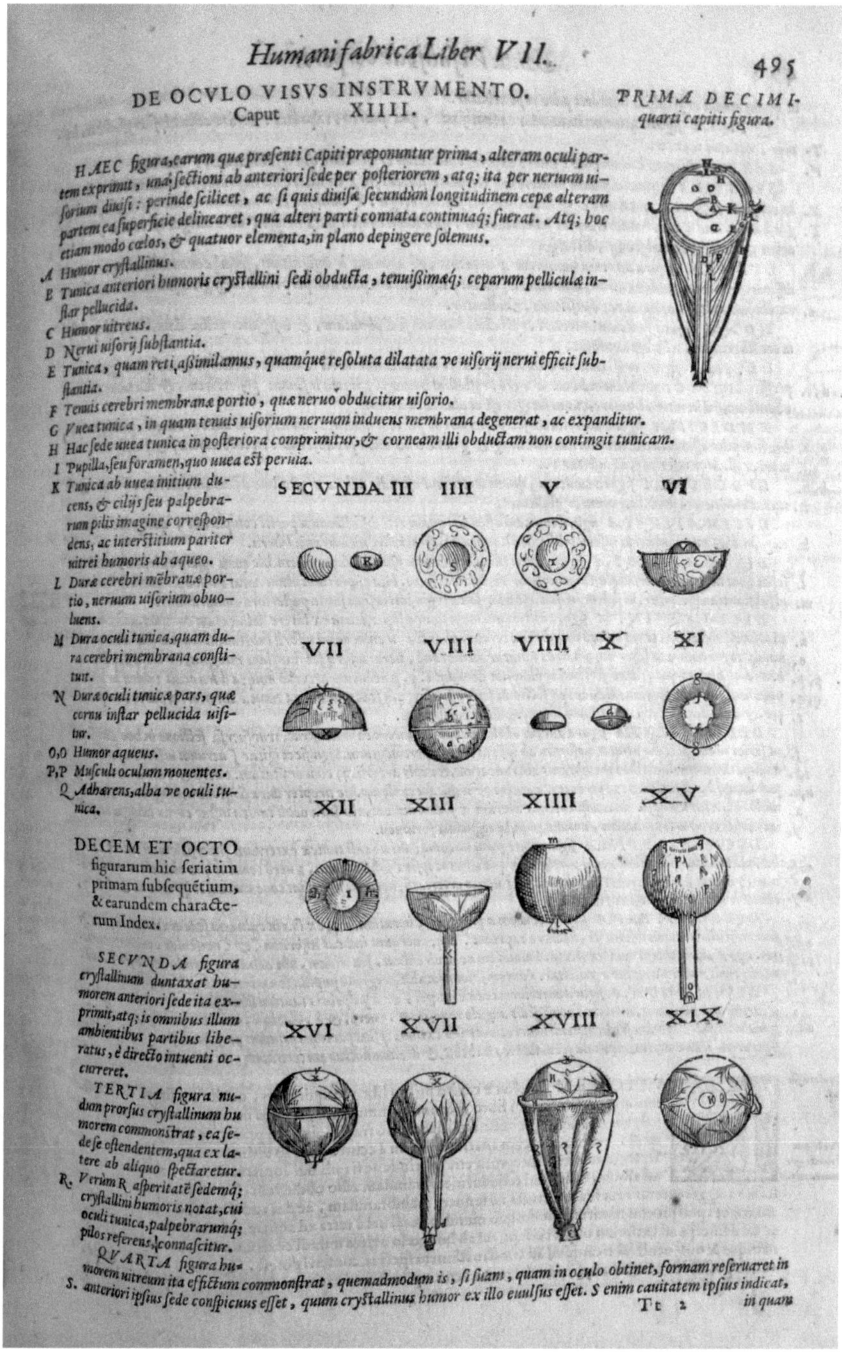

Fig. 9.3 Vesalius' depictions of the eye from a later, unauthorized edition. The images, though more crude, follow the originals closely. At the top is a cross-section of the eye with the crystalline humor at the center, which Vesalius likens to a picture of the heavens and the four elements. At the bottom is the "divine craftsman" sequence: reading top-left to bottom-right, we are given the composition of a living eye according to Nature's plan; reading in the reverse we are given instructions for dissecting the eye. Note the equal sizes of the aqueous (VII) and vitreous (VI) humors. (Vesalius 1568, 495.) (Courtesy Newberry Library)

deal of visual spirit present in the front chamber is lost upon death, to reconstruct his idealized configurations, despite his assertion that he has not yet been able to determine precisely how vision itself takes place (Baker 2019).

While Vesalius forcefully argued that anatomy and medicine should be based on direct visual and manual experience, his dissection of the eye is rudimentary by the standards of his time. His instructions for dissecting the eye are intended for a public demonstration; this highlights the lack of professional interest in the eye by learned surgeons and physicians and to the siloing of the disciplines dealing with vision and the eye. His instructions and his technique for public dissection are useless to determine the relative quantities of the humors or the location of the crystalline humor within the eye. Vesalius' method for manually rendering the eye into its parts was not designed to resolve questions that would later prove crucial for determining whether vision takes place by intromission or extramission, for determining the seat of the visual faculty in the eye (whether the *aranea*, the crystalline lens, the vitreous, or the retina), or indeed for understanding the usefulness of the parts in detail (Baker 2019, 2023).

Vesalius' images display Nature's rational composition of the eye while also cleverly representing dissection as the inverted sequence of this rational synthesis: read bottom-right to top-left, his woodcuts mirror his dissection instructions, which are written several chapters later. Moreover, Vesalius says that, in teaching students about the eye, he begins by drawing a diagram just like the one placed at the beginning of his chapter on the eye and afterwards dissects it with reference to the drawing (Vesalius 1543, 649). For the eye, at least, both his public and classroom dissections were meant to confirm the idealized account of the eye previously impressed on the minds of his audience; he used diagrams, together with Galenic pneumatic physiology (if not a Galenic extramissionist visual theory itself), to mitigate the potential chaos and confusion inherent in a public anatomical demonstration of the eye (Baker 2023).

Both the resolution of the eye into its constituent parts and its reassembly are theory-dependent here, even if both steps are somehow based on first-hand experience with dissected eyes. As elsewhere in the *Fabrica*, the order in which the body parts are treated in Vesalius's text does not mirror the order of dissection; instead, Vesalius prioritizes a clear understanding of the rational construction of the body and his audience's retention of that knowledge. This (along with the size and enormous cost of the *Fabrica*) limited its use as a dissection manual; smaller, pirated editions may have been used in such a capacity, but I am unaware of any evidence for this. We can say that, in the case of the eye (but not necessarily in his treatment of other parts of the body), a conceptual composition of the parts was prioritized; the material dissection was subsequently guided by these conceptual/theoretical results. Note, however, that Vesalius' images offered more than mere topological relationships: in a radical innovation that is easy to take for granted today, metrical properties—such as sizes and shapes—in his images depicted those of the living body parts they represent (Baker 2023, 200–228).

9.3 Varolio: Anatomy as Discovery vs. Anatomy as the Resolution of Final Causes

Costanzo Varolio (1543–1575) studied medicine at the University of Bologna and earned his doctorate in 1566, learning anatomy under Giulio Cesare Aranzio, an accomplished anatomist who studied the brain in particular.[6] He was extraordinary chair (i.e., the lowest rank) in surgery at University of Bologna from 1569 to 1572. From 1572 to 1573, he advanced to an ordinary chair in medical practice and held the responsibility for performing public dissections. Our knowledge of Varolio's short life is otherwise poor. Some reports have him joining the medical faculty of the papal university, and he may have been invited to become a physician or surgeon in the papal medical service. He is also reported to have successfully removed bladder stones from many patients in Rome, at the time a dangerous procedure that learned surgeons generally avoided (Siraisi 2013, 71–2).

Varolio has only two publications to his name. The first, *De nervis opticis*, is a letter to the famous physician Hieronymus Mercurialis (Girolamo Mercuriale, 1530–1606) together with Mercuriale's reply and Varolio's response. It contains two high-quality woodcuts depicting Varolio's new method for dissecting the brain, showing, among other things, his description of a region of the brainstem now known as the *pons varolii* (he believed that it was part of the cerebellum). Varolio was keenly interested in the origins of nerves. He used a new dissection method to show that the optic nerves originate behind the brainstem and argued that both ancient and recent anatomists were incorrect in the opinion that they enter the brain toward the front (Varolio 1573, 2r).

This argument is visually captured in his published images. A note introducing a detailed and skillfully executed woodcut (the first of two) reads (Varolio 1573, 17r),

> THIS FIGURE shows the base of the cerebrum denuded of the *calvaria* and the dura membrane, whose left part [i.e., the right side of the diagram] demonstrates the origin of the spinal medulla, as well as the seven pairs of nerves according to the common opinion of the Anatomists, the right part showing the true origin of the spinal medulla, and everything else which has been mentioned in this Letter, etc.[7]

The precise relationship between this woodcut and any original drawings in the letter is unclear (see Fig. 9.4). The above text and thus the image and the key to the image were by all indications written by Varolio, suggesting that (at least for publication) he invested significant work in his images and the accompanying key. It is possible that he had a hand in revising the letter for publication and that the woodcut was added at this later step. Note that this is a public letter, one that (along with the

---

[6] Background information on Varolio is based primarily on O'Malley (2008) and Andretta (2020).

[7] "HAEC FIGURA ostendit basim cerebri denudatam à calvaria, & dura membrana, cuius pars sinistra demonstrat originem spiinalis medullae, atque septem parium nervorum secundum communem Anatomicorum sententiam, pars dextra ostendit veram originem medullae, & aliorium, quorum in Epistola habita fuit menti, &c."

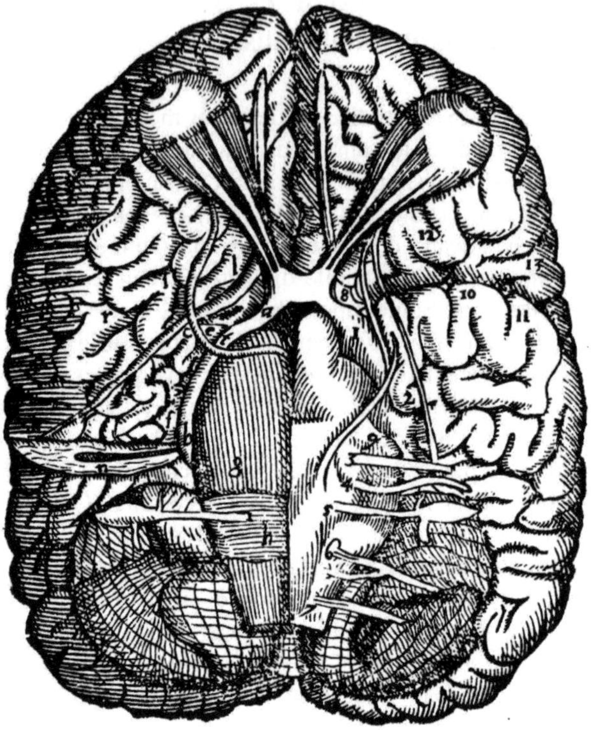

Fig. 9.4 On the left side of the page (i.e., the right hemisphere of the brain) is Varolio's new account; on the right is the traditional (roughly Galenic) account. In the key opposite this, we read, "a.b The whole part of the optic nerve, unknown to the rest of the anatomists, where b is the bending around of the previously mentioned nerve at the posterior part of the spinal medulla." (Varolio 1573, 17r–18v)

replies by the famous Mercuriale) was expected to be copied by other physicians, if not published (as it eventually was).

Dividing the brain into two halves allowed details to be compared and contrasted, perhaps even while the readers were performing a dissection to allow one to test the two accounts via *autopsia.* This visual technique is remarkable for the time, and it reflects the continually innovative use of medical images in this period.

Varolio's initial letter questions the traditional view of the number and functions of the ventricles of the brain. These ventricles were spaces within the cerebrum in which the animal spirits responsible for cognitive faculties were believed to reside. He gives a new description of the ventricles and suggests that a revised account of the locations of the cognitive faculties is necessary (Variolio 1573, 7v–8r). Traditional accounts had the optic nerves entering the front of the brain to transmit visual judgments to the front-most ventricle, where the common sense was thought to reside. Much of Varolio's account arises from his novel dissection method and observations made therefrom. The brain was traditionally dissected by removing the cap of the skull and then dissecting in layers from the top down. This was the primary technique employed by Vesalius; although Vesalius also dissected from the bottom up, he did so with the brain within the skull. Varolio (11r) writes,

> You are not unaware that the common method of dissecting the head, both by the ancients and the neoterics, was to begin with the upper parts, by first removing the upper part of the skull then the membranes, followed by the substance of the brain as far as the ventricles, and the rest of the parts, as you already well know. I, however—considering that most of the organs of the brain lie near the base of the head, and that the brain, by its weight (especially in the dead), compresses these organs between itself and the skull—I judged that this common method of dissection was subject to the greatest number of hinderances, and therefore that I should begin the dissection from the opposite side of the head, that is, from the base of the brain. By which method of proceeding each of its organs is observed so perfectly that nothing more seems to be wanting. But this method of dissection, as it is different from the common one, is also most difficult.[8]

Varolio then says that he used this manner (*ratio*) of dissecting the brain to demonstrate to others the truth of his findings (13v):

> Since, therefore, in the year of 1570 after Christ's birth, in the month of December (if I am not mistaken) within a private residence, I had shown to many students that the optic nerves originate from the posterior part of the spinal medulla (which I had observed before). And since those who had seen this had immediately reported it to everyone, the rest of the anatomists, who indeed thrive and flourish in this school, objected to me that what the students proclaimed was not true; in fact, that they had been deceived by me, for it was easy (they said) for me to deceive others, as someone who had been trained in dialectic. Thus, they gathered many people together, and acquiring human heads and dissecting in a communal manner they strove to persuade everyone of the falsity of my discovery. [...] There were notwithstanding many who did not see this, and who thus could not at all believe that it was true. Therefore in the year 1571, in April, given the opportunity of [several] bodies from which I was able to begin a public dissection of the human body in the Bologna Gymnasium, at the end of this dissection (the fabric of the head having been performed earlier), although the brain was shriveled, and because of the heat of the environment somewhat putrid, I once again made it so clear to everyone, that those who had seen it before were entirely reassured in this truth, and those who had not believed it prior both had peace in their soul and stood up for the truth.[9]

---

[8] "Non te latet communem modum administrandi caput tum antiquorum, tum neotericorum fuisse incipiendo a partibus superioribus, removendo primo sectione superiore calvariae partem, deinde membranas; & substantiam cerebri usque ad ventriculos, & reliquas partes, ut optime novisti. Ego verò considerans plurima cerebri organa extare propè basim capitis, & cerebrum suo pondere (praesertim in mortuis) ea. inter semetipsum, & calvariam comprimere iudicavi hanc communem rationem administrandi esse plurimus impedimentis obnoxiam, idcirco coleo ab opposita capitis parte adminstrationem inchoare nempe a basi cerebri, qua ratione procedendo singula eius organa ita absolutè observantur, ut nihil amplius desiderari posse videatur. Verum hic modus administrandi sicuti est a communi diversus ita etiam est dificillimus."

[9] "Quum [i]gitur anno a Christo ortu MDLXX. mense (ni fallor) Decembris intra privatos parietes multis scholaribus ostendissem nervos opticos nasci ex posteriori parte spinalis medullae (quod tamen prius ovservaveram[)], & quum qui hoc vidissent statim omnibus retulissent, reliqui Anatomici, qui sanè in hac schola plurimum vigent, & florent obiecerunt mihi non esse verum id, quod scholares praedicabant; sed eos fuisse a me deceptos, facile enim (dicebant) mihi erat alios

De nervis can be placed in the *epistolae medicae* and *observationes* epistemic genres (Maclean 2008; Pomata 2010a). On the *observationes* genre, Gianna Pomata writes that the genre arose in the late Renaissance and that its "rise and fortune was linked to the development of horizontal networks of exchange among European scholars" (Pomata 2010b, 197). She identifies the following features as typical: "the report of first-hand observation, an informal and agile format, a limited and yet ambitious goal—the communication of anatomical discoveries" (Pomata 2010b, 203). Novelty, the description of particulars without aiming toward general rules, and the stress on practice were all characteristic of the genre; the trend was toward a sort of "collective empiricism" (Pomata 2010b, 223).

Varolio in *De nervis* does not argue for a systematic method for anatomy, which accords with Pomata's account. Rather, a particular instance in which, for Varolio, the problems associated with death combined with the traditional way of dissecting the brain resulted in certain parts being both obscured and altered (if not destroyed). Varolio offers a new technique for dividing the brain into its parts, one that reveals new structures and connections. From this he implies that the current account of the cognitive faculties needs to be overhauled—by no means a small change—but this project is neither attempted nor outlined in his letters.

In contrast is Varolio's only other extant work, the posthumous *Anatomy, or the Resolution of the Human Body* (1591). Varolio addressed the work to Caesar Mediovillanus. The preface was written by Johannes Baptista Cortensius, the surgeon and anatomist behind the book's publication, and is addressed to Mercuriale. Cortensius says of Mediovillanus that he was a secretary to Pope Gregory XIII and that he was a "devoted student of the liberal arts, and especially of anatomical matters." (Varolio 1591, unnumbered preface.) He says that Caesar Mediovillanus received the manuscript from Varolio as a gift and that Mediovillanus urged Cortensius to print the work, which was Varolio's wish before his untimely death.

The book offers a thoroughly teleological anatomy, beginning with the end or final cause of the human being and then moving to analysis. That is, the body is first analyzed into the capacities necessary for the human body and soul to achieve its ends, which he calls the *methodus dissolvente,* and these, in turn, are resolved into the complex parts and then the simple ones necessary to sustain those capacities. Varolio writes that this form of anatomy, while it results in knowledge of the "bones, membranes, fibers, cartilages, vessels, nerves, and such like parts," does so only insofar as those structures serve the powers of the soul and thus conveys things that

---

decipere, utpotè qui in Dialectica esse exercitatus, & accipientes capita humana multos convocabant, modoque communi ea. administrantes falsitatem meae inventionis omnibus persuadere nitebantur, Ac si Dialecticae munus esset decipere sensum circa sensibilia, & ex eo quòd quispiam rem aliquam ignoret necessario inferatur eam non esse. Quum tamen multi essent, qui hoc non vidissent, & propterea id esse verum minime credere possent, ideo ubi anno MDLXXI, mense Aprilis data esset corporum occasio, unde publicam administrationem corporis humani in hoc Bononiensi Gymnasio aggredi possem, in fine eius administrationis (capitis fabrica prius declarata) quamvis cerebrum esset flaccidum, & propter caliditatem ambientis semiputridum, illud omnibus iterum adeo patefeci, ut qui prius viderant maxime in hac veritate confirmarentur, qui verò non crediderant & animo quiescerent, & veritatem consisterentur."

are worthy of a "noble man or a philosopher by nature (*ingenuus Philosophus*)" (Varolio 1591, 2). Of his method, he writes (2),

> In explaining these things, I shall proceed in such a way as to first propose the end, the highest of which has, thanks to the grace of God, chiefly been established in man; then by means of the Method of dissolution (from which these books take their inscription), collecting the number and quality of the of the powers of our soul. Next I will try and disclose to you in what way it was necessary for the human body to be constructed on account of this—that by postulating the order of natural things [it must be] constituted out of such and such a number of parts, neither more nor fewer, or in any other way disposed, so that starting from one part you may gradually perceive the necessity of all the others, down to the smallest and most secret things. Whereby, in addition to what was known to the rest of the writers, there are also many things that neither the common person nor any others have known before, as I hope you will see.[10]

Compared to *De nervis*, a very different sort of anatomy is here conveyed in a distinct anatomical genre whose readers are not primarily physicians or medical specialists. His invocation of "secret things," for example, seems calculated to appeal to a popular audience.

The final cause of human beings, according to Varolio, is to contemplate and understand the immense power (*virtus*) and divine essence of the craftsman who created us. This ultimate final cause is theological (though compatible with Galen), but the final causes of the complex parts, and perhaps even the simple parts, are also assumed in this work. That is, the teleological functions of the organs and systems of the body are not derived from the results of dissection or vivisection. In this conceptual analysis, neither a physical dissection, nor a *historia* of the parts—either gathered via first-hand dissection or compiled from other anatomical texts—are characterized as a resolution, dissolution, or analysis; the (largely traditional) account of what one ought to find in physical dissection is merely coordinated with the end products of a conceptual analysis, beginning with abstract final causes and ending with material descriptions.

Accordingly, his treatment of vision begins with the declaration that sight ranks highest among the five senses. After asserting his new account of the origin of the optic nerves, he begins with the most general property of the eye's interior—its transparency—arguing that this is necessitated by its object of vision—namely, color. For example, if the eye were yellow, we would always see yellow. Moving on to the parts, he writes that species of color are not impressed in air or water when rays pass through these substances and thus that the crystalline, which is the seat of vision, must be relatively thick for colors to be conspicuously received. This is all

[10] "In quibus explicandis ita progrediar, ut primum proponsito fine, cuius gratia Deus optime maxime hominem instituit: deinde Methodo dissolvente (à qua ipsi libri inscriptionem sumpserunt) numerum & qualitatem potentiarum animae nostrae colligam. Mox quomodo fuerit necessarium propter has humanum corpus ita construi, & id ordine rerum naturalium postulante, ex tot, nec pluribus aut pauciobus partibus, aut alia ratione dispositis, constitui, tibi aperite tentabo; ut ex unius partis principiis, opportunitate, gradatim necessitatem omnium aliarum, usque ad minimas atque intimas, percipere possis. Ubi praeter ea. quae reliquis scriptioribus nota fuerunt, multa etiam nec vulgata, nec aliis antea cognita, ut spero perspicies."

typical for the time. His resolution of the eye into its constituent parts proceeds entirely in this way, with the properties of the parts being necessitated by the physics of light, color, and perception imported from natural philosophy or optics. His physical descriptions are brief, general, and in no way give the impression that they were derived from *autopsia*. The sole image of the eye printed in the book depicts the activity of vision and the division of the visual field into the central (direct and clear) visual angle and the "confused" vision that takes place at larger angles; this idea is imported from mathematical optics.

Thus, one anatomist gives us two very different conceptions of anatomy, each constrained by a different audience. For the professional community of physicians, learned surgeons, and medical students, anatomy is a literal cutting up; prime importance is placed on his innovative artificial sectioning of the body, and he includes images that seem intended for the reader to compare, in an actual dissection, his account with the traditional one. For a popular audience interested first and foremost in humanity's relationship to God, by contrast, anatomy is a theoretical resolution of man's final cause: spiritual causes are conceptually analyzed to reveal material structure. Had he lived longer, Varolio might have reconciled these two approaches. In any case, he has left us two distinct anatomical methods, both of which might be labelled analytic.

9.4 Anatomy as Artificial Analysis of the Parts: Du Laurens on Anatomical Pedagogy

André du Laurens (Andreas Laurentius, 1558–1609) did indeed explicitly bring together the notions of material resolution/composition and conceptual composition in anatomy. Du Laurens was among the most prominent physicians and anatomists of the late sixteenth century, and his works on anatomy and medicine were extraordinarily influential for a time. He was a professor at Montpellier University and in the 1590s became a physician to Henry IV of France and Queen Marie de Médicis, then chancellor of Montpellier (while still remaining at court), and finally first physician to the king in 1606. Du Laurens is often described as an orthodox Galenist who defended Galenic views against Aristotelian ones.

Here, we will examine only du Laurens' early *Opera anatomica,* published in 1593 in Lyon and London (republished twice in 1595, in Frankfurt and in Hanau, all octavo format), although similar views, though expanded, can be found in his later *Historia anatomica* (in folio), first published in Paris and Frankfurt in 1600. His *Historia anatomica*, in particular, was an extremely popular textbook, and through this his classification of the methods and definitions of anatomy were widely disseminated. The *Historia* was reprinted throughout the seventeenth century in Latin, with translations into French and English, and his works were collected into an *Opera omnia*, first published in French in 1621 and Latin in 1627 (this contained the *Historia anatomica* but not the superseded *Opera anatomica*). The appeal of his anatomies was certainly their pedagogical utility: they were comprehensive, with largely up-to-date information; they thoroughly incorporated Greek terminology and adhered to Galen; and they provided a model for how to deal with anatomical/ philosophical controversies.

In his preface to the *Opera anatomica*, we read that there are two methods for learning and teaching anatomy—namely, analysis or genesis (Laurentius 1593, 8).

> Since, then, the utility and necessity of the anatomist is so great, you (lovers of medicine all) should devote painstaking and diligent work to its study, and neither dread nor let the difficulty of the art frighten you, for it is easy, believe me, if it is explained by way of order and method. Now there is a twofold method of learning and teaching anatomy, one [being] ἀναλύσεως [analysis] or resolution, which resolves the whole into its parts: the other γενέσεως [genesis] or composition, which composes the whole out of similar and dissimilar parts. Nearly all the anatomists have followed this. We will observe the analytical method in this history. For we shall cut (*secare*) the human body into four principal parts, the head, the thorax, the belly, and the limbs. We will cut these again more minutely, until we have reached the simplest [i.e. Galeno-Aristotelian homeomerous] ones, in each one continually observing the order of the section. Since, indeed, many things in anatomy are observed to be covered up [by other body parts], and very difficult, in order that nothing may be wanting regarding a perfect knowledge of this art, I will reveal and explain each one, then first describe the *historia* of the part, and immediately after I will make clear what is considered controversial.[11]

A similar statement with almost identical wording with respect to analysis and genesis may be found in his later *Historia* (du Laurens 1600a, b, 11). Du Laurens thus adopts the analytical method of teaching anatomy, which, for him, parallels the actual practice of dissection, making the text a suitable companion to anatomical demonstrations performed at medical schools.

In addition to the two methods described in the preface, in his first chapter, on "The Definition of Anatomy and the Parts," du Laurens states that anatomy also has two definitions (Laurentius 1593, 9–10):

> For it [anatomy] denotes either an action which is accomplished by the hand, or an attitude of mind and the most perfect action of the intellect. The former is said to be practical (πρακτικὴ), the latter theoretical (θεωρητικὴ): both are indispensable to the perfect physician towards the diagnosis, prognosis and therapy of the conditions. If you look at the prior meaning, anatomy will be defined as the artificial section of the external & internal parts. I call it an artifice which separates the parts from the parts in such a way that each one is seen wholly and not in any way mangled. The latter meaning will be defined as the science which

[11] "Cum itaque tanta sit Anatomes uitlitas & necessitas vos (φιλίατροι omnes) horror ut in eius stuido sedulam ac dilligentem operam impendatis, nec est quod artis difficultas vos terreat, facilis enim est, mihi credite, si modo ordiine ac methodo explicetur. Est autem discendae & docendae Anatomes methodos duplex, una ἀναλύσεως seu resolutionis quae totum in suas partes resolvit: Altera γενέσεως seu compositionis quae ex similaribus dissimilares ex his totum componit. Hanc ferè omnes anatomici sequunti sunt. Nos in historia hac anatomica analiticam methodum observabimus. Humanum enim corpus in quatuor praecipuas partes, caput, thoracem, ventrem, & artus secabimus. Has rursum minutius concidemus donec ad simplicissimas perventum fuerit in singulis ordinem sectionis perpetuò observantes. Quoniam verò multa in anatome observantur involuta, summéque ardua, ne quid ad perfectam huius artis cognitionem desideratur, singula patefaciam & explicabo, primúmque historiam partiunt describam, mox quae in ea. videbuntur controversa enodabo."

> examines and investigates the nature of each part. With the word "nature" however I include many things—substance, size, number, composition, connection, position, temperature, action, and use—of which only the first seven are demonstrated in the deceased in the sense of *autopsia*. The latter [action and use] are more visible in the living, and are judged both by reason and by sense.[12]

Here, anatomy is defined as either an art—dissection performed by an expert—or a science whose end is to understand the nature of each part; "nature" includes both material (sensible) properties as well as properties discovered via reason—for example, the parts' teleological functions. Despite calling his second definition of anatomy a "science," du Laurens emphasizes that the purpose or end of anatomy is the perfection of the art of medicine rather than knowledge for its own sake. That is, throughout the work, du Laurens focuses not on anatomy as a branch of natural philosophy but as a means of understanding the healthy human body—knowledge that is necessary for the sake of healing. This contrasts with the framing given by the Paduan anatomists, to which we shall turn next. Nevertheless, several of the controversies he tackles after his anatomical discussions are not strictly medical; some are purely philosophical.

The distinction that du Laurens makes between analysis and genesis is interesting, though not unprecedented: it points to something like developmental anatomy, or the manner and order in which nature generates and composes the human body. This implicit contrast between human artifice—dissection and vivisection—and nature's processes is made explicit by Helkiah Crooke, who in the course of translating the du Laurens on this distinction gives, "The other is called γένεσις or the way of composition, which of similar parts make dissimilar, and of these compoundeth the whole frame and structure. But we esteem this last not to be the way of *Art*, but of *Nature;* and therefore leave it to her who is onely able to performe it" (Crooke 1615, 19–20). This last sentence is not in the *Opera anatomica* or the *Historia;* I therefore take "we esteem" to indicate that this is Crooke's comment on du Laurens's distinction—something Crooke does frequently. Notably, neither du Laurens nor Crooke present analysis and genesis/synthesis as complementary steps for an investigator or author of a textbook. One is left to choose between the two organizational methods, rather than employ both in sequence.

It appears that analytical/resolutive method in anatomy may be theoretical (a science), practical (an art), or both—the last being perhaps the ideal, although how to integrated the first two is not explicitly mentioned. By contrast, the genetic/compositive method cannot be an art, and is therefore only theoretical, though perhaps

[12] "Nam actionem denotat quae manu perficitur, aut habitum animi & actionem intellectus perfectissimam. Illa ϖρακτικὴ haec ϑεωρητικὴ dicitur: utraque medico perfecto ad *διάγνωσιν*, πρόγνωσιν & ϑεραπείαν affectuum necessaria. Si priorem significationem spectes definietur anatome artificiosa partium externarum & internarum sectio. Artificiosam appello quae partes à partibus ita separat ut singulae integrae nec ullo modo lacerae videantur. Posteriore significato definietur scientia quae partis cuiusque naturam exquisitè rimatur & investigat. Naturae autem nomine multa complector, substantiam, magnitudinem, numerum, compositionem, connexionem, situm, temperiam, actionem, & usum: Ex quibus priora septem in demortuis Solo sensu αυτοψία demonstratur. Posteriora in viventibus magis sunt conspicua, & tum ratione, tum sensu iudicantur."

it ought to be based in part on observation of nature's processes. The synthetic/genetic approach is never fleshed out, however.

How does this understanding of the methods and definitions of anatomy impact du Laurens' treatment of the eye? Book 4, chapters 15 and 16 deal with the dissection and *historia* of the eye. Chapter 15 treats the eye as a whole, its place, shape, and temperament, including the *usus* or teleological function of the eye itself. He begins with a lengthy, humanistic discussion on sight, the objects of vision, and the eye as described by Aristotle, Theophrastus, Hippocrates, Alexander of Aphrodisias, biblical Hebrew writers, and others. He then writes, "I come to the history of this most noble organ, in which I observe these things: their use, shape, position, number, size, substance, temperament, connection, composition, and individual parts" (du Laurens 1593, 698–9).[13] Notably, he starts his history with the *usus in universam,* thus giving the whole eye along with its final cause before analyzing the eye conceptually alongside dissection instructions. The *usus* of the eye overall is dual: one is common to humans and brute animals—namely, so that the animal might survive and live better. On the second use, we read (du Laurens 1593, 699),

> there is another more divine use of the eyes, proper to man alone, the cognition of things, the contemplation of the invisible God through those things which are visible, and I would almost say happiness itself: for having received the species of heaven, by an acute intellect one is restored to the likeness of the workman.[14]

This echoes Plato's initial teleological account of the eye at 47a–c in the *Timaeus* (which influenced countless Neoplatonic, Arabic, and Christian writers, not to mention Galen): eyes exist so that humanity might observe the periods of the stars and planets, comprehend the intelligence that generated the heavens, and so bring order to our own souls. On the overall shape of the eye du Laurens also includes arguments from mathematical optics (du Laurens 1593, 699, 707–715), assuming the orthogonality condition for sight introduced by Ibn al-Haytham (Lindberg 1976). Thus, du Laurens leads with a teleological account of the eye before any dissection or analysis, and this teleological account relies upon both Christian theology (influenced by Neoplatonism) and accounts of the eye found in medieval optics. Note that, like many physicians in the sixteenth century, du Laurens holds that there is also an internal light in the eye, but he stops short of advocating a somewhat common infra-ocular extramission theory, according to which the incoming rays and rays sent out via the optic nerve meet in the crystalline humor; he adheres, overall, to a Peripatetic–perspectivist intromission theory.[15]

Chapter 16 covers the individual parts of the eye in the order revealed via dissection (the traditional order prior to Vesalius), starting from the muscles and then

---

[13] "Venio ad nobilissimi huius organi historiam, in qua haec spectando, usus, figura, situs, numerus, magnitudo, substantia, temperies, connexio, composito, partes singulae."

[14] "est divinior alter oculorum usus, soli homini proprius, rerum cognitio, Dei invisibilis contemplatio per ea. quae visibilia sunt, ipsaque penè dixerim, beatitudo: nam coeli specie recepta, intellectu acuto opifici suo simillimus redditur."

[15] For more on this intromission–extramission synthesis, see Vanagt (2010, 2012).

treating the tunics and then the humors. In this section, in contrast to his chapters on other parts of the body, du Laurens omits dissection instructions, merely giving the *historia* (in terms of temperament, texture, shape, size, and situation) and, immediately after, the combined action and *usus* of the part. The latter are not presented as arguments—that is, the actions and uses are not demonstrated from the results of du Laurens's physical and conceptual analysis together with observations and/or experiments. Nor are they explicitly imported from other disciplines, such as natural philosophy or optics. Instead, readers—primarily medical students, as du Laurens makes clear—must accept what du Laurens conveys on the basis of his authority and the authorities he had cited previously. In short, here, du Laurens indeed follows the analytic method for anatomy; actions and uses are not derived synthetically but are merely added for the sake of completeness. Moreover, his analysis (in contrast to other sections) is more theoretical than practical.

Compared with his predecessors, his account of ocular anatomy is far from cutting edge. His separate and thorough account of the controversies relating to vision—eight disputed questions lasting 24 pages, over twice the length given to the two chapters on anatomy—echoes the thirteenth-century polymath Pietro d'Abano's *Conciliator*. It is only here that du Laurens makes arguments, at times relying on up-to-date anatomical information to draw his conclusions. One controversy concerns the origin of the optic nerves, and while he does not cite Varolio, he argues for the latter's account based largely on observation (du Laurens 1593, 731–2).

While Laurens identifies analysis/resolution and genesis/composition as two different methods of learning and teaching anatomy, they are not seen as complementary or otherwise integrated. They are not, primarily, methods of investigation. Likewise, there is no explicit discussion of how (or whether) his two definitions of anatomy relate to his methods. In his discussion of controversies, he sometimes argues from anatomical evidence—but not always. Anatomy—as either analysis (be it physical or conceptual) or composition (be it natural or conceptual)—is not cast as a way to discover new structures, actions, or uses. This is confirmed by his actual account of the activities and final causes of the parts of the eye, where the structure of the eye as revealed via dissection is not systematically appealed to in order to determine the actions and uses of the parts of the eye.

9.5 Anatomy in Padua around the Turn of the Seventeenth Century

In contrast to all of this is the first official series of publications by Hieronymus Fabricius ab Aquapendente: *De visione, De voce,* and *De auditu* (Venice 1600).[16] In his statement concerning the order, method, and aim of anatomy at the beginning of *De visione*, we read (Fabricius 1600, 1),

[16] I have discussed Fabricius at length on these issues elsewhere. On him, I shall primarily summarize my previous work along with that of others.

> Our discussion will be in three parts. First we will reveal the entire workmanship and structure of the eye. Then we will proceed to the action of the eye, that is, to vision itself. Finally, we will contemplate the *utilitates* of the eye, not only according to the eye as a whole (*in universum*), but also [the *utilitates*] of the individual parts of the eye themselves. In general, moreover, we hunt all these things through dissection. Indeed, dissection (if one judges correctly) has the advantage that it makes evident not only those things which belong to the eye—that is, structure and history—but it also leads to an acquaintance with the actions or faculties, and finally it uncovers and declares the *utilitates*, of the eye. We begin, however, to dissect the eye exactly as it presents itself to sight.[17]

De visione, *De voce*, and *De auditu* were part of a larger project that he calls the "Theater of the Craftsmanship of the Whole Animal" (*Totius animalis fabricae theatrum*). It is concerned with carefully taking apart the body, scrutinizing the tangible properties of the simple parts, and then conceptually reassembling the body, arguing along the way for the actions and uses of the parts, organs, and systems based on the primary evidence from dissection. Although he does not rely on the terms, his project unites the notions of both resolution and composition, along with the manual/practical side of anatomy and the conceptual. Anatomy is, for Fabricius, a science aiming at causal knowledge of animals (particularly formal and final causes), but the science rests upon the art of dissection and sensible experience (and experiment, to some extent) with the parts as revealed by that art.

The procedures for dissecting the eye itself, moreover, are more carefully considered than given in most previous anatomies (see Fig. 9.5). Here, the naturalistic copper engravings provided by Fabricius, together with a chapter on how to anatomize the eye, clearly instruct the reader how to dissect the eye and what to observe; they seem intended both to memorialize what students would have seen in Fabricius' public dissections and to indicate to readers not in attendance how to dissect an eye properly (Baker 2023, 228–34).

As some, including Bigotti (2019, 63), have concluded, Aquapendente's anatomy in general is both analytical and synthetical, the latter in the sense of a rational reassembly of the sensible parts discovered via dissection. Regarding this conceptual recomposition of the body, by means of which the activities (*actiones*) of the eye's parts and then their purposes (*utilitates*) are determined, Fabricius cites optical or philosophical authorities but does not take their accounts for granted.[18] His anatomy of the eye is used to rethink the account of visual rays passing through the eye found in *perspectivist* optics. His anatomical findings were also used as a basis for philosophical disputation: he argues for a specific, novel intromissionist visual

[17] "Triparta erit nostra haec disputatio. Primò enim totus oculi fabricam structuramque patefaciemus. Deinde agemus de oculi actione, hoc est de visione ipsa. Postremò tum oculi in universum, tum singularum ipsius oculi partium utilitates contemplabimur. Haec autem omnia ferè per dissectionem venabimur. Dissectio enim (si quis recte aestimet) eum habet usum, ut tum ea., quae oculis insunt, hoc est structuram & historiam, manifestet: tum in actionis facultatisve notitiam deducat: tum denique oculi utilitates aperiat atque declaret. Incipiemus autem oculum dissecare, prout sese nobis offert aspectui."

[18] A more detailed account of the ideas in this paragraph and the next can be found in Baker (2014, 2016).

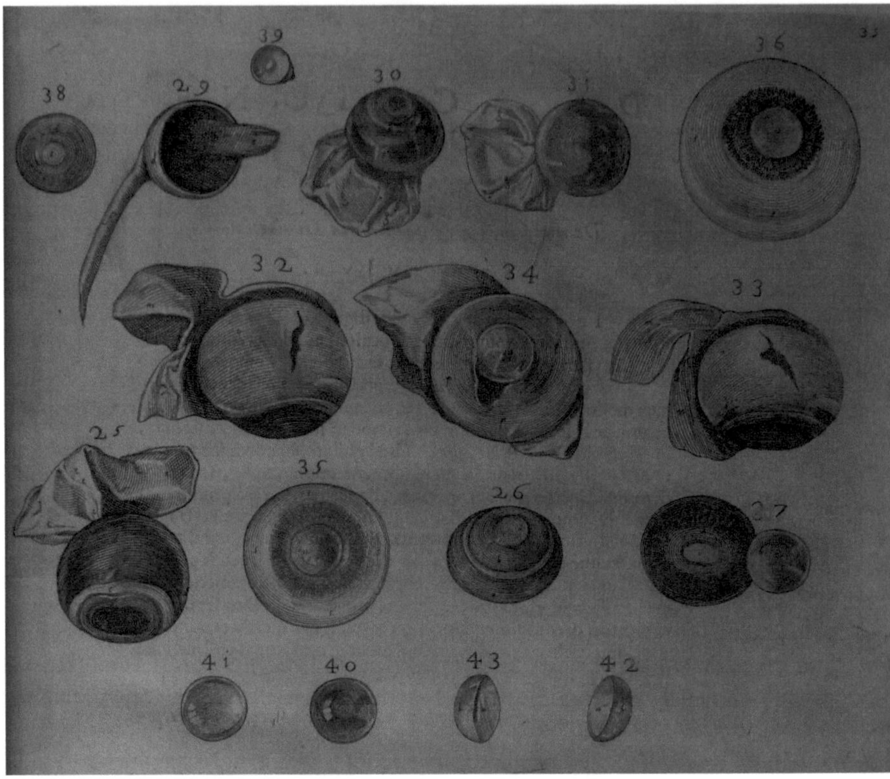

Fig. 9.5 Fabricius's visual account of the *dissectio* stage of ocular anatomy, or how to dissect the eyes of humans, cows, and sheep in order to reveal the shape, size, configuration, and texture of the parts (Fabricius 1600, 35).

theory based in part on the sizes and conformation of the humors; he argues that the visual faculty judges the appearances presented to the eye within the eyeball itself (rather than this judgment occurring in the common sense in the ventricles in the brain) based on anatomical details including the observation that there is no foramen in the optic nerve; and he even argues that color is an affection of light (a somewhat radical position at the time) based on details uncovered by dissection, such as the specific degree of transparency of the crystalline lens. Finally, in his account of the parts' usefulness, drawn from his meticulous *historia* of the parts of the eye, he argues against many positions given by Galen (and followed by many of Fabricius' contemporaries) on the purposes of the parts of the eye. In short, Fabricius uses the dual method of dissection and conceptual reassembly—all based, he claims, on the synthesis of the works of Galen and Aristotle—to overcome the problem of death in a way that does not merely import pre-established notions about the nature and processes of that living body. His method of investigating the dead body by skillful dissection, comparison of many animals, and sophisticated conceptual composition allows for a reevaluation of old accounts of structure, for new accounts of physiology, and indeed for new arguments about the workings of nature in general. In all this, he leans upon a Galeno-Aristotelian teleology, with the implicit argument being thus: because the eye is constructed rationally by Nature, the physics of light and color, along with the physics of visual perception and judgment, can be reverse-engineered through precise examination of all the sensible properties of parts of the eye generated by Nature, by observing how Nature organized these parts, and by discovering, via argumentation, deduction, and at times experiment, the activities and purposes of the parts in a living eye.

Crucial for Fabricius is the assumption that a dead eye, as revealed through numerous meticulous dissections, does not differ from a living one, in opposition to what Galen, Vesalius, and the perspectivists after Ibn al-Haytham claimed. Fabricius' eye does not contain copious amounts of spirt that dissipate upon death. He also takes for granted that the refraction of light within the eye can be determined by experimenting on the dissected humors themselves, something that conflicts with perspectivist optics. This led him to identify the crystalline humor as a burning lens, one of the most familiar and well-studied optical instruments prior to the invention of the telescope in the seventeenth century. The supposedly revolutionary notion that, apart from the faculty of vision itself, a dead eye behaves just like a living one is usually attributed to Kepler's 1604 *Paralipomena* (Shapiro 2008), but this notion was articulated by Fabricius 4 years prior to that and was also communicated in his lectures 10 years earlier (Baker 2019). How was Fabricius able to work around what we might call the theory-laden observations of Vesalius and Galen, along with the belief found in mathematical optics (which he read and cited copiously) that visual spirits actively managed the paths of rays in the eye? He had couched a few cataracts, and he noted that the procedure was bound to fail due to the closeness of the crystalline humor to the iris, but in this, he seems to be using his prior anatomical knowledge to caution physicians against performing ocular surgeries; he does not appear to be modifying or correcting his anatomical knowledge due to surgical experience (Baker 2017). We might simply point to his great skill and long experience, but these qualities were not unique to him.

9.6 Casserius: Anatomical Analysis United with Philosophical Synthesis

Finally, Fabricius' one-time student, then colleague, and later rival in anatomy and surgery in Padua, Casserius, presents a similar approach that couples a very precise and—in comparison to others, at least—theory-agnostic dissection of the body with a subsequent conceptual recomposition to determine the actions and uses of the parts of the body. In his first publication in 1601, *On the Anatomical History of the Organs of Voice and Hearing, Put Aright via Method and Industry, Set Down in Two Treatises*, and *Illustrated with Various Images in Copper* (hereafter *De vocis* for short), we read the following at the end of his preface:

> But, before I raise my hand from this paper, it seems that the reader should be forewarned that a definitive order of the things to be discussed is heeded; after the name [of the part] has been explained, the thing itself is revealed through the structure of the part, and from the final cause the usefulness is drawn out (though with the action given beforehand): whence anyone will easily unearth that a twofold methodology is to be continually followed without fail in the exposition of these [treatises] that deal with the parts of the human body. Of these methods the one can be plainly called active and operative history, which displays the body's fabric with great accuracy and produces so detailed an acquaintance with even the smallest parts that, as a result of the skillful dissection of these singulars, they can be separated out even on a living body. Such a method ought truly to be called the Anatomical Method, as Galen left excellent testament to this effect in the first passage in *De usu partium corporis humani*, Book I. [...] The second method (to which well-known men give the title that I have given to the first), altogether contemplative and intellective, belongs to the acuity of the mind alone, neglecting the work of the hands. It probes by means of the temperaments, that which follows [from temperament], and accidents, and pays out usefulness (*utilitates*). Of this, too, Galen made mention in the first book, chapter 9 of *De usu partium*.[19]

Note that Casserius here follows the distinction between *order* and *method* advocated by Zabarella (see Hattab, this volume). A twofold notion of the anatomical method is presented, although Casserius clearly feels that the label "anatomical" properly applies only to an "active and operative history" of the parts (the marginal index to the second "method" reads "A Most grave abuse of the anatomical method"). Both the operative and the contemplative methods, he notes, were described by Galen. While Casserius seems to distinguish between the speculative and the anatomical, he spends a great deal of time attempting to understand the actions and uses of the parts; in these sections, his arguments are generally based on the historical material gathered by the anatomical method. Thus, he does not shun the second approach but considers it to be philosophical. This, it appears, is the justification for his self-description on the title page as, in order, "philosopher, physician, anatomist." It is also worth noting Casserius' link between the anatomical method and vivisection and/or surgery, which we have not seen explicitly in our previous authors (on this, see also Klestinec 2011).

His 1609 *Pentaestheseion* includes a restatement of this in his short address to the reader. He says that he first manually uncovers the fabric of the body "according to the judgment of sense," and then, "progressing from that which falls under the

---

[19] "Sed antequàm, huic manum admoveam tabulae, praemonendus videtur lector, definitivam hic, observari rerum tractandarum fierem; quae explicato nomine, rem ipsam per structuram partis, eiusque à finali causa depromptas utilitates praemissa tamen actione patefaciat: Undè facilimum erit cuius eruere, duplicem sectandam perpetuò methodum in eorum expositione; quae circa humani corporis partes versantur. Earum altera historica planè Activa dici potest, & operatrix, quae fabricam accuratissimè pandit, & exquisitam adeò parit, vel minutissimarum particularum notitiam: ut ex hac singulae possint artificosa sectione, non lacerae, & illaesae ad vivum separari. Ita verè Anatomica Methodus dici debet, ut perpulchrè Galenus testatum reliquit primo de loc. Aff. I. cuius nominis gratum scio, vel gravissimis viris fuisse, & nunc esse, absum. Altera (quae prioris, apud illos insignitur titulo) contemplativa omninò, & intellectiva, solo mentis acumine, neglecta manuum ipera, ab iis quae insunt. Temperamentis, Consequentibus, Accidentibus, partium usus rimatur, & utilitates expendit; atque huius meminit Galen. I. de usu paritum cap. 9." Casserius, *De vocis auditusque organis*, 5.

sense to that which the mind contemplates, I first investigate the ACTIONS, then the USES of the parts in accordance with temperament, what follows [from temperament], and accidents"[20] (Casserius 1609, Preface). His account of the anatomical method is consistent across his two published works.

His fifth book in the *Pentaestheseion* treats vision and the eye, and here, Casserius multiplies the techniques for dissecting the eye: his dissection technique involves removing the layers of the eye in multiple ways—from the front, from the back, and from the side—to precisely determine the parts' relative locations and the connections between them. While the images have no doubt been cleaned up somewhat, I do not doubt that Casserius was able to dissect and present the parts of the eye as we see here, particularly given the praise heaped upon him by the medical students at Padua in recognition of his skill in dissection and surgery (Klestinec 2011). The images thus also depict his skill with the anatomist's knife. This comprises his account of the history of the eye, while his section on the actions of the eye he gives an exceedingly comprehensive philosophical treatise on disputed questions regarding light, color, transparency, vision, and related issues, during which he relies on the anatomical knowledge previously gathered as needed. Finally, his account of the uses of the parts is perhaps less innovative in its conclusions, but again, it is built on the foundation of the prior sections (Fig. 9.6).

Thus, Casserius' anatomical method is narrowly defined as an analytical, practical art, as well as a skill to be learned; nevertheless, a conceptual synthesis, by means of philosophical (i.e., syllogistic, demonstrative) reasoning, is necessary. While his goal is to understand the nature of vision itself by explicating the actions and uses of the parts of the eye, this aim is not strictly anatomical: the project requires a skilled anatomist who is *also* an adept philosopher.

9.7 Conclusion

We now return to the problem of how it might be possible to cut through the specific epistemic circle faced by anatomists in this period. How did our authors attempt this? For Varolio, natural divisions and connections were revealed through improved dissection methods via the comparison of traditional and novel procedures. This (ideally) allows one to judge which procedure is better but not necessarily which truly reveals the nature of the parts and their interconnections. Varolio's letter

[20] "Principio namque ad sensuum iudicium magis accedentem organorum FABRICAM cum suis requisitis explico, ut in propatulo fiat magnitudo, figura, numerus, situs, & substantiae proprietas. Quae causa fuit, ut non solùm hominis, verùmetiam variorum animalium partes in aes incidendas curaverim. Ab iis autem, quae sub sensum cadunt ad ea., quae mens contemplatur progressus primùm ACTIONES, deinde in temperamentis, consequentibus, & accidentibus consistentes USUS partium investigo."

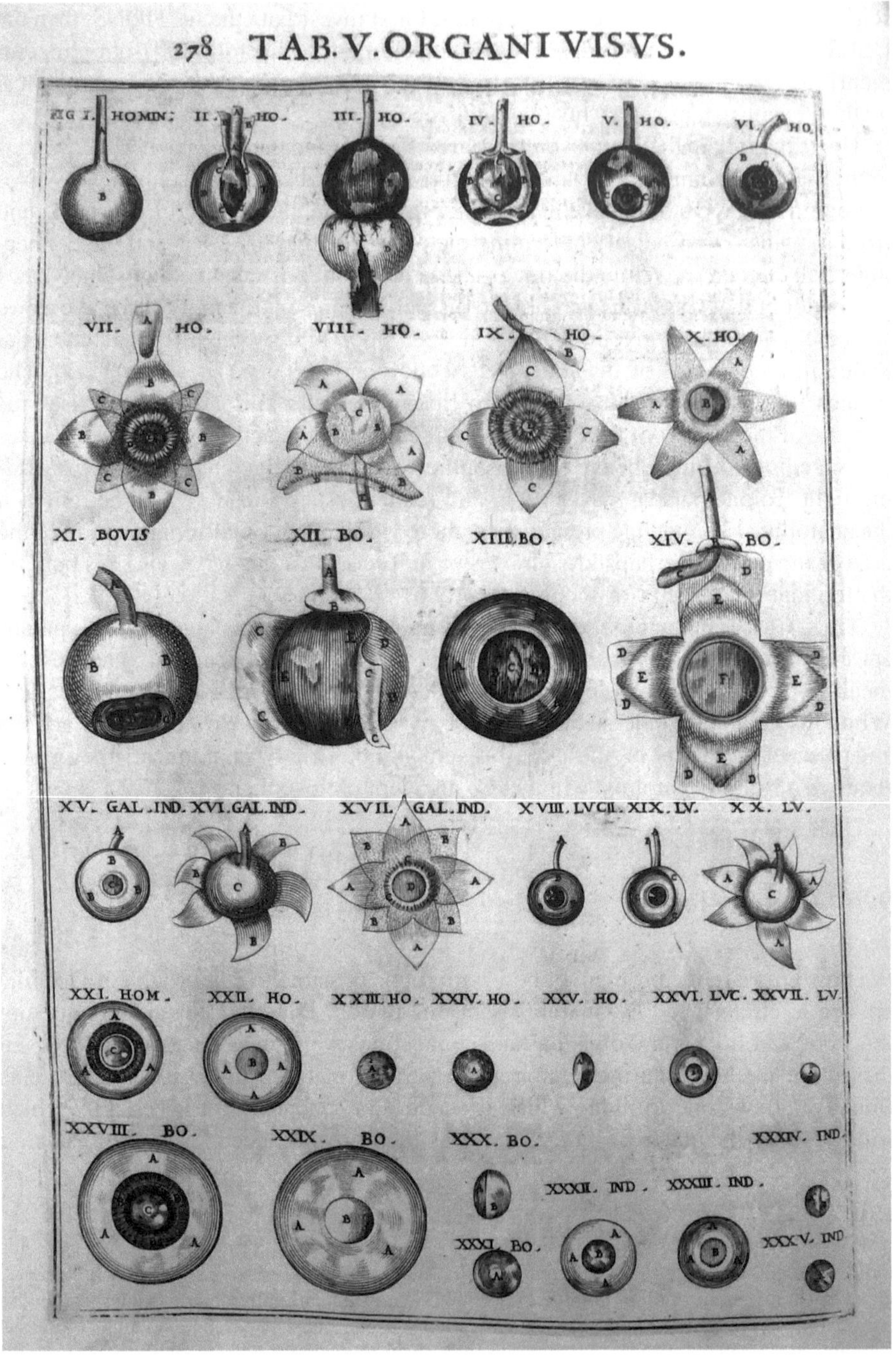

Fig. 9.6 Multiple methods of dissecting human, cow, and chicken eyes, all with the aim of understanding the animal capacity for vision through the actions and uses of the parts. (Casserius 1609, 278)

presents an anatomy that was necessarily incomplete: we may see it as a contribution to an ongoing, communal project. In his *Anatomia, sive de resolutione*, Varolio invokes external authorities—regarding the ends and actions of the organs—to implicitly break the circle; at times, necessity is invoked, particularly with regard to the actions and uses of the parts that constitute the organs. Bottom-level details, such as temperament, shape, size, location, position connection, and so on are also, it seems, simply declared authoritatively. In short, his method of dissolution is an order of exposition, followed here for the sake of capturing the interest and aiding the memory of the non-specialist.

Du Laurens gives us a fourfold categorization: an analytic art of analysis, a conceptual analysis, a natural genesis, and a conceptual synthesis. In all cases, knowledge is based on authority. As with Varolio's *De resolutione*, the epistemic circle is not tackled head-on: the method advocated by du Laurens is an order of exposition, selected as the best way to educate medical students. In neither the *De resolutione* nor the *Opera anatomica* is *methodus* intended as an active investigation of nature. In general, explicit discussions of analysis and synthesis in the anatomical tradition seem to be concerned with how knowledge may be ordered; one chooses, for the sake of pedagogy—that is, for conceptual clarity and ease of memorization—*either* analysis or synthesis.

In Padua, we see a deeper concern with the method of anatomy as an active investigation into nature, with the goals of the discovery of new parts of the body, determining the true activities of parts and organs, and more fully explicating the functions that parts and organs contribute toward the life of the animal. Notably, this method was also taught to students. This is, as has often been remarked, a philosophical turn. As some have suggested, this may also have been influenced by Zabarella's attack on medical humanism and his insistence that *ordo* is for the sake of pedagogy and *methodus* for the sake of knowledge.[21]

With Aquapendente and Casseri, anatomy emerges as both a resolution of the body via dissection and then a mental synthesis, via an investigation of the actions of the parts and, particularly, how the parts work together for the sake of the animal. We can identify a clear sense of material analysis together with a sophisticated conceptual synthesis from that empirical basis. In these works, we find an emphasis on the discovery of anatomical parts and structures, particularly corrections to previous accounts of structure but also new accounts of the parts' activities and purposes. Fabricius considers both analysis and synthesis to be anatomical—anatomy is at once manual and philosophical—while Casserius frames it as a combination of a manual anatomical method (skillful dissection and a determination of the

[21] This is already evident in Girolamo Capivaccio's (1523–1589) work. Anatomy is a twofold mental *habitus*, aimed to a certain extent at perceptible things but primarily at knowledge of the actions and uses of the body (Capivaccio 1593, 3). Anatomy is defined as an art that compares dissection, actions, and uses for the sake of understanding man's vital functions. *Ordo* concerns, primarily, the order of dissection (102–3). Capivaccio does not discuss analysis/resolution or synthesis/composition at any length, nor do we see in his treatise anything that might unequivocally be described as a sequence of analysis and synthesis. See also Bigotti (2020).

temperaments, sizes, shapes, connections, etc. of the parts) followed by a philosophical discussion (scholastic disputation regarding the actions and uses of the parts). For both, the epistemic circle is cut via a thorough and systematic determination of every sensible property of the parts, along with a belief that nature does nothing in vain. The problems posed by death are resolved on a case-by-case basis via manual skill, diligence, repetition, and comparative anatomy. In many cases, the anatomist's authority of experience cuts the circle. Moreover, since (in such teleological investigations) it is assumed that nature does nothing in vain, every sensible property of every part has a rational purpose and can be accounted for on the basis of a proper *historia* together with a great deal of careful argumentation. Thus, the belief in final causes offered further opportunities to make a cut in the circle, including the use of experiments to resolve certain questions (Baker 2016).

In the anatomical tradition, discussions of the "method of anatomy" provided a key locus for discussing analysis and synthesis, and such discussions of analysis and synthesis appear to be distinct from the four senses of analysis and synthesis as identified by Alan Shapiro (this volume). This is due to the disciplinary cohesion of anatomy and because of the distinct ways that the terms and concepts were deployed for the pedagogical ordering of an anatomical text or course of study. It was also potentially—though not invariably—because of the unique way that analytic and synthetic methods might be used to probe the nature of animal bodies. Shapiro also makes a distinction between decompositional analysis—be it material or conceptual—and regressive analysis, the latter being a search for causes. However, within the Galeno-Aristotelian framework of this period in particular, anatomy was understood as simultaneously decompositional and regressive, at least at Padua. It involves not only a literal dissection—a rendering apart of organs into their constituent elements—but also a search for causes via a sort of regressive method.

Take the eye as anatomized by Aquapendente. The purpose (final cause or *utilitas*) of the eye is obviously sight and, as most authorities said beforehand, the crystalline lens can be, at least provisionally, taken as the primary locus of vision in the eye. Aquapendente may be read as claiming that these two facts were only known imperfectly before a proper anatomical investigation. After dissection, the anatomical investigator must account for all of the sensible properties of the lens, not least its temperament, which is watery. The eye's temperament is watery, Aquapendente says, because only water can be condensed in several ways; note that Aquapendente is the first to make this argument. Thus, he must account for why the crystalline has the precise density that it has, with regard to *both* transparency/opacity and refractive power. He must also account for the shape of the crystalline lens. Aquapendente thus gives a twofold account of vision. First, images are received and retained in the crystalline humor on account of its thickness with respect to its reception of light and color. This then required, partly given that his dissections revealed that there is no inherent luminosity in the eyes of any animals, a determination of the precise relationship between light and color, which he offers. Second, he says that these impressed images are judged by the visual power sent forth through the optic nerves to the retina and then to the *aranea* surrounding the crystalline lens—again, via material connections discovered through a careful method of dissection. The refractive density of the lens, combined with its shape, meanwhile, is solely for the sake of perfecting vision: it helps strengthen these impressed images. The shape and relative difference in the refractive powers of the lens and the surrounding humors also serve a secondary purpose—namely, to ensure that light does not rebound from the retina causing images of the interior of our eyes to perpetually inhere in the crystalline lens (Baker 2016). From all this, and much more, Aquapendente may be understood as claiming that both what it means to see, and the precise actions and *utilitates* of all sensible properties of all parts of the eye—that is, their final causes—are known with precision only after this process of material analysis and conceptual synthesis.

It is tempting to trace a connection to Jacopo Zabarella's *regressus* method,[22] which Zabarella took to be the true method for natural philosophy. The Paduan anatomical method is at once practical, experimental, and philosophical; in contrast, Zabarella never uses practical or experimental examples in discussing his *regressus* method. Moreover, none of the Paduan anatomists mention the *regressus* method, or Zabarella, at all. While Zabarella's distinction between *ordo* and *methodus* seems to have influenced the Paduan anatomists, it is far more difficult to say that his *regressus* method did so. The approach adopted by Aquapendente and Casseri could certainly have come from elsewhere—for example, from a reading of Galen that differed from that of their contemporaries. Finally, much has been written on Zabarella's potential influence on Galileo and other seventeenth-century experimental philosophers, but I suggest that it is perhaps easier to connect these *novatores* to the Paduan anatomical tradition and its emphasis on anatomical resolution and theoretical recomposition.

This chapter might seem to imply a sort of teleological history: that sixteenth-century anatomical ideas were striving toward the perfection of the analysis/synthesis pairing found in turn-of-the-century Padua. However, regardless of the importance or influence of the latter, analysis and synthesis were not simply united hereafter in anatomy. Indeed, Paduan authors, to my knowledge, did not even use the terms. The careful consideration of anatomical method as *ordo* was crucial for professors; the analytic method, alone, was generally followed. As this volume demonstrates, pedagogy was one of the key contexts in the deployment of analysis and/or synthesis. If, in the history of science, we focus on the perpetual search for correct (or at least better) pedagogical methods, a profound—indeed existential—problem for every scientific discipline, then the Paduan understanding of "anatomical method" cannot be regarded as the perfection of analysis and synthesis in the early modern anatomical tradition.

[22] See Zabarella 2013.

References

Andretta, Elisa. 2020. VAROLIO, Costanzo. In *Dizionario Biografico degli Italiani*, vol. 98. https://www.treccani.it/enciclopedia/costanzo-varolio_(Dizionario-Biografico).

Baker, Tawrin. 2014. *Color, cosmos, oculus: Vision, color, and the eye in Jacopo Zabarella and Hieronymus Fabricius Ab Aquapendente*. PhD Dissertation. Bloomington: Indiana University.

———. 2016. Why all this jelly? Jacopo Zabarella and Hieronymus Fabricius Ab Aquapendente on the usefulness of the vitreous humor. In *Early modern medicine and natural philosophy*, ed. Peter Distelzweig, Benjamin Isaac Goldberg, and Evan Ragland, 59–88. New York: Springer.

———. 2017. The oculist's eye: Connections between cataract couching, anatomy, and visual theory in the renaissance. *Journal of the History of Medicine and Allied Sciences* 72 (1): 51–66.

———. 2019. Dissection, instruction, and debate: Visual theory at the anatomy theatre in the sixteenth century. In *Perspective as practice: Renaissance cultures of optics*, ed. Sven Dupré, 123–147. Turnhout: Brepols.

———. 2023. Images of the eye from Vesalius to Fabricius Ab Aquapendente: The rise of metrical representation in anatomical diagrams and the cross-fertilization of visual traditions. In *Reassessing epistemic images in the early modern world*, ed. Ruth Sargent Noyes, 219–239. Amsterdam: Amsterdam University Press.

Bartisch, Georg. 1583. *Ophthalmodouleia: das ist Augendienst*. Dresden.

———. 1996. *Ophthalmodouleia: That is the service of the eyes*. Trans. Donald Blanchard. Ostend: Wayenborgh.

Bigotti, Fabrizio. 2019. *Physiology of the soul: Mind, body and matter in the Galenic tradition of late renaissance (1550–1630)*. Age of Descartes Ser. Turnhout: Brepols.

———. 2020. Beyond Galen: Image, geometry, and anatomy from Vesalius to Santorio. In *Contre Galien: Critiques d'une autorité médicale de l'Antiquité à l'âge Moderne*, ed. Antoine Pietrobelli. Paris: Honoré Champion.

———. 2021. Logic, geometry and visualisation of the body in Acquapendente's rediscovered Methodus Anatomica (1579). *Medical History* 65 (3): 227–246.

———. 2023. Commenting on Aristotle with a knife: The heretical anatomies of Bassanio Landi. *Naturwissenschaften, Technik Und Medizin* 31 (1): 1–25.

Capodivacca, Girolamo. 1593. *De methodo anatomica liber*. Venice: Io. Baptista Ciottus Senenses.

Casserius, Julius. 1609. *Pentaestheseion, hoc est de quinque sensibus liber*. Venice: Nicolaus Misserinus.

———. 1601. *De vocis auditusque organis historia anatomica, singulari fide methodo ac industria concinnata tractatibus duobus explicata ac variis iconibus ære excusis illustrata*. Ferrara: Baldinus.

Crooke, Helkiah. 1615. *Microcosmographia: A description of the body of man together with the controversies thereto belonging. Collected and translated out of all the best authors of anatomy, especially out of Gasper Bauhinus and Andreas Laurentius*. London: William Iaggard.

Cunningham, Andrew, and Sachiko Kusukawa. 2003. *Gregor Reisch: The philosophic pearl (1503), and natural philosophy in the age of modern devotion*. London: Ashgate.

du Laurens, André. 1600a. *Historia anatomica humani corporis singularum eius partium*. Paris: Orry.

———. 1600b. *Historia anatomica humani corporis singularum eius partium*. Frankfurt: Theorodicus de Bry.

Edwards, William F. 1960. *The logic of Iacopo Zabarella*. PhD Dissertation. Columbia University.

Eyn Newes hochnützliches Büchlin. 1538. *Ein Newes hochnutzlichs Büchlin, von erkantnüs der kranckeyten der Augen, Sampt einer figur oder Anothomia eines augs*. Strasbourg: Heinrich Vogtherr.

———. 1539. *Eyn newes hochnützliches Büchlin und Anathomi eynes authgethonen augs*. Strasbourg: Heinrich Vogtherr.

Fabricius ab Aquapendente, Hieronymus. 1600. *De visione, voce, auditu*. Venice: Franciscus Bolzetta.

Fay, Isla, and Nicholas Jardine, eds. 2014. *Observing the world through images: Diagrams and figures in the early-modern arts and sciences*. Leiden: Brill.

French, Roger K. 1984. An origin for the bone text of the 'five-figure series'. *Sudhoffs Archiv* 68 (2): 143–156.

Heseler, Baldasar, and Ruben Eriksson. 1959. *Andreas Vesalius' first public anatomy at Bologna: 1540*. Uppsala: Almqvist & Wiksell.

Klestinec, Cynthia. 2011. *Theaters of anatomy: Students, teachers, and traditions of dissection in renaissance Venice*. Baltimore: Johns Hopkins University Press.

Kusukawa, Sachiko. 2012. *Picturing the book of nature: Image, text, and argument in sixteenth-century human anatomy and medical botany*. Chicago/London: University of Chicago Press.

Laurenza, Domenico. 2003. *La ricerca dell'armonia: rappresentazioni anatomiche nel Rinascimento*. Biblioteca di Nuncius. Firenze: L.S. Olschki.

Lennox, James G. 2018. Aristotle, dissection, and generation: Experience, expertise, and the practices of knowing. In *Aristotle's generation of animals: A critical guide*, ed. Andrea Falcon and David Lefebvre, 249–272. Cambridge Critical Guides. Cambridge: Cambridge University Press.

Lindberg, David C. 1976. *Theories of vision from Al-Kindi to Kepler*. Chicago: University of Chicago Press.

Maclean, Ian. 2008. The medical republic of letters before the thirty years war. *Intellectual History Review* 18 (1): 15–30.

Mikkeli, Heikki. 1992. *An Aristotelian response to renaissance humanism: Jacopo Zabarella on the nature of arts and sciences*. Helsinki: SHS.

Muller, Frank. 1987. Heinrich Vogtherr, Alias Heinricus Satrapitanus, Alias the 'Master H.S. with the Cross'. *Print Quarterly* 4 (3): 274–282.

———. 1997. *Heinrich Vogtherr l'Ancien: Un artiste entre Renaissance et Réforme*. Wiesbaden: Harrassowitz.

Murdoch, John Emery. 1984. *Album of science: Antiquity and the middle ages*. New York: Scribner.

O'Malley, C. D. 2008. Varolio, Costanzo. In *Complete dictionary of scientific biography*, vol. 13, 587–588. Charles Scribner's Sons. http://proxy.library.upenn.edu:2261/ps/i.do?p=GVRL&sw=w&issn=&v=2.1&it=r&id=GALE%7CCX2830904447&sid=googleScholar&linkaccess=abs

Paré, Ambroise. 1575. *Les oeuvres de M. Ambroise Paré*. Paris: Chez Gabriel Buon.

Pomata, Gianna. 2010a. Observation rising: Birth of an epistemic genre, 1500–1650. In *Histories of scientific observation*, ed. Lorraine Daston and Elizabeth Lunbeck, 45–80. University of Chicago Press.

———. 2010b. Sharing cases: The observationes in early modern medicine. *Early Science and Medicine* 15 (3): 193–236.

Raynaud, Dominique. 2020. *Eye representation and ocular terminology from antiquity to Helmholtz*. Hirschberg history of ophthalmology: The monographs 16. Amsterdam: Wayenborgh Publications.

Reisch, Gregor. 1503. *Margarita Philosophica*. Friburgi: Schott.

Shapiro, Alan E. 2008. Images: Real and virtual, projected and perceived, from Kepler to Dechales. *Early Science and Medicine* 13 (3): 270–312.

Shotwell, R. Allen. 2013. The revival of vivisection in the sixteenth century. *Journal of the History of Biology* 46 (2): 171–197.

Siraisi, Nancy G. 2013. *Communities of learned experience: Epistolary medicine in the renaissance*. Baltimore: Johns Hopkins University Press.

Tieleman, Teun. 2002. Galen on the seat of the intellect: Anatomical experiment and philosophical tradition. In *In science and mathematics in Ancient Greek culture*, ed. Lewis Wolpert, C.J. Tuplin, and T.E. Rihll, 256–273. Oxford University Press.

Vanagt, Katrien. 2010. Suspicious spectacles. Medical perspectives on eyeglasses, the case of Hieronymus Mercurialis. In *The origins of the telescope*, ed. Albert Van Helden, Sven Dupré, Rob van Gent, and Huib Zuidervaart, 115–28. History of Science and Scholarship in The Netherlands 12. Amsterdam: KNAW press.

———. 2012. Early modern medical thinking on vision and the camera obscura. V.F. Plempius Ophthalmographia. In *Blood, sweat and tears: The changing concepts of physiology from Antiquity into early modern Europe*, ed. Manfred Horstmanshoff, Claus Zittel, and Helen King, 569–593. Boston: Brill.

Varolio, Constanzo. 1573. *De nervis opticis nonnullisq: Aliis praeter communem opinionem in humano capite obseruatis*. Padua: Meiettus.

Varolio, Costanzo. 1591. *Anatomia, Sive de Resolutione Corporis Humani*. Frankfurt: Wechel.

Vesalius, Andreas. 1543. *De humani corporis fabrica libri septem*. Basel: Johannes Oporinus.

———. 1568. *De humani corporis fabrica libri septem*. Venice: apud Franciscum Franciscium Senensem, & Ioannem Criegher Germanum.

Zabarella, Jacobus. 2013. In *On methods*, ed. John P. McCaskey, vol. 2 vols. Cambridge (Mass.): Harvard University Press.

Tawrin Baker is an independent scholar, and from 2019 to 2023 was Visiting Assistant Professor in the Program of Liberal Studies and the History and Philosophy of Science at University of Notre Dame. His research focuses on the intersection of anatomy and medicine, natural philosophy, and mathematics in the early modern period.

Chapter 10
Taxis and Texture: Johann Daniel Major (1634–1693) on Spirits, Salts, and the Limits of Analysis

Vera Keller

Abstract Johann Daniel Major (1634–1693) made *taxis*, a revocable ordering of units, the main subject of his lifelong research across all domains. He derived this concept from experiments in color changes in liquid solutions, an area of research that was especially vibrant in his cultural and intellectual milieu. Major drew on Cartesianism to query qualitative views of color and color change but he disagreed with rational Cartesians that reason could suffice to counteract the impact of original sin on knowledge. Due to the strife and sickness debilitating matter and human thought (itself highly conditioned by matter), Major remained skeptical that it was possible to identify the most basic building blocks of matter and thus focused more on synthesis and the complex textures it produced than upon analysis. Understanding had to remain provisional as experimental knowledge developed. *Taxis* provided Major the framework through which a dynamic array of knowledge could be co-articulated with a dynamic array of nature.

Keywords Taxis · Taxonomy · Salts · Spirits · Color · Experiment · Cartesianism · Provisionality · Experiment · Analysis

10.1 Introduction: *Taxis,* Cartesianism and the Limits of Analysis

As this volume explores, the concepts of *analysis* and *synthesis* operated across many disciplines, from pedagogy and the arrangement of knowledge to chymical processes. This essay investigates one figure who made the connection between the arrangement of knowledge and the arrangement of corpuscles explicit by linking the disposition of material corpuscles to the texture of human wit (*ingenium*) and the actions of the animal spirits. As I have discussed elsewhere (Keller 2024), Johann Daniel Major (1634–1693), chair of medicine at the University of Kiel, made

V. Keller (✉)
Department of History, University of Oregon, Eugene, OR, USA
e-mail: vkeller@uoregon.edu

W. R. Newman, J. Schickore (eds.), *Traditions of Analysis and Synthesis*,
Archimedes 73, https://doi.org/10.1007/978-3-031-76398-4_10

taxis—a revocable ordering of units—the main subject of his lifelong research across multiple domains. Herein, I focus on Major's discussion of spirits, the most subtle form of matter. Investigation of the ways in which Major handled *taxis* on this scale—that is, on the scale of matter at the extreme limits of human knowledge and imagination—illuminates the interplay of chymistry and the human ability to know, according to Major.

Major did not regularly use the terms "analysis" and "synthesis". As the deployment, recall, and redeployment of individual units in varied encampments and military actions, *taxis* encompassed both coming together and breaking apart. While the ancient Greek *taxis* could be understood as the equivalent of the Latin *dispositio* in dialectics, its primary association was with the tactical arrangement of soldiers in a military context, exemplified by second-century works such as Aelian's *On Tactical Arrays of the Greeks* and Arrian's *Ars tactica* (e.g. Crassot 1619, 78). Military tactics demonstrated how the varied arrangement and motions of corpuscles alone could effect major actions without recourse to qualities. They offered a means of conceptualizing chemical reactions as dynamic clashes between arrays. Major even deployed the analogy of the signing of a peace treaty and the creation of a (temporary) alliance as a means of thinking about the formation of bonds following a reaction. Significantly, these military associations situated material clashes and alliances as agonistic and temporary.

Taxis offered Major a suitable way to imagine the state of postlapsarian nature as one embroiled in constant pain and antagonism. Major contended that, since the Fall, the most infinitesimal units of matter clashed with one another and with crasser structures in constant warfare and destructive erosion. Cumulatively, this friction injured the world and profoundly debilitated not only the soundness of the human body but also the knowledge that it could produce.

Major considered himself to be a follower of Descartes, but like some contemporaries such as Johann Bohn of Leipzig, he attributed greater roles to the material body than Descartes had ever intended (Scribano 2022, 235–57). He further differed from Descartes and many of his followers with respect to the significance of the Fall in human knowledge. Cartesianism offered others a means of overcoming "the limitations of the fallen intellectual faculties of Adam's seventeenth-century descendants," since the "Cartesian system was premised upon the assumption that both reason and the image of God in the soul were retained after the Fall" which made it possible for humans to acquire a "perfect science" through rational deduction (Harrison 2002, 239 and 246–7). By contrast, according to Major (1674), original sin had erased humanity's perfect knowledge, leaving behind just a faint memory of previous knowledge that merely incited the insatiable human desire to know.

Humans were impelled to pursue knowledge in various, foolish directions as their character and thought were shaped by infinitely varying material textures of the brain. There were, he wrote, "as many types of wits (*ingeniorum*) as there were temperaments, and as many temperaments as there are leaves on the branches of some huge tree, due to the special textures of atoms and an infinite variety of different placements, so that I seriously doubt whether it is possible to find in this whole world two or three humans whose inclination to learn and talent for philosophizing are exactly the same" (Major 1677a [A3]).[1]

Not only did molecular texture and arrangement differentiate individual approaches to philosophy, but the continuing friction of fallen materiality would further shift and distort thought. For instance, a gout-sufferer believed that a sword was sticking out of his foot. This showed that although "it is not always in the power of humans to achieve due to the confluence of rough salts in the affected area, it is of the utmost necessity to keep the mind calm even with gout, just as it is certain that it is impossible to philosophize well unless the mind is likewise first purged of all prejudice" (Hesporn and Major 1679, [D3r]).[2] Likewise, students entering the emotional arena of a dissertation defense would find it nearly impossible to preserve their equanimity. "Who in the entire human race was of such consummate moderation ... that they could step into a zone of competition where glory and ability were tested and remain unexcited by the impetus of the spirits and untouched by the force of that primitive ambition that already affected Adam in Paradise, so that they were not shifted at all from the equilibrium of indifference?" (Major 1681, [A3r]).[3]

Space does not allow here for a full discussion of how molecular *taxis* affected thought and thus the *taxis* or arrangement of disciplines themselves; I explore that question in depth in *Curating the Enlightenment.* Here, I emphasize only how the relation between debilitated materiality and cognition shaped Major's views on synthesis and analysis in two ways. First, it encouraged epistemic humility and spurred Major to resist claims that analysis had identified the simplest building blocks of matter. Instead, he emphasized material textures that could be accessed more easily by human experimentation. Second, it dramatized how important it was, nevertheless, to attempt to understand invisible material structures because of the effects they might have on shaping all thought and thus all knowledge. The weighty consequences of this topic, in turn, motivated Major to pursue an ambitious experimental agenda, even while working with pragmatic and temporary classifications of his experimental subjects. Major treated the category of simple bodies merely as a placeholder that would need to shift over time as a greater experimental knowledge of the *taxis* of nature shifted the *taxis* of knowledge.

[1] "Cum verò Ingeniorum tanta sit, quanta Temperamentorum; & Temperamentorum tanta, quanta Foliorum arboreorum, in unâ aliquâ magnâ Sylva, quoad Atomorum speciales texturas & positus infinities discrepantium, Varietas, ut seriò dubitem, an facilè inveniendi, in toto Terrarum Orbe, duo tresvè Homines sint, quoad Discendi Inclinationem, & Philosophandi Aptitudinem, exquisitissimè aequales. . . ."

[2] "Quod quidem plenè efficere, non semper in potestate hominum est, ob plura identidem Salia hispida, in Partem affectam confluentia: unum tamen maximè necessariis esse, Animum etiam in Podagrâ sedatum habere, tam certum est, qùam Impossibile coeteroquin est, rectè philosophari, nisi Animus itidem fuerit ab omnia ante omnia Praejudicio purgatus."

[3] "Eqcuis aut quotus in toto Hominum genere, utcunque existens excitati spirituum impetûs... adeò consummatae Moderationis est, ut praesentibus Arbitris in Arenam cum altero descendens, ubi Res Gloriae & Virtutis agitur, per vim primitivae illius in Paradiso afflatae Ambitionis, quae pars Veteris Adami dicitur, ultrò citroque quadantenus non deflectat ab Animi verè Indifferentis Aequilibrio... ."

10.1.1 The Origin of Major's Concept of *Taxis*

Major's immediate source for *taxis* was a discussion of color changes in liquid solutions, a topic afforded particular significance in his milieu. A Breslauer, Major studied at Wittenberg and Padua before taking up the first chair of medicine at the new University of Kiel, founded by Christian Albrecht, the Duke of Holstein-Gottorf. Major's first wife, Maria Dorothea Sennert, was the granddaughter of Daniel Sennert. After Major moved to Kiel, he developed close intellectual connections with the friends of Joachim Jungius in Hamburg, including Andreas Cassius (1604–1673). Major also drew upon the culture of material experimentation in regional courtly institutions, where experiments in metallic solutions and the colored glass they produced were at the forefront of European science.

The Gottorf court boasted a laboratory directed by Joel Langelott (1617–1680) and a glassworks operated by the Kunckel family, of whom Johannes Kunckel (1630–1708), a member of the third generation of glassmakers at Gottorf, is the best known. The Gottorf Duke's brother, August Friedrich, the Prince-Bishop of Lübeck, also entertained renowned physicians and chymists, such as members of the Cassius family, known for the "Purple of Cassius," a suspension of gold salts in a solution that turned a wonderful red color and was used medicinally as potable gold, as well as in the crafting of ruby glass. The color effects generated by these suspensions of minuscule particles facilitated a regional discussion of material texture and of its potential to explain various phenomena, such as color change, that might appear to be qualitative in nature.

In 1666, Major presided over the medical dissertation of Andreas Cassius Jr. (1645–1700) at Kiel; in a 1685 work on gold, Cassius Jr. later described his father's experiments on gold which Johannes Kunckel would also relate to his own productions of ruby glass (Cassius and Major 1666; Cassius 1685; von Kerssenbrock-Krosigk 2008). This was only one among the color transformations that Cassius Jr. described in that work. One preparation of gold could be made to appear black, green and citrine; another would appear white, then purple, then white, then carmine. A solution of gold added to a solution of tin would produce yellow, blue, and black, and, once air was admitted, would transform into an elegant ruby purple (Cassius 1685, 50, 74, and 106).

Major himself initially investigated color changes relating to medical infusions into blood. In 1665, he published a collection of letters from other scholars responding to his intravenous injections. One correspondent, Johann Daniel Horst, raised the issue of how the changing color of blood might be understood chymically and compared color changes in blood to the red crystals that could be created from a preparation of gold salts (Major 1665, 87). Major, in conversation with his friend Philipp Jacob Sachs, continued to discuss theories of color change in works from 1665 and 1667 (Sachs 1665, 402 and Sennert and Major 1667, [G]).

In another 1667 work discussing his blood injections, Major discussed the mixture of colors at length, using artists' materials such as paints that he mixed, or through experiments using liquid solutions, which, Major described having performed often as public spectacles during his anatomical demonstrations of the eye. These, he argued, indicated that changes in color resulted from the figure, motion and position of minimal particles in the liquid. He related these experiments to those color experiments described by Thomas Willis in the latter's *On Fermentation,* first published in Latin in 1659 (Major 1667, 220–1).

Willis recounted how a clear liquid could be poured into another clear liquid and produce a color, or how two colors could be combined to produce a clear liquid. This, Willis argued, demonstrated that no unique tinging body was added to the liquids to color them. Rather, the invisible atoms that made up the liquids rearranged in different formations, like soldiers shifting their stances and positions, so that light could enter in and reflect out in a different way. As Willis wrote, the "little Bodies or Atoms" "seems as an Army of Soldiers placed in their Ranks, who now draw into close Order, now open their Files and Ranks, now turn to the left, now to the right hand, as is diversely shown in the exercising of Tacticks, or the Art Military" (Willis 1659, 88; 1681, 48; Debus 1962). It was from this that Major drew his concept of *taxis,* a concept he would later apply to all forms of order.

Color changes in liquid solutions, Major argued, revealed that colors did not result "from any salt or sulfur," as a principle. Rather, the color of liquid solutions appeared to change due to rays of light being reflected in shifting ways as the spaces, shapes, and disposition of particles within the pores of the liquid moved. Like soldiers, the particles "contract or dilate, turn to the left or right, circle round, join, or depart, now on the flanks, now at the head, now throughout the entire corps, so, that seen from afar, they sort themselves through a change of order" (Major 1667, 227).

The effect of the distinct position of particles in giving rise to color phenomena could be shown further in the how colors could be "extinguished" or "resuscitated." Major drew one such example—that of making an ink appear and disappear—from Willis. When sublimated mercury was added, Major pointed out, it would turn the color of opal. The other example was Major's own. The volatile spirit of sal ammoniac added to water in a copper basin would cause the water to turn cerulean. If this were then poured in a glass, one could admire its transparent sapphire color "with great pleasure." Were one to then add a clyssus of antimony, the water would become clear again (Major 1667, 228).

As he later recalled, Major was performing these experiments that year in the court of Eutin, the seat of August Friedrich, the 21-year-old Prince-Bishop of Lübeck (1646–1705), whom Major began treating as his physician (Reinbacher 1998, 40). In an innovative experimental seminar that he delivered in his home, Major and his students recreated these experiments on August 20th, 1670. First, they examined the shifting colors of an opal closely as a means of demonstrating that "colors do not flow from some internal, or as they say, essential form of a body. Rather, the diverse articulations of particles of the same kind give rise to so many diverse phenomena." While gems of various colors and opals in particular were used medicinally based on a belief in the virtues of their essential forms, Major taught his students that gems were nothing but "natural glass" (Major 1670, "Specimen 1").

Major and his students further tested this idea by performing the "metamorphoses of colors" described in Thomas Willis' *On Fermentation,* producing an ink from two clear liquids, then extinguishing and reviving the color again and again by repeatedly adding oil of tartar followed by the spirit of vitriol. They also, as Major had described in his 1667 work, produced an opalescent effect by adding a sublimate of mercury. In his account of the seminar, Major further noted that he had previously performed these "metamorphoses, extinctions and resuscitations of color" at public events at both the University of Kiel and the court at Eutin (Major 1670, "Specimen 1").

In 1675, Major once again discussed color changes in his extensively annotated edition of the Lincean Fabio Colonna's work on ancient purple; among his dedicatees for this work was August Friedrich, Prince-Bishop of Eutin. This discussion hinted at yet another of the appealing features of color for thinking about analysis and synthesis, as Major grouped color combinations in much the same way as he categorized other mixed bodies. Colors could be simple or composed, a combination of composed and composed, or even further composed ("*decompositi*") in ways that made it very difficult to distinguish individual colors, as in the case of the iridescence on the neck of a duck or the tail of a peacock, in soap bubbles floating in the air, or a silk cloth woven from multicolored thread. A vast variety of such things might arise from infinitely mutable mixtures. However, Major contended, it was still possible to move backward in one's mind ("*in animo retrogredi*") from these superposed colors to the composed, and from these to the means of combining simpler colors, until one arrived at the three "fundamental"colors: yellow, red and blue (Major 1675, 33–4) that Boyle had identified among the "primary" colors (Boyle 1664, 219).

Boyle had based his identification of five primary colors on the practice of painters. White, black, red, blue, and yellow, variously "Compounded, and (if I may so speak) Decompounded, being sufficient to exhibit a Variety and Number of Colours, such, as those that are altogether Strangers to the Painters Pallets, can hardly imagine" (Boyle 1664, 220). Herein lies a difference between Major's "fundamental" and Boyle's "primary" colors. Whereas non-practitioners could "hardly imagine" how so few colors produced such variety according to Boyle, Major contended that it was possible to travel mentally back from the most mixed colors to three fundamental colors. Major ordinarily hesitated to attribute to the mind an ability to identify fundamental building blocks, but he did so in the case of colors.

10.2 Querying Analysis

10.2.1 Questioning Principles in the Publishing Circle of Schultze

Questions regarding the basic building blocks of matter, whether spirits or other principles, were not only entertained in academic seminars and court demonstrations but were also broached in works intended for a wider chymical audience of the kind to which the bookseller Gottfried Schultze (1643–1686) of Hamburg catered in works published between 1671 and 1673. This included a notable set of publications addressed to Gottorf court physician Joel Langelott (Newman 2018, 320).

In 1671, Schultze published the Latin translation of Robert Boyle's *Tractatus de cosmicis rerum qualitatibus* produced by the Kiel University professor and polyhistor Georg Morhof in 1671. Morhof then urged Langelott to address an account of his laboratory achievements to the Academy of the Curious about Nature, and Langelott did so, in an edition published in the Academy's journal as well as separately by Schultze in 1672 (Langelott 1672a, b). That same year, Schultze published a work addressed to Langelott by Hamburg physician David von der Becke, who sought to demonstrate how the volatilization of fixed salts could be explained through his understanding that so-called principles were in fact "not the ultimate constituents of bodies." Sulphur, spirits, and salts could all be reduced further into prior particles (Clericuzio 2000, 198 and von der Beke 1672). Schultze re-issued von der Becke's work in a collected volume that also contained Langelott's letter to the Academy and an epistolary work on the transmutation of metals that Morhof had addressed to Langelott (Schultze 1673).

In 1673, Schultze also published Rostock professor Sebastian Wirdig's account of a new system of medicine based on the study of spirits which Wirdig dedicated to the Royal Society of London and to which Morhof contributed a liminary poem. Wirdig was forced to issue a retraction following an investigation by the theologian and court preacher of Güstrow, but Major praised Wirdig's work frequently despite its suspected heresy (Wirdig 1673; Siricius 1684; Lengerken and Major 1676, [Dv]; Lehmann and Major 1685, 6). In 1673, Langelott then published with Schultze an edition of the recipes of Johann Tilemann, a professor of medicine at Marburg, concerning the solution of gold. In a prefatory letter to Leipzig professor Johann Bohn, Langelott boasted further of his philosophical mill (that he had already announced in his work of 1672a, b), which, by slowly breaking down materials into the "smallest particles (*minimas particulas*)" allowed for the production of potable gold and many other *arcana* (Tilemann 1673, [a8v]. In a letter to Jena professor Werner Rolfink, Major criticized Langelott's claims, saying that his mill was not as new nor as praiseworthy as Langelott had said (Major to Rolfink 1672).

Finally, in 1675, Johann Bohn published a letter to Langelott in which he argued against the concept of acids and alkalis, although he did so in Leipzig rather than in Hamburg. Bohn argued that those who sought to restrict the number of active principles had "brought out two gladiators into the arena, whose conflicts were supposed to provide the immediate and sole causes of generation, coagulation, dissolution, and in a word all natural phenomena" (Bohn 1675, 6). They could not explain everything, however; they were not the basic building blocks of things; and moreover, they had no good explanation themselves. Major approved greatly of this work of Bohn's, calling it "most elegant" (Lengerken and Major 1676, [F4v]).

In a dissertation on synthesis over which he presided at Leipzig, Bohn further argued that that which was called "coagulation" could be renamed a "concretion, combination or composition," that is, an agglomeration of many separated things as though weaving together one cloth from many threads (Heyse and Bohn 1680,

[C2]). When combined, bodies could take on apparently new natures, such as the combination of two clear liquids that could produce a red color, Bohn noted. However, many famous chymists were deceived when they dissolved concretions and thereby thought themselves to have produced simple bodies. They not only merely separated out bodies that were themselves also composites, but the dissolving menstruum often combined with the solute to produce a new synthesis (Heyse and Bohn 1680, [D4r-v]).

10.2.2 The Pragmatic and Provisional Identity of Simple Bodies

Like Bohn, Major proved skeptical of claims to have produced the basic building blocks of nature through analysis (1677a [Iir-v]). He acknowledges that he sometimes uses the term "element" pragmatically but asserts that he does not intend thereby to identify any particular material as the basic material building blocks of all other materials. He harshly criticized those, like Aristotle, who rush intemperately into such identifications.

> I should note that when I use the term element here, I mean nothing other than the material constituting any mixture. Not only do I not mean fire or air, water and earth as the simplest constituents of bodies according to the Aristotelians, but also [I do not mean] other textures, such as salt, sulfur, or mercury, that are commonly called principles, or whatever others that might be found under whatever other name. And these are either truly simple, or made of two or three simple things, so closely fused by their nature, that they cannot be reduced to a simpler form by any artifice or observed anywhere in that form. Of which condition salt is foremost, because through its properties and special ways of acting, it was noticed a long time ago that it was something different than the other principles of things, nor could it ever be found anywhere completely simple, because it would have in one case water, in another earth, and perhaps in a third some other substance as a companion. Thus, the distinguishing principles of this [salt] could perhaps be called middle bodies, that is because in comparison with truly simple bodies, they are composed, but in comparison with the mixts or the "further composed" (*decompositorum*), they are simple. If Aristotle had gained any knowledge of these through analysis of bodies, he doubtless would have given them titles too and enlarged the imperious quaternary of the elements.[4]

[4] "Quâ in Causâ, ut hoc praemoneam, notandum, me per Elementi vocabulum híc intelligere, quicquid ullo modo Materiam ullius Misti constituit: non Ignem videlicet solùm, Aërem, Aquam, Terram, tanquam Corpora Rerum constitutiva simplicissima, Peripateticis sic dicta; sed Principia Texturarum etiam reliqua, Sal, Sulfur, Mercurium, communiter postmodùm sic dicta, aut quaevis alia, quovis titulo alio incedentia: eaque item vel verè simplicia, vel ex simplicibus duobus aut tribus, ita arctè per naturam suam conflata, ut per artificium ullum ad formam simpliciorem redigi nequeant, aut in eadem uspiam observari. Cujus conditionis praecipuè est Sal. Quod quidem per Proprietates suas, & speciales agendi Modos, pridem innotuit, esse quid â Principiis rerum quibusvis aliis distinctum: nunquam tamen adeò simplex alicubi ultimò deprehenditur, quin in hoc casu Aquam, in alio Terram & forsan in alio iterum Substantiam aliam comitem habeat. Unde Principia istius Commatis, placet Media vocare, h. e. quae respectu verè simplicium, sint utcunque Composita; respectu Mixtorum verò aut Decompositorum, Simplicia. Qualium notitiam quoque si

While Major rejects the prior identification of any basic building block of nature, he does single out salts as one category that appeared to interact in a distinctive way with other forms of matter. Refusing to acknowledge that "salt" was a primary principle, he granted it an unusual "medium" status. As we will see, Major engaged in an ambitious program of testing the combinations of salts with various other materials and further synthesizing those syntheses together to attempt to reach some form of a more systematic understanding of what salts were.

10.2.3 Texture

Rather than focusing on the identity of the basic principles of matter, Major, like Robert Boyle, emphasized the efficacy of complex textures resulting from the syntheses of many other syntheses to achieve many visible phenomena in the world (Clericuzio 2000, 5). For Major, these complex textures applied to every form of organization imaginable. According to Major's oft-repeated phrase, any kind of unit could be simple, composed, or further composed (*decomposita*). Organs of the human body were not simple, endowed with a singular identifying function, but rather were composed and subsequently recomposed of various textures and capable of fulfilling various functions (Major 1677b, [A3r]). Similarly, the actions of the material soul were not simple, but were "composed and then further composed" (Hesporn and Major 1679, § 7).[5]

Major prided himself on having been the first to thoroughly treat the "difformity" of natural bodies—that is, the composition of bodies of more than one nature. This was a main theme in his various schemes for arranging collections and other knowledge repositories. In what he claimed as a new discipline of the "taxis of chambers," Major complained that objects in museums were not arranged according to the "rigor of natural science" (Major 1674, [C3r]).[6] In the model public museum that he opened in 1688, he showcased how the rigorous disposition of objects on the shelves of a collection could follow natural philosophical categorizations of matter. "Natural bodies are either Uniform or Difform, composed of two or more, which the natural philosophers have until now not brought into an appropriate, correct order," he argued (Major 1688, 24).[7] He claimed to have taken the shelf-labels in his museum directly from the categories of nature that he explored in a now lost natural philosophical manuscript. Quoting from this manuscript in his guidebook, he denoted various basic categories (water, salt, metal, earth, sulfur, stone, plant, and animals)

---

ex ullâ Analysi Corporum sibi parasset Aristoteles; tempestivè ac citra omne Dubium, eorum Titulis additis, imperiosum Elementorum quaternarium adauxisset."

[5] "Actiones animae materiales pleraeque non simplices sunt, sed compositae aut decompositae."

[6] „nach dem Rigor der Physicalischen Wissenschafft."

[7] "Natürliche Corper... sind entweden Uniforma (*Einfache*) oder Difformia, von zwey-oder mehrer zusammengesetzte; davon die Physici ... bis dato noch nicht/in gebühren-richtige Ordnung gebracht haben."

and identified difform natural bodies as varying compounds of these, such as, "water-salts, water-sulfurs, water-metals; earth-earths, earth-salts, earth-sulfurs, earth-metals, earth-plants or earth-animals; or salt-salts, salt-sulfurs, or salt-metals; or sulfur-animals.... or composed from many and then again further composed" (Major 1688, 24).[8] Noticeably, Major does not expend much effort in defining the terms such as "salt" or "sulfur" that constitute these hybrid categories. Rather than claiming knowledge of principles, Major treats simple bodies as pragmatic categories awaiting further investigation.

10.3 Spirits and Salts

10.3.1 The Suffering of the Giants

Major treated spirits at length in his 1676 *Suffering of the Giants.* Spirits were physical bodies of the most subtle "texture" that could be identified (Lengerken and Major 1676 [A4r]). By referring to spirits as having a texture, rather than being simple bodies themselves, Major refused to claim that spirits represented absolutely simple bodies. He confessed that he was aware of thus falling into an infinite regression of subtler and subtler parts. He treated spirits only as a pragmatic category. Spirits could be defined as the level of matter after which it was difficult to imagine a more subtle body (Lengerken and Major 1676, [A4v]).

Spirits were called by many names such as "Fires," "Lights," "Salts," "Archeos," "Balsamic Airs," and "Native Heat," but Major did not think their name was significant (Lengerken and Major 1676, [A4v]). He also refrained from debating whether spirits were of one kind or of many (Lengerken and Major 1676, [Br]). What Major specified as most interesting to him was not the basic or most subtle identity of spirits, but the way that their arrangements as fighting "factions" could offer new explanations for seemingly teleological physiology.

Just as the rebellious Titans were thrown down back into the body of Gaia by the gods, where they continued to tremble, destabilizing the earth's crust through earthquakes and floods, so too had original sin given rise to constant motion and friction in nature on the level of the spirits. Like the Titans beating against the womb of Gaia, spirits entrapped within grosser forms of matter constantly came into conflict with one another and with surrounding structures. Their sufferings gave rise to an attrition and debilitation "coeval with Adam and propagated through original sin to

[8] "Naturalia Naturalibus copulata, sunt vel Aqui-Salia, Aqui-Sulfura, aut Aqui-Metalla: Vel Terri-Terrae, Terri-Saliae, Terri-Sulfura, Terri-Lapides, Terri-Metalla, Terri-Plantae, aut Terr-Animalia; vel Sali-Salia, Sali-Sulfura, aut Sali-Metalla: vel Sulfur-Animalia... vel ex pluribus, aut plurimis composita, aut decomposita."

his posterity, yet not yet observed by physicians, as far as I know, nor treated fully by anyone" (Lengerken and Major 1676, [A2r]).[9]

Factions of spirits become combatants resembling gladiators, or "*Agones*, in a state of Disease approaching Death." Because they risked all in this fight, they waged battle "like two Athletes, fighting to the point of drawing blood" (Lengerken and Major 1676, [Bv]). Whereas others, like Major's much-admired Francis Bacon, discussed how the actions of spirits led to dryness, sickness, aging and death, for Major, all physiological processes, including generation and growth, resulted from the suffering of spirits within diseased materiality. Such was the regular state of the human: "there is no healthy material action in us that can be easily begun or concluded without some obscure disease interceding. In short, *We are never anything but morbidly healthy*." True health only applied to the body of Adam before the fall and to Christ with the Saints after resurrection (Lengerken and Major 1676, [A2v]. For all others, the study of how the "*taxis* of material particles" functioned in the body could moderate the ferocity of their disease, but not cure it (Lehmann and Major 1685, 3).

All actions of the material soul or animal spirits were the result of these fine fluids coming into friction with varying material textures in diverse structures. These actions of the animal soul were thus never simple, but many of them were composed out of two, three, or four things (Lengerken and Major 1676, [C2r]). To combat teleological understanding of the parts of the body, Major militated against assigning specific names to various processes in the body, located in specific organs and controlled by specific faculties. For instance, rather than nutrition, one might speak of augmentation or accretion (Lengerken and Major 1676, [C3r]).

Major also rejected all the terms used to describe various stages of digestion, which had proliferated, such as "chylification," "coction," "corruption," "fermentation," etc. They all insinuated some "transformation of a quality," while preferable terms such as "attrition, destruction, resolution, or separation," suggested merely a change of magnitude, shape, and number. He noted that some medical authorities conceptualized "*resolutio*" or "*analysis*" as the breaking up the union of the elements (Lengerken and Major 1676, [E2r]; Major cited Hofmann (1645, 232) who described this breakdown as though it were a return back into prime elements ("*quasi in prima elementa redigantur*"). Major instead preferred a language of separation that made no claims about the prime constituents of matter. "Just as all fermentation (natural or artificial) begins with the incipient dispersal of the subtle parts from the coarser ones, having been driven into sharp collisions and conflicts with the latter, it concludes with the total separation of both through random movements, so that they completely take off the garment of their former union" (Lengerken and Major 1676, [D3r-v]).[10] In this process of separation, the fermenting spirits do not escape the universal fate of suffering, as they must "dash against now this here and

[9] "coaevam tamen Adamo jam olim, & culpâ Peccati dein ad Posteros propagatam; imò nedum valdè observatam ab ipsis Medicis, quod sciam, aut plenè pertractatam ab ullo... ."

[10] "... quemadmodùm omnis Fermentationis (naturalis, vel artificio factae) Terminus proximus est, Portionum subtilium, cum crassioribus in acres Collisiones & conflictus actarum, ab iisdem crassi-

now that there," and "wrestle" with "very thick and resistant" textures, repeating "their blows upon it many times, before they reduce some portions of it to obedience" (Lengerken and Major 1676, [D3r-v]).[11]

Major addressed one of the main topics in period debates over the relationship between the qualitative and the mechanistic in corpuscularianism which, for many, centered on digestion and the nature of fermentation in particular in that process (Clericuzio 2012). Major appears to be taking Descartes' attempts to move the language describing physiology in more mechanistic directions even further. As Carmen Schmechel has explored, Descartes moved from "concoction" to "fermentation" as a less teleological means of describing physiological processes, but "fermentation" did not resolve the issues he had hoped it would (Schmechel 2022).

10.3.2 The Battle of Acid and Alkali

Major offered dramatic examples of chymical warfare in the interactions of acidic and alkali salts. When the salt of tartar (potassium carbonate), a weak base, was added to a strong acid, such as aqua fortis (nitric acid) or oil of vitriol (sulfuric acid), the two engaged in an "especially great battle, emitting sudden smoking explosions and heating up the glass." Soon, "innumerable bubbles rise up from the bottom to the surface of the fluid," but once "these motions have stilled, and the enemy forces have entered into a pact of alliance, then shining clear streaks and little rods coalesce," he said (speaking of the formation of potassium nitrate crystals). However, the "Sylviano-Helmontians" were wrong to claim that human physiology could be explained by the conflicts between these two types of salt (Lengerken and Major 1676, [F4r-v]; Major 1677a, [Gv]).

According to Franz de la Boë, or Sylvius (1614–1672) during digestion an effervescence resulted from the mixture of alkaline bile with acidic pancreatic juice. Quoting Johann Bohn's letter to Langelott, Major agreed with Bohn that the followers of Sylvius deployed the acid-alkali theory too dogmatically. For one thing, it was well known that acids often effervesced when added to other acids, so that if an effervescence was observed, it did not mean that one of the substances had to be alkaline. Furthermore, perhaps two effervescing substances might both be alkaline. Such a phenomenon was not yet known but was "there anyone who had done experiments of all possible modes of combination in bodies of every kind of texture?" (Lengerken and Major 1676, [F3r-Gr]).[12]

---

oribus incipiens Discessio; ultimus verô, utrarumque confusè motarum Separatio omnimoda, ut Veste pristinae Unionis exsuantur."

[11] "ut laboriosè proin nunc huc nunc illuc impingant" "plurimum crassae ac resistentis Texturae intùs inveniunt, cum quâ colluctentur, & opus habent, ictus in illam suos saepe repetere, antequam portiones illius aliquas redigant in obsequium."

[12] "Ecquis tamen per omnes possibiles Combinationum modos, in omnis generis Texturis Corporum, Experimenta fecit?"

10.3.3 Testing Salts

In a digression in the *Sufferings of the Giants*, Major touched upon how he and his students in his experimental seminar the previous August had undertaken an ambitious experimental program to test the combinations of many different kinds of salts. At this time, Kiel was embroiled, as was often the case, in the actual military engagements of the decades-long Northern wars. "Amid the unpleasant din of wars on every side," he and his students enjoyed a much more pleasurable scene of conflict as they "instituted certain experiments about the different kinds of salts, in order to perceive their actions & mute passions, motions, external figures, & changes of state, so as to reduce them to some more universal intellectual concept" (Lengerken and Major 1676, [H2r].[13]

For the "sake of experiments," Major and his students assembled multiple liquids, "aqueous, oily, and spirituous; odorous and odorless; sweet, acrid, more or less acid, or plainly corrosive, lixivial, sour, or bitter; clear and transparent or turbid and tinged; fixed and volatile; produced through coction, per deliquum, solution and distillation; hot, cold; *simple, composite and further composed* [emphasis mine]," as well as "many types of salts, acidic, basic, or neutral; artificial and natural, fixed and volatile, etc. and in this way we set up mixtures of bodies in order to see which kind of particles would be friendly or hostile, and which would be conform or difform one to the other" (Lengerken and Major 1676, [H2r-v]).[14] Notably, some of his liquids were already "composite" or "further composed," and he and his students sought to synthesize these further with salts to determine whether they would be "friendly" and produce an apparently unified body, or if they would appear to be "difform," that is, obviously composed of two or more parts. Major's language of friendship here picked up on his analogy between chymical bonds and temporary truces or pacts in the ongoing war of all matter.

---

[13] "ad universaliorem aliquem Conceptum Animi reducendas."

[14] "Assumseramus nimirum, Experimenti ergo, Liquores multiplices, Aqueos, Oleosos, Spirituosos; Olidos, inodoros; Dulces, acres, acidos plus minus, aut planè corrosivos, lixivos, austeros, amaros; Claros, & diaphanos, turbidos, aut tinctos; Fixos, volatiles; Coctiones, deliquio, solutione, ac destillatione ortos; Calidos, frigidos; Simplices, compositos, ac decompositos; nec non Salia una ac altera, Acida, lixiva, Enixa; Naturâ ac arte facta; Fixa ac volatilia, &c. & in istiusmodi Corporum Misturis constituendis, ad videndum, cujus generis particulae amicae vel inimicae, conformes aut difformes futurae essent alteris, non quidem curiosè nimis numerando emetiabamur Compositionum, ex binis, trinis, quaternis, &c. objectis, Modos omnes possibiles, quoniam fuisset (ut ex Arimeticis Additionum Multiplicationumque progressibus notum est) Varietas stupendi Numeri, vel Numorosa farrago Stupendae Varietatis hinc facilè exortura, operosior exsequutu, quàm aequè semper aut peromnia utilis: in medicori tamen, ac sat notabili Corporum varietate, ultrò citroque tractandâ occupatis, Vascula vitrea commoda incipiebant nobis deficere. Atque sic Successum avidi, moraeque impatientes, ac semidefatigati annotandis, memoriae ergò, Phaenomenis passim selectioribus, quorum unum saepè Misturae novae ulterioris atque iterum alterius, adornandae ansam dabat, non ad mensuram ampliùs ac pondus, hunc aut illum Liquorem, hoc aut illud Sal, cum aliis mista, huic vel illi Vitro, accuratè tegendo, inserere commodum exstitit... ."

Major hastened to confess that they had not been so stringent as to chart every single combination of two, three, four or more variables since that would soon give rise to an exponential progression of a stupendous number of combinations. He and his students pursued so many combinations that they eventually ran out of glasses and grew tired of carefully measuring and weighing "this liquid with that salt mixed with some others in this or that carefully covered glass" (Lengerken and Major 1676, [H2v]).

10.3.4 Describing Salts

In his *Erring Genius* of 1677, Major further explored what he wished for in a complete account of his investigation into salts. In this book, Major criticized the human capacity for error, including an error of his own mind—namely, his insatiable desire to have all the experiments in all areas of knowledge in the whole world recorded. His immoderate desire for a perfect experimental record stemmed from his fear that any general statement derived from experimental evidence might one day be disproven by some new experiment. Thus, in a position that Major acknowledged was futile, he desired first to know about all experiments before ever proposing a more general statement.

Others suffered from the opposite failing. Major criticized the "Praetorian edicts" issuing from renowned chymists like Glauber and Helmont. They had persuaded so many that all salts in the universe as a whole could be classified as either an acid, an alkali, or a composite of the two. However, they did not think to explain what the nature of an alkaline salt was in and of itself (Major 1677a, [G3v]. Rather, they defined an alkali merely in terms of an acid—as something that was not an acid but that invariably came into conflict with it. Bohn's very learned work had demonstrated the insufficiency of this acid-alkali theory. According to Bohn, "these two Hercules (namely, acid and alkali) do not necessarily always fight, so that without collision sometimes one of the powers degenerates into the other, and each is transformed as it were" (Major 1677a, [G4r]).

If the cause of these changes was to be found anywhere, it was most likely to be drawn "from the delicate part of Corpuscular Philosophy, which I call Phoranomic-Figuristic." In coining this term, Major referred to the highly complex and varied array of motion, shape, and number exhibited by particles. They might move in a "rectilinear, curvilinear, simple, reflex" manner. Their shape might be "irregular, orbicular, flat, smooth, hispid, long, short, sharp, obtuse, caudate, spicular, toothed, bifid, trifid, eel-like, helical, repand, annular, triangular, cubical, polygonal, pyramidal, or other." To account for all possible interactions between salts, therefore, the description of salts would thus have to record a vast array of substances, as Major and his students had already begun to investigate, as well as the many different permutations according to which they might be moving or disposed. Yet even this immense study of salts represented a tiny portion of the entire corpus of annals of experiments that remained to be written (Major 1677a, [H5r]).

10.4 Conclusion: The Order of Things

Major thought of synthesis and analysis as reversible processes. However, he criticized those who rushed to hypothesize about the prime constituents of matter. He treated so-called simple bodies as pragmatic rather than proven categories. The Fall embedded moral and epistemic waywardness in materiality itself. Invisible spirits tumultuously crashed against inhospitable surfaces, rendering all of shuddering matter, including that of the human brain, unstable and slipshod. Only experimentation into invisible chymical structures promised any means of improving the state of human knowledge by raising awareness of the conditions under which it was formed. Perforce, such experimentation would need to begin by framing merely temporary and pragmatic categories, which might be improved over time.

Theories of basic constituents of matter could be disproven, but the nature of primary, invisible constituents of matter could not be proven. For instance, color changes in liquid solutions showed that no particular entity caused color. The alternate addition of acids and alkalis that made color appear and disappear showed that color was an effect of reversible shifts in chymical structure. However, this experiment could not prove that acids and alkalis were simple bodies and prime constituents of matter; it seemed impossible to prove that one had ever reached the final stage of analysis.

Major felt on surer ground when dealing with synthesis, rather than analysis. By observing the syntheses of many types of salt, Major hoped to be able to reason back to the essential identity of a salt (that is, to "reduce them to some more universal intellectual concept"), just as he felt that it was possible to reason back to mental categories of three primary colors from the mixed colors in which they appeared (while denying those mental categories any physical reality as prime constituents of matter; red, yellow, and blue, like all colors, were still only the perceived effects of light refraction in different corpuscular structures). Major's perception of an error-prone, shifting knowledge encouraged pragmatic choices in selecting the basic categorical units (such as salt, sulfur, metal) that might be deployed for a time, similar to the arrangement of objects in a collection.

The curator of that collection would be obliged to question its categories continually as further research reorganized tactical fields. For example, a collection of salts formed for the purposes of researching what salt was would require continual additions, subtractions, and regroupings as some materials would be rejected in the end as not salts after all, or others might be grouped up temporarily as some unique form of salt, only later to be re-arranged with some other constellation of materials. The ordering of human mental conceptions of the categories of nature and the ordering of materials for experiments would need to remain in provisional and responsive articulation with one another.

Major thus apparently enunciated an empirico-pragmatic approach to principles long before such an approach became widely accepted in the middle of the eighteenth century (Best 2016, 46). However, he never wished scientific categories to remain merely pragmatic. As he and his contemporaries shifted experimental philosophy into the form of an academic discipline, they aimed to identify philosophically defensible terms for pragmatic categories used professionally by apothecaries and chymists. This was the criticism that Major and his colleague Johann Bohn launched at proponents of the acid-alkali theory. While these categories accounted for many observable phenomena in the laboratory, they had not been subjected to sufficient doubt, nor to the possibility of future discoveries, to be accepted philosophically (even within a probabilistic framework). Major aimed to establish academic practices of professionalized doubt and extensive testing that could establish experimental conjectures upon sounder footing, even as he acknowledged that absolute certainty would remain perpetually out of reach (Keller 2020, 2024).

Major deployed the term *taxis* for temporary orderings of both matter and of human-authored units of knowledge. The term itself would not, it seems, re-occur as a means for describing natural categories and arrangements until the early nineteenth century, when Augustin Pyramus de Candolle (1813) used *taxis* to coin a term for a dynamic and agonistic form of natural order, "*taxonomy*." Scholars who describe the eighteenth century as a taxonomic age might be surprised to learn that the term did not yet exist in that period. They may also ignore the ways that taxonomy, rather than a strict classification of a static natural order, was provisional and agonistic from the outset. The interaction between epistemological structures and natural structures permitted a view of *taxis* as a method of research, that is, as a means of ongoing investigation into the building blocks of nature and the ways in which they interacted with other forms and scales of order, including the categories of human knowledge itself.

Bibliography

Manuscript

Hamburg

Staats- und Universitätsbibliothek Hamburg Carl von Ossietzky.
Sup.e ep. 44, 114. Johann Daniel Major to Werner Rolfink, 30 October 1672.

Primary

Bohn, Johann. 1675. *Epistola ad virum … Joelem Langelottum … de alcali et acidi insufficientia pro principiorum seu elementorum corporum naturalium munere gerendo*. Leipzig: Hahn.
Boyle, Roberg. 1664. *Experiments and considerations touching colours*. London: Herringman.
Cassius, Andreas. 1685. *de Auro*. Hamburg Wolff.
Cassius, Andreas (*defendens*), and Johann Daniel Major (*praeses*). 1666. *De febre artificiali*. Kiel: Reumann.
Crassot, Jean. 1619. *Totius Philosophiae Peripateticae Corpus*. Vol. 1. Paris: Huby.

de Candolle, A.P. 1813. *Théorie Élémentaire de la Botanique*. Paris: Déterville.
de Lengerken, Hermann (*respondens*), and Johann Daniel Major (*praeses*). 1676. *De aerumnis gigantum in negocio sanitatis*. Kiel: Reumann.
Hesporn, Franz (*defendens*), and Johann Daniel Major (*praeses*). 1679. *Positiones medicae variae earumque praecipuae de podagra*. Kiel: Reumann.
Heyse, Ernst Gottfried (*respondens*), and Johannes Bohn (*praeses*). 1680. *Dissertationum chymico-physicarum secunda de corporum combinatione, seu concretione*. Leipzig: Hahn.
Hofmann, Caspar. 1645. *Institutionum medicarum libri sex*. Leiden: Huguetan.
Langelott, Joel. 1672a. Epistola. *Miscellanea Curiosa* 3: 96–106.
———. 1672b. *Epistola ad ... Naturae Curiosos*. Hamburg: Schultze.
Lehmann, Christoph (*respondens*), and Johann Daniel Major (*praeses*). 1685. *De moribundorum regimine*. Kiel: Reumann.
Major, Johann Daniel. 1665. *Joh. Danielis Horstii judicium de chirurgia infusoria*. Frankfurt: Fickwirt.
———. 1667. *Chirurgia infusoria*. Kiel: Reumann.
———. 1670. *Collegium medico-curiosum*. Kiel: Reumann.
———. 1674. *Unvorgreiffliches Bedencken von Kunst- und Naturalien-Kammern ins gemein*. Kiel: Reumann.
———. 1675. In *Annotationes, appended to Fabio Colonna, de Purpura*, ed. Johann Daniel Major. Kiel: Reumann.
———. 1677a. *Genius errans*. Kiel: Reumann.
———. 1677b. *Prospectus praeliminaris in theatrum profundae scientiae*. Kiel: Reumann.
———. 1681. *Programma quo... disputationem... de petechiis ... invitat*. Kiel: Reumann.
———. 1688. *Museum Cimbricum*. Plön: Schmidt.
Sachs, Philipp Jacob. 1665. *Gammaralogia*. Leipzig: Fellgiebel.
Schultze, Gottfried, ed. 1673. *Virorum clarissimorum epistolarum circa utilissima aliquot chymica experimenta*. Hamburg: Schultze.
Sennert, Johann Andreas (respondens), and Johann Daniel Major (praeses). 1667. *De lacte lunae*. Kiel: Reumann.
Siricius, Michael. 1684. *Victrix veritatis in censuris theologico-medicis de nova spirituum medicina*. Gustrow: Spierling.
Tilemann, Johann. 1673. In *Experimenta circa veras & irreducibiles Auri solutiones*, ed. J. Langelott. Hamburg: Schultze.
Tralles, Johann Christian (*respondens*), and Johann Daniel Major (*praeses*). 1677. *De malacia*. Kiel: Reumann.
von der Becke, David. 1672. *Epistola ad ... Joelem Langelottum*. Hamburg: Schultze.
Willis, Thomas. 1659. *Diatribae duae medico-philosophicae quam prior agit de fermentatione*. Hague: Vlacq.
———. 1681. *A medical-philosophical discourse of fermentation*. London: Dring.
Wirdig, Sebastian. 1673. *Nova medicina spirituum*. Hamburg: Schultze.

Secondary

Best, Nicholas W. 2016. What was revolutionary about the chemical revolution? In *Essays in the philosophy of chemistry*, ed. Eric Scerri and Grant Fisher, 37–59. Oxford: Oxford University Press.
Clericuzio, Antonio. 2000. *Elements, principles, and corpuscles: A study of atomism and chemistry in the seventeenth century*. Dordrecht: Springer.
———. 2012. Chemical and mechanical theories of digestion in early modern medicine. *Studies in History and Philosophy of Science, Part C* 43 (2): 329–337.
Debus, Alan. 1962. Solution analyses prior to Robert Boyle. *Chymia* 8: 41–61.

Harrison, Peter. 2002. Original sin and the problem of knowledge in early modern Europe. *Journal of the History of Ideas* 63 (2): 239–259.
Keller, Vera. 2020. Professionalizing doubt: Johann Daniel Major's observation 'On the Horn of the Bezoardic Goat,' curiosity collecting, and periodical publication. In *Institutionalization of science in early modern Europe*, ed. Giulia Giannini and Mordechai Feingold, 199–235. Leiden: Brill.
———. 2024. *Curating the enlightenment: Johann Daniel Major and the experimental century*. Cambridge: Cambridge University Press.
Newman, William. 2018. *Newton the alchemist: Science, enigma, and the quest for Nature's 'Secret Fire'*. Princeton: Princeton University Press.
Reinbacher, W. Rudolph. 1998. *Leben, Arbeit und Umwelt des Arztes Johann Daniel Major: Eine Biographie aus dem 17. Jahrhundert, mit neuen Erkenntnissen*. Linsengericht: Kroeber.
Schmechel, Carmen. 2022. Descartes on fermentation in digestion: Iatromechanism, analogy and teleology. *The British Journal for the History of Science* 55: 101–116.
Scribano, Emanuela. 2022. Powers of the body and eclipse of the soul. From Descartes on. In *Mechanism, life and mind in modern natural philosophy*, ed. Charles T. Wolfe, Paolo Pecere, and Antonio Clericuzio, 235–257. Cham: Springer.
von Kerssenbrock-Krosigk, Dedo, ed. 2008. *Glass of the alchemists: Lead crystal-gold ruby, 1650–1750*. Corning: Corning Museum of Glass.

Vera Keller, Professor of History at the University of Oregon, has published three monographs: *Knowledge and the Public Interest, 1575–1725* (Cambridge, 2015); *The Interlopers: Early Stuart Projects and the Undisciplining of Knowledge* (Hopkins, 2023), winner of the Phyllis Goodhart Gordan Book Prize from the Renaissance Society of America and *Curating the Enlightenment: Johann Daniel Major and the Experimental Century* (Cambridge, 2024). She is also the author of over forty articles and has edited six volumes and journal special issues. Her research has been supported by many grants and fellowships, including a Ryskamp Research Fellowship (ACLS) and a Guggenheim Fellowship.

Chapter 11
Phenomena and Principles: Analysis–Synthesis and Reduction–Deduction in Eighteenth-Century Experimental Physics

Friedrich Steinle

Abstract In the *Encylopédie*, d'Alembert hinted that "experimental physics," unlike mathematical sciences, had a specific epistemic goal, which he characterized as "reducing" the multitude of phenomena to a few "principal facts" that should serve as principles for "deducing" all others. However, he did not detail precisely how he understood that goal, the status of the principal facts, nor the process of "multiplying" the phenomena. Neither did Diderot, who supported d'Alembert's view. Herein, I present d'Alembert's ideas and examine a prominent and highly successful case of experimental research as illustration: Charles Dufay, whose work d'Alembert certainly knew. Dufay not only managed to impose a firm order on the bewildering phenomena of electricity but also presented his research using the same epistemological notions that d'Alembert would himself later use. Similar notions may be found in the earlier writings of Edme Mariotte, who defined "principles of experience" as the goal of experimental research. This may suggest a specific tradition of spelling out the age-old idea of a double pathway of empirical science—from particular experience to general statements, and back again to the level of experiences—in a detailed manner that differed considerably from how Newton would frame it in his famous notions of analysis and synthesis.

Keywords Eighteenth-century experimental physics · Principles · Analysis-synthesis · Reduction-deduction · Denis Diderot · Jean-Baptiste le Rond d'Alembert · Edme Mariotte · Charles Dufay

---

F. Steinle (✉)
Wissenschaftsgeschichte/History of Science, Institut für Philosophie, Literatur-, Wissenschafts- und Technikgeschichte, Technische Universität Berlin, Berlin, Germany
e-mail: friedrich.steinle@tu-berlin.de

W. R. Newman, J. Schickore (eds.), *Traditions of Analysis and Synthesis*, Archimedes 73, https://doi.org/10.1007/978-3-031-76398-4_11

11.1 Introduction

The idea that empirical science should have a double pathway—from the particular and direct experience to general statements, and from them back again to the level of experiences—extends as far back into history as the idea of systematic empirical research itself—that is, to Aristotle. He called the first direction *epagogé*, typically translated as "induction," but gave only a very brief explication of how this should be implemented. The four steps he mentioned—from perception (*aisthesis*) via memory (*mneme*) to experience (*empereia*) and finally to insight into the first principles (*episteme*)—remained very general (*Analytica posteriora* II, 19 or *Metaphysics* I, 1), and he did not spell out how the individual steps should be performed. For the second pathway, by contrast, the explanation or deduction of the particulars from first and general principles, he provided an elaborate system of syllogistic reasoning in the major parts of *Analytica priora & posteriora*.

While empirical research was practiced from Aristotle onward, in Hellenistic, Arabo-Islamic and Latin-Medieval science, including many cases of experimental work, the epistemological framework showed little change until the Early Modern period in European science.[1] With the claim of reforming or even revolutionizing science and its epistemology, allegedly new programs were formulated and implemented by authors and researchers such as Bacon, Galileo, and Descartes and institutions such as the early academies and, by the mid-seventeenth century, the enduring academies in London and Paris (not to forget the German-speaking 'Leopoldina'). While the basic idea of a double pathway between particulars and general statements still persisted, there were dramatic shifts in both the weights attributed to the two aspects and the specificities in which they were explicated. It is within this secular development that Newton's famous statement about analysis and synthesis should be understood. Here, he had presented "analysis" as the inductive pathway from the phenomenon to principles, and "composition" or "synthesis" as the reverse, thus drawing an analogy between epistemology, chemistry, and mathematics.

In this paper, I shall not dwell on Newton's statement so much, since Alan Shapiro's contribution to this volume offers a profound analysis. Rather, I wish to demonstrate that, from the seventeenth century onwards, there were several different attempts to articulate that double pathway in the context of the new and explicit emphasis on observation and experiment. In particular, I shall highlight one strand that differed strikingly from the Newtonian picture and terminology and was, I claim, of high historical relevance and visibility. However, it has been largely overlooked from the historiographical perspective, and while a full picture has still to be developed, I shall, at least, point out some clear cases that might inspire further research. I shall begin with two cases from the eighteenth century, before pointing

[1] This statement should be taken *cum grano salis*, because in research fields such as chymistry or medicine, particular epistemologies had been developed that did not fit the Aristotelian framework. For a pointed discussion, see Newman 1998, among others.

to their seventeenth-century roots, and in the end have a brief look to Newton's famous passage. In doing so, I hope to be able to add a new and scarcely discussed aspect to the historical understanding of the coupling of analysis and synthesis as epistemic procedures.

11.2 Experimental Physics Described in the Encyclopédie

I adopt the mid-eighteenth century as my starting point, with an illustrious author and an even more illustrious work: the *Encylopédie*, edited by d'Alembert and Diderot between 1751 and 1780. Many of the articles were written by d'Alembert himself, including those that treated the structural and epistemic character of the physical sciences. Among the key texts for this purpose are the lengthy introduction, the "Discours préliminaire" (d'Alembert 1751), and the article on experiment, called "Expérimental" (d'Alembert 1756).[2]

In these texts, d'Alembert consistently distinguishes sharply between two branches of physical science: On the one hand, he refers to physico-mathematical sciences ("sciences physico-mathématiques"), by which he means astronomy, falling bodies, statics, hydrostatics, mathematical optics, among others. On the other hand, he refers to what he calls "general and experimental physics" ("physique générale & expérimentale"), which includes fields such as heat and cold, magnetism, electricity, chemistry, and so on. He called Boyle the "father of experimental physics" and also mentioned Boerhaave and other physicians as examples.

The distinction did not merely point to different subject fields; rather, d'Alembert separated the two strands in several aspects. The role of mathematics, for example, differed considerably across the two, and the distinct strands also differed strikingly with respect to their certainty: while the degree of certainty in the 'sciences physico-mathématiques' came close to geometrical truths (carried along by mathematical procedure), that degree was much less in experimental physics. Most saliently in terms of my present purpose, he indicated profound methodological differences between the two, which became strikingly visible in the different roles of experiment and observation. In both strands, d'Alembert saw experiment as extremely important but ascribed different epistemic roles to it: In the 'sciences physico-mathématiques', only a small number of experiments were required, and they served primarily to produce (mostly quantitative) results that could then be compared with the theory's predictions. If differences occurred, they could be used to take more aspects than before into account, such as friction or air resistance ("Expérimental," 1756, 300). In experimental physics, by contrast, as many experiments as possible were required, with the goal of a systematic collection of a large field of phenomena (ibid.). To highlight the contrast, the differences are listed in Table 11.1.

[2] The heading's peculiar linguistic form was likely dictated by the fact that the French language does not have a specific word for what we call "experiment."

Table 11.1 Mathematical sciences and experimental physics in d'Alembert

Sciences physico-mathématiques	Physique générale & expérimentale
Astronomy, falling bodies, collision of bodies, statics, hydrostatics, mathematical optics, …	Heat and cold, magnetism, electricity, pneumatics, …
Mathematical	Typically non-mathematical
Certainty comes close to geometrical truth	Much less certainty
Few experiments, used to check the predictions of theory	As many experiments and observations as possible, with the goal of a "systematic collection"

While we might well be familiar with the first approach (in the "sciences physico-mathématiques") up to the present day (it resonates with the HD-approach in philosophy of science), the same cannot be said of the second use (in experimental physics). Hence, I shall focus on d'Alembert's "experimental physics" and investigate precisely what he had in mind with his characterization. I shall attempt to better understand what he meant by systematic collection and how one should arrive there.

Regarding how experimental physics should proceed, d'Alembert said the following:

> Thus, it is not at all by vague and arbitrary hypotheses that we can hope to know nature; it is by thoughtful study of phenomena, by the comparisons we make among them, by the art of reducing, as much as that may be possible, a large number of phenomena to a single one that can be regarded as the principle. Indeed, the more one reduces the number of principles of a science, the more one gives them scope. … This reduction which, moreover, makes them easier to understand, constitutes the true "systematic spirit." (*Discours préliminaire* 1751, vi, transl. 2009)[3]

Three points in particular warrant highlighting here. First, d'Alembert emphasized that it was necessary to proceed not by way of making arbitrary hypotheses but much more systematically. Second, and to specify how that might look, he emphasized the need to *compare* phenomena to one another and, with what might sound like a strange formulation, to "reduce a large number of phenomena to a single one that can be regarded as the principle." Finally, he emphasized the need to reduce the number of such principles as far as possible in further steps.

To illustrate his meaning, d'Alembert discussed the magnet and listed its general properties: the attraction of iron, the transfer of attractive power to iron, orientation towards the north, magnetic dip, etc. He emphasized that it would be desirable to sum up all those properties under a single, more general property. However, given that we do not yet know the origin of all those properties, he continued, our aim should be clear:

[3] Ce n'est donc point par des hypothèses vagues & arbitraires que nous pouvons espérer de connoître la Nature; c'est par l'étude réfléchie des phénomènes, par la comparaison que nous ferons des uns avec les autres, par l'art de réduire, autant qu'il sera possible, un grand nombre de phénomènes à un seul qui puisse en être regardé comme le principe. En effet, plus on diminue le nombre des principes d'une science, plus on leur donne d'étendue; …. Cette réduction, qui les rend d'ailleurs plus faciles à saisir, constitue le véritable esprit systématique.

> Since such knowledge and the necessary enlightenment concerning the physical cause [la cause physique] of the properties of the magnet are lacking, it would doubtless be an investigation most worthy of a philosopher to reduce, if possible, all these properties to a single one, while showing the liaison that they have with one another. ("Discours préliminaire," 1751, vi–vii, transl. 2009)[4]

Two points may be noted immediately here. First, d'Alembert drew a strong distinction between (ever more general) principles on the one hand and physical causes on the other hand. Second, the principles that he had in mind were just general accounts of the phenomena and were located on the same epistemic level as phenomena (which may differ from the level of the physical cause): he emphasized that the researcher should identify a single phenomenon to which all others could be reduced and that this phenomenon should henceforth count as a principle. It should be noted that we see here quite a specific meaning of the term "principle," a point to which I shall return later in greater detail.

On the question of how this may be achieved, d'Alembert became more specific:

> The only resource that hence remains to us in an investigation so difficult, ... is to collect as many facts as we can, to arrange them in the most natural order, and to relate them to a certain number of principal facts of which the others are only the consequences. (*Discours préliminaire*, 1751, vii, transl. 2009)[5]

In his article *Expérimental*, he said more about the procedure:

> It is the facts that the physicist must thoroughly search to know. He cannot multiply them too much; the more he has collected, the nearer he comes to their unity. His goal must be to put them in the order they allow, to explain the ones by the others as much as possible, and to form from them, if one can say so, a chain with as few missing links as possible. He will have done enough work then, having brought nature in good order. (*Expérimental* 1756, 301, my translation)[6]

In the second quotation, d'Alembert further specifies his claim in the first quotation above, with the phrase "to collect as many facts as possible and give them an order." He proposed to "multiply the phenomena" as much as possible, put them in order to form a chain, and to explain one with reference to another. This latter point resonates well with his earlier statement, when he had spoken of "reducing most of them to a small number of principal facts." However, several questions inevitably remain unresolved: what did "multiplication" of phenomena mean? What kind of "chain"

---

[4] Au défaut d'une telle connoissance, & des lumieres nécessaires sur la cause physique des propriétés de l'Aimant, ce seroit sans doute une recherche bien digne d'un Philosophe, que de réduire, s'il étoit possible, toutes ces propriétés à une seule, en montrant la liaison qu'elles ont entr elles.

[5] La seule ressource qui nous reste donc dans une recherche si pénible, quoique si nécessaire, et même si agréable, c'est d'amasser le plus de faits qu'il nous est possible, de les disposer dans l'ordre le plus naturel, de les rappeler à un certain nombre de faits principaux dont les autres ne soient que des conséquences.

[6] Ce sont-là les faits que le physicien doit sur-tout chercher à bien connoître: il ne sauroit trop les multiplier; plus il en aura recueilli, plus il sera près d'en voir l'union: son objet doit être d'y mettre l'ordre dont ils seront susceptibles, d'expliquer les uns par les autres autant que cela sera possible, & d'en former, pour ainsi dire, une chaîne où il se trouve le moins de lacunes que faire se pourra; il en restera toûjours assez; la nature y a mis bon ordre.

did he envisage? And how might the few "principal facts" be identified among the innumerable others? D'Alembert did not discuss these questions any further, nor can we turn to his own experimental practice for possible illustration, simply because he had none. Rather, for a more detailed understanding, it will likely be fruitful to consider research of others that he witnessed around him as pursuing experimental procedures or discussions that he then described in these terms.

Prior to doing so, I shall summarize several points that are relevant to the overall question. For experimental physics, d'Alembert described quite a specific epistemic approach. To recall, experimental physics, as he understood it, "is properly nothing but a systematic collection of experiments and observations" (*Discours préliminaire*, 1751, vii, transl. 2009).[7]

However, this collection had several specific characteristics: it was aimed at establishing principles rather than physical causes and did so by "multiplying" phenomena, and forming chains as close as possible. It aimed at identifying, among the phenomena, a few "principal facts," taking those facts as "principles," and "reducing" all other phenomena to these.

Despite the open questions, we see here an attempt to spell out a procedure that leads from the particular to the general in a quite specific way, or, put otherwise, a procedure that we might call—should we choose to use that terminology—epistemic analysis by way of induction. However, it should be emphasized that that procedure was regarded as constituting much more than dull, enumerative induction, and the items to which the induction process was directed were not the physical causes but the generalized facts or "principles" that were regarded as located on the same epistemic level as the phenomena.

As a side note, it is significant that d'Alembert was not alone in such a characterization. His long-standing companion Denis Diderot, in his *Pensées sur l'interprétaton de la nature*, of 1754 (sometimes regarded as the complement to d'Alembert's *Discours préliminaire*), gave quite a similar presentation of experimental physics that he clearly distinguished, much as d'Alembert did, from the mathematical sciences. Speaking of the procedure of experimental physics, Diderot emphasized the necessity of repeating and multiplying experiments since, as he emphasized, an individual experiment did not count much. He saw it as necessary to bring them in order by creating a chain ["chaine"] of experimental facts and to "multiply" experiments. Through such a procedure, a "reduction" of effects ["réduction des effets"] could finally be achieved (*Pensées* 1754, § 44, 116–120, my translation; see also Diderot 1999, 61). I shall not discuss those points in detail, but the quotes offered may suffice to illustrate the high degree to which Diderot shared d'Alembert's views on the proper approach to experimental physics. Regrettably, however, he also resembled d'Alembert in that he did not further explicate precisely what he meant by "multiplying" experiments and the "reduction" of effects.

---

[7] recueil raisonné d'expérience et d'obsérvations…

11.3 Experimental Physics in Paris and the Case of Dufay

What did d'Alembert find around him in Paris? In his article "Expérimental," he welcomed the fact that the university finally had created a chair of experimental physics (1756, 300–301). It was conferred on Jean-Antoine Nollet, who had worked broadly in experimental physics and published a six-volume course in that field. He had become best known for his original work in electricity and his (later) dispute with Franklin on those matters. However, Nollet's work contains few considerations as to how knowledge should be created. Moreover, his work is less systematic than d'Alembert's statements would suggest, providing no help towards understanding such ideas as multiplying, creating a chain, reducing, or establishing central facts.

This is not the case for one other very prominent experimental researcher with whom Nollet had collaborated for a time, often referred to as his mentor or teacher—in particular, in the field of electricity. Charles Dufay, intendant of the Royal Botanic Garden, member of the Académie Royale des Sciences, and well established in the Parisian academic scene, was broadly active in various fields of research, from botany and astronomy via dyestuffs and fire pumps to dew, fluid dynamics, and magnetism.[8] Of course, the *Jardin Royal* offered ample and extraordinary resources for all these activities. In a brilliant experimental series on luminescent minerals (Bolognese stones, as they were called), Dufay had demonstrated his interest in working with research fields that lacked even basic classifications (Daston 1997). His most famous and spectacular contribution, however, had been to electricity: he had worked on this field for over 5 years and published six memoirs that had spectacularly altered the entire science of electricity and raised broad interest all over Europe. It is safe to assume that d'Alembert was familiar with Dufay's printed memoirs, though it is unlikely that he met Dufay in person: Dufay had already died in 1739—that is, before d'Alembert had been employed at the Paris academy in 1741.

Within the bulk of his memoirs on electricity, Dufay inserted several methodological remarks on how empirical research should proceed. My claim that d'Alembert relied on Dufay in his idea of experimental physics derives from some striking resemblances between Dufay's statements and those of d'Alembert.

> It would mean attempting the impossible, would one look for the causes before one had discovered the mass of phenomena, ..., of which we have seen how they [the phenomena] derive from a small number of simple and invariant principles. (Dufay, 4. mém. 1734a, b, 477)

> Here we have the principles or, if one likes, the simple and primitive facts ("les faits simples & primitives"), to which all experiments on electricity can be reduced ("réduites") that so far are known. (Dufay, 6. mém. 1734a, b, 525)

What we see here is, first, a clear distinction between causes (the investigation of which must be postponed to a later stage) and principles (to which the phenomena can be "reduced" and the investigation of which must come first), with clear priority afforded to principles. Moreover, we see the idea that these principles can be

[8] For an overall account of Dufay's life and research, see Heilbron 1971.

exhibited by a small number of "simple" or "primitive facts" to which all other phenomena should be "reduced"—in other words, the idea that the principles from which the understanding of all phenomena should be derived were located on the same epistemic level as those phenomena themselves. In speaking of "causes," by contrast, Dufay pointed to other items, as I shall discuss below.

Again, those quotations leave many questions open. However, and in contrast to d'Alembert, Dufay was a prolific experimenter. Hence, to understand his meaning in greater detail, we may examine in detail some exemplary cases of his experiments and the way in which he drew consequences from them.

Dufay's electrical research can be understood only against the background of the state of electrical knowledge as Dufay found it in the 1730s. Numerous electrical experiments, often spectacular, had been performed, presenting a rich and sometimes wild variation. All those who had been dealing with electricity before and contemporaneously with him—for example, Guericke, Boyle, Hauksbee, and Gray—had been dealing with or even focusing on the question of what process lay behind the phenomena: a continuous stream of fluids, a vortex movement, particles attracting or repelling each other or the like? They aimed at discovering the causes, understood as being the processes, entities, or forces that were responsible for the phenomena. These causes were constantly taken for granted as lying beyond the visible realm—that is, on a different epistemic level than the phenomena.[9] However, such considerations were speculative and hypothetical, based on general (and diverse) views of nature's hidden workings. The Cartesian view (that the world was entirely filled with subtle fluids) and the Newtonian outlook (that minute particles attracted or repelled each another through empty space) were merely prominent examples. A wide range of proposals had been advanced, all of which could explain specific experiments and thus could claim some empirical success. However, none of them had received broad experimental support, and any decision between these views was to be made on the basis of theoretical preferences rather than empirical support. This was likely the mode of reasoning that d'Alembert sought to exclude in banning hypotheses.

When Dufay decided to turn his research to electricity,[10] he began with a systematic reading and careful summary of the state of research and, as a result, wrote the first ever 'history of electricity' (Dufay 1. mém. 1733a). Significantly, he focused almost exclusively on the empirical aspect without elaborating on the many 'theoretical' considerations that others had given, concerning the hidden processes behind the phenomena. This was also characteristic of his ensuing research agenda: the list of questions that he posed is most indicative (Dufay 2. mém. 1733b, 73)

- Which bodies/materials become electric by rubbing?
- Which bodies/materials become electric by transmission (contact or close approach)?
- Which bodies/materials enable or hinder the transmission of the electric virtue?

[9] For both a broad picture and a more detailed account, see Heilbron 1979, with many examples.

[10] Heilbron 1979, Chapter 9 provides a brief account of Dufay's research in electricity.

- Which bodies/materials get attracted more or less strongly?
- What factors (vacuum, air pressure, temperature, moisture, etc.) affect the strength of the electric effects?
- What is the relationship between electric effects and luminous effects of electric bodies?

The list clearly demonstrates that his task was to establish, for the first time, a solid foundation in a highly systematic way, and in this he succeeded: in John Heilbron's words, Dufay found a field full of weeds and left a well-ordered garden (Heilbron 1979, 252).

His approach to accomplishing this was not theoretical in the sense characterized above (i.e., in searching for causes) but was thoroughly and most broadly experimental. In light of the methodological reflections quoted above, it is worth examining precisely how he proceeded. In general terms, his procedure may be described as a systematic and very broad experimentation. The general guideline was the systematic and highly controlled variation of various parameters: materials, conditions, constellations, sizes, etc., with the aim of establishing general empirical rules. In some cases, he even invented new concepts to enable the formulation of such rules. In short, what we see may be labelled "exploratory experimentation" at its best (Steinle 1997, 2016, Ch.7).

For my present purpose, I shall illustrate the general point with a brief examination of his earliest enterprise: his analysis of the question of which materials could be electrified by rubbing (Dufay 1733b). Dufay was not the first to be interested in the question, and others had provided lists of materials that could or could not be electrified by rubbing. Clearly, Dufay was dissatisfied with these lists and began to systematically expand them with all the materials available to him. Given his environment at the Jardin Royal, numerous materials were at his disposal, and he tried glass, wood, wax, ivory, feathers, bones, hair, silk, wool, cotton, linen, sugar, paper, parchment, metals, pearls, magnets, sandstone, diamond, brick, animal fur, and many others. As a result, he formulated a bold conclusion: "All bodies, with the exception of metals and those bodies that cannot be rubbed, can be electrified by rubbing" (Dufay 1733b, 80).

It will be instructive to examine Dufay's procedure more closely here. Dufay formulated an 'all'-statement as a result of empirical research—that is, of induction. Furthermore, at first glance, it appears that this was just the simplest form of induction, namely enumeration. However, as has long been well known, induction by enumeration does not provide a valid conclusion but is invariably subject to empirical failure as new cases are tested. Had Dufay been epistemically naive? Based on a closer examination of his paper, I suggest, rather, that he had indeed been aware of the problem and pursued a procedure that allowed him to at least mitigate it. My argument is as follows: in his hundreds of experiments, Dufay not only varied and enumerated the materials but paid special attention to the conditions under which electrification succeeded particularly well. Beginning with the well-known advice that electric experiments were most likely to succeed when the environment was warm and dry, he went further and confirmed that the materials could be more easily electrified by rubbing when they were heated and dried. He took these conditions to the extreme in some cases; indeed, some materials that had initially been characterized as non-electrifiable were later revealed to be electrifiable if only they were sufficiently heated and dried. There exists a list in Dufay's hand, with clear experimental results, showing numerous instances of materials that were initially on the "negative" side but were subsequently deleted from there and show up on the "positive" side: examples include the entries "la porphyre" and "l'aimant."[11] The "negative list" was not entirely empty in the end, however. Dufay may simply have been less than assiduous in maintaining it, but one of the main entries that remained is 'the metals'—that is, the class of materials that Dufay named as the essential exception to his general rule. The procedure that we see here goes beyond enumerative induction: The experience that even those materials that first appeared as not producing electricity when being rubbed finally did so, provided that the conditions were strictly imposed (i.e., when they were heated and dried enough), encouraged and supported his conclusion that "all" materials—except one particular class—could be electrified by rubbing. I shall return to this striking case of induction later.

I have analyzed another line of Dufay's research elsewhere (Steinle 2006): his investigation of electric attraction and repulsion that resulted in his proposal—or discovery—of the twofold nature of electricity. Again, he had worked in a largely exploratory manner, but here he was aiming not at an "all"-claim but, rather, a law: the law of electric attraction and repulsion. It was the difficulties of that enterprise that finally led him to introduce the notion of two electricities as the only means that allowed the formulation of such a law that comprised all of his numerous experimental findings (Dufay 1733c).

Dufay thus worked intensely for several years and published his results in a series of six remarkable *mémoires* to the Paris academy. In one particularly striking case—the announcement of two electricities—he even sent the *mémoire* to the Royal Society, where it was translated and quickly printed in the *Philosophical Transactions* (Dufay 1734a).

This paper was particularly notable for his reflection on the status of his results. In the final section of his last *mémoire*, he summarized his achievements as follows: "Our research has not given us knowledge of the physical and primordial causes of electricity, but led to the discovery of several principles." He presented 16 of these principles, including the following:

1. All bodies, with the exception of metals and those bodies that cannot be rubbed, can be electrified by rubbing.
2. All bodies, with exception of the flame, become electric by transmission.
5. Bodies that are electric by themselves (*idio-électriques*) are least apt to transfer electricity over a distance, moist bodies are most apt.
9. There are two types of electricity, the vitreous and the resinous. Bodies charged by the same repel each other, those charged with the other attract.

[11] The document is kept in Archives de l'Académie des Science, Paris, Dossier Dufay, cahier "Electricité, par M.r DuFay"

11. Moist air destroys electricity and diminishes all effects.
12. Electric bodies exert their action also in the void.
14. All bodies with considerable electricity also shine with light.

At the end of the list, Dufay summarized with the following sentence:

> Here we have the principles or, if one likes, the simple and primitive facts ("les faits simples & primitives"), to which all experiments on electricity can be reduced ("réduites") that so far are known (Dufay, 6. mém., 1734b, 523–5).

It is worth noting that all of these principles or "simple facts" were generalized empirical statements, as we would call them, that Dufay put them center stage since all other experiments could be "reduced" to them, and that he kept that type of "reduction" carefully apart from the explanation by "physical and primordial causes."

11.4 Principles, Causes, and Experimental Approach

Although I cannot point to a "smoking gun" document, it appears highly plausible to assume that the procedures Dufay pursued in his electrical research illustrate and explicate not only his own epistemological positions but also those of d'Alembert: after all, d'Alembert used precisely the same words and insisted on the same distinctions. In recapitulating those positions, I shall highlight two points:

First, a sharp differentiation between principles and causes is evident. By causes, sometimes called "physical causes," both Dufay and d'Alembert meant those physical entities and processes that worked beyond the visible level and brought about the effects that could be seen as phenomena. In electricity, streams or vortices or particles were discussed; in chemistry, rather, corpuscles, like those that Boyle imagined; in optics, corpuscles as Newton had them; and so on. These causes were regarded as existing on a level that differed to that of phenomena, and in any case, they were not accessible to the senses. This specific understanding of causality was commonplace at the time (and for centuries to come).

When Dufay spoke of "principles," by contrast, he meant general statements, gained from the generalization of experimental findings. In some cases, he called them also "laws." I have also diagnosed such a use for d'Alembert, and this may present a suitable opportunity for a wider discussion.[12] The notion of "principle" has an immensely wide scope in history. In its first, epistemological meaning, the notion simply denotes a general starting point for deductions of various types. Other meanings have emerged, however, such as in chemistry, wherein the "principles" of a substance or drug are what make it efficacious—that is, as the cause of a specific property—a meaning that dates back at least to Paracelsus. In Joel Klein's contribution to this volume, this meaning emerges as central. Within that variety, the

[12] Many thanks to Jutta Schickore for drawing my attention to the need for such a discussion!

meaning used by Dufay and d'Alembert (as I claim) is quite specific: first, it is used only in an epistemological context, denoting the specific status of some propositions. While this is evident in Dufay's work, in d'Alembert's work it becomes clear when all the above quotes are taken together. Second, the specific status that they bear is that of a generalized phenomenon or an empirical law, in explicit contrast to what they regard as the physical cause. As I shall indicate in the next section, this understanding already had a visible tradition in France at their time.

The second point that I wish to emphasize concerns the pathway taken to arrive at those principles—that is, the experimental procedure and approach to reasoning. Dufay's work reveals a highly systematic procedure, characterized by a systematic and dense variation of many parameters. I argue that this should be taken as an explication of that which d'Alembert called "multiplying experiments" and "forming chains as close as possible." In this procedure, the experimental field should be explored as dense and as complete as possible, nothing should be left out, the researcher must not be selective. The following quote from Dufay highlights this point:

> [electricity] was up to now only known by some very complicated experiments, depending on bizarre circumstances, from which one could have particular judgements on certain questions, but in which nearly nothing of positive certainty was found. Today, it is perhaps a quality of matter in general, depending on invariable principles and subordinated to exact laws. (Dufay 6. mém. 1734b, 525–526, my translation)[13]

Here, Dufay contrasts his own, systematic procedure with the selective attitude towards experiments that he saw in those researchers before him. It was only his own, systematic experimental procedure that led to the principles and to certainty, while the selective procedure was mostly used in arguments about causes, and never led to any certainty. Again, it appears highly plausible that this was also d'Alembert's understanding.

Hence, in mid-eighteenth-century Paris, we see quite a specific understanding of the pathway from the particular to the general, concerning both the understanding of what type the general should be, what the empirical procedure looked like, and what type of experimenting and reasoning should be practiced. To be sure, and as the case of Nollet illustrates, that understanding was not the only one, and it is still an open task to determine how widespread it was.

Before going further, a general note on the notion of principle is appropriate. In a recent attempt to focus on that notion in the context of the Early Modern period, Peter Anstey synthesized several articles that illustrated the broad uses of 'principle' in various different fields: from mathematics to alchemy to experimental philosophy, from theology to metaphysics and jurisdiction (Anstey 2017a, b). In view of that puzzling variety, Anstey concluded that "there is no single or fundamental notion of what a principle is" in that period (2017a, b, introduction). While I agree

[13] '... jusqu'à présent n'étoit indiquée que par quelques expériençes très-compliquées qui l'avoient fait juger particuliére à certaines matiéres, & dépendante de circonstances bisarres, & dans lesquelles il ne se trouvoit presque rien d'assuré ni de positif. Aujourd'hui c'est peut-être une qualité de la matiére en général dépendante de principes invariables, assujettie à des loix exactes'.

to that general observation, one basic distinction might nevertheless be mentioned: in most cases, the term "principle" had an epistemological use, generally denoting those statements (of most varying types) on which, in an advanced stage of philosophy, deductions could solidly be based. In other domains—most prominently, chymistry—principles were understood as those constituents of matter that formed the basis of all changes and transmutations: Paracelsus' triad of salt, sulfur, and mercury was a famous but rather late example of a much longer tradition (Newman 2017). Given that broad range of meanings, the meaning that emerges in Dufay and d'Alembert appears relatively specific and sharp.

11.5 Seventeenth-Century Roots

The specific notion of principles that we see here gives rise to a historical outlook that goes further back. Within the Paris tradition of experimental physics, we find a striking and at the same time very prominent case.

In the early phase of the Paris academy, Edme Mariotte, one of its founding members, was its most prolific experimenter: Condorcet would later praise him as the one who brought experimental physics to France (Condorcet 1773, 49). Mariotte had worked on such diverse topics as collision of bodies, pneumatics (with the so-called Boyle–Mariotte law), hydraulics, colors, vision, pendulum and others. He not only practiced experimental research but also inserted, time and again, considerations of what could and should be achieved by such research.

What is significant for our purpose here is that he introduced, already in his early writings, a specific epistemological category for generalized empirical statements: the "principes d'éxperience," which we might translate as "principles of experience" or just "empirical principles" (Mariotte 1673, 179, 267, among others). As an early example, he mentioned the statement that weights, when they fell on a support, exerted a greater impact than when they were resting.

Briefly later, in his *Essay de Logique*, he presented a systematic account of those "Principes" (Mariotte 1678, reprint Picolet 1992: 89–90). He highlighted that for empirical research, those principles were central. They could not be based on intellectual grounds (such as in mathematics) but only on experience, and he called them "Maximes ou regles naturelles, ou principes d'expérience." As examples he mentioned both general statements, such as the rectilinear propagation of light, the unequal action of equal weights at unequal distances at the balance, the motion of iron towards magnets, and the heating effect of friction, and more specific regularities, such as the refraction of light from air into water towards the perpendicular or the proportionality of compression and weight of the air (later called "Mariotte's law" in France) (ibid., 91–3, 127, 132). Most of those empirical generalizations (as we would call them) were supported by experiment and not embedded in a larger theoretical framework. Mariotte knew well, moreover, that his contemporaries sometimes called statements of this type "laws of nature." In one case, referring to a relativity principle in collision (of Huygens' type), Mariotte even spoke explicitly of "un principe d'expérience, ou loy de la nature" (ibid., 99, cf. Mariotte 1686, 79). Already earlier, in presenting his famous proposition on pneumatics in 1676, he had introduced it as "une règle certaine ou loi de la nature" (Mariotte 1923, 8).[14]

It is important to note that the statement's reliability rested on experiments, carried out by Mariotte and the instrument maker Hubin, and that there was no explanation provided from principles such as the microscopic properties of air. Mariotte did indeed aim at such an "explication" in the long run, but admitted that, for the time being, it could not be achieved (ibid., 38). Opposing the idea that the expansion of air was caused by increasing distance between its particles, he favorized an analogy to the behavior of cotton wool, an idea likely taken from Pascal (without mentioning his name: ibid., 47–48). The status of the proportionality as a "principe d'expérience," however, remained unaffected by the lack of such an explanation.

What we see here in the last third of the seventeenth century is a specific approach to experimental reasoning and its goals that was considerably more epistemologically elaborate and differed significantly from what was evident in English experimenters, such as Boyle, during approximately the same period. Moreover, the similarities of what we see in Mariotte and later in Dufay are striking: they had the same specific understanding of what the first goals of experimental physics should be, and for that purpose made a sharp distinction between principles (call them empirical principles, viz. Mariotte's "principe d'expérience") on the one hand and physical causes on the other.

I have yet to investigate the possible historical pathway from Mariotte to Dufay in depth, but it is highly plausible that such a pathway existed. While Mariotte's publications originally appeared six decades before Dufay's, an edition of his collected works had been published in 1717—that is, shortly before Dufay commenced his experimental work. It is unlikely that Dufay failed to take notice of those works by his late colleague who was not only extremely prominent in general but was still authoritative on the topic of experimental physics. Mariotte might have served as a direct source and inspiration for Dufay's experimental approach and his epistemological framing. Even if this was not the case, there was likely a strong and highly visible epistemological tradition, from Mariotte onwards, taken up by Dufay and recognized by d'Alembert (who certainly was aware of the second edition of Mariotte's collected works in 1740) and others. This would also clarify why d'Alembert could refer to that tradition and its terminology without further explication: he could take it for granted that his readers, at least in France, understood what he was speaking about.

[14] On the problem of speaking of 'laws of nature' in that period, see Steinle 2008.

11.6 A Brief Look at Newton

The specific contours of such a tradition are clarified even further if we put it together and contrast it with what, only briefly after Mariotte, had been acknowledged as the proper mode of experimental reasoning in a highly prominent setting—namely, in England. In the 1706 Latin version of his *Opticks* (and in all further editions), Newton had already used the framework of analysis and synthesis to describe the reasoning in experimental philosophy (and might have been the first to do so). Below, I quote from the final version:

> As in Mathematicks, so in Natural Philosophy, the Investigation of difficult Things by the Method of Analysis, ought ever to precede the Method of Composition. This Analysis consists in making Experiments and Observations, and in drawing general Conclusions from them by Induction .... By this way of Analysis we may proceed from Compounds to Ingredients, and from Motions to the Forces producing them; and in general, from Effects to their Causes, and from particular Causes to more general ones, till the Argument end in the most general. This is the Method of Analysis: And the Synthesis consists in assuming the Causes discover'd, and establish'd as Principles, and by them explaining the Phænomena proceeding from them, and proving the Explanations. (Qu. 31 of 4. ed. *Opticks*, 1730)

The general structure outlined here was quite clear: analysis was defined as induction from the particular to the general and synthesis or composition seen as deduction. It is also clear that Newton used the "analysis–composition" couple to point to analogues in both mathematics and chemistry. Leaving the detailed study of Newton's position to Alan Shapiro (this volume), I shall highlight only two points:

First, the goal of analysis/induction was clearly defined: it was designed as a means of identifying the causes of phenomena. The crucial question, however—that of precisely what he meant by causes—was not treated by Newton in that passage. However, we may learn more here when we examine his works:

- In dealing with motions, causes were considered to be the mechanical forces, such as general gravitation.
- In optics, Newton took the rays of light and their properties to be the causes of phenomena: not only their refrangibility but also their "fits" of easy reflection or refraction and, of course, their power to generate light and color sensation. Rays of light, however, were never directly accessible but lay behind the level of phenomena.
- In pneumatics, he took the attractive or repulsive forces between the minute particles as the causes of the properties of the air, such as its elasticity in the form of Boyle's law.
- Finally, in electricity, he saw—as did most of his contemporaries—hidden processes of subtle matter as causes for the phenomena that could be produced by experiments.

This all aligns well the notion of causes that Dufay and others mentioned, viz. the understanding that causes were the properties of the smallest units of matter, typically inaccessible to direct experience. It should be noted that such a notion corresponded well with what mathematical "analysis" (or calculus) aimed at with fluxions or infinitesimals.

For the synthesis, then, these causes should be taken as principles from which the phenomena could be explained. Newton's notion of principle clearly conformed to the general meaning as the starting point for deductions.[15] The more specific type of principles that Dufay and d'Alembert had in mind (as generalized empirical statements) is not present in Newton's statement. Although he was most probably aware of Mariotte's 'principles of experience', he certainly did not adopt that meaning. Accordingly, principles did not generally differ from causes: once the cause had been detected, he argued, it should be taken as principle—that is, as the starting point from which the individual phenomena could be explained.

My second point concerns the procedure deployed to arrive at the causes. Newton did not explicate that procedure beyond the unspecific notion of "induction" from experiments and observations. Rather, he detailed how the results of such an induction, once obtained, could be further supported or rejected by experiments. However, he offered no procedural advice as to how that induction should be performed in the first place. Again, we might look, for illustration, at Newton's own practices. Here, we can see that his general ideas were often inspired by individual experiments, but not in a systematic way. Rather, he had some general views (such as the corpuscular views of light, matter, or electricity) already in mind and was asking how individual experiments would specify or modify those general assumptions. In any case, however, we see nothing comparable to what d'Alembert etc. spelled out as a systematic procedure of multiplying, chain-forming etc., neither in Newton's statements nor his practices. As such, his famous quote shows a strikingly different idea of experimental procedure and experimental reasoning than we saw in the Mariotte–Dufay–d'Alembert tradition.

11.6.1 Reduction-Deduction vs. Analysis-Composition

To summarize, I shall compile several keywords pertaining to the two modes of experimental reasoning found in the historical sources (Table 11.2).

This counterposition, rough though it might be, clarifies how distinctly the understandings of the two-way process between experimental phenomena and general statements differed by the mid-eighteenth century. It should be emphasized that these differences were intimately connected to what type of statement was ultimately aimed at: analysis, in Newton's sense, aimed at determining the properties of the smallest or elementary aspects (like light rays, corpuscles, fluids, or the like), much as mathematical analysis aimed at infinitesimals and chemical analysis aimed at properties of the ultimate particles of matter. Since these were not accessible to direct experience, hypothetical guesses were inevitable. Principles, by contrast, if

[15] A closer analysis is given by Kirsten Walsh (2017).

Table 11.2 Two traditions of understanding the double pathway between phenomena and general statements

	Reduction–deduction (Mariotte, Dufay, d'Alembert, ...)	Analysis–composition (Newton, many others, ...)
Epistemic goal	(Empirical) principles, *principes d'éxperience*	(Hidden) causes and forces
Epistemic procedure	Broad and systematic experimentation according to certain guidelines: "multiplying" phenomena, forming chains,	Inductive guesses from experiment and observation, not further specified
Mode of reasoning	Identifying a few 'simple facts', taking them as principles	Corroborating the hypotheses about causes by further experiments
Scope of experience	Comprehensive, leaving nothing out	Selective

understood in the Mariotte–Dufay–d'Alembert tradition, were generalizations of experience, and the way to arrive at them had to stay as closely as possible to a most dense field of experiments. The goals of induction and the inductive pathways were inseparably interwoven.

To be sure, I may have drawn the lines between the two approaches more sharply than is appropriate for all cases. My task was mainly to show that such different understandings existed, and this does not exclude that in some cases there might be intermediary cases and less clear terminology. I have indicated all this drawing on the examples of just a few prominent and highly visible researchers. To enhance the historical understanding of how broadly those approaches were shared and whether there were regional preferences or local traditions and so on, a far wider study that includes more experimenters from all over Europe is needed. Among other things, it would be of interest to identify where and when the term "analysis" was used in its epistemological meaning, as we see in Newton, and how it was related to the use in other fields such as chemistry or mathematics. A similarly broad investigation would be necessary for the term "principles" with all its variations. Such historical research would help us to identify epistemological traditions, similar to Jutta Schickore's work over several years now (see Schickore 2007 and 2017 as prominent examples), and this could in turn enrich our understanding of the few cases I have discussed here. Jutta Schickore (this volume) gives a profound view on the German-speaking landscape of the eighteenth century.

In conclusion, and to place these questions into their wider context, I wish to point out that such a counterposition went far beyond the eighteenth century. In nineteenth-century electromagnetism (with Ampère and Faraday, among others, see Steinle 2016) and optics (on Brewster as a striking example, see Steinle 2024), we find clear reflections of that counterposition, both in research practice and in methodological reflection. For the twentieth century, one might think of Einstein's famous distinction between "principle theories" and "constructive theories" in physics (Einstein 1919)[16]: by "constructive theories," he meant those that started with basic hypotheses—such as that of molecular motion—and aimed at deriving phenomena from them, while "principle theories" were based on general statements—"principles"—derived from experience and experiment. The degree to which that distinction matches the counterposition I have sketched in this paper remains to be determined, but already its basic structure indicates a significant resonance. Questions as to how the pathways that connect empirical or experimental findings with general statements and with the ultimate causes of things may be constructed have been challenging for all reflections on the empirical sciences.

Acknowledgments I wish to express my gratitude to the organizers of the workshops and the volume and to all participants and contributors for the enriching discussions. Special thanks go to Jutta Schickore for her detailed and thoughtful critique of the first draft of this paper, which helped me a lot in situating my claims, and to Bill Newman for carefully examining the final draft.

References

Anstey, Peter R., ed. 2017a. *The idea of principles in early modern thought: Interdisciplinary perspectives*. Routledge Studies in Seventeenth-Century Philosophy. New York: Taylor & Francis Ltd,.

———., ed. 2017b. Introduction. In *The idea of principles in early modern thought: Interdisciplinary perspectives*, 1–15. New York: Taylor & Francis.

Condorcet, Marie Jean Antoine Nicolas Marquis de. 1773. Eloge de Mariotte. In *Eloges des Académiciens de l'Académie Royale des Sciences, morts depuis 166, jusqu'en 1699*. Paris: Hotel de Thou.

d'Alembert, Jean-Baptiste le Rond. 1751. Discours préliminaire des éditeurs. In *Encyclopédie, ou Dictionnaire raisonné des Sciences, des Arts et des Métiers*, ed. Denis Diderot and Jean le Rond d'Alembert, Tome 1:(NA9), i–xlv. Paris.

———. 1756. Expérimental. In *Encyclopédie, ou Dictionnaire raisonné des Sciences, des Arts et des Métiers*, ed. Denis Diderot and Jean le Rond d'Alembert, t. 6, 298–301. Paris.

———. 2009. *Preliminary discourse. The encyclopedia of Diderot & d'Alembert collaborative translation project*. Trans. Richard N. Schwab and Walter E. Rex. Ann Arbor: Michigan Publishing, University of Michigan Library. http://hdl.handle.net/2027/spo.did2222.0001.083. Accessed 28 Jan 2023. Originally published as "Discours Préliminaire," Encyclopédie ou Dictionnaire raisonné des sciences, des arts et des métiers, 1:i–xlv (Paris, 1751).

Daston, Lorraine. 1997. The cold light of facts and the facts of cold light: Luminescence and the transformations of the scientific fact, 1600–1750. In *Signs of the early modern, part 2: 17th century and beyond*, ed. David Lee Rubin, vol. 3, 17–44. Studies in Early Modern France. Charlottesville: Rookwood Press.

Diderot, Denis. 1754. Pensées sur l'interpretation de la nature.

———. 1999. *Thoughts on the interpretation of nature and other philosophical works*. Manchester: Clinamen Press.

Dufay, Charles François Cisternai. 1733a. Premier mémoire sur l'électricité. Histoire de l'électricité. In *Histoire de l'Académie Royale des Sciences, avec les Mémoires de Mathématique & de Physique pour la même année*, 23–35.

[16] Many thanks to Theo Arabatzis and Giora Hon for directing me to that dichotomy. For analysis of the background to the distinction, see Giovanelli 2020, among others.

———. 1733b. Second mémoire sur l'électricité. Quels sont les corps qui sont susceptibles de l'électricité. In *Histoire de l'Académie Royale des Sciences, avec les Mémoires de Mathématique & de Physique pour la même année*, 73–84.
———. 1733c. Quatrième mémoire sur l'électricité. De l'attraction et répulsion des corps électriques. *Histoire de l'Académie Royale des Sciences, avec les Mémoires de Mathématique & de Physique pour la même année*, 457–477.
———. 1734a. *A letter from Mons. Du Fay, F. R. S. and of the Royal Academy of Sciences at Paris, to his grace Charles Duke of Richmond and Lenox, concerning electricity*. Translated from the French by T. S. M D. *Philosophical Transactions* 38, Nr. 431: 258–266.
———. 1734b. Sixième mémoire sur l'électricité. Où l'on examine quel rapport il y a entre l'électricité, et la faculté de rendre de la lumière, qui est commune à la plûpart des corps électriques, et ce qu'on peut inférer de ce rapport. In *Histoire de l'Académie Royale des Sciences, avec les Mémoires de Mathématique & de Physique pour la même année*, 503–526.
Einstein, Albert. 1919. What is the theory of relativity? *Times*, November 28, 1919.
Giovanelli, Marco. 2020. "Like Thermodynamics before Boltzmann." On the emergence of Einstein's distinction between constructive and principle theories. *Studies in History and Philosophy of Science Part B: Studies in History and Philosophy of Modern Physics* 71: 118–157. https://doi.org/10.1016/j.shpsb.2020.02.005.
Heilbron, John L. 1971. Dufay (Du Fay), Charles François de Cisternai. In *Dictionary of scientific biography*, ed. Charles Gillispie, vol. 4, 214–217.
———. 1979. *Electricity in the seventeenth and eighteenth centuries: A study of early modern physics*. Berkeley: University of California Press.
Mariotte, Edme. 1673. *Traité de la percussion ou chocq des corps, dans lequel les principales Regles de mouvement, contraires à celles que Mr Descartes, et quelques autres Modernes ont voulu establir, sont demonstrées par leurs veritables Causes*. Paris: Michallet.
———. 1678. *Essai de logique, contenant les principes des sciences, et la manière de s'en servir pour faire de bons raisonnemens*. Paris: Michallet.
———. 1686. *Traité du mouvement des eaux et des autres corps fluides*. Paris: Michallet.
———. 1923. Discours de la nature de l'air – De la végétation des plantes – Nouvelle découverte touchant la vue. In *Les Maitres de la Pensée Scientifique*. Paris: Gauthier-Villars.
Newman, William R. 1998. The place of alchemy in the current literature on experiment. In *Experimental essays – Versuche zum experiment*, ed. Michael Heidelberger and Friedrich Steinle, 9–33. Interdisziplinäre Studien/Interdisciplinary Studies. Baden-Baden: Nomos Verlag.
Newman, William R. 2017. Alchemical and chymical principles: Four different traditions. In *The idea of principles in early modern thought: Interdisciplinary perspectives*, 77–97. New York: Taylor & Francis.
Newton, Isaac. 1730. *Opticks, or, A treatise of the reflections, refractions, inflections and colours of light A treatise of the reflections, refractions, inflections and colours of light*. 4th ed., Corr. London: Printed for W. Innys.
Picolet, Guy, ed. 1992. *Edme Mariotte: Essai de logique, suivi de l'écrit intitulé "Les principes du devoir et des connaissances humaines" attribué à Roberval*. Corpus des oeuvres de philosophie en langue française. Paris: Fayard.
Schickore, Jutta. 2007. *The microscope and the eye. A history of reflections, 1740–1870*. Chicago: University of Chicago Press.
———. 2017. *About method: Experimenters, snake venom, and the history of writing scientifically*. Chicago: University of Chicago Press.
Steinle, Friedrich. 1997. Entering new fields: Exploratory uses of experimentation. *Philosophy of Science* 64 (Supplement): 65–74.
———. 2006. Concept formation and the limits of justification. "Discovering" the two electricities. In *Revisiting discovery and justification. Historical and philosophical perspectives on the context distinction*, ed. Jutta Schickore and Friedrich Steinle, 183–195. Archimedes. Dordrecht: Springer.

———. 2008. From principles to regularities: Tracing "laws of nature" in early modern France and England. In *Natural law and laws of nature in early modern Europe. Jurisprudence, theology, moral and natural philosophy*, ed. Lorraine Daston and Michael Stolleis, 215–231. Aldershot: Ashgate.

———. 2016. *Exploratory experiments: Ampère, Faraday, and the origins of electrodynamics*. Pittsburgh: Pittsburgh Univ. Press.

———. 2024. Controlling induction: Practices and reflections in Brewster's optical studies. In *In elusive phenomena, unwieldy things. Historical perspectives on experimental control*, ed. Jutta Schickore and William Newman, 105–124. Archimedes. Cham: Springer.

Walsh, Kirsten. 2017. Principles in Newton's natural philosophy. In *The idea of principles in early modern thought: Interdisciplinary perspectives*, 194–223. New York: Taylor & Francis.

Friedrich Steinle is professor emeritus of history of science at the Institute of History and Philosophy of Science, Technology, and Literature at the Technical University Berlin. His research interests include history and philosophy of experimentation, concept generation, and the history of the study of electricity and colors.

Chapter 12
Analysis and Induction as Methods of Empirical Inquiry

Jutta Schickore

What the sciences need is a form of induction which takes experience apart and analyses it Francis Bacon

Abstract This chapter traces the history of the notions of analysis and induction in German-language accounts of scientific inquiry between the mid-eighteenth and mid-nineteenth centuries. Many eighteenth-century scholars perceived analysis and induction as intertwined and tied to methods of empirical inquiry and discovery, as they had been perceived in the early modern period. Until well into the nineteenth century, the terms "(scientific) induction" or "inductive science" referred to any experience-based science (as opposed to a science based on first principles). Inductive science, thus understood, encompassed the making, evaluation, and testing of hypotheses—even of hypotheses about unobservable things—and "reduction" rather than "deduction" was its opposite. While the Kantian understanding of analysis influenced the notions of induction and scientific method, it did not wholly alter their interpretation. After Kant, formal logic and philosophy gradually became independent, but a more applied philosophy of empirical inquiry also persisted, linked to scientific practice while rooted in older ideas about analytic and inductive methods. Several authors recommended evaluating and testing hypotheses by deriving testable consequences but did not consider hypothesis testing to be a strictly deductive process opposed to, or more powerful than, induction. Inductive inquiry and the hypothetico-deductive (H-D) method are both integral to the inductive sciences, according to these pragmatic theories of empirical inquiry.

Keywords Scientific method · Inductive science · Practical logic · Analysis · Discovery

J. Schickore (✉)
Department of History and Philosophy of Science and Medicine, Indiana University, Bloomington, IN, USA
e-mail: jschicko@iu.edu

W. R. Newman, J. Schickore (eds.), *Traditions of Analysis and Synthesis*, Archimedes 73, https://doi.org/10.1007/978-3-031-76398-4_12

12.1 Introduction

In 1934, philosopher of science Karl Popper published *Logik der Forschung* (translated as *The Logic of Scientific Discovery*), a seminal account of the scientific method. In the English edition, he introduced his approach as follows: "According to a widely accepted view—to be opposed in this book—the empirical sciences can be characterized by the fact that they use 'inductive methods', as they are called. According to this view, the logic of scientific discovery would be identical with inductive logic, i.e., with the logical analysis of these inductive methods. It is usual to call an inference 'inductive' if it passes from singular statements (sometimes also called 'particular' statements), such as accounts of the results of observations or experiments, to universal statements, such as hypotheses or theories" (Popper 2002, 3).

Popper equated induction with inductive *logic*, which he famously sought to replace with "deductive *logic*." A philosophical theory of scientific method was a theory of "the methods of deductive testing" (Popper 2002, 10). Deductive logic is the opposite of the "inductive logic" of scientific inquiry, whereby Popper insinuated that the latter was an impoverished notion of scientific method. Popper used the technical term "analysis" in the context of logical or linguistic analysis.

Readers familiar with early modern accounts of empirical inquiry will be immediately struck by the difference between Popper's notion and the famous methodological statement from Newton's *Opticks* that several contributors to this volume invoke: "As in Mathematicks, so in Natural Philosophy, the Investigation of difficult Things by the Method of Analysis, ought ever to precede the Method of Composition. This Analysis consists in making Experiments and Observations, and in drawing general Conclusions from them by Induction, and admitting of no Objections against the Conclusions, but such as are taken from Experiments, or other certain Truths" (Newton 1718, 380). For Newton, the method of analysis encompassed experiment, observation, *and induction*, and its goal was to discover the causes—eventually, the "most general" causes—of the phenomena we experience.[1]

Francis Bacon also considered analysis to be integral to induction. In the preface to the *Novum Organum*, he explained, "By far the biggest question we raise is as to the actual form of induction, and of the judgement made on the basis of induction," adding, "What the sciences need is a form of induction which takes experience apart and analyses it, and forms necessary conclusions on the basis of appropriate exclusions and rejections" (Bacon 2000, 17).

Statements such as Popper's shaped the philosophical understanding of scientific methodology, knowledge generation, the certainty and limits of knowledge, and the relationship between philosophy and the sciences and logic for most of the twentieth century. They also served as the lens through which philosophy of science was

[1] Alan Shapiro reminded me that the first edition of the *Opticks* of 1706 did not mention induction. For a close examination of the development of Newton's view of analysis, see Shapiro's contribution to this volume.

viewed prior to the twentieth century. Like Popper, twentieth- and twenty-first-century philosophers of science have often portrayed Newtonian and Baconian approaches as "inductive *logic*" and as deficient, praising their nineteenth-century antecedents as advocates of hypothetico-deductivism who were capable of overcoming these deficiencies. John Stuart Mill's main work is typically characterized as a *Logic of Induction*. Robert Butts also calls William Whewell's philosophy a "logic of induction"; others, however, have portrayed Whewell as an early advocate of the hypothetico-deductive (H-D) method of theory testing (Achinstein 2004).

Viewed through such a lens, the image of nineteenth-century philosophies of science becomes distorted and Newton's statement becomes obscure. Until well into the nineteenth century, the term "induction" had a considerably broader scope than the quotation from Popper's book suggests. The terms "(scientific) induction" or "inductive science" broadly referred to any science built on experience (as opposed to a science based on first principles), but they did not imply that inductive science properly conceived must proceed strictly through inductive inferences, let alone through induction by enumeration. By contrast, inductive science encompassed the making, evaluation, and testing of hypotheses, even of hypotheses about unobservable things.[2] Furthermore, if "induction" could be said to have an opposite, it was "reduction" rather than "deduction." This understanding of inductive science aligned more closely with the Baconian and Newtonian methodologies than with Popper's.

Moreover, many scholars prior to the nineteenth century framed their accounts of scientific method in terms of "analysis" rather than "induction." The process of discovery—empirical inquiry—was often described as a method of analysis or resolution. The concept of analysis resonated with several ancient traditions, including those described in Niccolò Guicciardini's, Helen Hattab's, and Alan Shapiro's contributions to this volume.

In this paper, I trace the history of the notions of analysis and induction in German-language accounts of scientific inquiry between (roughly) the mid-eighteenth and mid-nineteenth centuries. Numerous scholars in philosophy and the empirical sciences addressed these topics, many of them illustrious, prolific, and sometimes opaque writers, including Christian Wolff, Kant, Schelling, Johann Heinrich Lambert, Johann Friedrich Herbart, Schopenhauer, and Hegel. My account is thus rather condensed, as my aim is to offer some signposts for orientation in the complex history of theories of empirical inquiry in the eighteenth and early nineteenth centuries.[3] This chapter thus complements Hattab's, Guicciardini's, and Shapiro's chapters on pre- and early modern traditions and Steinle's on eighteenth-century developments, although it has a somewhat different emphasis. Hattab and Guicciardini are more concerned with rationalist, Cartesian and Leibnizian strands of thinking about method. Like Steinle, I examine how the notions of analysis and

[2] Laura Snyder has demonstrated that Whewell proposed what she calls a "discoverer's induction" (Snyder 1997). This is convincing, but her point may be more broadly applied, as we shall see.

[3] For an overview, see Tonelli (1976).

synthesis are deployed in experimentalist traditions. Unlike Steinle, who explores French physics, I focus on the German corpus—specifically, on the many German-language textbooks on "applied" or "practical logic" that were published in the decades surrounding 1800.[4] Contrary to that which the Popperian lens suggests, in the first and second thirds of the nineteenth century, both notions—analysis *and* induction—continued to exist in accounts of applied logic; they were not equivalents, and both existed in formal and non-formal guises until the early twentieth century.

I begin with a very brief review of those pre-modern and early modern notions of induction and analysis that are especially important for the later traditions I am investigating. For more in-depth discussions, I refer the reader to Hattab's, Guicciardini's, and Shapiro's chapters. Like Hattab and Guicciardini, I stress the diversity of meanings associated with these terms. I show that for many eighteenth-century scholars who wrote on applied logic and methodology, analysis and induction were closely tied together, and tied to methods of empirical inquiry and discovery, just as they were in the early modern period. We shall also see that the eighteenth-century understandings of "analysis" and "induction" were rather eclectic and infused with miscellaneous pre- and early modern ideas on these topics. Then, I shall highlight several key features of Kant's influential re-casting of "analysis" as a formal method of analysis of cognition (i.e., of the basic elements of thought and intuition). I describe the shift in the understanding of "analysis" and the transformation in the relations between logic, philosophy, and the sciences that Kant helped bring about.

While the Kantian understanding of analysis affected the notions of induction and scientific method, it did not completely change their interpretation. Several nineteenth-century German philosophers indeed followed Kant and confined "analysis" to formal and conceptual questions. Others, however, sought to distinguish a formal and a non-formal notion of analysis in empirical inquiry. Formal logic and formal philosophy gradually became independent fields of study. Alongside these formal disciplines persisted a more applied philosophy of empirical inquiry, which remained closely tied to the practice of science and was rooted in older ideas about analytic and inductive methods. I shall demonstrate that these notions continued to inform the discussions until well into the nineteenth century. The terminology was also debated, as some scholars, often influenced by the British tradition, dispensed with the term "analysis" altogether and merely used "inductive science" in their accounts of scientific inquiry. Others insisted that "analysis" was the appropriate term to use, while others still continued to use both.

As I shall demonstrate in the concluding sections of this essay, the term "inductive science" continued in use as a general term for empirical inquiry in major scientific works, such as Matthias Jakob Schleiden's 1848 edition of *Botanik als induktive Wissenschaft* and the 1840 translation of Whewell's history of science as *Geschichte der inductiven Wissenschaften*, as well as in discussions about the status

[4] These books remain largely unexplored in the history of philosophy of science.

of empirical psychology vis-à-vis philosophy. Most importantly, these nineteenth-century authors did not equate "inductive science" with "inductive logic," as Popper did in the 1930s. Nor did they simply advocate the H-D method of theory testing. On the contrary, they espoused a broad understanding of inductive science as encompassing both logical inferences and methodological strategies, including strategies for observation and experimentation, hypothesis formation, hypothesis testing, and evaluation.

12.2 Early Modern Notions of Analysis, Induction, and Discovery

In today's philosophy of science, scientific method is typically divided into two main forms of reasoning—inductive and deductive—whereby inductive reasoning is understood as ampliative, deductive reasoning as truth-preserving. The two main strands of philosophy of science are inductive-statistical approaches and H-D approaches to confirmation and test of hypotheses. Analysis (philosophical analysis), on the other hand, is understood as a general method of conceptual clarification, not specifically tied to scientific method, certainly not understood as a method of gaining empirical knowledge. As the contributions to this volume show, the pre-modern and early modern notions of analysis, synthesis, and induction were markedly different.

In a lucid discussion of the pre-modern history of "analysis," Hans-Jürgen Engfer has demonstrated that early modern thought was informed by several different understandings of analysis: the analytic or analytic–synthetic method of ancient geometry, the method of regression, and mathematical analysis.[5] Engfer highlights that the geometric model of scientific method emerged from early interpretations of Euclidean geometry, notably the two influential interpretations by Proclus and Pappus. Both interpretations are complex, ambiguous, and themselves difficult to interpret.[6] For the purposes of this essay, it is important to note that according to Proclus's interpretation, Euclid's *Elements* exemplify a synthetic method. According to Pappus, however, Euclidean methods are analytic. Understood as synthetic method, the geometrical method as modeled by Euclid's *Elements* proceeds as a deduction of conclusions from true, evident premises—an unfolding of the rich consequences of the premises. Understood as analytic or analytic–synthetic method, the geometrical method proceeds from something posited by ascending to premises and principles, on the basis of which that which is posited can in turn be demonstrated.

[5] Engfer (1982) also identifies a fifth tradition, going back to Raymundus Lullus, that influenced early modern linguistic theories. I shall not discuss this strand here, although traces of it may be seen in eighteenth-century books on applied logic.

[6] Besides Hattab, Guicciardini, and Engfer, see also Hintikka and Remes (1974).

In the *Nicomachean Ethics*, Aristotle proposed a derivative notion of analysis that applied to deliberations regarding means and ends.[7] The relevant passage reads as follows:

> And we deliberate not about our ends but about the things which contribute to our ends; a doctor does not deliberate whether to heal, or an orator whether to persuade, or a statesman whether to create good order in society, nor do any of the rest deliberate about their end; rather, having posited their end, they consider how and by what means they can achieve it. And if it appears that it can come about in several ways, they consider by which it will be best and most easily achieved, and if it can be achieved by only one, how it will come about by means of that and by what means that in turn will come about, until they reach the first cause, which is the last thing to be discovered. The deliberator seems to investigate and analyse in this way as one does a geometrical problem (it seems that not every investigation is deliberation, e.g. mathematical investigations, but all deliberation is a sort of investigation), and the last step in the analysis is the first in bringing about the result. (Book III, 3., Taylor and Taylor 2006, 22)

This notion of analysis also informed subsequent writings on method—for instance, the Galenic understanding of medical practice[8]; and much later, Ernst Mach still framed his understanding of analysis in such terms, as we shall see in the concluding section of this chapter.

These Greek texts leave several questions open. Is the geometrical method as a whole analytic or synthetic, or does it comprise both processes? Which of the two processes is associated with discovery (the generation of knowledge)—analysis, the intuitive grasping of premises, or synthesis, the unfolding of consequences from premises? Is synthesis the strict reversal of each step of the analysis, or might it proceed along different steps? Are geometrical and mathematical, logical, and physical analyses the same or distinct from one another? Discussions regarding how these Greek sources might best be interpreted continue today. The noteworthy point here is that the ambiguity of the Greek texts and the inconsistent terminology used in their interpretations contributed to the terminological diversity in eighteenth-century understandings of method.

An influential early modern understanding of analysis, which was particularly important for theories of scientific inquiry, may be traced back to Zabarella's account of scientific method.[9] For Zabarella, the most important kind of resolution was "demonstratio ab effectu," which demonstrates causes from effects. This resolution is possible because any phenomenon that we observe—an effect, something that is familiar to us—also "contains," in some manner, its cause. Therefore, an analysis of the thing present can reveal the cause (understood as a property). This constitutes a discovery that leads us to causes or principles. These, in turn, serve as the basis for the demonstration of the effect. According to Zabarella, this process is not circular given that initially, we know the phenomenon indistinctly—that is, we

[7] See Oeing-Hanhoff (1971, 238–240).

[8] On the Aristotelian–Galenic tradition of anatomy, see also Tawrin Baker's contribution to this volume.

[9] On this, see Engfer (1982, 90–94); see also Hattab (2014, 2021).

know only that it exists. Once we know its cause, we know the phenomenon distinctly and understand its nature.

Zabarella notably acknowledged that between discovery and the demonstration of the effect, there is a third phase of consideration, whereby the putative discovery is probed for plausibility.[10] He further emphasized that the method of regressus was specific to empirical inquiry. The mathematical regressus or analysis differed from the scientific in that the regressus proceeds to principles that are known to us (Engfer 1982, 95).

The history of theories of scientific inquiry is further complicated by the fact that in pre-modern understanding, scientific inquiry includes the demonstration of knowledge to others—that is, communication and instruction in an orderly fashion. Notably, for Zabarella, method is associated not only with the study of cause and effect but more generally with order—that is, with the arrangement of things to be treated in teaching and learning. According to him, the most effective order of teaching and learning is synthetic (Hattab 2021, 90).

This broad understanding of empirical inquiry is reflected in the definition of method in the influential Port Royal Logic.[11] Here, method is defined as "The art of arranging a series of thoughts properly, either for discovering the truth when we do not know it, or for proving to others what we already know, can generally be called method." The authors, Antoine Arnauld and Pierre Nicole, further distinguish two kinds of method, "one for *discovering truth*, which is called *analysis*, or the *method of resolution*, and which may also be termed the *method of invention*; and the other for explaining it to others when we have found it, which is called *synthesis*, or the *method of composition*, and which may be also called the *method of doctrine*" (Arnauld and Nicole 1861, 303). Analysis is then defined as consisting in "the attention we give to that which is known in the question we wish to resolve, ... the whole art being to derive, from this examination, many truths which may conduct us to the knowledge of what we seek" (Arnauld and Nicole 1861, 313).

Bacon's and Newton's writings about scientific inquiry echo these older understandings of scientific method. Bacon's *Novum Organum* was conceived as a logic responding to Aristotelian syllogistic logic (Cassan 2021, 256–7). In the final aphorism of the book, Bacon reminded his readers that

> in this *Organon* of ours we are dealing with logic, not philosophy. But our logic instructs the understanding and trains it, not (as common logic does) to grope and clutch at abstracts with feeble mental tendrils, but to dissect [*persecet*] nature truly, and to discover the powers and actions of bodies and their laws limned in matter. Hence this science takes its origin not only from the nature of the mind but from the nature of things; and therefore it is no wonder if it is strewn and illustrated throughout with observations and experiments of nature as samples of our art (Bacon 2000, 220).

[10] "And so, the first procedure, which is from effect to cause, having been performed, before we go back from it [i.e., the cause] to the effect, it is necessary that there intercede some third intermediate effort by which we are led into distinct knowledge of that cause, which was known only confusedly" (McCaskey 2013, 378).

[11] The authors of the Port Royal Logic did not build directly on Zabarella but on Descartes, who, in turn, drew on Zabarella.

Bacon thus conceived this "inductive logic" as a tool with which properly to connect experience and knowledge. It incorporates formal and psychological aspects and relies on the material–conceptual process of dissection.[12] For Bacon, induction was a complex methodological practice of working from experience or empirical data to more general ideas and propositions. As such, induction required analysis—exclusions and dissolutions, as Bacon called it. Notably, for him, empirical data are not sensations or that which is empirically given but *experiences*—that is, complexes of sensations, as in the Aristotelian tradition.

Bacon's account of how facts should be gathered is provided in *Of the Advancement and Proficience of Learning* and exemplified in his natural history writings.[13] In Book V of this work, Bacon carefully lays out different uses of "the Faculties of the Mind of Man," first distinguishing logic and ethics. Logic is further divided into "the Arts, of Invention; of Iudgment; of Memorie; and of Tradition" (Bacon 1640, 217–18).[14] The arts of discovery and invention comprise the discovery of arguments and the discovery of arts, the latter of which, according to Bacon, is both the most important and "wanting" (220). The discovery of arts, again, comprises two parts—"literate Experience" and "a New Organon" (220).

Chapter II of Book V is devoted to literate or learned experience itself (to that which Bacon called the "hunt of Pan"). This chapter details Bacon's methodology of inquiry. Learned experience is either pursued "in the light" of axioms or following some direction and order or by feeling out one's way in the dark. The method or direction "chiefly proceeds; either by *variation of the experiment*, or by *Production of the Experiment*, or by *translation of the Experiment*; *or by inversion of the Experiment, or by compulsion of the Experiment, or by Application of the Experiment, or by the Copulation of the Experiment, or else by the lots and chance of the Experiment*. And all these are limited without the termes of any *Axiome* of *Invention*. For that other part of the *New Organ* takes up and containeth in it all the *Transitions of Experiments into Axioms, or of Axioms into Experiments*" (Bacon 1640, 226).[15] These are all processes of discovery—processes of extending knowledge, of making it more exact.

A well-known passage from Newton resonates with both pre-modern and Baconian ideas about scientific inquiry in characterizing analysis as a "cause-revealing" procedure and thus as regressus: for Newton, analysis proceeds "in general from Effects to their Causes, and from particular Causes to more general ones, till the Argument end in the most general." The method of analysis "consists in making Experiments and Observations, and in drawing general Conclusions from them by Induction." For Newton, as for Bacon, induction proceeds by drawing

---

[12] Bacon did not make an explicit distinction between the material and conceptual decomposition of things.

[13] See Rusu (2013), Jalobeanu (2011), Jalobeanu (2015).

[14] This quotation shows that for Bacon, demonstration to others (transmission) was an integral aspect of the scientific method.

[15] Bacon described and exemplified these strategies in detail.

inferences from experiences. Induction presupposes the analysis—decomposition—of these experiences, as it does for Bacon.[16]

In sum, a multitude of interpretations of "analysis" existed in the pre- and early modern period. As a procedure or method, analysis can mean decomposition, reduction, or identifying the means for realization. These procedures may be applied to numbers (in math), problems and figures (in geometry), sentences and concepts (in logical reasoning), and things (in empirical inquiry). The procedures or methods can proceed from the known to the unknown or less known (ampliative—"discovery"); from the less known to the known (demonstration, clarification); or from the known to the known (systematic ordering, which is truth-preserving). The procedures or methods connect causes and effects, parts and wholes, particulars and generals, or principles and consequences.

Early modern scholars or scholarly traditions put these notions together in various combinations and permutations. They discussed sometimes the formal (mathematical and logical) and sometimes the scientific traditions of inquiry. The resulting interpretations and explications of the terms "analysis" and "synthesis" are highly diverse. Often, though not invariably, analysis is presented as the method of discovery—as a way of finding new things, notably the causes of what is known to us in experience. Often, but not always, synthesis is interpreted as a form of demonstration—in the sense of justification as well as of instruction. Some scholars, notably Zabarella, clearly distinguished between the method used in geometry and mathematics and that used in the natural sciences. Others, notably Leibniz and Descartes, sought to devise a "mathesis universalis" that encompassed all science, mathematics, and logic.

"Induction" typically refers to a process of working through specific instances to see where they lead. Induction, thus understood, relies on analysis. It has become clear that early modern interpretations of "analysis," "induction," and related terms differ considerably from our modern understanding of these terms. Pre- and early modern scholars often assumed that decomposition reveals causes, principles, or essences. In the modern sense, decomposition denotes breaking a complex thing apart into *particulars*, often literally disassembling things into smaller constituents. For many early modern scholars, by contrast, decomposition entailed the decomposition of a particular thing into more *general* elements. Elements are more general insofar as each element may be a part of many different, complex things. Most importantly, when early modern authors used the term "induction," its scope was not confined to enumerative induction. In fact, Bacon declared that this form of induction was impoverished ("childish," Bacon 2000, 17). Induction "is poor if it reaches the principles of the sciences by simple enumeration without making use of exclusions and dissolutions, or proper analyses of nature" (Bacon 2000, 57). The understanding of induction or inductive science was broad, even broader, in fact, than eliminative induction. Inductive science was regarded as a science based on

[16] See, e.g., Bacon (2000, 57).

experience—as opposed to a science that emphasized first principles. Inductive science, thus understood, comprises and relies on "proper analysis of nature."

12.3 Eighteenth-Century Works on Applied Logic

Recent work on the history of logic has revealed not only the continuities between early modern writings and pre-modern logic but also the fact that many early modern logics were not confined to formal discussions of inferences.[17] Rather, they encompassed investigations of the mind and mental operations as well as discussions surrounding habits of studying and how empirical research might best be organized. Discussion of these themes persisted throughout the eighteenth century. But as in the preceding decades, the terminology continued to be polysemous and in flux.

Consider, for instance, the entries "analysis" and "analytic method" in Johann Zedler's multi-volume encyclopedia of the arts and sciences. According to Zedler's authors, in general, analysis means "a division, dissection of a thing, body, or speech." Logical analysis—concerned with speech—leads "the thoughts of other people, as well as our own, to the principles from which they came" (Zedler 1732a, 38). Analysis "is for the *Chymicis* a dissolution, dissection, or reduction of a body into its *prima principia* or in its original essence [*Wesen*]."[18] In anatomy, however, analysis is the explanation of the use of a body part through the means of its realization (the Aristotelian–Galenic understanding of seeking out the means by which a goal is realized): "The *Anatomici* by contrast understand it [analysis] as explanation offered of each part of the body, whereby it is not only itself distinctly shown but its use and other circumstances that are necessary [*nötig*] to know are demonstrated" (Zedler 1732a, 38).

Zedler's authors also covered the analytic method in "mathesis" and algebra. In mathesis, one employs analytic methods "when one divides a project [*vorhabende Sache*] or question into its parts or circumstances, and dissects, and regards those separately and against each other, and investigates their causes and reasons, until one reaches step by step the origin and first main reason, such that one can show with certainty the nature and entire composition and can provide a thorough answer to the question posed." The authors borrowed the Cartesian framework, explaining that the analytic method comprised naming, equation, reduction, and construction (Zedler 1732b, 39).[19]

I shall not discuss the complex history of the concept of analysis in the contributions to seventeenth- and early eighteenth-century philosophy here.[20] Rather, I shall

[17] See Cassan (2021) for a recent overview.

[18] This is similar to Zabarella's regressus in empirical inquiry, as described in the previous section.

[19] I thank Niccolò Guicciardini for drawing my attention to the Descartes connection.

[20] See Tonelli (1976) and Engfer (1982) for discussion.

skip forward to the later eighteenth-century books on practical or applied logic. Written primarily as textbooks for introductory philosophy courses at universities, these books introduced future researchers as well as future lawyers, doctors, pastors, state officials, and other professionals to the methods of empirical inquiry. These texts employed the terminology of analysis and induction to illuminate the roles of observation and experiment for knowledge generation.

The accounts that these books present draw on the various traditions of thought about analysis and piece together bits from the works of Zabarella, Bacon, and other early modern scholars—notably, Leibniz, Locke, and Christian Wolff.[21] While the authors sometimes explicitly referred to these illustrious predecessors, they often left their sources unnamed.

In this section, I shall demonstrate how these texts helped to transmit early modern thought about analysis and induction to the later eighteenth and early nineteenth centuries. The authors of applied logic books used portions of previous accounts for their own purposes of instructing budding scientific practitioners and vocational students on how to make empirical inquiries. Given the existence of multiple different notions of analysis, synthesis, and induction, readers would find considerable diversity in the presentations, but all authors pursued similar goals—namely, elucidating how best to organize experiences and how to investigate cause–effect relations. Theories of judgment were deemed important for this purpose but not sufficient to account for all the various aspects of scientific inquiry. The authors of applied logic books placed considerable emphasis on the explication of how learned or literate experience might best be gained.

As before, some authors used the Greek terms in their accounts, while others favored Latin terms. Still others used both, and many used the German cognates—separating [*trennen*], dissecting [*zergliedern*], dividing [*teilen*], putting together [*zusammensetzen*], and so on (e.g., Ebeling 1785, 62). Those who did discuss induction contrasted it with "reduction," not "deduction" (e.g. Beseke 1786), and they certainly did not understand induction narrowly as enumerative induction, as simple generalization over particulars.

Johann Gottlieb Melchior Beseke voiced his disdain for those who espoused a narrow, formal understanding of induction. He commented that "the common philosopher" [*der gemeine Philosoph*] used the two terms "inducing" and "realizing" in such a way that "inducing" meant "inferring an unknown truth from experience" and "realizing" meant "explaining and proving a known truth from experiences" (Beseke 1786, 48). For instance, based on repeated exposure to dying people, the common philosopher would induce that all humans are mortal. That induction would be realized via the repeated experience that people die. Beseke had no patience for such simplistic philosophical exercises, instead praising those "true philosophers" who gained knowledge from actual empirical inquiry. The physiologist (a true philosopher).

[21] In addition, eighteenth-century German scholars were attuned to the Scottish common-sense philosophers, see Kuehn (1987).

> shows from the structure and growth of the body, its vessels, muscles, cartilage and bones, that dying is natural for humans; he infers it with the spirit of observation and shrewdness from the inner structure of the human body, pronounces this from concepts and experiences as true recognized judgment with a reliability to which level a common philosophy, this previous dwarf science [*Zwergenwissenschaft*], this unmanly creature, cannot reach. The true philosopher has the anatomical knife, the syringe, infusions, magnifying lenses, telescopes, microscopes, fishing nets, insect traps, herbals, crucibles, tongs, files, hammers—fire and water, always in his hand, and lets them guide him, as by subservient spirits, on his way, and when it gets dark, being led by their torch (Beseke 1786, 49).

In this perspective, induction—inductive science, science based on experience—constitutes a broader process of empirical discovery. Induction thus understood, Beseke implied, is much more valuable than an exercise in inferential reasoning might be. It is the process through which scientific knowledge can be gained.

Virtually all authors writing during this period emphasized that in the context of empirical inquiry, induction was typically incomplete and therefore did not lead to absolutely certain conclusions. They offered guidelines as to how these inferences might be made more secure—for instance, by drawing analogies to better-known phenomena or by seeking out more cases.

Some authors framed their accounts of empirical inquiry in terms of analysis, in a way that was reminiscent of Zabarella's understanding of analysis as *regressus*. For instance, in his treatise on experience in medicine, the physician and philosopher Johann Georg Zimmermann stated that the analytic method "ascends from the effects to the causes" (Zimmermann 1777, 429). Those authors who employed German cognates to characterize empirical inquiry also focused on the study of causes and effects. Irrespective of the precise terms they used, all offered explanations of induction, analysis, or *Zergliedern* that included discussions of empirical inquiry, of observations, trials, and hypothesis formation. They assumed, like Bacon and Newton, that experiences must be analyzed—decomposed—to initiate the process and that experiments and observations were important tools in this endeavor.

Before Beseke launched his attack on the "common philosophers" and their sterile notions of induction and reduction, he explained the terms via the difference between observation and trial. Observations begin with sense impressions and proceed to thoughts. The technical term [*Kunstwort*] for this process was "inducing." Trials, by contrast, begin with "thoughts of the soul"; this way of "descending" from thoughts to sensations was called "realization or reducing" (Beseke 1786, 48).

In his *Practical Logic* of 1764, which incorporates ideas from the works of John Locke and Christian Wolff, Justus Christian Henning interspersed his discussion of (incomplete) induction with discussions of how both the empirical basis of induction and the ampliative inference itself might be secured. He aimed to explain how "one could discover the cause of an effect through experience."[22] He advised that we must consider the cause of a change in "one individual real thing," which necessitated investigating if "our senses had perhaps betrayed us." Moreover, we must note "all circumstances that directly preceded or simultaneously existed with the effect

[22] Hennings' discussion is similar to Wolff (1713, 66–67).

or change of the thing." He described the procedure meticulously. One should take every single circumstance by itself and see if, through removing it,

a) either a change happened in the thing and in the effect we had perceived. If so, then

aa) either a complete change in the effect occurs, that is, the previous effect, which we noticed as the circumstance was still connected to the thing, disappears entirely.
bb) Or through the removal of that circumstance only a partial change in the thing or the perceived effect occurs. In both cases the circumstance must be regarded as a cause [*Grund*] of the effect. […]

b) or not. Here I can infer that the circumstance did not contribute to the effect and therefore is not a cause of it (Hennings 1764, 116–17)

Furthermore, one should connect the opposite circumstance with the thing of which I wish to discover the cause of the effect. Here remains

α) either the same effect, which shows that the previous circumstance posited does not have an influence on the effect and therefore is not a cause of it.
β) or the opposite effect is generated. Then I am all the more justified in taking my previously supposed circumstance as a cause of the effect (Hennings 1764, 117).

Like many eighteenth-century authors of applied logic books,[23] Hennings did not confine himself to describing the steps of a reasoning process (such as eliminative induction) but included a long note discussing the practical difficulties that one might encounter in the attempt clearly to distinguish circumstances from one another or to manipulate them individually, referring to experiments on blood transfusions and the conclusions that one could or could not draw from them. The emphasis that these authors placed both on the element of creativity in induction and the complementary concern about the possibility of error in scientific investigations distinguishes them markedly from Leibniz' insistence on the ideal of mechanized reason, as described in Guicciardini's chapter.

Virtually all books on applied logic include extensive discussions of practices of empirical inquiry, observation, trials, and hypothesis making and test. The suggestions that they offer are similar, and they frequently refer to ongoing scientific research (albeit citing different examples). Beseke, for instance, devoted a substantial portion of his treatise to rules for making observations and trials.[24] These rules describe, among other things, how to organize experiences in such a way that causes may be established as securely as possible. Similarly, Heinrich Ebeling offered

---

[23] Christian Leduc has described in detail Johann Heinrich Lambert's account of empirical inquiry (Leduc 2021). Lambert's *New Organon* was published in the same year as Hennings' practical logic (Lambert 1764). Lambert's empirical epistemology is not as unique in eighteenth-century German scholarship as Leduc claims.

[24] The rules are discussed in two places, first in the part "rules for correct sensation" (which draws heavily on Locke) and again in the part "means for extending our knowledge".

several rules for the proper use of one's common sense to arrive at distinct, correct, and reliable knowledge. The fourth of these rules is, "Be careful and cautious in your inferences; to this end, observe and make trials" (Ebeling 1785, 43). Ebeling's long explication of this rule ranges over basic points of word use and forming judgments (in the general sense of associating a predicate with a subject) and the different kinds of errors that one might commit in judging. Attention, observation, and trial were crucial in preventing these errors. Observation is "nothing but sustained attention to particular specific things, with a specific purpose, namely to discover the causes of the effective powers and all the circumstances on which a success depends" (Ebeling 1785, 54). Trials are necessary because one may "not be able to observe a thing like one wants, one must put it in a position and in the circumstances in which it can be observed according to purpose" (Ebeling 1785, 56). Ebeling explained in detail how the success of these trials could be ensured, for instance, by eliminating all circumstances that might unduly influence the trials and by repeating the trial often and with many variations.

Readers of the German translation of Jean Senebier's prize essay on the art of observation (Senebier 1776a, b, translated by the Göttingen researcher and professor Johann Friedrich Gmelin) would find in Part V of the work, entitled "The Observer as Interpreter of Nature" discussions of both induction and analysis, placed in different sections. In both sections, the discussion covered formal and material aspects of the processes. Senebier defined induction as an inference from the particular to the general and further explained that such an inference would proceed from the knowledge of effects to the knowledge of causes (Senebier 1776b, 314). Like Hennings, Beseke, and others, Senebier portrayed inductive inquiry as a complex process involving reasoning and practical action. For the induction to be useful, he argued, "one must consider each of the occurrences, which are part of it, separately, tear them out of nature for some time to see them better when they are isolated, put them back into nature, and not to draw the inferences they suggest unless one has perceived separately that, which could be said or denied of it" (Senebier 1776b, 315). Induction could also lead the way if "one suspects that a phenomenon has several causes, one would soon, through destroying one cause after the other, find out which is the true one, or which are the ones that contribute to its generation" (Senebier 1776b, 317).

Senebier's discussion of analysis reiterates several of the points made in the section on induction. Using the analytic method, "one gradually dissects a composite thing and, in going through all the parts of which it consists, comes to the simplest one" (Senebier 1776b, 357). Like induction, physical analysis is "going backwards from the effects to the causes" but simultaneously also "from the composite to the parts, to the way in which the composition happened, to the knowledge of the variations [*Schattierungen*] through which this composite being has gone, in order to get from the first moment of its development to the point at which it now is, in order to discover through the nature of these compositions all relations between this being with everything else" (Senebier 1776b, 358). The example for analysis is Bonnet's and Haller's empirical work on generation, which is noteworthy because the study of generation and development "in reverse," as it were, is clearly different from a physical decomposition of a living organism into parts. Rather, one would observe the unfolding of a seed or germ; and the analytic procedure would then be a conceptual process working backward through the stages in which they were observed.

There are other instances of authors characterizing analysis and synthesis in such a way that they are not mirror images. Zimmermann explained that in using the synthetic method, "effects are further determined and the events described such that one emerges from the other and thus can understood and demonstrated better" (Zimmermann 1777, 430). The synthetic method did not simply repeat the same steps of analysis in reverse order.

Like some pre-and early modern scholars, the authors of works on applied logic often did not clearly distinguish between material division into parts and conceptual division of properties or features. Consider Ebeling's discussion of composition and separation, which moves smoothly from things to concepts and back again. By dividing and putting together things and their parts or features, one could discover new things, both by chance and often

> when we cautiously and with the appropriate attention put together or separate different things and consider them in various relations [*Verhältnisse*]. The main thing is that one first knows by their properties the things that one wishes to separate or combine, that one observes properly, and obtains an exact complete and correct insight. To discover the thoughts and the concepts belonging to an object the most secure means is to develop and determine all concepts of the main sentence, dissect the properties one after another and to think about them until one has arrived at simple, in themselves clear concepts of sensations, and clear knowledge of each thing, proper separation and dissection of its parts and cautious connection is generally the most secure means of discovery [*erfinden*] (Ebeling 1785, 61–2).

Such an equation of concepts and things was common.

It was also common for the authors of applied logic books to include sections on methods of persuasion and instruction. In this context, the concepts of analysis and synthesis return as methods of ordering, demonstration, and teaching empirical knowledge. Opinions were divided as to whether instruction would be more effective if the analytic or the synthetic method were used, whether both methods should be mixed, or whether different methods should be used for different audiences and purposes.

It is difficult to do justice to all facets of the debate, particularly given that the authors had different understandings of "analytic" or "synthetic" to begin with. Senebier, for instance, found the analytic method advantageous for discovery and also very useful for instruction (Senebier 1776b, 367). Others, as we have seen, described the analytic method as a method of discovery and the synthetic method as a method of instruction. Still others distinguished between instruction to the uninitiated and scholarly discourse (Hennings 1764, 178) or between mathematical instruction and instruction in the empirical sciences (Basedow 1765, 89), and so forth. The important point to note is that the concepts of analysis and synthesis continued to play a crucial role in reflections on discovery and demonstration.

12.4 Kant's Notions of Analysis and Induction

Among the German-speaking scholars, the publication of Kant's critical philosophy in the 1780s significantly influenced discussions about scientific methodology, the scope of knowledge, and the nature and status of metaphysics and logic in relation to the sciences. Kant's work completely transformed the interpretations of analysis and synthesis, drawing a distinction between analytic judgments, which express only what is already contained in the premises or in our thinking, and synthetic judgments, which extend our knowledge. This stands in opposition to, for instance, the position expressed in the passage from the Port Royal Logic that I quoted above.[25]

An in-depth examination of the intricacies of Kant's theoretical philosophy lies beyond the scope of the present discussion. But, given that many nineteenth-century scholars on scientific methodology referred to Kant's writings (in one way or another), it is worth reviewing some key features of his understanding of analysis and synthesis. Kant drew several new distinctions that had implications for the older views on logic and scientific method. His novel explication of analytic and synthetic judgments addressed the problem of how our knowledge relates to the outside world and the associated problem of whether we can acquire new knowledge (i.e., make discoveries) without empirical input (i.e., a priori synthesis as method of discovery).

While many early modern scholars did not distinguish between the material and the conceptual—that is, between causes and reasons, consequences and effects, components and properties—Kant drew an explicit distinction between the analysis or decomposition of matter and the analysis or decomposition of cognition. In his *Critique of Pure Reason*, he mobilized the chemical concept of the decomposition of matter to explain his approach, which he presented as an analogous cognitive process. In Kant's philosophical framework, the process of analysis or decomposition targets the cognitive conditions of knowledge exclusively. In critical philosophy, analysis and analytic judgments concern the mind and cognition rather than the external world; analytic judgments are non-empirical, because they concern only the mind—either concepts or empirically givens. Therefore—and in sharp contrast to the Port Royal Logic (or Newton, for that matter)—analysis cannot "discover" anything—at least, not in the sense of an empirical discovery.

Kant also drew a distinction between "observing" the human mind as it works within all the limitation and conditions of life and exposing the rules of thinking (of what the mind "does" —namely, judging). Only the latter, the rules of thinking, are within the scope of logic: "Were we to take principles from psychology, that is, from the observations on our understanding, we should but see how thinking goes on, and how it is under the various subjective impediments and conditions; this would consequently lead to the knowledge of merely contingent laws. In logic, however, the

[25] In his announcement of the 1805 prize question on the method of analysis at the Berlin Academy of Science, the writer and publisher Friedrich Nicolai chided Kant for having created confusion regarding the terms "analysis," "synthesis," and "regressus" (for details, see Engfer 1982, 32–3).

inquiry is after not contingent, but necessary rules; not how we think, but how we are to think" (Kant 1819, 13–14).

Kant's conception of logic was thus narrower than that of those who composed applied logic works. For Kant, logic did not and could not have an empirical component. Meanwhile, he distinguished two kinds of logic, pure or formal logic and transcendental logic, which discusses the conditions of the possibility of thinking that is concerned with objects [*auf einen Gegenstand bezogen*].

Finally, among the key questions Kant raised was whether synthesis could be achieved a priori such that *new* knowledge could be gained—new knowledge without empirical input. In the Kantian understanding, discovery was synthesis or composition, the bringing together of elements that had been uncovered through analysis. These might include elements found in the description of our experiences or elements found in the analysis of our cognition. The latter type of synthesis would be a priori. For Kant, a priori discovery—making synthetic judgments a priori —was possible, but analytic empirical discovery did not exist.

12.5 Nineteenth-Century Developments

While Kant's critical philosophy was extremely influential, it did not fully change the conversation about analysis, synthesis, and induction in the nineteenth century. Kant's new philosophical position and his complete overhaul of older notions of logic, analysis, and synthesis stimulated intense critical discussion, not least because it had implications for the traditional order and scope of scholarly disciplines. Not everyone was immediately on board with his views. Many scholars interpreted and re-interpreted Kantian philosophy, praising or disdaining it.[26] In 1805 and 1808, the Berlin Academy of Science issued two prize questions on the method of analysis in the philosophical sciences (in contrast to mathematics) and the relation between analytic and synthetic methods, which attests to the consternation that Kant's philosophy caused at the time.[27]

Beyond academic philosophy, the broader pragmatic tradition of examining the features of experience-based—"inductive"—inquiry still prevailed. Contributors to this strand of discussion often had training in philosophy as well as scientific training and expertise in botany, psychology, chemistry, and other fields. They hoped to say something that had practical relevance to the nature of empirical inquiry. During

[26] Publications on all aspects of Kantian thought proliferated in the German lands and beyond, and clearly these cannot be covered in detail here. The discussion surrounding Kant's philosophy and its significance and impact continues to this day. For an overview, see Köhnke (1993). See also Friedman and Nordmann (2006), Beiser (1987), and the various articles related to Kantian thought in the *Stanford Encyclopedia of Philosophy*.

[27] The prize-winning author G. S. Francke and Johann Christoph Hoffbauer were both working through Kant's philosophy, albeit with markedly different results, and not all their comments were praise (Francke 1805; Hoffbauer 1810).

these discussions, various issues were gradually sorted out and separated, including the difference between analyzing things and analyzing thoughts, the relation between analysis and induction, and formal and non-formal interpretations of induction.

As an early commentator on Kant, Gottlob Ernst Schulze was a significant voice in the discussions. In 1810, he published a logic textbook that went through several editions. It included a detailed discussion of the "two methods in the sciences," the "dissecting (analytic) and the composing (synthetic, adding one insight to another)" (Schulze 1810, 174). Schulze characterized thinking according to the analytic method as "resembling the process of understanding in proving through induction and analogy" (Schulze 1810, 175). This process was complex; it was not a simple generalization based on enumeration but rather a comparison of specific facts to identify similarities and differences. From the similarities thus discerned, one could proceed to more general knowledge claims. Schulze noted that this process was characteristic for all human inquiry; it was thus often termed the "heuristic" method. It was by no means a formal, algorithmic procedure, as no general rule was in place to govern its execution. Schulze pointed out that this process "leaves a lot of room for the mind for free movements and allowed imagination and memory to support the endeavor of the mind"—in fact, sometimes it involved "genius" (Schulze 1810, 176). It is evident that these scholars had diverged far from the Leibnizian ideal of mechanized reasoning.

Schulze also noted that because the analytic method moves "from the grounds to what is grounded" it should not have been called a regressive method (Schulze 1810, 176). We witness here a gradual move away from the pre-modern idea that the components produced by analysis are *general* elements of things. For Schulze, analysis begins with *concrete* parts, particular components. Analysis begins with "seeking out the particular [*Besonderes*], from which must proceed the speculation gradually leading to the general" (Schulze 1810, 177). That process would typically fall into oblivion once one had found the more general truths, leading to disregard for the importance of the analytic method (Schulze 1810, 177).

Schulze characterized the synthetic method as the systematization of knowledge, whereby both general and specific truths would be organized into a system. Still, the process of subsuming required familiarization with the specific; otherwise, one would not know whether it could be subsumed under the general concept (Schulze 1810, 179). Here, Schulze launched a long complaint about the tendency to build systems without having sufficient knowledge about specific phenomena (Schulze 1810, 185).

In 1828, the Jena philosophy professor and future director of the museum of mineralogy and zoology Carl Friedrich Bachmann distinguished analysis and synthesis as follows: Analysis begins with effects, consequences [*Folgen*], wholes—particular things [*Besonderes*]—and proceeds to causes, principles, elements—the general. Synthesis, in turn, begins with causes, principles, elements, the general, and proceeds to effects, consequences [*Folgen*], wholes, particular things (Bachmann 1828, 358).

Bachmann's explication of the terms is noteworthy in that it encompasses all the disparate notions of analysis and synthesis—the material, the pragmatic, the logical, and the epistemological. Some philosophers—notably, Arthur Schopenhauer—had already argued that these notions should be kept separate. In his 1813 treatise on the fourfold root of the principle of sufficient reason, Schopenhauer chided his fellow philosophers for erroneously equating the relationship between reason and consequence in judgments with cause–effect relations in the world (Schopenhauer 1813).

In the 1830s, several works on scientific inquiry were published in Britain that would subsequently be regarded as milestones of the history of philosophy of science, John Herschel's *Discourse on the Study of Natural Philosophy*; John Stuart Mill's *System of Logic*; and William Whewell's two multi-volume treatises on the history and philosophy of the inductive sciences. They all provided broad accounts of "inductive" scientific inquiry, which did not reduce induction to a formal step in reasoning. Moreover, in each of these accounts, analysis played a role as part of inductive inquiry.

These works were quickly published in German translation. Consequently, the term "*induktive Wissenschaft*" gained currency in German-language discussions. Some German commentators, such as the botanist Matthias Jakob Schleiden, adopted the term for their own writings on methods of empirical inquiry. In 1842, Schleiden published a handbook entitled *Principles of Scientific Botany* [*Grundzüge der Wissenschaftlichen Botanik*]. In its third edition, published 1848, the first part of the book carries the subtitle "Botany as an Inductive Science." In the long methodological introduction to the original edition of the work, Schleiden outlined the methods of inductive botany. This closely resembled the methods of empirical inquiry that English authors had described, so much so that the English translator chose to omit this part.[28] Subsequently, "inductive" methods were also favorably discussed in other scientific fields, including geology (von Cotta 1867, 377, citing Liebig), economy (Weisz 1871), and geography (Dronke 1885). "Inductive science" continued to be understood broadly as an umbrella term for any science that is grounded in experience. Within this broad understanding, inductive science encompassed observation and experiment, hypothesis formation, and hypothesis testing.[29]

In the context of psychology or empirical studies of the soul [*Erfahrungsseelenkunde*], discussions were particularly lively, not least because Kant's philosophy had significant implications for the status of this field as a science. In this context, the appropriateness of the terms "analysis" and "induction" as labels for psychological investigation was a theme. The discussion about the nature, status, and limits of psychology was wide-ranging and engaged numerous scholars,

[28] A footnote explains, "As general introductions on the principles involved in scientific inquiry, we have in our own language two admirable works, —Sir John Herschell's [sic] "Discourse on the Study of Natural Philosophy," and Professor Whewell's "Philosophy of the Inductive Sciences"" (Schleiden 1969, iii).

[29] The understanding of "analysis" as dissection lives on in sensory physiology, for instance in Ernst Mach's *Analysis of Sensations* (Mach 1886). In Mach's case, at least, there are explicit connections to the older notion of analysis; see below.

from academic philosophers like Herbart to empirical researchers like Hermann Helmholtz. As in earlier decades, opinions were divided, approaches differed, and the authors disagreed about the appropriate terminology for philosophy of science.

Philosopher Christian Martin Julius Frauenstädt, editor of and commentator on Schopenhauer's works, insisted that "inductive" was the more appropriate label for the empirical method. Frauenstädt approvingly cited Schopenhauer, who had (in Frauenstädt's interpretation) regarded philosophy as a science that was "founded on the entirety of experience and is thus empirical science" (Frauenstädt 1854, 83). Indeed, Schopenhauer had explicitly proposed the use of "induction" for the analytic method in empirical science, as it "goes from the facts, the specific, to the principles, the general, or from the consequences to the reasons; the other [the synthetic] the reverse: for the traditional names are inappropriate and express the matter badly" (Schopenhauer 1844, 121).[30]

Other philosophers, including the Königsberg philosopher Johann Karl Friedrich Rosenkrantz, who had studied with Hegel, complained in a contribution to the *Allgemeine Monatsschrift für Literatur* on "psychology as a natural science" that the use of the term "induction" instead of "analysis" was due to the influence of British writers:

> If one thinks of the method of the natural sciences as a specific one, it would have to be the *analytic* one, which is currently called the *inductive*. How cuddly are we Germans with foreign countries! Since Whewell wrote the history of inductive science and now since Mill gave an inductive Logic with relation to natural science and with examples from Liebig's chemistry, many of us dance around this golden calf, exhibited for devotion. The inductive method is the new salvation! But the inductive method is nothing but the analytical, which goes from the problem through setting out the determinations found in the object of knowledge, through comparison and grouping of the related according to instances, as Bacon called it, to a subsumptive inference, which can itself be threefold and of which the induction as such is only *one* of its forms (Rosenkranz 1850, 162).

Rosenkranz approvingly cited his Leipzig colleague, the mathematician and philosopher Moritz Wilhelm Drobisch, who prefaced his 1842 *Empirical Psychology According to the Method of the Natural Sciences* with the pronouncement that he aimed to present a psychology "without reference to mathematics, through mere disinterested observation, dissection, comparison and combination of the facts of our inner experience" (Drobisch 1842, III).

A terminological trend may be discerned, however. As many academic philosophers continued to engage with Kant's work and as formal logic became a more specialized subject independent from philosophy of science, the term pairing "induction–*reduction*" gradually came to be replaced by "induction–deduction." The notion of deduction as the opposite of induction also entered the practice-oriented discussions about proper methods and methodologies of scientific inquiry.

[30] This passage is an addition in the second edition. In the first edition, the relevant passage is less concerned with terminology and more with the question of what the distinctive features that make an endeavor "scientific" are—namely, the grounding in experience, not certainty (Schopenhauer 1819, 98–102).

Chemist Justus Liebig's 1865 speech at the Munich Academy of the Sciences is indicative of the changes in the conceptualization of inductive science and methodologies of inquiry that happened during the nineteenth century. The speech was subsequently published in a booklet entitled *Induction and Deduction* (emphasis added). In it, Liebig compared the "inductive" and the "deductive" researcher. He set the theme by characterizing inductive research as research relying on enumerative inductive inferences, an interpretation of induction he associated with Aristotle. He further characterized the entirety of early modern science as inductive and as an art. Early modern researchers would operate on the level of visual experience, they would "inductively" combine various visual experiences into a whole. Liebig did not suggest that this kind of research was mindless, as the combination of experiences required imagination [*Einbildungskraft*], but the point was that—according to Liebig—early modern researchers were simply not interested in explanations or in establishing causes.

Modern researchers, by contrast, would pursue their experiments in light of specific ideas about the product they wanted to make: "the discovery of a new fact or reaction, to which can be connected the idea of a thing hitherto unknown, yet useful or important for industry or life, is sufficient to raise the conviction of its existence in many individuals" (Liebig 1865, 23). Both the understanding and the imagination were equally necessary in science, but understanding ruled in modern science: "induction guided by imagination is intuitive and creative, but indeterminate and excessive; deduction guided by the understanding analyzes and limits, and is determinate and measured. One of the essential characteristics of deductive research in the natural science is the measure, and the ultimate end of its work is an unchanging quantitative expression for all properties of things, for processes and phenomena" (Liebig 1865, 23).

This passage is remarkable because it misrepresents the early modern and eighteenth-century accounts of scientific inquiry in a way that is reminiscent of Popper's statement, with which we started. It contrasts induction and deduction rather than induction and reduction and reduces induction to enumerative induction. In the text, however, Liebig offered a more informal characterization of empirical research, although he did not go into detail about how experimental and observational practice should be organized. Ernst Mach would later praise Liebig not because Liebig had drawn attention to the importance of deduction in scientific inquiry but because Liebig emphasized the creativity that was required to perform scientific work.

12.6 Analysis and Synthesis in Knowledge Generation

I shall conclude this account with a brief glance at two books on logic and method, both published around 1900 by influential figures on the Austro–German scientific scene: Wilhelm Wundt's *Logic* (published in 1880) and Ernst Mach's *Knowledge and Error* (1905).

The second volume of Wundt's *Logic* begins with a general introduction to the methods of scientific inquiry.[31] Wundt's methodology applies to all sciences, from physics to chemistry and from history to psychology. The first section of the introduction covers analysis and synthesis, and the third section discusses induction and deduction (Sect. 12.2 concerns abstraction and determination). According to Wundt, analysis and synthesis are the "most general form" of investigation, on which two pairs of composite methods are erected: abstraction and determination and induction and deduction (Wundt 1883, 1).

Wundt presented analysis as a three-step operation of thinking and doing. Its first step is elementary analysis—that is, the decomposition of a phenomenon into its constituent phenomena [*Theilerscheinungen*] without regard for the relationships between these components. Wundt explicitly noted that this analysis could be perceptual, introspective, or mediated by scientific instruments, historical documents, or statistics (Wundt 1883, 2).

The second step is causal analysis, which is the decomposition of phenomena into components with regard to the causal relations of the parts. The predominant method of causal analysis is the experimental method, the "intentional isolation" of particular elements from the complex facts to be investigated, the systematic variation of the relations between the elements, their removal, or a quantitative change of the element itself—for example, an increase in size or intensity (Wundt 1883, 4). If practical manipulations of this sort are impossible, the experimenter must resort to drawing analogies with already known phenomena.

The third step is logical analysis, "the decomposition of a complex fact into its parts with regard to the logical relations of the latter" (Wundt 1883, 5). Wundt exemplified this step with the mathematical equation of a curve or the formulation of a law of nature on the basis of a causal analysis of phenomena in experiment or trial.

In the context of this volume, it is particularly noteworthy that Wundt then explicitly distinguished two kinds of synthesis, "reproductive synthesis," the simple reversal of the analysis, and "productive synthesis," the new and independent combination of the elements found in analysis. Unlike many of his predecessors who had valued the probative force of reproductive synthesis so highly, Wundt thought that this form of synthesis was of "relatively limited value" precisely because it was done mainly in the interest of checking the results of the analysis. Productive synthesis was much more valuable because it was constructive; it went beyond the analytic investigation. Wundt exemplified the constructive power of synthesis with a nod to Newton: The analysis of sunlight, Wundt explained, led to the discovery of the spectral colors as their elements, and their synthesis produced white. But "a modification of the procedures simultaneously suggests itself, whereby the path of a mere reproductive synthesis is abandoned" (Wundt 1883, 8).

---

[31] In the mid-1870s, Wundt briefly held the professorship in "inductive philosophy" at the University of Zurich, Switzerland, before he took up a more permanent post at Leipzig, Germany.

Like analysis, synthesis involved elementary, causal, and logical steps, whereby causal synthesis was a key element of experimental procedures. Causal analysis was "not only applied when an analytic result had to be confirmed through a reversal of the investigative process but also often brought about complex phenomena through new combinations of elementary conditions" (Wundt 1883, 8). However, because the operation of productive synthesis required relatively simple processes or elements, it was only of limited use in the historical, social, and human sciences with their complex and convoluted facts.

Analysis and synthesis were part and parcel of inductive science. For Wundt, as for many of his predecessors, the inductive method was the method of scientific investigation, broadly understood. The inductive method "seeks to limit the interpretation of facts through the manifold varying use of the analytic and the synthetic method" (Wundt 1883, 22). The method hypothetically assumes a possible interpretation as valid [*wirklich geltend*], unfolds its consequence, and tests these consequences empirically. In this way, "various hypotheses can be investigated successively such that one is eventually left with that which recommends itself most through its agreement with the facts" (Wundt 1883, 22). Wundt thus advanced a position that twentieth-century philosophers of science would characterize as "hypothetico-deductivism." Unlike these philosophers, and in line with the eighteenth- and early nineteenth-century scholars, Wundt described this method as "inductive" and included general characterizations of the practical, experimental processes by which hypotheses, including causal hypotheses, could be generated.

Mach's epistemological treatise *Knowledge and Error* provides a fitting endpoint to my story. Published at the beginning of the twentieth century, it is infused with traditional notions of analysis, synthesis, and induction. Mach utilized these long-standing ideas about scientific inquiry for his theory of knowledge, invoking Euclid in the characterization of the analytic and the synthetic method. According to the analytic method, he explained, we proceed "starting from the result and working back to admitted premisses"; according to the synthetic method, we proceed "starting from admitted premisses and working forward to the result." The "indirect" method demonstrates that the contradictory of the result is impossible. The three methods are not only methods of inquiry but can also be used "in demonstrating what is already known" (Mach 1976, 188).

The analytic method—the method of knowledge generation or invention—regresses from the given thing, the "secure facts", to the conditions that might have brought it about. Mach emphasized that the analytic method could not be formalized; it was a considerably less determined task than the deduction of consequences from a theory.[32] Mach illustrated the analytic, knowledge-generating procedure with several examples from the history of early modern and modern physics. He reiterated several times that performing an analysis was not a formal, algorithmic procedure but involved creativity and imagination; it was vague and indeterminate and

[32] The German original explicitly characterizes this inference as "regressive, analytic, indeterminate inference" (Mach 2002 [1926], 235). The English translation misleadingly reads: "starting from the given, secure facts and inferring back to the indefinite conditions" (Mach 1976, 173).

thus highly challenging. It succeeded "only in tentative steps" (Mach 1976, 197) with the help of hypotheses, which may be erroneous or irrelevant.

Mach approvingly cited Liebig, praising him for having correctly characterized analysis as an art form (Mach 1976, 236). In another passage, he illustrated the analytic method or process of discovery or invention[33] in a more Aristotelian spirit as a technical operation of deliberating about and organizing the means to a given end:

> If you want a tree trunk laid across a stream in order to walk over, you imagine the problem solved: by considering that the trunk must be dragged into place, but first the tree must be felled and so on, you tread the path from the sought for to the given, which in actual construction of the bridge he has to traverse in the opposite direction, reversing the sequence of operations. This is a case of very ordinary practical thinking. Most great engineering inventions seem to rest on this process, insofar as they were not gradually provided by chance but rapidly called into existence by spontaneous effort (Mach 1976, 191).

Like the older approaches to scientific methodology and like Wundt, Mach characterized knowledge generation as complex. On occasion (as in the example above and in other examples from the history of physics) he characterized the discovery process as a pragmatic process that included engaging with materials (e.g., Mach 1976, 269). Mostly, however, he characterized the process as a "cognitive [*psychisch*] operation," explaining that "the mental operation by which new insights are gained, which is usually called by the unsuitable name 'induction', is not a simple process but a rather complex one. Above all it is not a logical process, although logical processes may figure as auxiliary intermediate links. Abstraction and the activity of phantasy does the main work in the finding of new knowledge" (Mach 1976, 235–6). Mach noted that it was once common to call this complex process "induction," but he implied that he found that terminology misleading.

For Mach, uncovering cause–effect relations—stable connections between phenomena and events—was still one of the main goals of empirical inquiry. Unlike his more optimistic predecessors, he emphasized the complexity and apparent irregularity of events. In fact, according to Mach, the constant *change* between regularity and irregularity that we encounter as we pursue our practical goals makes us wonder which associations are regular and constant and which are accidental, and when we look closely, we almost invariably find "that the so-called cause is only one of a whole set of conditions that determine the so-called effect; so that according to which of these have been noticed or overlooked, the condition in question may differ greatly" (Mach 1976, 204).

[33] The German "Erfindung" has a double meaning, at least as it appears in the texts prior to 1900—it means designing a new object, but it also means "finding" or discovering a piece of new knowledge.

12.7 Conclusion

Historians of philosophy have long considered the middle decades of the nineteenth century to be the period of the rise of the H-D method, with Popper as one of its most vocal twentieth-century advocates, and of the simultaneous downfall of "Baconian" science.

The contours of an alternative picture have emerged from my account. First, it has become clear that the notion of "Baconian" science is useless as a historiographical category. "Baconian" in what sense? In the sense that rules for experimentation are provided, as in *De Augmentis?* In the sense that rules for reasoning from facts are provided, as they are laid out in the *Novum Organum*? In the sense of a demonstration of a particular matter theory, as, again, in the *Novum Organum*? In the sense that the collection of facts is promoted, as in Bacon's project of natural history? In the sense that "fructiferous" experiments are described, as in *Sylva Sylvarum*? Depending on how we understand what is distinctive about "Baconian" science, eighteenth- and nineteenth-century philosophies of science either exemplify a Baconian approach, contrast with it, or deal with something wholly different. Moreover, as we have seen, other scholars also developed comprehensive accounts of empirical inquiry; it would thus be misleading to label the entire tradition of theorizing empirical inquiry "Baconian." German scholars also drew on Christian Wolff, who, in turn, owed something to Robert Boyle.

Second, throughout the eighteenth and well into the nineteenth century, theories of scientific inquiry were significantly concerned with the practice of inquiry, specifically with the question of how to establish cause–effect relations. Causal inquiry was understood to comprise hypothesis making, hypothesis evaluation, and hypothesis testing, as well as the making of observations and experiments. The latter were often portrayed in terms of analysis—a conceptual-material process of revealing epistemically, ontologically, or physically prior things. Philosophers writing about applied or practical logic wanted to offer comprehensive accounts of empirical inquiry that were useful for the practice of research and for its future practitioners.

Third, prior to the nineteenth century, inductive and analytic methods were often intertwined and linked to discovery, as in Newton's famous rule. Induction and deduction were not considered opposites; if induction had an opposite, it was "reduction." "Analysis" and "induction" slowly separated around 1800, not least through the influence of Kantian philosophy and the ensuing rise of academic philosophy and formal logic. Analysis gradually moved out of theories of discovery and became separated from inductive science. In philosophical theories of thinking after Kant, an inversion of the concepts of analysis and synthesis took place. While early modern writers often (but not always, as we saw) regarded analysis and induction as methods of discovery, modern philosophers commonly considered analysis as a process of clarification and associated synthesis with discovery. But that did not happen overnight, as evidenced by prize essays, the mid-nineteenth-century discussions about methods of empirical inquiry, and the theories of scientific method expounded by Wundt and Mach.

If the British philosophers publishing in the 1830s and 1840s did exert an impact on the German discourse during the nineteenth century, it was in matters of terminology. With the translation of Mill's and Whewell's works and publications by Anglophiles such as Schleiden, the label "inductive science" became increasingly common in the German discussions about the methods of empirical inquiry. Nevertheless, the spirit of the older, broad, pragmatic accounts of methodology persisted in nineteenth-century scientific methodologies.

This, then, is the fourth point: practice-oriented theories of inquiry lived on outside formal logic and foundational theoretical philosophy. These theories were considerably more sophisticated than twentieth-century critics of "inductive" or "Baconian" methodologies suggested and were in fact quite similar to the H-D approaches that those critics recommended. The nature of empirical inquiry—of "inductive science" broadly understood—remained a theme in various scientific fields—for instance, in discussions about the true nature of psychology but also in specific sciences, such as botany. But these theories were no longer called "applied logic." Moreover, while many nineteenth-century scientists (Liebig and Helmholtz, among others) gave popular lectures or wrote popular science books, these outreach activities were aimed at the broader public and were no longer considered integral components of scientific method.

Several eighteenth- and nineteenth-century authors recommended evaluating and testing hypotheses by deriving testable consequences; nevertheless, they did not regard the derivation of testable consequences as a narrowly deductive inference that would be opposed to, or epistemically more powerful than, induction. Rather, hypothesis testing was understood to be a creative, non-algorithmic process and an integral part of the inductive sciences.[34] Only in the formal sciences (and in Kant's philosophy!) would the deduction of consequences from principles be considered a formal process. In the empirical sciences, where one may not know the mechanisms by which a cause generates an effect, predictions (of effects) are generated tentatively and stepwise. That insight, well familiar to eighteenth- and nineteenth-century scholars, had to be regained by philosophers of science after Popper.

Today, the concept of analysis has lost the connection to discovery that it once had. The all-encompassing notion of inductive science has been narrowed down to the formal notion of inductive inference, and we usually talk about "inductive logic" and not about "inductive science" in the broader sense of "empirical science". But only by keeping in mind the earlier meanings of induction and analysis will we be able to understand past and present discussions about scientific inquiry.

The broader interpretation of "inductive science" was alive and well during the nineteenth century. Inductive science begins with the immediately experienced, but that does not mean that it proceeds by drawing enumerative inferences. Many authors recommended evaluating and testing hypotheses by deriving testable

[34] In fact, many of the discussions about whether or not nineteenth-century scientists (notably Darwin) were advocates of the H-D method are superfluous if we keep in mind that "the method of hypothesis" and "inductive methodology" are not opposites.

consequences as legitimate elements of methods of empirical inquiry.[35] But they did not regard hypothesis testing as a strictly deductive process that is opposed to, or more powerful than, induction. While enumerative inductive inferences and formal deductions logically occupy different statuses, inductive inquiry and H-D methodology are not in opposition to one another. Rather, both are integral aspects of the inductive sciences, according to pragmatic theories of empirical inquiry. That is the tradition of scientific methodology that Popper ignores.[36] The magnification of the justificatory, hypothesis-testing step in scientific methodology unnecessarily impoverishes scientific methodology, and it took many decades for philosophy of science to recover from this move.

Acknowledgments I am most grateful to the participants of the Sawyer Seminar working group "Rigor: Control and analysis in historical and systematic perspectives" at Indiana University Bloomington, especially Claudia Cristalli, Bill Newman, and Evan Arnet, for the numerous valuable and inspiring discussions we had over the past few years. I thank the contributors to the Sawyer conference and workshop on analysis and synthesis for their stimulating papers and Niccolò Guicciardini, Helen Hattab, Alan Shapiro, and Bill Newman for their comments on this chapter.

References

Achinstein, Peter, ed. 2004. *Science rules. A historical introduction to scientific methods*. Baltimore: Johns Hopkins University Press.

Arnauld, Antoine, and Pierre Nicole. 1861. *The port-royal logic, translated from the French, with introduction, notes, and appendix*. 5th ed, 1662. Edinburgh: James Gordon.

Bachmann, Carl Friedrich. 1828. *System der Logik*. Leipzig: Brockhaus.

Bacon, Francis. 1640. Of the advancement and proficience of learning. Oxford: Printed by Leon Lichfield printer to the University, for Robert Young and Edward Forrest.

———. 2000. *The new organon*. Cambridge: Cambridge University Press.

Basedow, Johann Bernhard. 1765. *Theoretisches System der gesunden Vernunft*. Altona: In Commission bey David Iversen.

Beiser, Frederick. 1987. *The fate of reason. German philosophy from Kant to Fichte*. Cambridge, MA: Harvard University Press.

Beseke, Johann Gottlieb Melchior. 1786. *Versuch einer praktischen Logik oder einer Anweisung den gesunden Verstand recht zu gebrauchen*. Leipzig: Müllersche Buchhandlung.

Cassan, Elodie. 2021. Introduction: Logic and methodology in the early modern period. *Perspectives on Science* 29 (3): 237–254. https://doi.org/10.1162/posc_e_00367.

[35] Laura Snyder rightly pointed out that Whewell advocated a broad notion of "inductive" science, which *included* instructions for hypothesis-testing. This broader understanding of inductive science is entirely in line with the broader ideas of induction and analysis that are presented in the applied logic books. It would be misleading, as she noted, to subsume Whewell's ideas under the "H-D method."

[36] His immediate predecessors, notably Mach, knew that the derivation of consequences from hypotheses was not simply a "deductive" step. Popper's critics, such as Lakatos, who chided Popper for narrowing hypothesis testing to a logical step and ignoring its complexity, thus revived earlier ideas about empirical inquiry.

Drobisch, Moritz Wilhelm. 1842. *Empirische Psychologie nach naturwissenschaftlicher Methode*. Leipzig: Leopold Voss.
Dronke, Adolf. 1885. *Die Geographie als Wissenschaft und in der Schule*. Bonn: Eduard Weber's Verlag.
Ebeling, Heinrich Matthias Friedrich. 1785. *Versuch einer Logick für den gesunden Verstand*. Berlin und Stettin: Friedrich Nicolai.
Engfer, Hans-Jürgen. 1982. *Philosophie als Analysis: Studien zur Entwicklung philosophischer Analysiskonzeptionen unter dem Einfluss mathematischer Methodenmodelle im 17. und frühen 18. Jahrhundert*. Vol. Bd 1 *Forschungen und Materialien zur deutschen Aufklärung. Abteilung II, Monographien*. Stuttgart-Bad Cannstatt: Frommann-Holzboog.
Francke, G.S. 1805. *Über die Eigenschaft der Analysis und der analytischen Methode in der Philosophie*. Berlin: Johann Friedrich Unger.
Frauenstädt, Julius. 1854. *Briefe über die Schopenhauer'sche Philosophie*. Leipzig: Brockhaus.
Friedman, Michael, and Alfred Nordmann, eds. 2006. *The Kantian legacy in nineteenth century science*. Cambridge, MA: MIT Press.
Hattab, Helen. 2014. Hobbes's and Zabarella's methods: A missing link. *Journal of the History of Philosophy* 52: 461–486.
———. 2021. Methods of teaching or discovery? Analysis and synthesis from Zabarella to Spinoza. In *History of universities*, ed. Mordechai Feingold, 85–112. Oxford: Oxford University Press.
Hennings, Justus Christian. 1764. *Praktische Logik*. Jena: Christian Heinrich Cund.
Hintikka, Jaakko, and Unto Remes. 1974. *The method of analysis: Its geometrical origin and its general significance*. Dordrecht; Boston: Reidel.
Hoffbauer, Johann Christoph. 1810. *Ueber die Analysis in der Philosophie*. Halle (Saale): Hemmerde und Schwetschke.
Jalobeanu, Dana. 2011. Core experiments, natural history and the art of experientia literata: The meaning of Baconian experimentation. *Society and Politics* 5: 88–103.
———. 2015. *The art of experimental natural history. Francis Bacon in context*. Bucharest: ZETA books.
Kant, Immanuel. 1819. *Logic*. London: Printed for W. Simpkin and R. Marshall, Stationers Court, Ludgate Street.
Köhnke, Klaus Christian. 1993. *Entstehung und Aufstieg des Neukantianismus. Die deutsche Universitätsphilosophie zwischen Idealismus und Positivismus*. Frankfurt am Main: Suhrkamp.
Kuehn, Manfred. 1987. *Scottish common sense in Germany, 1768–1800: A contribution to the history of critical philosophy*. Vol. 11 McGill-Queen's studies in the history of ideas. Kingston [Ont.]: McGill-Queen's University Press.
Lambert, Johann Heinrich. 1764. *Neues Organon oder Gedanken über die Erforschung und Bezeichnung des Wahren und dessen Unterscheidung vom Irrthum und Schein*. Vol. I. zitiert nach dem Nachdruck, Hildesheim 1965. Leipzig: Wendler.
Leduc, Christian. 2021. Conducting observations and tests: Lambert's theory of empirical science. In *What does it mean to be an empiricist?: Empiricisms in eighteenth century sciences*, ed. Siegfried Bodenmann and Anne-Lise Rey-Courtel, 215–234. Cham: Springer.
Liebig, Justus. 1865. *Induction und deduction*. Munich: Königliche Akademie.
Mach, Ernst. 1886. *Beiträge zur Analyse der Empfindungen*. Jena: Gustav Fischer.
———. 1976. *Knowledge and error: Sketches on the psychology of enquiry*. Dordrecht/Boston: Reidel.
———. 2002 [1926]. *Erkenntnis und Irrtum*. Reprint of the 5th ed. Berlin: Parerga.
McCaskey, John P., ed. 2013. *Jacopo Zabarella. On methods, books III–IV, On regressus*, vol. II. Cambridge, MA/London: Harvard University Press.
Newton, Isaac. 1718. *Opticks. Or, a treatise of the reflections, refractions, inflections and colours of light*. 2nd ed., with additions. London: Printed for William Innys at the Westend of St. Paul's.
Oeing-Hanhoff, Ludger. 1971. Analyse/synthese. In *Historisches Worterbuch der Philosophie*, ed. Joachim Ritter, 232–248. Basel: Schwabe Verlag.
Popper, Karl. 2002. *The logic of scientific discovery*. London/New York: Routledge.
Rosenkranz, Johann Karl Friedrich 1850. Ueber die Psychologies als Naturwissenschaft. Allgemeine Monatsschrift für Literatur 1: 157–169.

Rusu, Doina-Cristina. 2013. *From natural history to natural magic: Francis Bacon's Sylva sylvarum*. Dissertation: Radboud University.

Schleiden, Matthias Jacob. 1969. *Principles of scientific botany. Or, botany as an inductive science*. New York/London: Johnson Reprint Corporation. 1849.

Schopenhauer, Arthur. 1813. *Ueber die vierfache Wurzel des Satzes vom zureichenden Grunde: eine philosophische Abhandlung*. In Commission der Hof- Buch- und Kunsthandlung: Rudolstadt.

———. 1819. *Die Welt als Wille und Vorstellung: vier Bücher, nebst einem Anhange, der die Kritik der Kantischen Philosophie enthält*. Leipzig: F.A. Brockhaus.

———. 1844. *Die Welt als Wille und Vorstellung*. Vol. II. 2nd ed. Leipzig.

Schulze, Gottlob Ernst. 1810. *Grundsätze der allgemeinen Logik*. Göttingen: Vandenhoeck & Ruprecht.

Senebier, Jean. 1776a. *Die Kunst zu beobachten*. Trans. Johann Friedrich Gmelin. Leipzig. 1775.

———. 1776b. *Die Kunst zu beobachten. Zweyter Band*. Trans. Johann Friedrich Gmelin. Leipzig. 1775.

Snyder, Laura J. 1997. Discoverers' induction. *Philosophy of Science* 64: 580–604.

Taylor, C.C.W., and C.C.W. Taylor. 2006. *Aristotle: Nicomachean ethics, books II–IV: Translated with an introduction and commentary*. Oxford: Oxford University Press, Incorporated.

Tonelli, Giorgio. 1976. Analysis and synthesis in the XVIIIth century prior to Kant. *Archiv für Begriffsgeschichte* 20: 178–213.

von Cotta, Bernhard. 1867. *Die Geologie der Gegenwart*. Leipzig: Verlagsbuchhandlung von J. J. Weber.

Weisz, Béla. 1871. Die Nationalökonomie und ihre Methode. *Jahrbücher für Nationalökonomie und Statistik*: 148–160.

Wolff, Christian. 1713. Vernünftige Gedancken von den Kräften des menschlichen Verstandes und ihrem richtigen Gebrauche in Erkäntniß der Wahrheit. *Halle (Saale)*.

Wundt, Wilhelm. 1883. *Logik. Eine Untersuchung der Principien der Erkenntniss und der Methoden wissenschaftlicher Forschung*. Zweiter Band: Methodenlehre. Stuttgart: Ferdinand Enke.

Zedler, Johann. 1732a. Analysis. In *Grosses vollständiges Universal-Lexicon aller Wissenschaften und Künste*, 35–36. Halle/Leipzig: Zedler.

———. 1732b. Analytische Methode. In *Grosses vollständiges Universal-Lexicon aller Wissenschaften und Künste*, 38–39. Halle/Leipzig: Zedler.

Zimmermann, Johann Georg. 1777. *Von der Erfahrung in der Arzneykunst*. Neue Auflage ed. Zürich: Orell, Gessner, Füesslin und Compagnie.

Jutta Schickore is Ruth N. Halls Professor of History and Philosophy of Science and Medicine at the Department of History and Philosophy of Science and Medicine, Indiana University (Bloomington). Her research interests include philosophical and scientific debates about scientific methods in past and present, particularly debates about (non)replicability, error, and negative results.

Chapter 13
From Chemical Analysis to Analytical Chemistry in Germany, 1790–1862

Peter J. Ramberg

Abstract The phrase "chemical analysis" has historically had two meanings: first, the long-standing, traditional definition of decomposing substances into their simpler, possibly ultimate, components and, second, the development of methods for identifying the proximate components contained in unknown mixtures and materials. The second meaning straddles the elusive border between theoretical and applied chemistry and has been present since antiquity. However, by the mid-nineteenth century, this second meaning of "chemical analysis" had become established as a new area of chemistry called "analytical chemistry." Based on searches of key phrases using Google Ngrams and within digital texts, this essay traces the appearance of key concepts in analytical chemistry in German textbooks between the late eighteenth century and the appearance of Karl Fresenius' *Zeitschrift für analytische Chemie* in 1862. It argues that analytical chemistry had become a subdiscipline by the middle of the nineteenth century and has the characteristics of a second-order activity, characterized by its focus on devising trustworthy and reliable techniques for the identification of unknown substances.

Keywords Analytical chemistry · Chemical analysis · Nineteenth-century chemistry textbooks · Chemical testing · Elements

13.1 Introduction

The twin concepts of analysis and synthesis have been central to chemistry for centuries, and linked closely to the belief that there are only a few fundamental elementary substances, atoms, or building blocks that make up all material substances. Indeed, as Joel Klein argues elsewhere in this volume, it was the development of experimental techniques of analysis and synthesis in the seventeenth and eighteenth

---

P. J. Ramberg (✉)
School of Science and Mathematics, Magruder Hall, Truman State University, Kirksville, MO, USA
e-mail: ramberg@truman.edu

W. R. Newman, J. Schickore (eds.), *Traditions of Analysis and Synthesis*, Archimedes 73, https://doi.org/10.1007/978-3-031-76398-4_13

centuries that led to a hierarchical conception of chemical combination and made chemistry autonomous. However, until the late eighteenth century, chemists had not agreed on a single operational definition of either analysis or synthesis—indeed, one of the conceptual difficulties is in unambiguously determining whether an analysis or a synthesis has occurred—until they adopted Lavoisier's quantitative criteria based on gravimetry and the operations of the laboratory. These new definitions created a different path by which chemists to maintain their focus on analysis and synthesis by establishing stoichiometrically defined proportions and identifying several new elementary substances through the nineteenth century.

During the nineteenth century, chemists continued to adapt and reconfigure the meanings of both analysis and synthesis. Analysis became, in its traditional sense of identifying fundamental components, the quantitative determination of atomic formulas using Daltonian atomic weights, while in organic chemistry, it was transformed into the concept of chemical structure. Synthesis assumed a variety of uses and meanings, particularly in the context of the rise of organic chemistry during the middle of the century, but all of those meanings were variations on its traditional meaning as the construction of materials from simpler components (Russell 1987; Rocke 2010; Jackson 2014; Jackson 2023).

This essay concerns another, lesser-known trajectory of chemical analysis in the early nineteenth century that is reflected in the title: the emergence of a new branch of chemistry known as "analytical chemistry." The traditional phrase, "chemical analysis," refers to a type of analysis, whereas "analytical chemistry" refers to a kind of chemistry that has a specific subject area, methods, and aims. "Analytical chemistry" is today an exceptionally large area of chemistry whose core concern is the creation of methods and instruments for detecting, identifying, and quantifying unknown compounds in materials. Curiously, "chemical synthesis" does not have a counterpart area called "synthetical chemistry," although chemists have been and are still avidly interested in developing synthetic methods.[1] In short, when we follow the trajectory of chemical analysis into the nineteenth century, we find that chemists continued to regard analysis as central to their discipline in the traditional way but that it was also transformed along a second, parallel path into analytical chemistry, created by a consciously reflexive consideration of the reliability of the methods and tools for identification available to chemists.

This area of chemistry appears to have been well formed by the middle of the century. In July of 1861, Karl Remigius Fresenius, director of the chemical laboratory in Wiesbaden, wrote to his mentor Justus Liebig that a new journal dedicated to analytical chemistry had become "not only useful, but absolutely necessary."[2] Fresenius had found that existing chemistry journals, especially the *Annalen der Chemie*, had become dominated by articles on organic chemistry, and so after "long

---

[1] As Catherine Jackson has argued, A.W. Hofmann created what he called "synthetical chemistry" as a means of probing the constitution of organic molecules, but the term did not become widespread (Jackson 2014, 2023).

[2] "nicht allein für nützlich, sondern für absolut nothwendig," Fresenius to Liebig, July 21, 1861, in Poth 2007, 68.

and careful consideration," Fresenius solicited contributions for his new journal (Poth 2007, 68). His requests proved fruitful and the first volume of the *Zeitschrift für analytische Chemie* appeared in 1862, with 519 pages—approximately 125 pages per quarterly issue—of both original research articles and abstracts from the literature. The first article was by Robert Bunsen and Gustav Kirchoff on their new spectroscopic analysis of the alkali metals, and other articles were by Friedrich Mohr, Heinrich Rose, Otto Erdmann, and Georg Lunge. Fresenius' *Zeitschrift* was the first specialized journal dedicated to a subspecialty of chemistry, predating the more famous *Zeitschrift für physikalische Chemie,* founded by Wilhelm Ostwald and J. H. van't Hoff in 1887.

Fresenius did not write an introduction or preface to the new journal that would justify it or explain its scope. It may have been the case that he did not believe it was warranted. By 1862, he had already completed 11 editions of his extremely popular *Anleitung zur qualitativen chemischen Analyse* and four editions of the corresponding volume on quantitative analysis. Both books were the market leaders among the many different textbooks and handbooks in analytical chemistry that emerged during the first half of the nineteenth century. Elsewhere, Fresenius also emphasized the importance that higher-order concerns about methods would have for chemistry as a whole, writing that "it doesn't take much effort to show that all the great developments in chemistry are linked directly or indirectly with new or improved analytical methods."[3] In other words, analytical chemistry lies at the heart of chemistry itself—all chemists are analytical chemists. However, the new journal, the multiple editions of Fresenius' textbook, and his perceived need for a new journal, along with the dozens of other textbooks of analytical chemistry that appeared in the decades before 1862, suggests that by 1862, "analytical chemistry" had become a distinct area of chemistry.

As I shall argue here, the emergence of analytical chemistry can be traced to the period between 1790 and 1830, when the phrase itself entered the chemical literature alongside the classical term "chemical analysis." The phrase "chemical analysis," as I shall use it here, has historically had two meanings: first, the traditional definition of decomposing substances into their simpler, possibly ultimate components and second, the development of methods for identifying the proximate components contained in unknown mixtures and materials. There was also a significant overlap and ambiguity between the two meanings since chemists were often unable to determine whether they had identified proximate or ultimate components. The second meaning straddles the elusive border between theoretical and applied chemistry and has been present since antiquity in the methods used for assaying metals and detecting alterations of pharmaceuticals (Newman 2000 and Totelin, this volume). By the eighteenth century, chemical analysis was applied to minerals, pharmaceuticals, mineral waters, forensic analysis, and many other practical areas in which the identity, purity, and quality of material substances are commercially

[3] "Ohne Mühe lässt sich nachweisen, dass alle grossen Fortschritte der Chemie in mehr oder weniger direktem Zusammenhang stehen mit neuen oder verbesserten analytischen Methoden" (Fresenius 1897).

important. As I hope to demonstrate here, by the early nineteenth century, this particular kind of chemistry dealing with the second meaning of "chemical analysis" had become "analytical chemistry," defined by the practitioners who had also marked out its concepts and methods.

What follows is a very preliminary step toward understanding the emergence of analytical chemistry by the examination of published texts to determine where and how chemists began to use these terms. To keep the number of sources manageable, I have restricted my analysis to books in German. I began with a Google Ngrams search for the term "analytical chemistry" (*analytische Chemie* in German) and two concepts that appear to be central to the emerging subject: "reagent" (*Reagens* or *Reagenz*), and "qualitative and quantitative analysis" (*qualitative* and *quantitative Analyse*). The results of these searches resulted in 34 German textbooks that cover analytical chemistry in some form, published between 1790 and the establishment of Fresenius' *Zeitschrift* in 1862. All the books examined were those available on Google Books or HathiTrust, which in most cases also facilitates a reasonably thorough search of the text for particular words or phrases. My discussion adheres closely to the published texts without pulling back to consider the broader context of the texts' authors, use, or publication, although I shall offer some observations on the broader trends and contexts in the conclusion.

13.2 Secondary Literature

Compared with the history of atomism, organic chemistry or physical chemistry, the emergence of analytical chemistry has received little historical treatment, with the exception of three works that have served as my starting point. The standard comprehensive history of analytical chemistry is Ferenc Szabadváry's *History of Analytical Chemistry* (Szabadváry 1966). Szabadváry begins in antiquity and ends in the early twentieth century with Mikhail Tsvett's invention of chromatography. While rich in detail, Szabadváry did not construct an overarching narrative about the emergence of analytical chemistry as a subject area within chemistry. Rather, his work is a survey of the invention of various analytical techniques, and, intentionally or not, agrees with Fresenius that all chemistry is analytical chemistry.

The other major historical work of note in analytical chemistry is Ernst Homburg's landmark 1999 *Ambix* paper on the role of analytical chemistry in the development of the German chemical profession. Homburg explained the rise of the chemical profession by elaborating several trends that converged during the 1820s (Homburg 1999). The first trend was the emergence of smaller, portable laboratories for chemical analysis in the field that made chemical testing simpler. The second trend was the state's involvement in creating public health regulations concerning pharmaceuticals, water purity, and food adulteration, which created the need for trained and certified chemists to carry out analyses. This resulted in pharmaceutical-chemical schools of analytical chemistry, especially in France and Prussia, with the emergence of new courses in "analytical chemistry" and their incorporation into the university, initially by Friedrich Stromeyer in Göttingen but followed elsewhere in the post-Napoleonic era.

Although both Szabadváry and Homburg provide an excellent starting point, neither explicitly define the concept of "analytical chemistry," although both recognize that the term emerged in the late eighteenth century as a variation on "chemical analysis." Similarly, both mention in passing the concept of the "chemical reagent," tracing the term's origin to Torbern Bergmann, who, according to Homburg, "deserves the title of 'father' of analytical chemistry because he perfected the use of reagents into an analytical methodology…" (Szabadváry 1966, 73; Homburg 1999, 4). Bergmann's role in the development of analytical chemistry, however, remains ripe for further study, and below I shall further flesh out the history of the term "reagent," placing Bergmann more fully in the context of both the earlier eighteenth century and what would follow in the nineteenth century.

William Jensen is among the few historians to have suggested that analytical chemistry was established early on as a specialty within chemistry, emerging during the first part of the nineteenth century. Like Szabadváry and Homburg, however, he does not examine contemporary definitions of the field. Jensen argues that it found its origins, together with inorganic and physical chemistry, in mineral chemistry and metallurgy (Jensen 2003, 126–7, 131). According to Jensen, analytical chemists focused on techniques, with little interest in the underlying chemistry and physics behind their tests, until Wilhelm Ostwald's 1894 text, *Die wissenschaftliche Grundlagen der analytischen Chemie*, in which Ostwald grounded the results of tests on the new theory of ionic dissociation and equilibrium (Jensen 2003, 153–4; Jensen 2017, 221).

13.3 Analytical Chemistry

We can begin with the phrase "analytical chemistry" itself. An Ngrams search for "analytische Chemie" suggests that the earliest use in German was in a 1787 work by the young Alexander von Humboldt on muscle and nerve fibers, where in a footnote he used it in the traditional sense:

> Our analytical chemistry gives correct information about the specific constituents of bodies and their quantitative proportions; but we are still far behind in the tricks for probing the relative masking (*Umhüllung*) of the [nature of] the elements.[4]

The term next appears in 1788 and 1789, in announcements by Sigismund Hermbstaedt in Berlin for his new private course in practical chemistry for aspiring pharmacists. The *Bibliothek der neuesten physisch-chemischen, metallurgischen, technologischen und pharmazeutischen Literatur* of 1789, for example, contains a

[4] Unsere analytische Chemie giebt über die spezifischen Bestandtheile der Körper und ihre quantitativen Verhältnisse richtige Aufschlüsse; in den Kunstgriffen aber, die relative Umhüllung der Elemente zu prüfen, sind wir noch weit zurück (Humboldt 1797, 128).

"Report of a chemical boarding facility for youth who want to become practical chemists." The courses included physics, mineralogy, pharmacy, *materia medica*, and analytical chemistry

> which I understand as special guidance for chemically analyzing unknown substances (*Körper*); where simultaneously fire assaying (*Probirkunst*), and metallurgical chemistry are practically worked through.[5]

It seems clear that already here, Hermbstaedt considered "analytical chemistry" to be a broad category, encompassing both assaying and metallurgical chemistry. At roughly the same time, Johann Trommsdorf (Private instructor in Erfurt) and Johan Friedrich Göttling (Professor in Jena) advertised *Probierkabinette* ("testing" or "assaying" cabinets) for sale. These *Probierkabinette* were physical, portable cabinets that contained various solutions and reagents for testing in the field. Both Trommsdorf and Göttling also wrote companion books that explained how such cabinets could be used (Göttling 1790; Trommsdorff 1801).

In 1798, Wilhelm August Lampadius at the Freiberg Mining Academy announced a course entitled "Analytical Chemistry," and in 1801, published the first textbook that used the phrase, his *Handbuch zur chemische Analyse der Mineralkörper*. He wrote that the book

> concerns chiefly the decomposition of mineral substances, and since the *aim* of this work remains only the analysis of these substances, I have preferred to name this occupation *analytical* chemistry.[6]

Two items from this passage are particularly noteworthy. First, Lampadius' explicit goal is the traditional chemical goal of decomposition or analysis of minerals, and so he continued to define "analytical chemistry" in the traditional sense. Second, the title of Lampadius' book itself contains the phrase "chemical analysis," which is consistent with his definition of analytical chemistry. Szabadváry cited Lampadius' book as one of the first—if not the first—textbook in analytical chemistry, although it is clear that Lampadius focused on mineral analysis and continued to use the traditional definition of analysis.

Between 1800 and 1820, the phrase continued to appear but was not used universally. It did not appear, for example, in Carl Friedrich August Hochheimer's 1792

[5] "Nachricht von einer chemische Pensionanstalt für Jünglinge, die sich zu praktischen Chemikern bilden wollen. … hierunter verstehe ich die besondere Anleitung, noch unbekannte Körper chemisch zu analysiren; wobey zugleich die Probirkunst, und metallurgische Chemie, praktisch durchgearbeitet wird." "Nachricht von einer chemische Pensionsanstalt fur Jünglinge, die sich zu praktischen Chemikern bilden wollen," *Bibliothek der neuesten physisch-chemischen, metallurgischen, technologischen und pharmazeutischen Literature* volume 2 (Berlin 1789, 253). *Allgemeine deutsche Bibliothek*, volume 84 (Berlin 1788, 620). All of these dates correlate with the earliest known use of the phrase reported by Homburg 1999.

[6] Lampadius' course was published in the *Intelligenzblatt der Allgemeine Literatur-Zeitung* (Jena and Leipzig) October 6, 1798, 1132. "Wir werden uns hier vorzüglich mit der Zerlegung der Mineralkörper beschäftigen, und da der *Zweck* dieser Arbeit immer nur Analysis der genannten Körper ist, so habe ich diese Beschäftigung vorzugsweise *analytische* Chemie genannt" (Lampadius 1801, 1). Lampadius's emphasis.

two-volume work on chemical mineralogy, or in Martin Klaproth's comprehensive and influential multivolume work on mineral analysis begun in 1795 (Klaproth, 1795–1810; Hochheimer 1792).[7] Göttling would go on to use the phrase in his substantially updated 1802 version of his book on how to use his commercially available *Probirkabinett,* but it appears only in the preface and is not defined for the reader in any specific way (Göttling 1802). Heinrich Kopp's 1805 text noted that "the perfection of mineralogy and the art of separation in recent times has raised the analysis (*Zerlegung*) of minerals (*Fossilien*) to a special branch of applied chemistry" but did not refer to it specifically as analytical chemistry.[8] In 1808, Johann John continued to use the phrase "chemische Analyse" (John 1808).[9] Karl Stahlberger mentions his "preference" (*Vorliebe)* for analytical chemistry in the preface to his 1819 text but otherwise does not define it (Stahlberger 1819).

In 1814, August Schulze Montanus (1782–1823) published a treatise on assaying materials that he revised for a second edition in 1818 and again in 1820. Montanus appears to have been active in Berlin, but further details about his life and career have proven elusive. His book is not mentioned by Szabadváry, Homburg, or Jensen, but its existence in three editions suggests that it was popular throughout the 1820s. In the preface to the second edition, Montanus describes himself as a physicist but one who often needed "pure assaying materials (*Prüfungsmittel*) or reagents, and with the lack of outside help, I felt compelled to prepare them myself."[10] Montanus proceeded to write 564 pages of "elementary instruction for beginners and less learned laboratory workers (*Laboranten*)." Montanus' focus was on the reagent (more on this later) and the application of analytical methods beyond metallurgy, though he did not use the phrase "analytical chemistry." In 1830, Wilhelm Lindes, also in Berlin as instructor at the *Königliche Realschule*, completed a revised fourth edition of Montanus' work. Lindes kept the emphasis on reagents but dropped the mineral water analysis section and added a section on the blowpipe, adapted from Berzelius' recent publication on the subject (Lindes 1830). The importance of Lindes' edition is that Fresenius would recommend it in 1841 as a detailed introduction to analytical techniques.

If this survey is representative, up until about 1820, the term "analytical chemistry" was known and used but not universally or consistently. This would change in 1821, when the Kiel chemist Heinrich Pfaff published his *Handbuch der analytischen Chemie für Chemiker, Staatsärzte, Apotheker, Oekonomen, und Bergwerks*

[7] There is little information on Hochheimer, who, judging by his numerous books, was active in Leipzig as a prolific author and scholar in many areas.

[8] "Die Vervollkommnung der Mineralogie und Scheidekunst in neuern Zeiten hat erst die Zerlegung der Fossilien zu einem besondern Zweige der angewandten Chemie erhoben," (Kopp 1805). Kopp was the father of the prominent mid-nineteenth century chemist Hermann Kopp (Rocke and Kopp 2012).

[9] John was a student of Klaproth and located in Moscow.

[10] "reiner Prüfungsmittel oder Reagentien, und sahe mich, in Ermangelung fremder Aushülfe, genötigt, mir solche selbst zu bereiten" (Montanus 1818, iv). I have been unable to locate a copy of the 1814 edition. Montanus also wrote on surveying (Montanus 1819).

Kundige, which put "analytical chemistry" directly in the title. Pfaff noted that a new book in this area of "practical chemistry" was long overdue, because Lampadius' textbook was no longer sufficient, and Montanus' text was "grossly ignorant in several places, and generally very deficient." Nonetheless, "no area of chemistry," wrote Pfaff, "is more diligent (*emsiger*) than analytical chemistry," carried out by the best chemists of the day, including Klaproth, Vauquelin, Berzelius, and Stromeyer. "Analytical or decompositional chemistry," Pfaff continued,

> is a specific part of applied chemistry, insofar as it has a specific purpose, namely the determination of the specific composition of substances (*Stoffe*) and the basic substances that compose them. Since chemistry has been correctly characterized according to a higher, more general notion as the science of the forces of affinity of matter and of the laws according to which these forces, in their interplay with all other forces of nature, bring about (*hervorbringen*) and establish the composition and decomposition, the formation and destruction, of all kinds of substances, the earlier narrower definition of it [chemistry], expressed by the name *Scheidekunst*, can still be applied to analytical chemistry as a specific branch of chemistry.[11]

Pfaff noted that analytical chemistry was both a practical and scientific skill (*Kunst*) that "does not confine itself to the determination and representation of the ultimate elements (*letzten Grundstoffe*), but should also teach how to depict the proximate components (*nähern Bestandtheile*) of mixtures (*Zusammensetzungen*), or at least to detect their presence."[12] One year later, Pfaff followed up this book with a second volume that completed the list of analytical techniques for inorganic compounds and added a section on the analysis of organic compounds. A second edition would follow in 1824.

Following Pfaff's book, more texts would include "analytical chemistry" directly in the title. In 1829, Heinrich Rose published one of the most influential compendia of identification techniques in analytical chemistry, the *Handbuch der analytischen Chemie*, which would go through many editions over the next 30 years. Rose's aim was to create a "guide *(Leitfaden)* for chemical-analytical investigation," but it was not so much a textbook as a handbook of methods for experienced chemists without

---

[11] "... die an mehreren Stellen grobe Blößen gibt, und überhaupt sehr mangelhaft ..."

"Kein Theil der Chemie ist emsiger bearbeitet worden, als die analytische Chemie."

"Analytische oder zerlegende Chemie ist ein besonderer Theil der angewandten Chemie, soferne der Zweck derselben ein bestimmter ist, nehmlich die Ausmittlung der bestimmten Zusammensetzung der Körper und der Grundstoffe aus welchen sie bestehen. Seitdem man die Chemie nach einer allgemeinen und höhern Idee als die Wissenschaft von den Verwandtschaftskräften der Materie, und den Gesetzen, nach welchen diese Kräfte in ihrer Wechselwirkung mit allen übrigen Kräften der Natur die Zusammensetzung und Zersetzung, die Bildung und Zerstörung der Körper aller Art hervorbringen und begründen, richtig karakterisirt hat, kann die frühere engere Definition derselben, die durch den Nahmen Scheidekunst ausgesprochen war, noch ihre Anwendung auf einen einzelnen Zweig derselben, die analytische Chemie, zulassen" (Pfaff 1821, 224, vi, 1).

[12] Die Aufgabe der analytischen Chemie schränkt sich aber nicht bloß auf die Ausmittlung und Darstellung der letzten Grundstoffe ein, sondern sie soll auch lehren, die nähern Bestandtheile der Zusammensetzungen darzustellen oder wenigstens ihrem Daseyn nach auszumitteln … (Pfaff 1821, 3).

discussing any techniques or giving any definitions (Rose 1833). While Rose would revise this work for several later editions, he never added any introductory material. Several later authors, including Fresenius, remarked that Rose's book was comprehensive and detailed but boring and inadequate for students.

In his 1830 edition of Montanus' textbook, Lindes also added "analytical chemistry" to the title but did not define it explicitly. In his 1835 textbook, Heinrich Wackenroder (director of the pharmaceutical institute at Jena) used the phrase "chemical analysis" in the title, and "analytical chemistry" in the text but, again, did not define it explicitly (Wackenroder 1836). Adolf Duflos' 1835 handbook of pharmaceutical-chemical practice, which contained standard techniques of qualitative analysis, did not mention either "chemical analysis" or "analytical chemistry" in the title or the text proper (Duflos 1835).

The appearance of Pfaff's text thus appears to have marked a turning point in the development of analytical chemistry, since the phrase "analytical chemistry" seems to have come into broader use after its publication. However, Pfaff seems to have been alone in explicitly defining the term. The reason for this lack of explicit definitions through the 1830s remains difficult to discern and requires additional analysis of the texts and their purposes. It may be that the phrase had become sufficiently common that practicing chemists were already familiar with the term and did not need it introduced. This was likely the case for Rose's text, which was written for practicing chemists, but not for some of the other texts written for students, such as Lindes, who also did not define the term.

In writing the first edition of his textbook, Fresenius also followed this pattern. The 1841 edition was a short 82 pages, with a brief introduction and a list of various testing procedures (Fresenius 1841). Fresenius wrote it for himself while he was a student in Bonn,

> Because for a long time I was not fortunate enough to do chemical analyses under the guidance of a teacher, but was completely confined to myself in their execution, I had a special opportunity to recognize the difficulties which the beginner, left to his own devices, almost inevitably encounters, despite the excellent instructions from H. Rose, Duflos and other masters.[13]

Fresenius initially wrote the book for his own use to distill the huge amount of material to a manageable level, to clarify the actual processes used for identification, and to remove any frequent misconceptions in earlier texts. Like most of his predecessors, he did not provide any definitions of the fundamental concepts behind analytical chemistry.

Fresenius continued to use his manual after moving to Giessen, where he provided his students with it for mineral analysis, but by 1843, he had expanded the work to 248 pages and added an extensive introduction. Liebig provided a preface,

---

[13] Da ich längere Zeit nicht das Glück hatte, mich unter der Leitung eines Lehrers mit chemischen Analysen zu beschäftigen, sondern in ihrer Ausführung ganz auf mich selbst beschränkt war, so bot sich mir besondere Gelegenheit, die Schwierigkeiten zu erkennen, welche dem sich selbst überlassenen Anfänger, trotz der trefflichen Anleitungen von H. Rose, Duflos und andern Meistern, fast unvermeidlich entgegentreten (Fresenius 1841, v).

considering it "a very convenient preparation (*Vorschule*) for using Professor Rose's excellent handbook."[14] Like Pfaff and other earlier authors, Fresenius recognized the interdependency of general theoretical chemistry and analytical chemistry. "Chemistry," wrote Fresenius,

> as is known, is the science that teaches us about the materials (*Stoffe*) that compose our Earth, their composition and decomposition, generally their behaviors with one another. A specific division of chemistry is designated with the name analytical chemistry, insofar as it has a specific aim, viz., the decomposition (analysis) of compound bodies (*Körper*) and the determination of their components.[15]
>
> Although chemical analysis draws on general chemistry and cannot be practiced without knowing it, analysis must on the other hand, also be considered as a main pillar on which the entire edifice of the science (*Wissenschaftsgebäude*) rests, for it is of almost equal importance for all parts of chemistry, theoretical as well as the applied, and its usefulness? for doctors, pharmacists, mineralogists, rational agriculturalists, engineers and others does not require explanation.[16]

The success of Fresenius' book spurred the publication of other new texts in analytical chemistry that would follow Fresenius' lead and define the area for the student in almost identical ways. Ludwig Posselt, a Privatdozent in Heidelberg, and Eduard Schweizer in Zürich argued that "analytical chemistry teaches the methods by which one has to proceed to investigate the composition of any compound (*Körper*)."[17] Georg Städeler defined "qualitative chemical analysis," as

> general information … obtained first about the nature of the compounds and their most important components by means of the dry way (a preliminary test); the bodies are then dissolved and an investigation is subsequently done using the wet way.[18]

---

[14] "sehr zweckmässige Vorschule für die Benutzung des trefflichen Handbuches vom Professor H. Rose." (Liebig, preface to Fresenius 1843).

[15] Die Chemie ist wie bekannt, die Wissenschaft, welche uns die Stoffe, aus denen unsere Erde besteht, ihre Zusammensetzung und Zersetzung, überhaupt ihr Verhalten zu einander kennen lehrt. Eine besondere Abtheilung derselben wird mit dem Namen analytische Chemie bezeichnet, insofern sie einen bestimmten Zweck, nämlich die Zerlegung (die Analyse) zusammengesetzter Körper und die Ausmittelung ihrer Bestandtheile verfolgt (Fresenius 1843, 1).

[16] Obgleich sich nun die chemische Analyse auf die allgemeine Chemie stützt und ohne Kenntnisse in derselben nicht ausgeübt werden kann, so muss sie andererseits auch als ein Hauptpfeiler betrachtet werden, auf dem das ganze Wissenschaftsgebäude ruht, denn sie ist für alle Theile der Chemie, der theoretischen sowohl, als der angewandten fast von gleicher Wichtigkeit und der Nutzen, den dieselbe dem Arzte, dem Pharmaceuten, dem Mineralogen, dem rationellen Landwirth, dem Techniker und Andern gewährt, bedarf keiner Auseinandersetzung (Fresenius 1843, 2).

[17] "Die analytische Chemie lehrt die Methoden kennen nach denen man zu verfahren hat, um die Zusammensetzung irgend eines Körpers zu erforschen" (Posselt 1846).

"Die analytische Chemie lehrt die Methoden kennen, durch welche man die Zusammensetzung von Körpern zu erfahren im Stande ist" (Schweizer 1848).

[18] Bei qualitativen Analysen sucht man sich zunächst durch eine Prüfung auf trocknem Wege (Vorprüfung) allgemeine Aufschlüsse über die Natur der Verbindungen und ihre wesentlichsten Bestandtheile zu verschaffen; man bringt darauf die Körper in Lösung und lässt eine Untersuchung auf nassem Wege folgen. (Städeler 1857, 3). Städeler's *Leitfaden* appeared in multiple editions through the remainder of the nineteenth century and was one of the chief rivals to Fresenius' text.

Finally, Carl Rammelsberg's *Leitfaden für die qualitative chemische Analyse* (1860) was another attempt to make Rose's handbook suitable for beginners and offered what was perhaps the simplest definition:

> The task to be solved by analytical investigations is in general, twofold: 1. Which components does the substance (*Stoff*) in question contain? and 2) in what quantities are these components present?[19]

In summary, the history of the appearance of "analytical chemistry" may be divided into three phases. The first began with Lampadius, who used the phrase but as another way of describing the traditional concept of chemical analysis. The term would continue to make its appearance in the literature, but it would not be used consistently until Pfaff's text in 1821. During the second phase, between 1821 and the appearance of the second edition of Fresenius' text, the term appears to have been in common use but not always defined for the reader. Following the second edition of Fresenius' text, the concept appears to have become stabilized, and it became necessary to include a formal definition for the purpose of training students.

13.4 Reagents

According to Jensen and Szabadváry, the concept of a chemical "reagent" originated with Torbern Bergmann in his *Chemical and Physical Essays* (1788), but the term and the concept of a "reagent" predates Bergmann by at least a century. It had appeared in its Latin form (*reagentia*), for example, as early as Johann Becher's *Physica Subterranea* (1669):

> With weighing alone, as they do in saltworks, it is impossible to determine the quality of waters, even though the quantity may be evident; and even if whatever body may have a diverse quantity, quality cannot be found from quantity, because, the relationship [*proportio*] of quantity with quality remains unknown to us up to this day, except insofar as the quality is known to us in a general sense, as in the case of the metal crown whose quality Archimedes discovered through quantity. Yet the more skilled physicists and chemists determine [*probent*] waters and mineral species by means of reagents [*reagentia*], for common salt precipitates a solution of silver, galls turn black if vitriol lies hidden in water, lye rejects sulfur, and so on.[20]

---

[19] Die Aufgabe, welche die analytische Untersuchung zu lösen hat, ist im Allgemeinen eine zweifache: 1. welche Bestandtheile enthält der zu untersuchende Stoff? und 2. in welcher Menge sind diese Bestandtheile vorhanden? (Rammelsberg 1860).

[20] Sola vero ponderatione, ut in salinis faciunt, aquarum qualitatem invenire, impossibile est, licet quantitas exinde pateat, et licet quodlibet corpus diversam quantitatem habeat, ex quantitate tamen qualitas reperiri nequit, quod quantitatis cum qualitate proportio, in hunc usque diem nobis incognita sit, nisi, quoad qualitatem subjectum nobis in genere notum est, ut in corona metallica, cujus qualitatem ex quantitate Archimedes invenit, licet quoque peritiores Physici & Chymici aquas, & species minerales per reagentia probent, nam sal commune preaecipitat solutionem lunae, si vitriol in aqua lateat, denigrabitur Galla, si sulphur, lixivium respuit &c (Becher 1669, 192). I thank Bill Newman for bringing this passage to my attention and for translating the text.

Here, Becher uses "reagent" casually, clearly assuming that the reader knows what reagents are, which suggests that the term has an even earlier history.

An Ngrams search reveals that "reagent" was used in German texts as early as 1723 by two unknown authors and again in 1749.[21] All three of these authors continued to use the Latin form (*reagentia*) in their texts, as indicated by the use of Latin typeface in the midst of the German Fraktur.[22] The term's earliest usage in German occurs in 1771 by Peter Simon Pallas, who used it casually in a footnote to describe an analysis of lake water: "The water has behaved in the following ways with the usual reagents."[23] Pallas then described each of the reagents used and the results.

It appears, then, that Bergmann neither coined the term nor the concept behind it in his *Chemical and Physical Essays*. He did, however, use the word "reagentia" extensively throughout the original Latin version of the *Essays*, and it may have been Bergmann's usage that influenced later chemists to use the term and concept more broadly, perhaps also in English, though this is uncertain.[24] Further Ngrams searches reveal similar definitions in the late eighteenth century, and the term is already common in the German chemical literature, appearing for example, in the first volume of Lorenz Crell's *Beyträge zu den chemischen Annalen* (1785), in a review article by a chemist named Struve, "On the reagents and their use in the

[21] Searches were done using the two different spellings of "reagent" in German: *Reagens* and *Reagentien* are the modern singular and plural forms, whereas the older spelling is *Reagenz* and *Reagenzien*. The only difference between the two terms is the change in spelling that occurred around the turn of the nineteenth century, so in what follows I have combined the searches for both spellings into a single analysis.

[22] Unknown author, *Aurea catena homeri. Oder eine Beschreibung von dem Ursprung der Natur und natürlichen Dingen* (Frankfurt and Leipzig, 1723, 64); Unknown author, "Specimen Chymicum de Diagnose Rerum mixtarum per Reagentia, oder von Untersuchung der Mineralischen Wasser, als da sind warme Bäder, Sauer-Brunnen, und Salz Quellen," *Sammlung von Natur- und Medizin- wie auch hierzu gehörigen Kunst- und Literatur-Geschichten*, vol 23 (Breslau 1723); Walthiere 1749, 24.

[23] "Das Wasser hat sich mit denen gewöhnlichen Reagenzien folgender massen verhalten" (Pallas 1771, 107).

[24] According to the OED, the earliest use of the term "reagent" in English is in Edmund Cullen's 1784 translation of Bergman's *Physical & Chemical Essays*, but a search of Cullen's translation reveals only one use of the term" reagent," in a footnote on page 125 of the translation (corresponding to pages 92–3 of Bergmann's original Latin text), containing the passage cited by the OED. Importantly, the note is Cullen's commentary, not a translation of Bergmann's text (In the note, Cullen mentions Guyton de Morveau, who is not mentioned by Bergmann in the corresponding Latin text). Curiously, when Cullen translated Bergmann's *Essays* he consistently translated "reagentia" as "precipitant." This is the source of some disagreement about Bergmann's use of the term "reagent." Following Cullen's translation, Jensen notes that Bergmann used the term "precipitant," while Szabadváry, translating Bergmann directly, said that Bergmann used the term "reagent." Cullen's use of "reagent" in the single location suggests that Cullen was familiar with the term, but for some reason chose not to use it in the translation. "Precipitant" is also a useful term, but it is unclear how common it was in English at the time and does not include all types of positive reactions, such as color changes that occur without precipitation. Given this analysis, the first full use of "reagent" in English remains unknown. (Jensen 2017; Szabadváry 1966, 73; Bergmann 1788).

analysis (*Zerlegung*) of Mineral water" (Struve 1785, p. 97).[25] Struve defined "reagent" as follows:

> According to the meaning of the word, a reagent must, for example, in mineral water, produce a reaction (*Gegenwirkung*), the appearance of which reveals the substances (*Stoffe*), which are mixed with water.[26]

Struve then described Bergmann's various reagents (*Reagentien*)—for example, litmus (*Lackmus*), turmeric (*Gilbwurz*), and tincture of oak gall (*Gallapfeltinktur*). In his 1790 guide for using his *Probirkabinett*, Göttling writes that the cabinet should contain the necessary "*reactive substances*" (*Reagentia*) that have become recently known in chemistry."[27]

The earliest German textbook to use the term appears to be the 1800 edition of Hermbstaedt's *Systematischer Grundriß der allgemeinen Experimentalchemie zum Gebrauch bey Vorlesungen* (1800). This book includes a section entitled "On the reactive substances (*gegenwirkende Mittel*), or reagents," in which Hermbstaedt wrote,

> The perceptible changes that specific different bodies, but especially their constituent parts, bring about in relation to one another when they are subjected to mutual contact, serve the chemist to infer the existence of certain components within them. By a suitably arranged preliminary examination of this kind, as well as the most precise observation of their outcomes, the chemist prepares for the actual decomposition of the body and the release (*Entstehung*) of its component parts. Substances that are able to bring about such changes to a high degree and in such a way that the results are decisive and can be perceived by the senses are called reactive substances (*Reagentia*).[28]

These three early uses of the concept of reagent reveal the different ways in which chemists tried to explain what the term "reagent" meant as it gained traction in German texts. For Struve, a reagent produces a "Gegenwirkung" with the materials in the unknown. Göttling and Hermbstaedt both use "gegenwirkenden Mittel." The

[25] "von den Reagentien und ihrem Gebrauche bey der Zerlegung der Mineralwasser "(Struve 1785, 97).

[26] Nach der Bedeutung des Worts muß ein Reagens, z. B. im Mineralwasser, eine Gegenwirkung hervorbringen, deren Erscheinungen die Stoffe, welche zur Mischung dieses Wassers kommen, zu erkennen geben (Struve 1785).

[27] Zu dem chemischen Probircabinette, welches ich im September 1788 ankündigte, und welches jener Ankündigung zu folge, die vorzüglichsten, durch die Scheidekunst in neuern Zeiten bekannt gewordenen gegenwirkenden Mittel (Reagentia), die bey der chemischen Zerlegung der Körper auf dem nassen Wege unumgänglich nothwendig sind ... (Göttling 1790, iii). Göttling's emphasis.

[28] Von den gegenwirkenden Mitteln, oder Reagentien. Die in die Sinne fallenden Veränderungen, welche specifisch verschiedene Körper, vorzüglich aber ihre Mischungstheile gegen einander veranlassen, wenn sie einer wechselseitigen Berührung unterworfen werden, dienen dem Chemiker dazu, auf das Daseyn gewisser Bestandtheile in ihnen zu schliessen, und sich durch eine zweckmäßig angestellte vorläufige Prüfung solcher Art, so wie der genauesten Beobachtung ihrer Erfolge, auf die wirkliche Zergliederung des Körpers, und die Entwickelung seiner Mischungstheile vorzubereiten. Stoffe, die dergleichen Veränderungen in einem hohen Grade, und so zu veranlassen vermögen, daß die Erfolge entscheidend und in die Sinne fallend sind, werden gegenwirkende Mittel (Reagentia) genannt. (Hermbstaedt 1800, 106).

most common translation of *Gegenwirkung* is "countereffect," and so "gegenwirkenden Mittel" would then become "counteracting substance." Neither of these translations is wholly satisfactory, since the reagent does not have a negating or countereffect, but is *revealing* something to the senses, as noted by Hermbstaedt. As Struve, Göttling and Hermbstaedt and others would further emphasize, these "reagents" produce a sensible effect—a change in odor, taste, color, or a precipitate, when the "agents" in the unknowns met the "reagents." Therefore, a more accurate translation of "Gegenwirkung" is simply "reaction," and "gegenwirkende Mittel" would then translate to "reactive substance." This is not entirely satisfactory in all cases, but it seems to be the closest to what these chemists were struggling to describe as the Latin word made its way into the German chemical literature.

After Hermbstaedt's 1800 text, similar definitions of "reagent" would appear in nearly all the textbooks that I examined. In 1802, Göttling defined reagent using the German term:

> In chemistry, the name reactive substance (*gegenwirkende Mittel*) is understood to be those chemical products, which by a color change, a precipitate that often differs in color, or by another phenomena that is easily observed, quickly indicates the presence of a component found by chemical investigation.[29]

Lampadius' first use of "reagent" was not in the context of its definition but in the context of the reagent's purity. "The analyst" wrote Lampadius,

> must be familiar with the preparation and use of the chemical aids (*Hülfsmittel*) that are used in analysis. These must be of the greatest purity. The results of so many imperfect analyses are simply due to the use of impure reagents.[30]

A considerable portion of Lampadius' book was devoted to describing how pure reagents could be made consistently. In a later section on "The instructions for preparation and testing of reagents," Lampadius defined the term "reagent" itself:

> By reagents we understand all the chemical aids (*Hilfsmittel*) that the analyst must apply, both in the wet and in the dry way, to discover the presence of any component of a mineral substance (*Mineralkörper*). These aids demand that they work definitively, so that there is no doubt remaining in the indicators obtained regarding the presence of a substance (*Körper*).[31]

---

[29] Man versteht in der Chemie unter dem Namen gegenwirkende Mittel diejenigen chemischen Produkte, welche durch eine Farbenveränderung, einen oft an Farbe verschiedenen Niederschlag, oder durch eine andere, leicht in die Augen fallende Erscheinung, die Gegenwart eines durch chemische Untersuchungen aufzufindenden Theils schnell anzeigen (Göttling 1802).

[30] Der Analytiker muß ferner die Bereitung und Untersuchungsweise der chemischen Hülfsmittel, welche bey der Analyse gebraucht werden, verstehen. Diese müssen von der größten Reinigkeit sein. Wie viel unvollkommene Analysen verdanken wir nicht bloß der Anwendung unreiner Reagentien (Lampadius 1801, 7).

[31] Anleitung zu der Zubereitung und Prüfung der Reagentien. Unter Reagentien werden hier alle die chemischen Hilfsmittel verstanden, welche der Analytiker sowohl auf dem nassen als auch auf dem trocknen Wege anzuwenden hat, um die Gegenwart irgend eines Bestandteiles in dem Mineralkörpern zu entdecken. Von diesen Hilfsmittel verlangt man, daß die bestimmt wirken, so daß bey den durch dieselben erhaltenen Anzeigen für das Dasein eines Körpers kein Zweifel übrig bleibe (Lampadius 1801, 53).

Montanus not only included "reagent" in the title of his text but also provided a lengthy definition for the reader:

> what has been said up to this point about chemical relationships, about the combination and decomposition of different substances that can be explained by this, etc. is entirely based on the nature and use of *chemical reagents*.
>
> *Reagents* are those substances which are used to examine the components of a substance (*Körper*) created by nature itself or by artificial preparation, or to decompose a body whose components are already known, or in many cases can only be guessed. They are called reagents because of the mutual influence they exert on other substances and, vice versa, are also experienced by them. In German, the words *Gegenwirkung* (mutual activity) or *Rückwirkung* (reactivity) would best describe their meaning.[32]

Montanus' attempt to define reagents for the reader reflects the difficulty of expressing what precisely a reagent is. From the context, Montanus means that reagents are mutually interacting substances, and so in this case *Gegenwirkung* is better translated as "mutual activity." Montanus' choice of *Rückwirkung* as a synonym of *Gegenwirkung,* would indicate that reagents "react to" the presence of a substance in the unknown and therefore have a certain empirically determined "reactivity" to certain substances.

In his 1821 textbook, Pfaff offered one of the clearest definitions of reagent as a "simple or compound substance, brought into chemical interaction with any other substance, that indicates the presence of this substance, by displaying a noticeable change to a sensory organ, especially the eye, but also smell [or] taste, in a distinctive way."[33] Pfaff went on to argue for the centrality of "reagent theory" for analytical chemistry:

> The theory of reagents (*Lehre der Reagentien*) forms the main foundation of analytical chemistry, and in certain respects it can be called its introductory part, as it constitutes an independent whole by itself, but can at the same time be regarded as a propaedeutic to analytical chemistry. A precise knowledge of the language of reagents and their manifold relationships ... facilitates the analyst's important task and prepares him anew for these tasks,

[32] Auf das, was von chemischer Verwandtschaft, von der daraus zu erklärenden Vereinigung und Zerlegung der verschiedenen Stoffe etc. bis hierher angeführt wurde, gründet sich nun ganz und gar die Natur und der Gebrauch der *chemischen Reagentien*.

Reagentien heißen solche Substanzen, deren man sich bedient, um die Bestandtheile eines durch die Natur selbst, oder durch künstliche Zubereitung entstandenen Körpers zu untersuchen, oder durch einen Körper, dessen Bestandtheile bereits bekannt sind, in vielen Fällen aber nur vermutet werden, zu zerlegen. Reagenzien heißen sie in Betracht der wechselseitigen Einwirkung, die sie auf andere Substanzen ausüben, und hinwiederum auch von diesen—vice versa—erfahren. *Gegenwirkung, Rückwirkung* möchte ihre Bedeutung in deutscher Sprache am besten bezeichnen (Montanus 1818, 22–3). Montanus' emphasis.

[33] Unter einem chemischen Reagens verstehen wir im allgemeinen irgend eine einfache oder zusammengesetzte Substanz, welche, in chemische Wechselwirkung mit irgend einer andern Substanz gebracht, durch eine bemerkbare Veränderung für irgend ein Sinnorgan, besonders das Auge, aber auch für den Geruch ... den Geschmack ... von eigenthümlicher Art das Vorhandensein dieser Substanz anzeigt (Pfaff 1821, 29).

> yet without completing it himself, as it requires a knowledge of the actual analytical operations and their appropriate sequence.[34]

Pfaff and several other authors noted that, in principle, any substance can serve as a reagent, but only a few offered quick and decisive results. Every substance that produces a chemical effect can be used as a reagent.[35]

In the first edition of his textbook, Fresenius assumed a general knowledge of chemistry, reagents, and instruments but recommended Lindes and Carl Winkelblech's *Elemente der analytische Chemie* (1840) for this information (Fresenius 1841, vi).[36] For the second edition, Fresenius himself provided definitions for the reader, dividing the book into four major points: operations, reagents, knowledge of the behavior of bodies with reagents, and a systematic course of investigation (Fresenius 1843, 3–4). "The substances (*Körper*)," Fresenius described,

> that indicate the presence of other substances by any conspicuous appearance, are called, in consideration of their mutual action, reactive substances, or reagents.[37]

From what I have seen so far, nearly all textbooks define the concept of a chemical "reagent" for the student in a similar way. Similarly, many authors note that all compounds are potentially reagents, but the best ones produce a quick effect that is detectable by our senses. Furthermore, all reagents must also be as pure as possible and prepared exactly according to the directions given, or they will give false positives.

[34] Die Lehre von Reagentien bildet die Hauptgrundlage der analytische Chemie, und sie kann in gewisser Hinsicht der propädeutische Theil derselben genannt werden, da sie für sich selbst ein unabhängiges Ganzes bildet, zugleich aber auch als Vorschule für die analytische Chemie betrachtet werden kann. Eine genaue Kenntniß der Sprache der Reagentien, ihrer mannigfaltigen Verhältnisse, in Folge welches Anzeigen des Daseyns der einfachen Grundstoffe sowohl, als der zusammengesetzten Substanzen geben, und zur Scheidung der ersteren dienen, erleichtert dem Analytiker sein wichtiges Geschäft und bereitet ihm neu demselben vor, doch ohne ihn selbst noch zu vollenden, wozu eine Kenntniß der eigentlichen analytischen Operationen, und ihrer zweckmäßigen Folge auf einander, erforderlich ist. (Pfaff 1821, 28).

[35] Hermbstaedt, for example, wrote in 1880: Jeder materielle Stoff im Weltraum, er sei einfach, oder gemischt, kann in gewisser Hinsicht als ein solches Reagens angesehen werden, und es kommt nur auf dessen zweckmäßige Anwendung am gehörigen Orte an. Eine vollständige Uebersicht solcher Reagentien, findet man in einigen davon besonders handelnden Schriften. Ihre Zubereitung, so wie ihre einzelne Anwendung werde ich in der Folge am gehörigen Orte einschalten (Hermbstaedt 1800, 106).

[36] I have been unable to locate a copy of Winkelblech.

[37] "Die Körper nun, welche die Gegenwart anderer durch irgend auffallende Erscheinungen anzeigen, nennt man, in Betracht ihrer wechselseitigen Einwirkung, gegenwirkende Mittel, Reagentien. "(Fresenius 1843, 19).

13.5 Qualitative and Quantitative Analysis

The third major concept of analytical chemistry to consider here is qualitative and quantitative analysis—in brief, what materials are contained in a sample (qualitative), and how much of that material is contained in it (quantitative). Previously, we saw how Becher in 1669 had already distinguished between these two forms of analysis, mentioning that measurements of quantity cannot typically result in conclusions about a material's quality (meaning identity). An Ngrams search for "qualitative Analysen" reveals that the phrase appears in one very brief mention in 1816 but sees more widespread use only in the 1820s. This correlates with the texts that I have found before 1820, none of which has the phrase "qualitative analysis," to judge from a search for that term in the texts.

An Ngrams search for "quantitative Analysen" yields some mentions from as early as 1805, in connection with the analysis of mineral waters (Graf 1805, XXIII). This also correlates with the texts I have found, where the earliest use of "quantitative Analyse" is in John's *Chemisches Laboratorium* from 1808 which appears in the context of investigation of metals and minerals (John 1808, 181). However, the most common early appearance of the phrase is in connection with the analysis of mineral waters, where the amounts of the dissolved minerals were important to know for medical and commercial purposes (Coley 1990). A section of Montanus' book is devoted to the "Investigation of mineral water according to its qualitative and quantitative components," in which he wrote

> Since the seventeenth century, outstanding chemists have most diligently endeavored to research qualitatively and quantitatively the constituents of select (*vorzüglich*) mineral waters. There are many results of this sort. Among many others, Hoffmann's book, *Systematische Uebersicht von Gesundbrunnen und Bädern* (1815), lists 242 mineral waters in the countries of the German Staatverein, according to their quantitative and qualitative components.[38]

Pfaff uses the term once:

> The main goal of analytical chemistry is the determination of the fundamental substances (*Grundstoffe*) of bodies in the most exact and complete way, according to their peculiar nature (qualitative analysis) and according to the amount in which each of the basic substances enters the composition of the body (quantitative analysis).[39]

---

[38] Die ausgezeichneten Chemiker haben sich in den nähern Zeiten seit dem Ende des 17ten Jahrhunderts eifrigst bemüht, die Bestandteile der vorzüglichen mineralischen Wasser qualitativ und quantitativ zu erforschen. Angaben dieser Art hat man jetzt viele; unter andern findet man in der Schrift: *Hoffmann's Systematische Uebersicht von Gesundbrunnen und Bädern* etc. (Berlin, 1815) 242 mineralische Wasser in den Ländern des deutschen Staatenvereins nach ihren quantitativen und qualitativen Bestandteilen aufgeführt (Montanus 1818, 385).

[39] Die analytische Chemie hat zu ihrem Hauptzweck, die Grundstoffe der Körper nach ihrer eigenthümlichen Beschaffenheit (qualitative Analyse) und nach der Menge, in welcher jeder der Grundstoffe in die Zusammensetzung der Körper eingeht (quantitative Analyse) auf das genaueste und vollständigste zu bestimmen (Pfaff 1821, 2).

The second volume of his text (1822) included extensive directions for performing a quantitative analysis of the components of mineral waters, beginning with the quantities of dissolved gases, followed by detection of dissolved minerals. Pfaff found such analyses essential for understanding the composition of surrounding geological features to get at the inner processes taking place within the Earth.

All texts emphasize the relationship between qualitative and quantitative analysis. The latter cannot be performed unless the former is completed first. Wackenroder (1835) illustrates this:

> Quantitative chemical analysis is so intertwined with qualitative analysis, and is so deeply rooted in the latter, that upon a little reflection, what is most essential and most important of the former emerges from qualitative chemical analysis.[40]

In addition to the concepts of qualitative and quantitative analysis, there is also the development of standardized methods for achieving these goals. We have already seen how reagents fulfilled the goal of simply identifying the nature of components, and this concept was well established by the early nineteenth century. The goals of quantitative analysis could be fulfilled by two major methods: gravimetric and volumetric. Gravimetric methods were the first to be adopted and nearly all textbooks that treat mineral water analysis describe techniques for determining the quantities of gas and minerals dissolved in the sample of mineral water. The gases could be systematically boiled out of solution and collected. Soluble minerals were then precipitated by reagents, followed by filtration and careful weighing. These techniques were well established by the 1820s.

Volumetric analysis, or titrimetry, has a more complicated history and was the last standard method to be adopted before 1862, at least in Germany. Volumetric methods were first developed in France in connection with the production of soda, chlorine, and sulfuric and hydrochloric acids for the rising soda and bleaching industry. Industrial use of soda and bleach required tests for their strength and concentration, and titrimetry—the addition of a solution of known concentration to the sample with an indicator to show when all the material had reacted—was a relatively quick and simple method for testing. These techniques were first developed by chemists in eighteenth-century France, and by the early nineteenth century, the techniques had been expanded to include many kinds of compounds, culminating in the work of Joseph Gay-Lussac in the 1820s, who made titrimetry more convenient, rapid, and accurate (Szabadváry 1966, 197–227; Crosland 1978, Chapter 9).

German chemists were slow to incorporate these volumetric techniques. Pfaff does not mention titrimetry, and Rose mentions it only briefly (Szabadváry 1966, 237). As late as 1846, Fresenius also hesitated to endorse titrimetry, writing,

> Sometimes solutions are measured, especially in applied industrial analysis ... but it is difficult to use the apparatus required to obtain accurate results in important analyses, therefore

[40] Die quantitative chemische Analyse ist so innig mit der qualitativen verschlungen, und wurzelt so tief in der letzteren, dass sich schon bei einiger Reflexion das Wesentlichste und Wichtigste der ersteren aus der qualitativen chemischen Analyse gleichsam aufdringt (Wackenroder 1836, ix).

> it is preferable to use the balance rather than this method (Fresenius in Szabadváry 1966, 237).

This hesitation can be attributed to two factors. Szabadváry suggests that German chemists were largely unfamiliar with the technique, and titrimetry became popular in Germany only after the publication of two comprehensive books on the subject by Karl Heinrich Schwarz in 1853, and Friedrich Mohr in 1855 (Schwarz 1853). Importantly, both had learned titrimetry outside of Germany, Schwarz at the *École Polytechnique*, and Mohr in Britain and America. Homburg suggests that cultural forces were also a factor, as titrimetry had long been associated with industrial analysis, and used only in that context, and academic German chemists did not see it as a reliable method in an academic context (Homburg 1999).

Nevertheless, the detailed descriptions of the accuracy of titrimetry given by Schwarz and Mohr would eventually convince German chemists of titration's precision and convenience, and this technique was deemed worthy of greater attention. In the third edition of his textbook (1853), Fresenius had changed his mind:

> Although volumetric analytical methods were also used earlier, they were rather isolated and were employed more for technical content determination than for scientific (*wissenschaftliche*) research. However, there is currently a trend toward also using volumetric analysis in scientific investigations. With the assistance of volumetric analysis, the aim is to achieve significantly faster results than would be possible by using weight-based analytical determination methods but without compromising accuracy.[41]

13.6 Analytical Chemistry and Rigor

Two aspects of "rigor" are under consideration here: one for qualitative analysis, the other for quantitative. For a proper qualitative analysis or simple identification of a mixture's components, particular standards were established for identification by the known effect produced by a reagent, either a color change, precipitate, or a change in taste or smell. This rather large set of tests, which ran to several hundreds of pages in most of the textbooks I examined, was well known by the 1830s, and it was Fresenius who developed a successful rational stepwise scheme for narrowing down the list of possible compounds in an unknown sample. The methods of qualitative analysis were not "exact," but they were "rigorous" in terms of the results' certainty, in so far as the reagents were pure and the technique followed properly.

The methods of quantitative analysis were "exact" in a measurable, numerical sense, in that they state precisely how much of a given component was in an unknown mixture. Generally speaking, quantitative analysis could only follow a

[41] Benutzte man auch früher maassanalytische Methoden, so standen solche doch ziemlich vereinzelt da und wurden mehr bei technischen Gehaltsbestimmungen als bei wissenschaftlichen Untersuchungen angewandt, während jetzt das Streben der Zeit dahin geht, auch bei letzteren mit Hülfe der Maassanalyse, unbeschadet der Genauigkeit, ungleich rascher zum Ziele zu kommen, als dies bei Anwendung gewichtsanalytischer Bestimmungsmethoden möglich ist (Fresenius 1853, xi).

successful qualitative analysis—that is, after the identity of the components was known. Gravimetric methods required precipitation, filtration, drying, and careful weighing of the material, considering the weight of the filter paper or its ash after combustion. Chemists could determine very small quantities. For example, in 1818, Montanus noted the quantities involved in mineral analysis:

> It is easy to overlook the fact that this type of investigation will become increasingly difficult when more substances are in a [sample of] water and when the quantities of water are smaller. Often the dissolved substances are scarcely 1/6000th of the weight of water.[42]

Titrimetry could be equally precise, by using a solution of precisely known concentration to react with the unknown. Stoichiometric calculations would then give the amount of the unknown in the sample.

The prominent textbooks that I have examined above cultivated this rigor as a desired virtue for the emerging analyst. The reliability of analytical results in either qualitative or quantitative analysis was highly dependent on the reagents' purity, and the confidence in any result was only as good as the ability to know that their reagents were pure and that their technique was sound. Lampadius emphasized persistence and patience:

> *Tenacity* in work is a quality that is essential for every practicing natural scientist, particularly for an analyst. Many analyses (*Zergliederungen*) in my early days of engaging in analysis failed simply because I could not wait until the decomposition of the mineral (*Fossil*) was properly completed through reagents, filtration, recaustitizing and other similar processes! Those who cannot wait for weeks and months to obtain a specific result would be better off not starting analytical work at all.

He then extolled the virtues of complete laboratory work:

> It is hardly necessary to note that *impartiality* and a *love for truth* should justly be commended. It is far from my intention here to target anyone with these remarks. But whoever entertains the idea of wanting to conclude analyses on paper should consider how much he harms science (*Wissenschaft*) by his self-centeredness. He analyses (*zergliedert*) not to convince himself, but to be regarded as an analytical chemist (*Analytiker*) by the chemical public.[43]

---

[42] Man übersieht leicht, daß diese Art von Untersuchung immer um so schwieriger werden müsse, je mehr Stoffe sich in einem Wasser befinden, und je geringer die Quantitäten derselben sind. Oft betragen die aufgelösten Substanzen kaum 1/6000 vom Gewicht des Wassers (Montanus 1818, 244).

[43] *Beharrlichkeit* bey der Arbeit ist eine Eigenschaft welche zwar jedem ausübenden Naturforscher, vorzüglich aber dem Analytiker notwendig ist. Wie manche Zergliederungen sind mir anfänglich bey meiner Beschäftigung mit der Analyse bloß darum verunglückt, weil ich es nicht erwarten konnte bis die Zerlegung des Fossils durch das Reagens oder die Filtration, das Aussüßen und dergleichen Arbeiten mehr gehörig beendet waren! Wer nicht Wochen und Monate lang warten kann um ein bestimmtes Resultat zu erhalten, der fange lieber die analytischen Arbeiten gar nicht an.

Unpartheylichkeit und *Wahrheitsliebe* sollte man billig anzuempfehlen gar nicht nöthig haben. Auch sei es fern von mir hier durch diese Bemerkung jemand treffen zu wollen. Wem es aber einfällt die Analysen auf dem Papier zum Theil beendigen zu wollen, der bedenke wie sehr er der Wissenschaft durch seine Eigenliebe schadet. Er zergliedert nicht um sich zu überzeugen, sondern

Pfaff offered several "rules" (*Regeln*) for the successful use of reagents in pursuit of "unambiguous" (*unzweideutig*) results. He noted the reagent's concentration, and the effect (*Rückwirkung*) of non-aqueous solvents. The sample may include other substances that might affect the test for a given substance. The temperature of the reagent and the necessary time required to produce a positive test could affect the results. The analyst should ensure the reagents' "utmost purity" (*grössten Reinheit*) and note carefully the different possible textures of various precipitates as well as knowing how to decide which reagent to use first when the components of the sample are completely unknown (Pfaff 1821, 21–8).

Montanus also emphasized the importance of a reagent's purity and of following the written procedures:

> … the reagents must be chemically pure, meaning free from foreign impurities; otherwise, multiple complex results will be obtained, but [with] no definite indications of a certain substance. The preparation of each of the chemically pure reagents should therefore be noted clearly and concisely as required by the purpose of this book, so that enthusiasts (*Liebhaber*) of the subject are enabled to prepare at least the principal testing agents with little effort from raw materials. The task is not easy, but with persistence and some experimental skill, everyone will succeed, except for a few cases that can be difficult even for the most experienced chemists. Potassium ferrocyanide, the principal reagent for metals, especially for iron dissolved in solutions, has not yet been prepared in a pure form free of iron content […]. In these difficult cases, one must be content with the most purified preparation of the reagent that is possible.[44]

In his 1847 introduction to quantitative analysis, Fresenius elevated these virtues of patience, practical skill, and knowledge of reagents and techniques to a moral issue for the practicing chemist. In this lengthy passage, he exhorted the reader that

> *Skill must unite with knowledge.* This principle applies in general to all applied sciences; but if it deserves special mention in any of them, it is in quantitative analysis. Armed with the most thorough knowledge, one cannot determine how much common salt is in a solution unless one can pour a liquid from one vessel into another without splashing or a drop running down the rim of the vessel, etc. The hand must acquire the ability to perform with care and skill the operations involved in quantitative analysis, an ability which can only be acquired by practice.

---

um im chemischen Publik als Analytiker zu gelten (Lampadius 1801, 13–14). Lampadius' emphasis.

[44] Zu diesem Behuf müssen aber die Reagentien chemisch rein d.h. frei von fremdartiger Beimischung seyn, weil widrigenfalls mehrfältige verwickelte Resultate, aber keine bestimmten Anzeigen auf einen gewissen Stoff erhalten werden. Die chemisch reine Darstellung der Reagentien soll daher bei jedem derselben mit Deutlichkeit und Kürze, welche der Zweck dieses Buches befiehlt, angemerkt werden; damit Liebhaber dieses Gegenstandes in den Stand gesetzt werden, sich wenigstens die vorzüglichsten jener Prüfungsmittel mit wenig Aufwand aus den rohen Materialien selbst zu bereiten. Die Sache ist nicht leicht, aber mit Beharrlichkeit und einiger Geschicklichkeit im Experimentiren wird sie gewiss jedem gelingen, einige wenige Fälle ausgenommen, welche auch den geübtesten Chemikern genug zu schaffen machen. Das blausaure Kali, das vorzüglichste Reagens auf Metalle, besonders auf Eisen in Flüssigkeiten aufgelöst, hat bis jetzt noch nicht ganz rein von Eisengehalt dargestellt werden können. (Klaproths chem. Wörterb. I. p. 406.) In solchen schwierigen Fällen muss man sich mit der möglich reinsten Darstellung des Reagens begnügen (Montanus 1818, 24–6).

Knowledge and skill must complement the desire to strive honestly for truth and the utmost conscientiousness. Anyone who has dealt with quantitative analysis to any degree knows, especially at the outset, that there are occasional cases where there are doubts about whether the result will be accurate, or where it is certain that the result will not be very exact. Sometimes a little has been spilled, sometimes there is a loss through decrepitation [roasting or calcining a substance], sometimes there are doubts about whether there was a mistake made in weighing, sometimes two analyses do not quite agree. In such cases it is a question of having the conscientiousness to immediately do the work again. Anyone who does not have this willpower, who shies away from trouble where the truth is at stake, who gets involved in guesswork and conjecture where the aim is to obtain positive certainty, must be denied the ability and profession to carry out quantitative analyses just as well as if he lacked knowledge or skill. Anyone who does not have full confidence in his own work, who cannot swear by his results, may go ahead and analyze for practice, but be careful not to publish his results as reliable or apply them. It would not be to his advantage, and it would be to the detriment of science.[45]

All of these passages suggest that training students as analysts required that the discipline necessary for patient, diligent work be instilled in them along with an acute awareness of how error-prone chemical analysis can be, to acquire the necessary confidence in their results, which could have significant commercial, legal, medical, or economic consequences. Analysts required knowledge of theoretical chemistry (mainly stoichiometry), practical skills in transferring and weighing materials, and a thorough knowledge of both the proper preparation and use of reagents. In Fresenius' words, this would avoid "guesswork and conjecture" and create "positive certainty" in the results.

---

[45] *Mit dem Wissen muss das Können sich vereinigen.* Dieser Satz gilt im Allgemeinen bei den gesammten angewandten Wissenschaften; wenn er aber bei irgend einer insbesondere hervorgehoben zu werden verdient, so ist es bei der quantitativen Analyse der Fall. Mit den gründlichsten Kenntnissen ausgerüstet, ist man nicht im Stande zu bestimmen, wie viel Kochsalz in einer Lösung ist, wenn man nicht eine Flüssigkeit aus einem Gefäß in ein anderes gießen kann, ohne dass etwas wegspritzt oder ein Tropfen am Rande des Gefässes hinabläuft u. s. w. Die Hand muss sich die Fähigkeit erwerben, die bei quantitativen Analysen vorkommenden Operationen mit Umsicht und Geschick auszuführen, eine Fähigkeit, welche einzig und allein durch praktische Uebung erworben werden kann.

Das Wissen und Können muss das Wollen, das redliche Streben nach der Wahrheit, die strengste Gewissenhaftigkeit ergänzen. Jeder, der sich nur einigermaßen mit quantitativen Analysen beschäftigt hat, weiß, dass sich, besonders am Anfange, zuweilen Fälle ereignen, in denen man Zweifel hegt, ob das Resultat genau ausfallen wird, oder in denen man gewiss ist, dass es nicht sehr genau ausfallen kann. Bald ist ein wenig verschüttet worden, bald hat man durch Decrepitation einen Verlust erlitten, bald zweifelt man, ob man sich im Wägen nicht geirrt habe, bald stimmen zwei Analysen nicht recht überein. In solchen Fällen handelt es sich darum, dass man die Gewissenhaftigkeit habe, die Arbeit alsobald/alsbald noch einmal zu machen. Wer diese Selbstüberwindung nicht hat, wer Mühe scheut, wo es sich um Wahrheit handelt, wer sich auf Schätzen und Muthmassen/Muthmaßen einlässt, wo es die Erlangung positiver Gewissheit gilt, dem muss Fähigkeit und Beruf zur Ausführung quantitativer Analysen eben so gut abgesprochen werden, als wenn es ihm an Kenntnissen oder Geschicklichkeit gebräche. Wer seinen Arbeiten selbst nicht volles Vertrauen schenken, wer auf seine Resultate nicht schwören kann, der mag immerhin zu seiner Uebung analysiren, nur hüte er sich, seine Resultate als sicher zu veröffentlichen oder anzuwenden, es dürfte ihm nicht zum Vortheil, der Wissenschaft aber würde es nur zum Nachtheil gereichen (Fresenius 1847, 3–4). Fresenius' emphasis.

13.7 Conclusion

In 2007, Jan Frercks and Michael Markert argued that during the first decade of the nineteenth century, chemists had "invented" the concept of theoretical chemistry (*theoretische Chemie*) by reconfiguring the central ideas of chemistry in the textbooks of the period (Frercks and Markert 2007). If my analysis here is correct, during roughly the same time, chemists—at least in Germany—also "invented" analytical chemistry, and by 1830, "analytical chemistry" had already coalesced into a distinct area within chemistry. However, a great deal more may be said about this process beyond my textual analysis. A full study of the emergence of analytical chemistry in the early nineteenth century, even confined to Germany, would require a book-length treatment and would need to include other factors, including the establishment of teaching positions, journals, and professional societies.

We can, however, take a small step back, look at the wider context of the texts examined above, and note at least two factors deriving from trends in late eighteenth-century chemistry that led to this new area. First, as Homburg has argued, by 1830, chemists had become professionalized because of the commercial and legal need for chemical analysis and new analytical techniques. This need was met by the growing number of private and university laboratories for training chemists in these new techniques. While the later dye industry is credited with the large-scale professionalization of organic chemists, the first major source of jobs for trained chemists emerged during the first half of the century with the application of analytical techniques to serve commercial needs.

Second, as Ursula Klein has demonstrated, the eighteenth-century culture of chemistry encompassed a broad spectrum of activities, from artisanal skills on one end to theories and analytical knowledge on the other, but all shared equipment and goals, resulting in considerable crossover—for example, pharmacists moved into natural inquiry, and academics became involved in practical matters related to mining and pharmaceuticals (Klein and Lefèvre 2007; Klein 2008, 2020). During the eighteenth century, chemists in Germany also benefited from a general increase in moral support and justification for their discipline as a legitimate profession, as reflected, for example, in the new salaried chairs for chemistry at universities and at the Berlin academy (Hufbauer 1982). In other words, the developing qualitative and quantitative techniques of the late eighteenth century in industrial, mineralogical, pharmaceutical, agricultural, and cameral contexts provided a ready means for chemists to serve broader cultural, political, and commercial needs.[46]

As part of the process of carving out this professional niche, therefore, chemists created the concept of "analytical chemistry" itself as the branch of chemistry that deals with the identification and quantification of unknown substances. As Pfaff wrote, "analytical chemistry" took on the mantle of the older terms *Scheidekunst*

---

[46] While the concept of "cameral chemistry" is intriguing, it remains under-researched to date. For the example of Swedish cameral chemistry in the eighteenth century, see Fors 2014. An example in early nineteenth-century Germany is Hermbstaedt 1808.

and *Probirkunst* to designate the array of methods used to assay the purity of various substances and determine the components of various materials and solutions. Analytical textbooks predate textbooks in organic chemistry by a wide margin, beginning with Lampadius in 1801 and possibly before.[47]

These trends raise further questions that lie well beyond the scope of this essay. First, there is the shift from the very old term *Probirkunst*, the specific term for assaying, to the broader term "analytical chemistry." The inclusion of water analysis and the many other commercial needs for testing may have necessitated a more general term. Second, the early nineteenth-century German texts that I examined also skew heavily toward Prussia and northern German states, particularly the Berlin circle of chemists, which included Hermbstaedt, Klaproth, and Humboldt, who began his career as a Prussian mining official (Klein 2012). We could also add Montanus and Lindes to the list of Berlin chemists. There is also the broader Prussian circle of Lampadius, Göttling, and Trommsdorff.[48] Only a broader institutional and biographical study could determine the importance of Prussia and Berlin.[49] Moreover, within the Berlin circle of chemists itself, Klaproth's role in the emergence of analytical chemistry requires further study. His multivolume text on mineral chemistry is cited extensively in nearly all works written in the 30 years after its appearance, and Klaproth was among the most distinguished of these early analytical chemists. Curiously, however, he did not use the phrase "analytical chemistry" and used the terms "reagent" and "*gegenwirkende Mittel*" only very sparingly throughout all five volumes (Klaproth 1795).

The concept of "reagent" itself also requires further historical analysis. It became central to qualitative analysis and analytical chemistry generally and had already become commonplace in German by the end of the eighteenth century. Yet, as I noted above, the term appears to have had a long history of common usage in Latin well before 1800, dating at least to the seventeenth century. The concept is connected with the exceedingly long-standing tradition, dating from antiquity, of assaying minerals and testing pharmaceuticals for purity and authenticity (Newman 2000 and Totelin, this volume). The long history of these identification techniques and the origin and use of "reagentia" in Latin is yet to be written, but beginning in the late eighteenth century, German chemists imported the word "reagentia" into German and adapted it to match both the goals of the professional analytical chemist and the new array of elements defined under Lavoisier's criteria. By the early nineteenth century, the word "reagent" itself had been transformed into a German word but was defined by its German equivalents.

---

[47] Given that textbooks on metallurgical chemistry and technical chemistry had already emerged in the eighteenth century, an argument may be made for their own emergence as distinct areas within chemistry. For example, see Gellert 1750 and Gmelin 1786. I thank Ernst Homburg for this suggestion.

[48] Pfaff was German but located in the northern city of Kiel, which before 1864 was part of Denmark (Kragh and Bak 2000).

[49] I thank Ernst Homburg (personal communication) for raising these important questions.

Significantly, as we have seen earlier, some early texts refer to the "theory of reagents" or to "reagent theory." An Ngrams search for "Reagentienlehre" reveals the earliest appearance in an 1811 Nürnberg *Realschule* program, in the description of a course on chemistry.[50] In Giessen, Wilhelm Zimmermann (Justus Liebig's predecessor) offered a course on "Reagentienlehre" in the fall of 1822.[51] According to Ngrams, the usage begins to rise around 1820, peaks around 1832, and appears only sporadically thereafter up to 1900, when it disappears.[52] This phrase did not become commonplace, although a review of Fresenius' 1843 book notes that given

> the great number of works on analytical chemistry and on the theory of reagents (*Lehre von Reagentien*), which have appeared recently, one would think that it would be unnecessary or at least not urgent, to publish yet another work of this kind.[53]

A more thorough search of the literature might identify more examples, but the idea of a "reagent theory," even if the term itself did not catch on, suggests a new conceptual foundation for analytical chemistry. The key transition here seems to be the conversion of a much older, well-known term into a broader conceptual scheme.[54]

The consistent—albeit short-lived—use of the phrase "reagent theory" suggests it had become a central concept in analytical chemistry, in the same way that "isomerism" or "structure" would become central to organic chemistry, or "equilibrium" in physical chemistry. The twin concepts of quantitative and qualitative analysis also served to define the discipline. Further examination of the textbooks would be necessary to identify and explore other central concepts. Among these would be the rise and fall of the blowpipe (present only in some of the texts I have seen), the "grouping" of various elements by systematic testing, and the concept of wet and dry methods.

In light of the factors I have described here—the creation of a professional identity, of new publications, of a core set of defined concepts and techniques—analytical chemistry appears to have emerged as a distinct subdiscipline of chemistry in the first third of the nineteenth century, well before the appearance of the better-known subdisciplines of organic and physical chemistry. However, two significant

[50] 1911. *Verzeichniß sämtlicher Schüler der Königlichen Real-Studienanstalt zu Nürnberg*. Nürnberg: Gehald.

[51] 1822. *Großherzoglich Hessisches Regierungsblatt*, September 1822, 423.

[52] In addition to the Nürnberg and Zimmermann courses, at least six other texts that use the phrase *Reagentienlehre* were published between 1820 and 1830.

[53] … man hätte glauben sollen, daß der großen Zahl von Schriften über analytischen Chemie und über die Lehre von Reagentien, welche in der neueren Zeit erschienen sind … es unnötig oder wenigsten nicht dringend gewesen wäre, noch eine Schrift dieser Art zu veröffentlichen. … Diese Besorgnis [p. 47] die vorliegende Schrift als etwas Überflüssiges zu erkennen, wird aber fast beim ersten Blick schon gehoben, wenn man die von Dr. Fresenius veröffentlichte Arbeit etwas ins Auge faßt und man überzeugt sich bald, daß diese Anleitung zur qualitativen chemischen Analyse dem trefflichen Werke von Heinrich Rose recht gut zur Seite stehend und als eine sehr zweckmässige Vorschule desselben betrachtet werden kann (Review in the Gelehrter Anzeiger (Munich), in Poth 2007, 46–7).

[54] By the end of the nineteenth century, the concept of a reagent had also become widespread in organic chemistry for the identification of specific functional groups (Jackson 2017).

developments emerge aside from the creation of a new area of chemistry. The first relates to the topic covered by Bernadette Bensaude-Vincent's paper in this volume, in which she questions the assumed symmetry and reversibility of analysis and synthesis. Bensaude-Vincent traces two historical traditions—the intellectual tradition, founded by Lavoisier, which mapped logical and linguistic analysis onto material composition (a building-block version of chemistry), in which matter is conceived as fundamentally passive. The alternative pragmatic tradition continued of the older tradition of *mixtio,* wherein matter is active and does not consist of passive building blocks. According to Bensaude-Vincent, these two versions of synthesis and analysis continue to co-exist and complement one another. The emergence of analytical chemistry, a variant of chemical analysis, suggests another sort of asymmetrical relationship of analysis and synthesis that began to appear in the early nineteenth century. That is, as analytical chemistry emerged, it was pursued independently of and without regard to synthesis.

The second broad characteristic to note about the emergence of analytical chemistry relates to the multiple discussions in the literature about the purity of reagents, the reliability of techniques, and the need for discipline among its practitioners. In other words, analytical chemists consciously drew attention to shoring up the potential weaknesses in identification techniques. Jutta Schickore (2007) has described the emergence of this kind of reflexive analysis in the context of creating reliable tools and methods in microscopy, at roughly the same time as the emergence of analytical chemistry. Schickore has described this activity as the creation of "second-" or "higher-order" discussions and reflections about techniques. The development of analytical chemistry appears to follow a similar process, in which chemists created a set of methods and tools (but not in the literal sense of instruments like microscopes) that covers chemistry in its entirety. Analytical chemistry is another manifestation of this kind of second-order thinking, in which chemists solidified and codified the epistemic functions of existing techniques in textbooks, manuals, and technical papers, creating a sophisticated methodology. Looking at analytical chemistry in this light explains Fresenius' attitude toward analytical chemistry noted in the introduction—all chemists are analytical chemists, and analytical chemists are not creating any essentially "new" chemistry but are refining identification techniques based on known reactions.[55]

A consideration of analytical chemistry as part of a higher-order form of reflexive thinking about techniques also helps to explain why the formal emergence of analytical chemistry, although explored in the literature, has largely gone unnoticed as a subdiscipline of chemistry that predates organic and physical chemistry. The sole exception to this is Jensen's argument, detailed above, that analytical chemistry emerged from metallurgy and mineral chemistry well before other organic and physical chemistry.[56] Although my analysis supports Jensen's view, it does not sup-

[55] I am grateful to Jutta Schickore for pointing out this similarity.

[56] Homburg's study makes the emergence of analytical chemistry as a subdiscipline implicit, but he does not explicitly argue for this point.

port Jensen's claim that it was only in the 1890s that analytical chemistry found its theoretical foundations. Many of the authors of the textbooks examined herein grounded their methods in the principles of general chemistry. Many introductions include extensive discussions of affinity theory (*Verwandschaftslehre*) to provide the theoretical basis for how the reagents worked to identify compounds. Textbooks are organized by the known chemical elements and their compounds, with a composition based on the new theories of stoichiometry and multiple proportions. Fresenius, in particular, integrated the practice of analytical chemistry with general chemistry. Textbooks were not merely focused on technique and its refinement, with "little or no interest in unraveling the underlying chemistry and physics," as Jensen claims (Jensen 2003, 157).

Nevertheless, although Jensen correctly claims that analytical chemistry had emerged very early in the nineteenth century, it has gone largely unnoticed in general histories of nineteenth century chemistry.[57] Why might this be the case? The answer lies, perhaps, in another curious asymmetry, that between analytical chemistry texts and general chemistry texts of the same period. As I have just noted, many of the analytical texts provide a thorough discussion of general chemical theory to ground the testing methods.

The same does not seem to be true for general chemistry textbooks in this same period. That is, while general texts grew and often branched into separate volumes dealing with inorganic and organic chemistry, textbook authors generally did not see the need for a third volume of analytical chemistry. This could be explained by the attitude that general chemistry itself incorporated analytical chemistry—as Fresenius argued, all chemists were analytical chemists to some degree. In his monumental textbook, for example, Berzelius did not include a separate volume on analytical chemistry but did mention reagents in the entry on "analysis" in the tenth volume's encyclopedia of chemical operations and terms (Berzelius 1841, 27). Berzelius did not write a separate entry for "analytical chemistry" but did pen entries for "organic analysis" and "inorganic analysis." The vast *Handwörterbuch der reinen und angewandten Chemie,* started by Liebig in 1842 and continued by Hermann Kolbe, does not have an entry for "analytical chemistry," and for "reagent" the reader is directed to the brief article on "reaction" (Liebig et al. 1842, 1854).

One final example from Friedrich Wöhler is insightful. In 1849, he anonymously published a small book for his own students entitled *Beispiele zur Übung in der analytischen Chemie*. This book includes no table of contents, definitions, or introduction but is simply a list of techniques. As Wöhler wrote to Liebig, it was a "cookery book" for beginners to save the time that would be spent explaining everything in person. Another version would follow, also anonymous, because, as he wrote to Liebig, "anyone can put a book like this together" (Keen 2005, 340).[58] Certainly not all chemists agreed, but Wöhler's attitude, if representative, is insightful. The

[57] This unfortunately includes my own recently edited volume (Ramberg 2022).

[58] A.W. Hofmann translated Wöhler's list of techniques into English, and Oscar Lieber translated them for an American press. Lieber also added an extensive introductory chapter of techniques (Wöhler 1852, 1854).

absence of analytical chemistry volumes in general chemistry texts, the subject of most historical studies, may have tended to make it less visible, because those chemists who wrote general chemistry textbooks did not perceive the need for a separate category of analytical chemistry. However, the critical mass of practitioners and literature became sufficiently large for a major new journal to appear in 1862. Perhaps this "invisibility" of analytical chemistry in general chemistry texts accounts for why the early emergence of analytical chemistry as a subdiscipline of chemistry has not been formally recognized.

Acknowledgments Many thanks to Bill Newman and Jutta Schickore for organizing this project and for their many helpful comments, particularly on the finer points of translation, and to the participants of the 2021 Zoom conference and 2023 workshop. I also thank Alan Rocke and Ernst Homburg for their many useful comments on an earlier draft of this essay. I am grateful to Google Books and the Hathitrust for making this kind of research on nineteenth century texts both possible and extraordinarily accessible. Unless noted, all translations from the German are my own, and although I am somewhat reluctant to admit it, I am also grateful to Google Translate and ChatGPT for helping me to get started on occasion.

References

Becher, Johann Joachim. 1669. *Physica subterranea.*

Bergmann, Torbern. 1788. *Physical and chemical essays*. Trans. E. Cullen. London: Murray.

Berzelius, Jakob. 1841. *Lehrbuch der Chemie*. Trans. Friedrich Wöhler. Reutlingen: Macken.

Coley, Noel G. 1990. Physicians, chemists, and the analysis of mineral waters: "The most difficult part of chemistry". In *The medical history of waters and spas*, ed. Roy Porter, vol. 34, 56–66. London: Wellcome Insitute.

Crosland, Maurice P. 1978. *Gay-Lussac, scientist and bourgeois*. New York: Cambridge University Press.

Duflos, Adolph Ferdinand. 1835. *Handbuch der pharmaceutisch-chemischen Praxis, oder Anleitung zur sachgemässen Ausführung der in den pharmaceutischen Laboratorien vorkommenden chemischen Arbeiten, richtige Würdigung der dabei stattfindenden Vorgänge, und zweckmässigen Prüfung der officiellen chemischen Präparate*. Breslau: Max and Komp.

Fors, Hjalmar. 2014. *The limits of matter: Chemistry, mining and enlightenment*. Chicago: University of Chicago Press.

Frercks, Jan, and Michael Markert. 2007. The invention of *theoretische Chemie*: Forms and uses of German chemistry textbooks, 1775–1820. *Ambix* 54: 146–171.

Fresenius, Carl Remigius. 1841. *Anleitung zur qualitativen chemischen Analyse*. Bonn: Henry & Cohen.

———. 1843. *Anleitung zur qualitativen chemischen Analyse*. 2nd ed. Braunschweig: Vieweg.

———. 1847. *Anleitung zur quantitativen chemischen Analyse*. 2nd ed. Braunschweig: Vieweg.

———. 1853. *Anleitung zur quantitativen chemischen Analyse*. 3rd ed. Braunschweig: Vieweg.

Fresenius, Heinrich. 1897. Zur Erinnerung an R. Fresenius. *Zeitschrift für analytische Chemie* 36: III–XVIII.

Gellert, C.E. 1750. *Anfangsgründe zur metallurgischen Chymie, in einem theoretischen und praktischen Theile nach einer in der Natur gegründeten Ordnung*. Leipzig.

Gmelin, Johann Friedrich. 1786. *Grundsätze der technischen Chemie*. Halle.

Göttling, Johann Friedrich. 1790. *Vollständiges chemisches Probir-Cabinet zum Handgebrauche für Scheidekünstler, Aerzte, Mineeralogen, Metallurges, Technologen, Fabrikanten, Oekenomen und Naturliebhaber*. Jena: Mauke.

———. 1802. *Praktische Anleitung zur prüfende und zerlegende Chemie*. Jena: Mauke.
Graf, Johann Baptist. 1805. *Versuch einer pragmatischen Geschichte der bayerischen und oberpfälzerfälzischen Mineralwasser*. Munich.
Hermbstaedt, Sigismund Friedrich. 1800. *Systematischer Grundriß der allgemeinen Experimentalchemie zum Gebrauch bey Vorlesungen und zur Selbstbelehrung beym Mangel des mündlichen Unterrichtes, nach den neuestes Entdeckungen entworfen*.
———. 1808. *Grundsätze der experimentellen Kammeral-Chemie für Kammeralisten, Agronomen*. Berlin: Forstbediente und Technologen.
Hochheimer, Carl Friedrich August. 1792. *Chemische Mineralogie, oder vollständige Geschichte der analytischen Untersuchung der Fossilien*. Leipzig: Barth.
Homburg, Ernst. 1999. The rise of analytical chemistry and its consequences for the development of the German chemical profession (1780–1860). *Ambix* 46 (1): 1–32.
Hufbauer, Karl. 1982. *The formation of the German chemical community, 1720–1795*. Berkeley: University of California Press.
Humboldt, Alexander. 1797. *Versuche über die gereizte Muskel- und Nervenfaser nebst Vermuthungen über den chemischen Process des Lebens in der Thier- und Pflanzenwelt mit Kupfertafeln*. Berlin: Rottmann.
Jackson, Catherine. 2014. Synthetical experiments and alkaloid analogues: Liebig, Hofmann, and the origins of organic synthesis. *Historical Studies in the Natural Sciences* 44: 319–363.
———. 2017. Emil Fischer and the ‘art of chemical experimentation’. *History of Science* 55: 86–120.
———. 2023. *Molecular world: Making modern chemistry*. Cambridge, MA: MIT Press.
Jensen, William B. 2003. *Philosophers of fire: An illustrated survey of 600 years of chemical history for students of chemistry*. Cincinnati: Oesper Collections.
———. 2017. Remembering qualitative analysis: The 175th anniversary of Fresenius’ textbook part I. *Educacion Quimica* 28: 217–224.
John, Johann Friedrich. 1808. *Chemisches Laboratorium. Oder Anweisung zur chemischen Analyse der Naturalien. Nebst Darstellung der nöthigsten Reagenzien*. Berlin: Maurer.
Keen, Robin. 2005. *The life and works of Friedrich Wöhler (1800–1882)*. Nordhausen: Verlag Traugott Bautz.
Klaproth, Martin Heinrich. 1795–1810. *Beiträge zur chemischen Kenntniss der Mineralkörper*. 5 vols. Berlin: Rottman.
Klein, Ursula. 2008. Not a pure science: Chemistry in the 18th and 19th centuries. *Science* 306: 981–982.
———. 2012. The Prussian mining official Alexander von Humboldt. *Annals of Science* 69: 27–68.
———. 2020. *Technoscience in history: Prussia, 1750–1850*. Cambridge, MA: MIT Press.
Klein, Ursula, and Wolfgang Lefèvre. 2007. *Materials in eighteenth-century science: A historical ontology*. Cambridge, MA: MIT Press.
Kopp, Heinrich. 1805. *Grundriß der chemischen Analyse mineralischer Körper*. Frankfurt: Hermann.
Kragh, Helge, and Malena M. Bak. 2000. Christoph H. Pfaff and the controversy over voltaic electricity. *Bulletin for the History of Chemistry* 25: 83–90.
Lampadius, Wilhelm August. 1801. *Handbuch zur chemischen Analyse der Mineralkörper*. Freyburg: Crazischen Buchhandlung.
Liebig, Justus, J.C. Poggendorff, and Friedrich Wöhler, eds. 1842–1854. *Handwörterbuch der reinen und angewandten Chemie*. 6 vols. Braunschweig: Vieweg.
Lindes, Wilhelm. 1830. *Versuch einer ausführlichen Darstellung der Lehre von den chemischen Reagentien. Zunächst als Vorbereitung auf das Studium der analytischen Chemie*. Berlin: Eichhoff und Krafft.
Montanus, August Schulze. 1818. *Die Reagentien und deren Anwendung zu chemischen Untersuchungen nebst zwei ausführlichern Abhandlungen über die Untersuchung der mineralischen Wasser und die Prüfungen auf Metallgifte*. Berlin: Flittner.

———. 1819. *Systematisches Handbuch der gesammten Land- und Erd-Messung mit ebener und sphärischer Trigonometrie*. Berlin: Rücker.
Newman, William R. 2000. Alchemy, assaying, and experiment. In *Instruments and experimentation in the history of chemistry*, ed. Frederic L. Holmes and Trevor H. Levere, 35–54. Cambridge, MA: MIT Press.
Pallas, Simon Peter. 1771. *Reise durch verschiedene Provinzen des Rußischen Reichs*. St. Petersburg.
Pfaff, Christoph Heinrich. 1821. *Handbuch der analytischen Chemie für Chemiker, Staatsärtzte, Apotheke, Oekonomen, und Bergwerks Kundige*. Altona: Hammrich.
Posselt, L. 1846. *Die analytische Chemie tabellarisch dargestellt*. Heidelberg: Winter.
Poth, Susanne. 2007. *Carl Remigius Fresenius (1818–1897): Wegbereiter der analytischen Chemie*. Stuttgart: Wissenschaftliche Verlagsgesellschaft.
Ramberg, Peter J., ed. 2022. *The Bloomsbury cultural history of chemistry in the nineteenth century*. London: Bloomsbury.
Rammelsberg, C.F. 1860. *Leitfaden für die qualitative chemische Analyse*. Berlin: Lüderitz
Rocke, Alan J. 2010. *Image and reality: Kekulé, Kopp, and the scientific imagination*. Chicago: University of Chicago Press.
Rocke, Alan J., and Hermann Kopp. 2012. *From the molecular world: A nineteenth-century science fantasy*. New York: Springer.
Rose, Heinrich. 1833. *Handbuch der analytischen Chemie*. Berlin: Mittler.
Russell, Colin A. 1987. The changing role of synthesis in organic chemistry. *Ambix* 34: 168–180.
Schickore, Jutta. 2007. *The microscope and the eye: A history of reflections, 1740–1870*. Chicago: University of Chicago Press.
Schwarz, Karl Heinrich. 1853. *Praktische Anleitung zur Maaßanalyse*. Braunschweig: Vieweg.
Schweizer, Eduard. 1848. *Praktische Anleitung zur Ausführung quantitativer chemischer Analysen*. Chur: Wassali.
Städeler, Georg. 1857. *Leitfaden für die qualitative chemische Analyse unorganischer Körper*. Zürich: Orell.
Stahlberger, Karl. 1819. *Sammlung chemischer Reagentien für gerichtliche Ärtzte, Pharmazeuten und Liebhaber der Chemie*. Vienna: Gerold.
Struve. 1785. Von den Reagentien und ihrem Gebrauche bey der Zerlegung der Mineralwasser. *Beträge zu den chemischen Annalen.*
Szabadváry, Ferenc. 1966. *History of analytical chemistry*. London: Pergamon.
Trommsdorff, Johann. 1801. *Chemisches Probierkabinett oder Nachricht von dem Gebrauch und den Eigenschaften der Reagentien*. Erfurt: Henning.
Wackenroder, Heinrich. 1836. *Anleitung zur chemischen Analyse unorganischer und organischen Verbindungen, nebst Beiträgen zur genaueren Kenntniss des Verhaltens und der Anwendung der Reagentien bei analytisch-chemischen Untersuchungen*. Jena: Cröker.
Walthiere, Balthasari. 1749. *Neue Beschreibung des Halts von dem weltberühmtesten Pfäfferser-Mineral Wasser*. Zug: Schäll.
Wöhler, Friedrich. 1852. *The analytical chemist's assistant: A manual of chemical analysis, both qualitatitve and quantitative*. Trans. Oscar Lieber. Philadelphia: Henry Baird.
———. 1854. *Hand-book of inorganic analysis. One hundred and twenty-two examples, illustrating the most important processes for determining the elementary composition of mineral substances*. Trans. Hofmann, August Wilhelm. London: Walton and Maberly.

Peter J. Ramberg is Professor of History of Science at Truman State University in Kirksville, Missouri, where he teaches history and philosophy of science and chemistry. His research interests include broadly the history of modern chemistry, but especially the institutional and conceptual history of chemistry in nineteenth century Germany. He is the editor of the recent *The Bloomsbury Cultural History of Chemistry in the Nineteenth Century* (2022), and author of *Chemical Structure, Spatial Arrangement: The Early History of Stereochemistry, 1874–1914* (Ashgate, 2023).

Chapter 14
Questioning the Symmetry Between Analysis and Synthesis in Chemical Practices

Bernadette Bensaude-Vincent

Abstract This paper interrogates the general assumption of the symmetry between analysis and synthesis. It argues that the emphasis on symmetry between analytic and synthetic operations proceeds from a reconceptualization of the empirical tradition of chemical analysis under the aegis of mathematical and philosophical notions. Given that the experimental analyses and syntheses are viewed as mere translations of a mode of reasoning onto the material realm, I refer to this approach as the *intellectualist* tradition. I highlight the contrast between this intellectualist approach, which describes analysis and synthesis as two distinct and successive operations, and more *pragmatic* approaches to analysis and synthesis as practical arts that emphasize the *synchrony* between the two processes rather than their *symmetry*.

Keywords Chemistry · Synthesis · Étienne Bonnot de Condillac · Chemical language · Lavoisier · Decomposition and recomposition · Synthetic biology · Bachelard · Rational materialism

This paper is concerned with the ways in which analysis and synthesis have been coupled in the discourses about science. Irrespective of whether they are conceived of as mental or material processes, they are often described as symmetrical operations. In mathematics, for instance, analysis proceeds from consequences to principles, while synthesis proceeds from principles to conclusions. As such, they are construed as two reversible operations—two pathways of reasoning. In chemistry, analysis proceeds from wholes to parts, and synthesis assembles parts into wholes. In both domains, analysis and synthesis have been conceived of as akin to mirror images of one another, and it is because of this affinity that they have been adopted as core methods for advancing knowledge in modern sciences. So prevalent was this tendency to couple analysis and synthesis that it came to be regarded as the optimum method for demonstrating truths in natural philosophy by the end of the eighteenth century. For instance, in the second preface to the *Critic of Pure Reason*,

B. Bensaude-Vincent (✉)
Professeure émérite, Université Paris 1-Panthéon-Sorbonne, Paris, Cedex 05, France
e-mail: bvincent@univ-paris1.fr

W. R. Newman, J. Schickore (eds.), *Traditions of Analysis and Synthesis*, Archimedes 73, https://doi.org/10.1007/978-3-031-76398-4_14

Immanuel Kant clearly suggested that the "light" spread by experimental philosophy derived partially from the reversibility of the analytical and synthetic procedures in Georg-Ernst Stahl's operations of metals reduction and calcination:

> When Galilei experimented with balls of a definite weight on the inclined plane, when Torricelli caused the air to sustain a weight which he had calculated beforehand to be equal to that of a definite column of water, or when Stahl, at a later period, converted metals into calcx, and reconverted lime into metal, by the addition and subtraction of certain elements; a light broke upon all natural philosophers. (Kant 1787, Preface to the second edition)

In a similar vein, Lavoisier set up a spectacular public experiment in February 1785 to demonstrate the composition of water by separating and recombining oxygen and hydrogen (Daumas and Duveen 1959). So deeply engrained in our mental framework is the view of analysis and synthesis as symmetric processes that provide clear and robust demonstrations of the nature of things that it is taken for granted. This framework is particularly familiar to historians of chemistry, a science that was traditionally defined in terms of the separation and combination of the constituent elements of bodies. In particular, Paracelsus' followers developed a framework for chemistry—spagyria—aimed at separating the constituent principles—the *tria prima*: salt, sulfur, and mercury—of bodies and recombining them after a process of separation from their impurities into the original substance (Pagel 1982; Kahn 2007). Van Helmont conducted quantitative experiments of analysis and synthesis, a method that is not dissimilar to Lavoisier's balance sheet method (Newman and Principe 2002). The various uses of analysis and synthesis in chemistry have been the subject of numerous historical studies,[1] most of which have been concerned primarily with the *ontological* status of elements separated by analysis: Were they speculative entities or tangible substances close to our sulfur and mercury? Were they viewed as the primary constituents of matter or merely as simple, provisionally un-decompound substances? (See Clericuzio 2000; Kim 2003; Klein and Lefevre 2007.) Other historical studies focus on *methodological* issues: what physical operations were performed for extracting the principles: distillation (fire analysis) or solution and precipitation (wet analysis) (Debus 1967; Holmes 1971). However, historians of chemistry rarely address the question of how analysis and synthesis have been articulated in various traditions. Are they symmetric operations resembling mirror images? Are they two successive or synchronic processes? It is essential that such questions be addressed to raise *epistemological* issues concerning the heuristic power and the probative force of analysis and synthesis.

In this paper, I shall interrogate the general assumption of the symmetry between analysis and synthesis. While from a logical point of view it is easy to understand the symmetry of this pair of operations and its probative force, the issue is rendered more complex when the material processes of decomposition and recomposition are considered. I shall address this question from the perspective of historical epistemology rather than from a logical point of view by focusing on several episodes—taken mainly though not exclusively from the history of French chemistry—over the

[1] See, for instance, Newman and Principe 2005 and *Ambix* special issue 61(4) (2014).

longue durée. My purpose here is not to use the analysis and synthesis pairing as a reliable indicator to trace changes or continuities or to bridge early modern and modern chemistry.[2] Rather, my aim is to unravel the epistemic and ontological presuppositions embedded in the claim that analysis and synthesis are symmetric, reversible processes.

I shall argue that the emphasis on symmetry between analytic and synthetic operations proceeds from a reconceptualization of the empirical tradition of chemical analysis under the aegis of mathematical and philosophical notions. Given that the experimental analyses and syntheses are viewed as mere translations of a mode of reasoning onto the material realm, I refer to this approach as the *intellectualist* tradition. I shall highlight the contrast between this intellectualist approach, which describes analysis and synthesis as two distinct and successive operations, and more *pragmatic* approaches to analysis and synthesis as practical arts that emphasize the *synchrony* between the two processes rather than their *symmetry*. When analysis and synthesis are described as arts or practical operations upon and with materials, they are praised for reasons other than their probative force, and analysis and synthesis may be decoupled. I do not wish to effect a kind of paradigm shift whereby the intellectualist model would replace the pragmatist model and reject it in the obscure tradition of premodern chemistry. Rather, I wish to demonstrate that the two approaches remain vital and are not necessarily mutually exclusive.

I first characterize the pragmatic model of chemical analysis and synthesis that prevailed in the mid-eighteenth-century French tradition of chemistry by pointing to the distance that separated the logical model of analysis and this art of chemists. In Sect. 14.2, I present the intellectualist model of analysis and synthesis as the product of a fusion of the mathematical and the pragmatic models by Étienne Bonnot de Condillac's "metaphysics of language," which inspired Lavoisier's chemical revolution. I then proceed to consider the views of synthesis to determine whether the two models have a future. In Sect. 14.3, I describe Marcellin Berthelot as a follower of Lavoisier who applied his intellectualist approach to synthesis, and I suggest that several twenty-first century synthetic biologists are also following this model. The subsequent sections focus on the relationship between the two contrasted approaches with respect to whether they are wholly antagonistic or capable of being reconciled. In Sects. 14.4 and 14.5, I identify and characterize two cases of composition of the two models by Gaston Bachelard, a philosopher of science, and Roald Hoffmann, a famous chemist. To further explore the antagonism between the two views of synthesis, Sect. 14.6 surveys the controversy that surrounded the emergence of nanotechnology in 2000.

---

[2] On this, see Klein and Ragland 2014.

14.1 Analysis as Art: The Pragmatist Model of Diderot's *Encyclopédie*

The entry "Analysis" in the first volume of the *Encyclopédie* edited by Denis Diderot and Jean Le Rond d'Alembert is subtitled "understanding, reason, philosophy of science, natural science, pure mathematics, literal arithmetic or algebra" (Diderot and d'Alembert 1751, 400–03). It runs to a total of four pages, of which one and a half describe analysis in algebra, with one paragraph describing its use in grammar; two pages are dedicated to analysis in logic, with one paragraph on literature and a mere eight paragraphs on analysis in chemistry. Analysis is thus primarily presented as a mental operation, an art of reasoning and of solving puzzles in mathematics. Remarkably, the small section on chemistry at the end of the entry states that analysis encompasses chemistry in its entirety but is largely disconnected from the preceding sections. The various meanings are simply juxtaposed. For d'Alembert, the mathematician whose signature is attached to the section on algebra, analysis is a highly powerful approach to solving mathematical problems by reducing them to equations exemplified in algebra. In grammar, analysis is the resolution or simplification of a whole into its parts; in logic, analysis is the royal road to the discovery of truth, the method for unveiling the simples that provide the starting point for any search for truth. This Cartesian rationalist method appears to be shared by empiricist logicians, since this subsection notes that it is better defined as a method for tracing the origins of our ideas, developing their generation. By contrast, Paul-Jacques Malouin, the author of the brief section on chemical analysis, describes analysis as an "art," a difficult art that requires special skills, lengthy training, and "*du métier*." While Malouin claims that "analyzing bodies or resolving them into their component parts is the principal object of the art of chemistry," he conveys a very poor image of the achievements of this method.

> It is difficult to know by analysis the composition & properties of things; one must be learned & experienced in Chemistry, to separate the principles which compose bodies, & have them as they are naturally, in order to be able to say what they are. (Malouin 1751; my translation)

Analysis by fire of vegetable matters fails to account for their virtues or properties: it resolves venomous plants into the same principles as healthy plants because the fire destroys the active principles. The analysis of mineral waters is a turf for charlatanry, for people who abuse the credulity of men. Analysis has no real cognitive power because it requires both theoretical knowledge and practical skills. "To be able to speak knowledgeably about waters and the principles of which they are composed, one must not only be well versed in Chemistry, but even very skilled at it."

So striking is the contrast between the praise of mathematical or logical analysis as a method of reasoning and the depreciation of chemical analysis that historian William Albury characterized this dual view of "analysis" as "intellectual schizophrenia" (Albury 1972, 62). Indeed, Gabriel-François Venel, who succeeded Malouin and was responsible for all the entries relating to chemistry subsequent to Volume 1, presented a considerably more positive image of analysis but he also characterized analysis as an "art" divorced from the method of reasoning. In using the then-fashionable contrast between ancients and moderns, he clearly distinguished fire analysis as belonging to "ancient chemistry" from solution analysis "discovered by modern chemists." In the entry on "Chymistry," he even claimed that chemists no longer used fire analysis:

> Because modern Chemists have discovered a better method, separation by dissolution, they have moved away from the earlier method. And since today's science is sufficiently advanced to measure the movement of all the reactive agents excited by heat in compound bodies, we can examine them using distillation caused by the violence of fire as easily as we can propose a chemical problem in the manner of Geometers and with the same degree of usefulness. (Venel 1753, §100)

Venel despised fire analysis as too violent and destructive. He agreed with Boyle and Boerhaave that it cannot yield any reliable knowledge about the nature of the constituent principles, but he blamed them for behaving as ancient chemists who adopt a physicist's approach to matter while ignoring the art of modern analysis. The entry "Distillation" praises this method for its economic interest while minimizing its philosophical significance for advancing knowledge (Venel 1754). Distillation is good for preparing highly valued products, such as essential oils for pharmacy and other arts, but it generates erroneous views on the constitution of substances. Only when "fire is administered according to the art," when the artist uses mediations and proceeds carefully to retrieve a sequence of products, is it possible to reassemble the original substance and learn something about its nature. Put otherwise, resynthesis is required to confirm that a genuine analysis into constituents has taken place. This article gives a lengthy description of the categories of bodies that can be submitted to distillation and insists on the art of "governing fire." It culminates in a list of practical rules regarding the choice of vessels, their size, and about how and where fire should be applied. By contrast, the entry "Menstrual analysis" describes the method of dissolution as a sure way to scrutinize the chemical nature of compounds. Venel praises dissolution analysis for two reasons: first, he claims that it is applicable to all categories of bodies including metals, plants and animal tissues; and second it is a gentle and gradual process that releases the principles in order of their composition: secondary principles will be divided into first principles or elements. Although Venel adopts Stahl's hierarchical view of the composition of matter (from simple bodies, to mixtures, compounds, supercompounds, etc.,) in the entry "Chymistry,"[3] he is less concerned with the identification of the ultimate elements than with the gradual extraction of and the access to the immediate principles (*principes prochains*) of plants by dissolution. Analysis in mid-eighteenth-century French chemistry was essentially praised as a method used by physicians and pharmacists to isolate the immediate principles of plants for making medicines or to denounce charlatans who sell adulterated or inferior medicines, thus threatening the public confidence in chemists. Analysis was associated with an

[3] Venel 1753, §70: "The bodies belonging to each of these kingdoms are distinguishable one from another by their simplicity or by their degree of mixing. They are simple and their character is essential relative to the ways in which chemists undertake to examine them."

order of nature that was divided into three kingdoms—mineral, vegetal, and animal—and that was practically oriented toward chemical arts (Simon 2002). Most chemists, Venel in particular, insisted that the art of analysis demands considerable care, skill, and mediation to isolate the various levels of principles. Only if these are applied does it allow a perfect demonstration.

> If a Chemist manages to gather back together in an orderly fashion all those principles that he has separated in the same fashion, and if he manages to recompose a body that he has broken down, he has attained a true demonstration of Chemistry. And indeed, the art of Chemistry has reached this level of perfection in several essential areas. (See syncresis). (Venel 1753, §72)

Unfortunately, Venel never wrote the entry "Syncresis" as he was too busy for writing all chemistry entries of the last volumes of the *Encyclopédie* during the 1760s. However, the terms "diacresis" and "syncresis" were still in use in the mid-eighteenth century—not least in an unpublished treatise on chemistry written by Jean-Jacques Rousseau, who attended Guillaume-François Rouelle's public lectures at the Jardin du Roy in Paris in the 1740s, while tutor to Dupin de Francueil, the son of Claude Dupin, a wealthy tax-collector.[4] The unfinished manuscript, entitled *Institutions chymiques*, relied heavily on Rouelle's oral lectures, which disseminated Joachim Becher's and Georg-Ernst Stahl's doctrines in France (Rappaport 1958, 1960). Rousseau presented them as the founders of modern chemistry. It was Becher, Rousseau claimed in Section III, who made decomposition and composition the core concepts of chemistry. The next section raises a lexical puzzle. The titles of the first two chapters of Section IV are "Analyse ou synchrèse" (Chapter 1), and "De la diagrèse ou composition" (Chapter 2), and analysis is treated as a synonym of synthesis and diagresis as synonym of composition.

Was Rousseau inconsistent in his exposition of these two fundamental operations of chemistry? The term "diagrèse" derived from ancient Greek *diakrisis*, which means "separation." As such, we would expect to see it used as a synonym of analysis. The term "*syncrèse*," from the Greek *synkrisis*, means assemblage or combination.[5] The terms "analysis" and "synchresis" should thus be regarded as antonyms rather than synonyms, and the same may be said for the terms "diagresis" and "composition."

However, closer examination of the chapters' contents reveals that Rousseau was not inconsistent. Both chapters treat analysis and synthesis as synonymous. The dissolution of salts features in the chapter "Analysis or synchresis." Rather than being described as a separation, it is interpreted as the union of a solvent to a solute. It results from the attraction of the same to the same; salt is dissolved in water because it contains a lot of water in its crystals. The theoretical explanation of such unions

---

[4] For a broader view of Rousseau's practices of science, see Bensaude-Vincent and Bernardi 2003.

[5] The eighteenth-century meaning of the term "synchresis" differs from its current use. Nowadays, it is a compound term associating "synchronization" and "synthesis." In acoustics, it refers to a phenomenon produced by simultaneous sound and image. They blend with each other so that it becomes difficult to separate them.

lies in the notion of "latus." All mixts present different sides (latus), and so the *menstrue* (dissolvent) attacks the aqueous side of salt because it is of the same nature.

> If we pay attention to what happens during dissolution, we cannot doubt that the dissolvent is intimately united with the dissolved body so that, after this union, they become one homogeneous body that must be considered as a single substance. (Rousseau 1999, 301)

The synchresis is the generation of a new mixt or homogeneous body through an intimate union with the dissolvent. It can only be decomposed when the dissolvent is precipitated by a third body presenting a greater affinity with the dissolved one. Rousseau noted that while chemists often used dissolution to purify salts (to separate them from impurities), dissolution was above all used to create useful compounds (Rousseau 1999, 302). Put otherwise, analysis is the result of a synthesis, from both a theoretical perspective within the affinity framework, and from a practical perspective.

In the subsequent chapter "On diagresis or composition," Rousseau continues the review of chemical operations that are simultaneously both separation and union. He describes the reduction of metal calxes to obtain the metal in its metallic state as a synthesis in which the calx combines with phlogiston.[6] In Rousseau's view, fire was not primarily an instrument of analysis but also the major instrument of the synthesis of mixts. This chapter begins with a definition of synchresis:

> Chymical synchresis consists of new mixts so that two substances that are intimately mixed and blended together compose a third of a strong union different in nature from each of those that composed it and where none of them is recognizable anymore. (Rousseau 1999, 305)

This definition implicitly connects chemical synthesis with the Stahlian notion of mixt advocated by Venel in the *Encyclopédie*'s entry "Mixt & mixtio." Stahl clearly distinguished between mixts and aggregates. Mixtio is the complete union of two bodies with no retention of the parts in the mixt. It results from the union of principles, while the aggregate results from the union of integral parts.

> The mixt or compound chemical bodies are formed by the union of various principles, water & air, earth & fire, acid & alkali, &c. They differ essentially in this from the aggregates, aggregates or molecules which are formed by the union of similar or homogenous substances. (Venel 1765a, b)

A mixt is made of two principles, and its properties differ from those of its constituent principles. In an aggregate, by contrast, the whole and the part are of the same nature. The difference between mixts and aggregates is not in the *degree* of composition but in the *nature* of union or combination.[7] Aggregation is a mechanical union, a juxtaposition of units. Whether it is understood in terms of hooking corpuscles or Newtonian attraction, it refers to the general properties of masses and

[6] The term "combination," which is the subject of Chap. 3, is defined as any operation by which principles united in the same quantity and proportions can form different compounds. The examples developed in this chapter are fermentation and vitrification. See Rousseau 1999, 323.

[7] For further details on Stahl's reconceptualization of the Aristotelian notion of mixt, see Bensaude-Vincent 2009b.

their movements. Mixing, by contrast, requires the qualitative diversity and individuality of the constituents, creating new homogeneous bodies from heterogeneous elements.

To sum up this section, in mid-eighteenth-century France, chemical analysis was essentially viewed and praised as an art rather than as a mental operation. This art requires both the work of skilled artists and the spontaneous affinities of chemical substances. Given the joint agencies of chemists and materials, analysis and synthesis could be understood as two *correlated rather than inverse*, *simultaneous rather than successive* operations. Within the affinity theory underlying Rouelle's lectures, composition was a generic concept encompassing distinctive types of union, such as aggregation and mixtion or chemical combination. The latter was primarily viewed as an operation of *exchange* between constituent and active principles from which there emerged a new compound with properties that differed from those of its constituent principles. The arts of chemical analysis and synthesis consisted in securing the appropriate experimental conditions to generate compounds with properties of interest for practical purposes.

14.2 The Intellectualist Model and the Reform of Chemical Language

As a historian and philosopher of chemistry, I have been intrigued by the significance of the heroic figure of Lavoisier as the founder of modern chemistry and the correlative view of chemistry as languishing in the obscure alchemical tradition that preceded his chemical revolution (Bensaude-Vincent 1993). While professional historians of science over several generations have demonstrated that so-called premodern chemistry was a well-established academic science and a booming investigative enterprise with a rich experimental tradition, grounded on a strong conceptual and theoretical basis, the founder myth is extremely resilient, and still vivid in chemistry communities (Holmes 1989; Bensaude-Vincent 1996).

One influential interpretation of the chemical revolution has focused on Lavoisier's notion of composition. Robert Siegfried viewed the chemical revolution as the emergence of the compositional paradigm, beginning with Lavoisier's work and culminating in Dalton's atomic theory (Siegfried and Dobbs 1988; Siegfried 2002). In this paradigm all substances whether they be natural or manmade in a laboratory are to be defined by the nature and proportion of their components. However, the assumption of such a paradigm shift must be nuanced in view of the power conferred on analysis and synthesis with respect to understanding the nature of chemicals in so-called premodern chemistry and the importance of the notion of degrees of composition between the simple and the compound. In particular, Mi Gyung Kim argued that the view of chemical elements as simple substances distinct from the ultimate elements of matter that she named the "analytic/philosophical ideal" was commonplace in seventeenth- and eighteenth-century chemistry textbooks (Kim 2001). As John Powers has argued convincingly, Lavoisier simply followed this tradition, which had been widely disseminated by Hermann Boerhaave's *Elementa Chemiae* in the eighteenth century, inviting chemists to abandon the search for the ultimate principles of matter (Powers 2014).

I contend that one major reason that Lavoisier is still celebrated as the founder of modern chemistry by professional chemists is the way in which he embedded the analytic ideal in the chemical language that he co-constructed with several French academic chemists and presented before the Académie Royale des Sciences de Paris in 1787. The reform of chemical language provided Lavoisier with an opportunity to establish a more philosophical order of nature, detaching chemistry from natural history and setting aside the practical orientation of pharmaceutical chemists (Simon 2005).

The project of a global and systematic reform of chemical language played a key role in stabilizing and reinforcing the "analytic-philosophical" ideal.[8] The exchanges between eighteenth-century chemists all over Europe coupled with an intense activity of translations emphasized the defects of the names inherited from a long tradition of chemists apothecaries and metallurgists became particularly visible. Many names were used to designate one and the same substance. Moreover, new names were needed for novel, recently identified substances, such as cobalt and vanadium, named after Swedish deities, and the crowd of "aeriform fluids" isolated and characterized in pneumatic science. Torbern Bergmann in Sweden and Joseph Macquer in France made timid attempts at inventing systematic names for substances that were recently identified, such as gases, or classified, such as salts and minerals (Bergman 1784).[9] Louis-Bernard Guyton de Morveau, a chemist and lawyer from Dijon, who was in charge of the chemistry dictionary for the *Encyclopédie méthodique* and in correspondence with various international chemists, was extremely receptive to the invitations to build up a universal and systematic language for chemistry. In 1782, he initiated a tentative project aimed at reforming the chemical nomenclature based on the general principle already at work in the botanical nomenclature established by Carl Linnaeus that denominations "must be, as far as possible, in conformity with the nature of things." For a chemist embracing the commonplace "analytic ideal," the "nature of things" was the expression of their composition. He steadily based his reform on firm principles, such as "the name of a chemical compound is clear and exact only as far as it recalls the component parts by names in conformity with their nature" (Guyton de Morveau 1782). He thus forged simple names for simple substances and compound names for chemical compounds, which express their composition. When the composition is uncertain, Guyton added, a meaningless term is preferable. His reform, like earlier attempts, was clearly designed to reach a consensus among European chemists. He shared the draft of his project with chemists of the Paris Academy of Science, and they

[8] On this reform of the chemical language the standard reference remains Crosland 1962. See also Dagognet 1969; Bensaude-Vincent and Abbri 1995.

[9] See also Smeaton 1954; Crosland 1962.

collectively worked out systematic names for not only all known chemical substances but also new substances yet to be discovered.

The memoir published in April 1787 and entitled *Méthode de nomenclature chimique* contains an introduction by Lavoisier, an exposition of the basic principles by Guyton, and two dictionaries of synonyms. Lavoisier embedded the new language in a broad metaphysical framework borrowed from Étienne Bonnot de Condillac's *Logic* (Guyton de Morveau et al. 1787). Thanks to this philosophical patronage, Lavoisier managed to merge the chemical process and the mental operation of analysis.

Following Condillac, Lavoisier presented analysis as a mental process, an art of reasoning inspired by nature that enabled children to form complex ideas by association of simple sense data. Condillac entitled his last treatise *Logic or the First Developments of the Art of Thinking* in response to the influential *Logic or the Art of Thinking* published by Antoine Arnault and Pierre Nicole in 1662 (Condillac 1780).[10] Condillac's core claim is that the method of reasoning is taught by nature itself, as it is exemplified in the knowledge acquisition process exhibited by children. Ideas are formed in the human mind according to analytical logic: the association of simple sensorial data generates primary ideas that gradually lead by association to increasingly complex and abstract notions. Analysis also forms a mental operation that consists in sequentially displaying the elements that are perceived simultaneously by our senses to form ideas. Analysis examines in succession that which is given simultaneously by sensations. Condillac explained it with the help of a metaphor: I suppose a viewpoint from a castle in the countryside. At first glance, I do not grasp all the simple elements that make up this landscape. In reviewing them successively, the mind introduces an order based on the relationships between objects and thus recomposes a global image.[11] For Condillac, analysis is a twofold process of decomposition and recomposition. It is the "universal method" of discovery that can be applied to sense data, to ideas, or to objects. "In the art of reasoning, as in the art of calculating, everything is reduced to compositions and decompositions, and one should not believe that these are two different arts" (Condillac 1780, 413). This "art of reasoning" includes the double move from the whole to the parts and from the parts to the whole. To attain perfect knowledge of a machine, one must decompose it and study each part separately before reassembling them in the same order.

However, Condillac's *Logic* merits its status as a landmark in the history of analysis for another reason: as William Albury argues, Condillac overcome the "intellectual schizophrenia" visible in the entry "Analysis" of the first volume of Diderot and d'Alembert's *Encyclopédie* (Albury 1972, 62). He managed to reconcile the two meanings of decomposition and algebra thanks to the introduction of language. He assumes that ideas must be connected with signs and that the art of reasoning

[10] On Condillac's influence on Lavoisier, see Albury 1972; Beretta 1993; Bensaude-Vincent 2010, 49–65.

[11] As William Albury notes the idea of individual things is formed through the relations that the mind creates between them (1972, 67).

presupposes a language. Condillac then boldly claims that every language is an analytic method and that "the art of reasoning is reduced to a well-made language." For Condillac, algebra is the most perfect language because it operates at the sole level of signs and relies on relations of magnitude. It is the analytic method *par excellence*. It permits the equation of unknown quantities with a known value, thus alleviating the burden of the long chain of reasons leading to the solution of complex questions. Condillac thus managed to unify the algebraic method and the method of decomposition that coexisted in d'Alembert's article.

The connection between analysis and language first attracted Lavoisier's interest in Condillac's *Logic*. If languages are not just a system of signs expressing ideas and images, if they are true analytical methods, methods for proceeding from the known to the unknown, then remaking the language is remaking science. The modest reform of language outlined by Guyton de Morveau was thus metamorphosed into a radical subversion of the science itself by the analytical method. Referring to Condillac's "metaphysics of language," Lavoisier assumed that words, facts, and ideas were, so to speak, three faces of a single reality.

> The perfection of the nomenclature of chemistry, considered in this respect, consists in rendering the ideas & facts in their exact truth, without suppressing anything of what they present, especially without adding anything to it: it must be only a faithful mirror, for, we cannot repeat it too often, it is never nature nor the facts that it presents, but our reasoning that deceives us. (Guyton de Morveau et al. 1787, 14)

This means that the binomial names of compounds formed by the juxtaposition of two words referring to two components (e.g., lead oxide or sodium chloride) are just symbolic transcriptions of their true nature. Just as ideas in the human mind are formed by the addition of elementary sense data, chemical compounds are supposedly formed by the simple addition of two elements. The *ratio essendi* and the *ratio cognoscendi* merge. Given that the names are supposed to follow nature, the nomenclature constitutes more than a lexicon. It is a "method of naming" rather than a nomenclature. It provides a program for constructing the names of substances yet to be discovered. Condillac's analysis makes it possible to anticipate the experimental analyses performed by chemists in their laboratory. While we do not actually know what all natural bodies are made of, we may conjecture, inspired by Condillac, that they are made by composition or addition, from simple to compound. This is the "logic of nature," it will be the logic presiding over chemical language.

To build up a systematic language, the four French reformers started from the 33 simple bodies identified in 1787 as residues of material analyses. Their language is a mirror image, in the strict sense that the words are the inverted image of the operations carried out in the laboratory and in the human mind. It thus follows Condillac's view of "natural logic" in two different ways. First, it reflects the true nature of chemical compounds. Whether they are mineral, plant, or animal in origin, they are supposedly formed by two simple substances or two radicals acting as elements. Thus, the new language of chemistry is rooted in a specific view of composition as the addition of two elements (or groups of elements) (Crosland 1962). The authors of the systematic language tacitly assumed that the nature and the proportion of the components are sufficient to define a compound and determine its properties entirely. By ignoring the puzzle of emerging properties in chemical reactions, they legitimized the organization of chemistry textbooks along one single logic: from simple to compound substances.

Therefore, Lavoisier claimed that the new language would bring about a "revolution in chemistry teaching," and two years later, he presented his *Traité élémentaire de chimie* as the natural outcome of the reform of language. In the preliminary discourse, Lavoisier explicitly mentioned Condillac as a kind of mentor who guided his reorganization of chemistry. He compared the genesis of ideas in children's minds with the learning of chemistry and decided to proceed from the known to the unknown, which, in his view, also meant proceeding from the simple to the complex (Lavoisier 1789, xx). In other words, for Lavoisier, these two processes are one and the same: analysis proceeds from the known/simple to the unknown/complex. In this respect Lavoisier subverted the early modern understanding of analysis as described in this volume by Helen Hattab. The "simple" is the known, in Lavoisier's view, whereas analysis proceeded from the "known to us" (that is, the complex/composite/specific) to the "unknown to us" (that is the general/simple) in Zabarella's *De Methodis*. Synthesis proceeding in the reverse direction was the preferred method for teaching.

Lavoisier claimed that the strict application of the rule from the known/simple to the unknown/complex would distinguish his textbook from its antecedents; that his *Elements of Chemistry* would be the first truly elementary textbook accessible to beginners (Lavoisier 1789). In fact, the simple-to-compound order was by no means revolutionary since it had prevailed in the exposition of chemistry for a few decades, but it was identified as "synthetic order." Antoine Baumé, for instance, had already adopted the order "from simple to compound and from compound to more compound."[12] Far from subverting the traditional organization of chemistry textbooks according to the three realms of nature, Lavoisier's order rather legitimized the natural history order. Like Baumé and Fourcroy, Lavoisier based his classification of vegetable matters on the results of analysis, but he did not retain the results of solvent analysis that were so highly valued by physicians and pharmacists for practical purposes. Lavoisier was only concerned with elementary analysis by fire.

Moreover, Lavoisier's claim that there was no distinction between the two processes—from the simple to the complex and from the known to the unknown—was already manifest in Pierre-Joseph Macquer's *Elements of Theoretical Chemistry*. Playing on the ambiguity of the term "element," he tacitly assumed that what is elementary in the order of substances was also elementary in the order of knowledge.

> Assuming that my reader knows no chemistry, I plan to lead him from the simplest of truths, which requires the least knowledge, to compound truths which require more. This order obliges me to start by treating the simplest substances that we know and that we look upon as the elements of which the others are composed, because knowledge of the properties of these elementary parts leads naturally to the discovery of those of their different combinations. And contrariwise, knowledge of the properties of compound bodies requires that we

[12] Baumé (1773, t. I, xii–xiv, quoted from p. xivl).

> be already familiar with that of their principles. The same reasoning obliges me when dealing with the properties of a given substance, not to speak of those of any other substance of which I have not spoken. (Macquer 1753, xvi–xvii, my translation)

If Lavoisier was simply following a common, established view when he claimed that the simple in nature was also simple for human understanding, how might we account for his prominent status in the chemists' community? Lavoisier changed the meaning of the chemists' routine experiments of decomposition and recomposition by using algebra as a model to reorganize chemistry along this analytical logic. As Marco Beretta rightly noted, his "epistemological generalization" transformed an ordinary laboratory operation into a central principle, the foundation of chemistry (Beretta 1993, 201).

Lavoisier reconfigured the concept of analysis by applying Condillac's connection between analytical reasoning, language, and algebra to chemistry. In the compositional paradigm, chemistry is refocused on the quest for simples.

> The principal object of chemical experiments is to decompose natural bodies, so as separately to examine the different substances which enter into their composition. By consulting chemical systems it will be found that this science of chemical analysis has made rapid progress in our times [...] Thus as chemistry advances toward perfection, by dividing and subdividing, it is important to say where it is to end; and these things we at present suppose simple may soon be found quite otherwise. (Lavoisier 1789, 176–7)

Analysis is the top priority, and it must be practiced in Condillac's sense as a method for proceeding from the known to the unknown modeled on algebra.

Lavoisier's algebraic view of composition transformed his experimental practice because he soon tied the analytical method to quantification in algebra. Just as mathematicians reach the solution of problems by simply rearranging quantitative data, chemists can use quantitative data to grasp the unknown, thanks to equations. Chemists should, above all, be concerned with weighing the quantity of inputs and outputs of chemical reactions, to make a balance sheet by addition and subtraction of quantities of matter. It is a special kind of algebra, since the terms on either side of the equation are individual material entities rather than letters standing for unknown commensurable numerical values. However, Lavoisier overlooked the difficulty in encouraging an idealized view of chemical analysis. In the course of the chapter on the decomposition of plants in wine fermentation in his *Elements of Chemistry*, Lavoisier focused on equal weights when he articulated the famous principle "Rien ne se crée."

> We may lay it down as an incontestable axiom, that, in all operations of art and nature, nothing is created; an equal amount of matter exists both before and after the experiment; the quality and quantity of the elements remain precisely the same; and nothing takes place beyond changes and in the combination of these elements. Upon this principle the whole art of performing chemical experiments depends. We must always suppose an exact equality between the elements of the body examined and those of the products of its analysis. (Lavoisier 1789, 129–30)

In chemistry, analysis should be based on the careful determination of the quantitative proportion of the constituents through weighing. As Trevor Levere (1992) has argued, Lavoisier's instruments followed the same logic. Thanks to the vogue for precision instruments encouraged by public experimental demonstrations and his personal revenues, he was able to use precision balances and gasometers made by skilled craftsmen. He translated Condillac's analytic method into the balance sheet method, guided by the assumption that all reactions would include a true equality or equation between the principles of the body under examination and those regained by analysis.

Lavoisier grounded the demonstrative power of analysis and synthesis on this algebraic reconceptualization of chemical composition, based on the axiom that the whole is the sum of its parts. Like many earlier chemists, he was acutely aware that the use of analysis to identify chemical principles was subjected to several objections. Both analysis and synthesis are open to skepticism when they are performed in isolation. Together, however, they provide a solid foundation upon which to establish the truth.

Lavoisier acknowledged that the decomposition of water does not necessarily give access to its constituent principles. He also assumed that the synthesis of water was not sufficient to provide a demonstration of the composition of water after performing a synthesis of water on June 24, 1783, because it provides only qualitative results. Consequently, he felt the need to perform a large-scale two-day experiment demonstrating the analysis and synthesis of water in February 1785. The narrative of this experiment suggests that it was more a theatrical performance than a precise demonstration, but its symbolic power convinced Claude-Louis Berthollet and several other chemists to adopt Lavoisier's theory (Daumas and Duveen 1959).[13]

To sum up this section, Lavoisier did not introduce the analytic ideal in chemistry, but he provided a philosophical legitimization for this shared ideal, and this philosophical detour had a tremendous impact on the identity of chemistry. It helped to establish chemistry as an ambitious and autonomous science, distinct from the natural history framework. Indeed, Lavoisier's famous definition of elements as final results of decomposition deprived the elements of their ontological status as universal constituents of matter. However, the abandoning of metaphysical ambitions—which was already commonplace in the eighteenth century, as mentioned above—was compensated by the additional meanings conferred on the empirical operations of analysis and synthesis. Laboratory experiments not only offered access to simple substances but they also shaped a view of nature as the expression of the analytic logic of the human mind, since chemical compounds came to be viewed as resulting from the additive union of two principles. Lavoisier shaped an identity of chemistry around the assertion of a triple parallel between the *ratio operandi*, the *ratio cognoscendi*, and the *ratio essendi*. His table of simple substances provided the alphabet of chemistry as a language mirroring a nature that is shaped by the logic of the human mind. The prevalence of the operational criterion of simplicity (the *ratio operandi*),[14] along with the decomposition of several substances

[13] See also Bensaude-Vincent 1993, 184–5.

[14] The prevalence of the operational approach ("*ratio operandi*") must be nuanced since not all of the 33 simple substances listed by Lavoisier were simple residues of experimental attempts at

that Macquer and Baumé had considered to be elements, inspired Antoine de Fourcroy's claim in 1800 that chemistry had conquered its autonomy. Thanks to a classification of its own, based on the nature and proportion of the constituent principles, chemical science would soon be emancipated from natural history and its reference to the realm of nature (de Fourcroy 1800, vol. I, xxxiij–xxxv). The criterion of composition underlying the reform of the language of chemistry at the end of the eighteenth century thus favored the claims that chemistry had become an autonomous and teachable science.

As a result of Lavoisier's reconceptualization of chemical analysis, the distinction between compounds based on the modes of union and separation that prevailed in Diderot's *Encyclopédie* was gradually abandoned in favor of the dichotomy between simple and substances compounds by the end of the eighteenth century. The interest of chemists shifted from the qualitative distinction between various modes of composition to seek out more powerful means of decomposition. Strikingly, the distinction between two types of union (mixtion and aggregation) that was key to distinguishing chemistry fundamentally from medicine in Stahlism and from mechanics in Diderot's *Encyclopédie* had disappeared from most chemistry textbooks by the end of the eighteenth century. The old notion of mixt inherited from Aristotle and reconceptualized by Stahl was eliminated from the language of chemistry following the controversy that opposed Louis Proust and Claude-Louis Berthollet in the early nineteenth century. The issue at stake was whether chemical compounds result from the union of components in fixed proportions, or in a continuous manner. The controversy was closed without winners or losers by the linguistic decision: the compounds without fixed proportions would be named "mixtures" and the compounds formed in fixed proportions "combinations." The study of the latter, which we now call "stoichiometric compounds," almost exclusively preoccupied the chemists during half of the nineteenth century.

Is it the case, then, that the pragmatic view of analysis was wholly abandoned, struck into obsolescence by Lavoisier's attempt to reshape chemistry according to a broad metaphysics? Chemists can hardly dispense with the practical aspects of analytical methods that Lavoisier set aside. Hence, the emergence of analytical chemistry in the nineteenth century, presented in this volume by Peter Ramberg, as a special branch of chemistry that crosses the boundary between theoretical and applied chemistry. In the sections that follow, I shall demonstrate that both the pragmatist and intellectualist models had a future in chemistry in the nineteenth and twentieth centuries chemistry and that they occasionally merged into composite models or entered into competition with one another.

---

decomposition. Under the heading 'elements belonging to the three realms of nature," the first class of five elements including light and caloric, oxygen, nitrogen and hydrogen.

14.3 Synthesis as a Way of Knowing Through Making

Among the followers of Lavoisier's intellectualist approach to chemistry was Marcellin Berthelot. In 1867, he published *La chimie organique fondée sur la synthèse*, a bestselling volume that underwent many printings under the title *La synthèse chimique*. This book is famous for one sentence in particular that synthetic chemists still like to quote:

> Chemistry creates its object. This creative faculty, similar to that of art, distinguishes it essentially from natural or historical sciences. The latter have an object given in advance and independent of the will and action of the scientist: the general relations which they can glimpse or establish are based on more or less probable inductions, sometimes even on simple conjectures, whose verification cannot be pursued beyond the external domain of observed phenomena. These sciences do not have their own object. Also they are too often condemned to an eternal impotence in the search of the truth, or must be satisfied to possess some scattered and often uncertain fragments of it.

On the contrary, the experimental sciences have the power to realize their conjectures. [...They] pursue the study of natural laws, creating a whole set of artificial phenomena which are their logical consequences. In this respect, the procedure of the experimental sciences is not without analogy with that of the mathematical sciences. (Berthelot 1893, 275–6; my translation).

In emphasizing the creative power of synthesis, Berthelot did not exactly brand synthetic chemistry as a promising force of production that would provide technological solutions to social and economic issues. In this book, rather, he praised it for its capacity to advance knowledge.[15] Like Lavoisier and Malouin before him, he considered that analysis has a limited cognitive power. Like Boyle and others, he argued that analysis cannot lead to a true knowledge of the nature and proportion of the components of a body because it creates artifacts. Decomposition, when it is sufficiently controlled, generates a host of new organic principles that were not made by living organisms, that are alien to nature. Ironically, for Berthelot, the products of analysis are more artificial than synthetic products. Synthesis, by contrast, truly extends our cognitive capacities because it proceeds from a hypothesis and has the power to test them by making substances. For Berthelot, synthesis is the materialization of a conjecture, of an idea. In this respect, the synthetic chemist acts like a mathematician, as the creative power of synthesis extends beyond the realm of real substances to the realm of the possible according to nature's laws.

> Not only can it create phenomena, but it also has the power to form a multitude of artificial entities similar to natural ones, and sharing all their properties. *These artificial entities are the instantiated images of abstract laws*, that [chemistry] seeks to know [...] Without leaving the sphere of legitimate ambition, we can hope to conceive the general types for *all possible substances* and create them; we can, I claim, hope to recreate all the substances that have been developed since the very beginning and form them in the same conditions,

[15] By the end of the century, Berthelot nevertheless promised that in the year 2000 chemistry would provide food and energy for all. See Berthelot (1894) and Dam (1894).

according to the same laws, using the same forces that nature put into effect to do so. (Berthelot 1893, 276–7)

Thus, chemical synthesis is first and foremost a way of knowing through making. Berthelot's conviction that making is true understanding is related to his skepticism about structural formulas and his opposition to chemical atomism (Rocke 2001). It relies on the assumption that understanding an object means going through its genesis rather than describing or visualizing its molecular structure. To know something is to experience how it comes into being through the assembly of its parts or out of the combination of two constituent elements. Its true nature is disclosed through action, it consists in a précis of the synthetic process. Therefore, Berthelot considered that 'generative formulas' better described organic substances than their hypothetical molecular structure. In his view the right formula of benzene was (C_2H_2) (C_2H_2) (C_2H_2), or tri-acetylene because when heated to 600 °C, acetylene produced a liquid containing traces of benzene, which he isolated in turn by fractional distillation.

Despite his strong emphasis on the creative power of synthesis, Berthelot did not challenge the symmetry of analysis and synthesis. He adopted Lavoisier's intellectual approach as he viewed synthesis as essentially the material implementation of a mental process from simple to compound. Berthelot conceived synthesis as a gradual process leading step by step from the element to more and more complex compounds. This is evidenced by the vast program of synthesis described in the conclusion of *La synthèse chimique*. The program included four steps: (1) with carbon, and hydrogen, making hydrocarbons, the keystones of the edifice; (2) with these binary compounds, creating the ternary substances (alcohols); (3) then the union of alcohols with acids generates ethers or aromatic essences; the same alcohols united with ammonia yield amines and vegetable alkalis; and united with oxygen they yield aldehydes or organic acids; (4) finally, these organic acids combined with ammonia produce amides (for example, urea).

In practice, Berthelot was unable to realize his grandiose plan, never progressing further than step 1 with his synthesis of acetylene. His view of the synthesis of organic compounds was based on a rational view of nature as an edifice constructed from the bottom up through the gradual assemblage of bricks. In his genetic perspective, synthetic compounds are basically the materialization of a mental process. They are sort of traces, signatures of an experimental process of production rather than molecular structures.

Berthelot's view of synthesis as a materialized mental process is still a model for numerous bioengineers working in synthetic biology. In the 2000s, the champions of this new branch of biology, whose name inspires a rapprochement with synthetic chemistry, liked to quote Richard Feynman's statement "What I cannot create I do not understand." Like Berthelot, they praise synthesis as a way of knowing through making. Biology so far based on observation and analysis, was among those sciences if which Berthelot wrote that they "do not possess their object." Thanks to the synthesis of biological bricks, biologists also become capable of verifying their conjectures. In particular, a striking analogy may be drawn between Berthelot's program of gradual synthesis and the BioBricks approach developed by Drew Endy, Jay Keasling, and Rob Carlson, who worked hard to establish synthetic biology as a discipline.[16] In their view, synthetic biology proceeds from the fusion of two worlds: molecular biology, which provided access to the building blocks of life, and computer technologies. Endy advocates a step-by-step approach, moving from independent modules or BioBricks, to devices, and finally to systems. Berthelot, the champion of synthetic chemistry and Endy, the champion of synthetic biology share the conviction that the rational simple-to-compound method of design is the key to success. Although Endy derived three guidelines—standardization, decoupling, and abstraction—from computer engineering rather than from chemistry, he clearly aims at a rational design of micro-organisms, just as Berthelot aimed at the rational design of organic molecules (Endy 2005). In both cases, the synthesis proceeds from well-characterized, standard, independent parts, separated from their milieu or environment. The resulting whole is nothing but the sum of its parts. There is no emergence, no dynamic interactions between the building bricks.

Just as Berthelot's grandiose program of gradual synthesis did not result in many industrial synthetic compounds, over the past 20 years, the BioBricks program has been unable to fulfill its promises of disruptive innovation. It is not to say that rational methods of synthesis are never fruitful. Actually, a majority of the new compounds synthesized in the late nineteenth century were predicted by an intellectual framework. Rational methods of synthesis are fruitful provided that they do not focus exclusively on the simple-to-compound pathway, thus overlooking the role of interactions and exchanges between atoms and molecules.

14.4 Synthesis and Rational Materialism

Berthelot's conception of synthesis as a creative process inspired the French philosopher Gaston Bachelard. As a philosopher trained in chemistry, he advocated a new form of materialism based on his interpretation of the practices of synthesis in contemporary chemistry. Bachelard's "rational materialism" features another variation on the intellectualist model in the sense that analysis and synthesis are first and foremost seen as creations of the mind. However, it also instantiates some aspects of the pragmatist model.

In line with Berthelot, Bachelard insisted that chemistry constructs its object and requires the production of artifacts. Chemistry that is ruled by the "spirit of synthesis" typically transforms the *fictitious* into the *factitious*.

> One must bring into existence bodies that do not exist. As for the ones that do exist, the chemist must, in a sense, remake them in order to endow them with the status of acceptable purity. This puts them on the same level of 'artifice' as the other bodies created by man. (Bachelard 1953, 60; my translation)

---

[16] https://biobricks.org/; see also Campos 2009; Bensaude-Vincent 2009a.

The molecular structures imagined by structural chemists are fictitious. They are hypothetical or predicted on the basis of theoretical knowledge. They become factual material entities tin the laboratory through a sequence of practical operations involving a lot of work and technical instrumentations. It is the power of materialization, of transforming speculative entities into actual matters of fact, that distinguishes chemistry for Bachelard. Both synthesis and analysis belong to "phenomenotechnics," a technological production of pure artificial substances mediated by instruments. Unlike Berthelot, however, Bachelard explicitly challenged the symmetry between analysis and synthesis.

> The spirit of materialist synthesis, in the proportion where it is not the strict opposite of the spirit of analysis, corresponds to a phenomenological attitude to be studied closely, to be characterized by total and eminent positivity. Too often when one reflects on the relationship between synthesis and analysis, one is content with viewing a dialectic of union and separation. This is to forget an important nuance. In fact, the process of synthesis is, in modern chemistry, the very process of invention, the *process of rational creation* by which the rational plan of an unknown substance is posed, as a problem, to the realization. (Bachelard 1953, 61; my translation)

Bachelard emphasized that the power of realization requires intensive labor with technical operations performed by chemists with the help of instruments. However, materials are also at work. Chemical substances or materials are not made of bricks passively waiting to be arranged. Bachelard clearly stated that the analytical ideal where the simple explains the compound is obsolete. The form and properties of a compound are dependent on interatomic "forces of composition."

> Not only does the carbon atom not *possess* the shape of a tetrahedron, but in its solitude, in its *own being*, it does not have the potentiality of an exact distribution of valences, with the angles indicated in the geometry of the tetrahedron. This angular distribution depends on the other atoms (or groups of atoms) that are offered to the carbon atom to constitute a chemical molecule. (Bachelard 1953, 198; my translation)

For Bachelard, molecular structures are nothing like rigid solid scaffolds. Rather, they are flexible because they depend on their surrounding, neighboring atoms. Bachelard's emphasis on the inventive power of synthesis thus relies on a notion of composition that restores the *dunamis* of the elements and the role of atomic and molecular interactions. His view of composition radically breaks with the analytical model promoted by Lavoisier. Chemistry, he claims, is an "intermaterialism." Bachelard outright refuted the two alternative visions of a compound as a fusion of elements in a mixt or as the juxtaposition of atoms. In his view, their antagonism instantiates the "naïve materialism" of alchemists that he opposed to the "learned materialism" based on the observation of the actual experimental practices of modern chemists (Bachelard 1953, 188; my translation). For Bachelard, the symmetrical view of analysis and synthesis rests on an ideal notion of the ultimate components (atoms, molecules) treated as abstract entities detached from their material environment.

14.5 Synthesis as Artful Design

Most remarkable is Roald Hoffman's composition of the two models. Hoffman more radically disrupted the symmetry between analysis and synthesis. A chemist and poet, Hoffmann collaborated during the 1960s with Robert B. Woodward, a synthetic organic chemist. Together, they devised rules based on a principle of conservation of molecular orbital symmetry in the products of organic reactions to explain the mysterious mechanisms of pericyclic organic reactions. The Woodward–Hoffman (W–H) rules—acknowledged with a Nobel prize in 1985—are a precious tool in the hands of synthetic chemists to anticipate whether concerted reactions are permitted or forbidden.

Of the duo who co-authored the rules, Woodward was the skilled experimentalist, an expert in the total synthesis of complex natural compounds who had received the Nobel Prize for Chemistry in 1965. His synthesis of vitamin B12 in 1972 was a major feat that he achieved alongside Albert Eschenmoser and a team of dozens of chemists. Hoffmann, a physical chemist, was the theorist, responsible for performing the calculations on the basis of the molecular orbital theory. However, far from praising the rational approach to synthesis, he praised synthesis as a form of art. Indeed, it is no longer art in the sense of craft; rather, it is art in the sense of a creative activity mobilizing various resources, molecular orbital theory and valence-bond theory, semi-empirical rules like W-H rules, computers, imagination, curiosity and a dynamic academic milieu.

Hoffmann likes to describe synthesis as a game strategy requiring a lot of anticipation, imagination, and skills. He himself mobilized an arsenal or resources in the process of discovery of the W-H rules (Seeman 2015). He used knowledge of symmetry in quantum chemistry derived from group theory to expand computations from planar aromatic compounds to 3-D structures. To extend Hückel theory, he did not follow the algebraic pathway from the simple to the complex, but rather he moved from the complex to the simple using his previous research on boron compounds to simpler small ring pericyclical reactions (Seeman 2022, 7). He also relied on visualization of the geometrical structures. He included considerations from photochemistry and aromaticity. He interacted with synthetic chemists to examine the role of potential intermediates in organic reactions: "he was flying through all of organic chemistry, applying eHT [expanded Hückel theory] calculations to whatever interesting molecules came his way – and everything that came his way was interesting" (Seeman 2022, 32).

A few years later, in a popular publication entitled "In praise of synthesis," Hoffmann distinguished three types of synthesis: elemental, industrial, and planned (Hoffman 1995). His description of elemental synthesis resembles a caricature of Berthelot's synthesis of acetylene:

> You take a substance A, perhaps an element, perhaps a compound, mix it with substance B, beat it with heat, light, zap it with an electrical discharge. In a puff of foul smoke, a flash,

> an explosion, out pop lovely crystals or desired substance C. This is a comic book stereotype of chemical synthesis. (Hoffman 1995, 94–5)[17]

Hoffman considers that this approach is simply a chimera, given that the vast majority of chemical compounds are being synthesized by chance as much as by design, in "that limbo between serendipity and logic."

By contrast, industrial syntheses are carefully planned, but their strategy is mainly driven by the competitive pressure to reduce cost. Hoffmann considers that industrial syntheses result from a compromise between the imperatives of cost and safety, to which environment should be added. Although this combination may lead to ingenious inventions optimizing the industrial process and products, industrial syntheses do not inspire Hoffmann's view of synthesis as an art.

The third variety—the syntheses planned in academic milieus exemplified in the design of an unnatural product such as cubane—is described as an elegant strategy analogous to the creation of a chess problem. Cubane, the end-product, results from a clever sequence of moves based on rules and aimed at mating the most recalcitrant opponent, nature (Hoffman 1995, 103). The art of synthesis essentially revolves around paying attention to the conditions of reaction at each step:

> Each reaction might be composed of five to twenty distinct physical manipulations: weighing out reagents, dissolving them in a solvent; mixing, stirring, and heating; filtration, dessiccation; and so on. A step might take an hour or a week. And the scheme does not include the laborious and ingenious analytical chemistry required to identify those intermediate molecules. (Hoffman 1995, 101)

Like early modern chemists, today's synthetic chemists are more concerned with what is going on in the vessel, with the practical conditions of reaction, than with the control of products. They are concerned with the monitoring of spontaneous molecular interactions rather than with demonstration of the true nature of things.

Like Bachelard, however, Hoffman stresses that synthetic practices demand a lot of bookish knowledge. Synthetic chemists must know all the processes and reagents already described in the literature to trigger a reaction with a catalyst or protect a chemical bond in an intermediate product. They thus create a scenario out of a set of possible routes opened up by the rules and laws of quantum chemistry, thermodynamics, and physical chemistry.

Hoffmann thus instantiates another approach to composing the two models that differs from Bachelard's rational materialism. Synthesis is no longer described as the material realization of human reasonings or as the outcome of a rational design; rather, it involves a complex game or partnership with materials and nature's laws.

[17] See also 101–06.

14.6 Diverging Ontological Assumptions

As Philipp Ball has noted, "It is becoming increasingly clear that the debate about the ultimate scope and possibilities of nanotech revolves around questions of basic chemistry" (Ball 2003). The controversy that surrounded the emergence of nanotechnology in 2000 is helpful in clarifying the ontological assumptions underpinning the contrast between the two antagonistic views of synthesis as either a method of rational design or an "art."

The rational design approach is focused on the control of the process of assembly from the simple to the compound, from the bottom up, from atoms to molecules, supramolecules… The key to success in assembling the parts from the nanoscale to the meso and macroscales is to take advantage of the molecular machinery designed by nature in living cells. Because biological evolution has produced selective assemblers like ribosomes and proteins, the convergence of nanotechnology and biotechnology is inevitable.

This is the message delivered by Eric K. Drexler, a champion of nanotechnology, who advocated "molecular manufacture." In his bestselling book entitled *Engines of Creation*, he describes chemical synthesis as a primitive and dirty design method and wonders how chemists, lacking molecular hands with which to put the parts in the right place for assembly, can achieve successful synthesis:

> Chemists have no direct control over the tumbling motions of molecules in a liquid, and so the molecules are free to react in any way they can, depending on how they bump together. Yet chemists nonetheless coax reacting molecules to form regular structures such as cubic or dodecahedral molecules, and to form unlikely-seeming structures such as molecular rings with highly strained bonds. (Drexler 1986, 13)

For Drexler, the synthesis of complex natural compounds is a kind of miracle akin to making a toy car by stirring the parts together in a box for a few hours to get them to assemble in the correct order. In his view of a molecular manufacture, the parts will be assembled from the bottom up by all-purpose, universal assemblers taken in the protein machinery. Drexler presents them as "molecular hands" manipulating nano-objects, just as children's hands manipulate Lego bricks and placing them wherever they need to go to perform the desired function.

Drexler's metaphor of "molecular hands" elicited strong criticism from the chemistry community. In particular, Richard Smalley, who was honored with a Nobel prize for the self-assembly of nanotubes and who later started a nanotube production facility at Rice University, objected that it is impossible to manipulate atoms. Molecular fingers would obviously take up too much space and prevent the closeness needed for reactions at the nanoscale (the "fat fingers" problem). Inevitably, manipulators would adhere to the atom that was being moved, making it impossible to move a building block to the desired location (the "sticky fingers" problem) (Smalley 2001). For many materials chemists, Drexler's proposal to force chemical reactions by placing the reagents in the right position is futile. Their "art" of synthesis consists in making the components spontaneously converge in the right location and assemble into larger molecular units without *any external intervention*.

Synthetic chemists do not manipulate the molecules. Neither do they rely on the genetic program to build molecular assemblers, because the components move by themselves.

In a popular book entitled *Soft Machines*, Richard Jones noted that the chemistry of life is nothing like Drexler's molecular manufacture. Atoms and molecules are not static, solid, or rigid blocks or bricks; rather, they are constantly moving in a liquid milieu. Jones insisted on the role of Brownian motion, van der Waals forces, and entropy in the molecules' self-assembly and the production of nanoscale and larger structures. In other terms, to the mechanistic model of the advocates of a rational synthesis from the bottom up, synthetic chemists oppose a dynamic model of self-assembling molecules. Smalley emphasized the importance of the inner dynamics embedded in carbon atoms in his Nobel Lecture entitled "Discovering the Fullerenes":

> The discovery that garnered the Nobel Prize was the realization that the carbon makes the truncated icosahedral molecule, and larger geodesic cages, all by itself. Carbon has wired within it, as part of its birthright ever since the beginning of this universe, the genius for spontaneously assembling into fullerenes. (Smalley 1996)

"Elemental synthesis," from the bottom up, would be impossible without the assumption of "a genius" located in carbon atoms. This animist metaphor emphasizes that the agentivity of material components is a necessary condition for achieving the synthesis of complex natural compounds. Matter can no longer be viewed as a passive receptacle upon which information is imprinted by a rational designer from the outside because self-assembly rests on spontaneous reactions between materials. Molecules have an inherent activity, an intrinsic *dunamis* allowing the construction of a variety of geometrical shapes (helix, spiral, etc.).

This art of synthesis is instantiated in supramolecular chemistry and soft chemistry, which perform syntheses at room temperature using a wide range of molecular interactions. Instead of using covalent bonds like traditional organic chemists, supramolecular chemists use weak interactions, such as hydrogen bonds, Van der Waals and electrostatic interactions. They use microfluidics and surfactants to produce self-assembled monolayers that, in turn, allow them to move from the atomic and molecular levels to the meso- and macroscales. The building blocks can assemble themselves to form supramolecular structures and beautiful molecular machines, such as the rotaxanes.

Supramolecular chemists develop an art of synthesis that has nothing to do with reverse analysis. They mix haphazardly components in a vessel and wait to see what happens, as Drexler mockingly noted. This is precisely the challenge of their art of synthesis. They do not intervene in the process to assemble the parts. Neither do they use nanorobots mimicking the universal assemblers taken from biological evolution. They delegate the task of assemblage to the molecules themselves and rely on their capacities for spontaneous self-assembly. As Harvard chemist George Whitesides noted in 1995,

> A self-assembling process is one in which humans are not actively involved, in which atoms, molecules, aggregates of molecules and components arrange themselves into

> ordered, functioning entities without human intervention… People may design the process, and they may launch it, but once under way it proceeds according to its own internal plan, either toward an energetically stable form or toward some system whose form and function are encoded in its parts. (146)

This does not mean that synthesis is an obscure process or that chemists do not understand what they are doing; on the contrary, to plan their synthetic protocols, they mobilize the resources of thermodynamics, physical chemistry, and molecular biology and even plant and marine biology when they take inspiration from nature. Their practice of synthesis is based on three assumptions that clearly distinguish them from the conventional practices of chemical synthesis based on analytic logic.

First, whereas Drexler's rational bottom-up design was based on a mechanical model of atoms and molecules, for supramolecular chemists, molecules are dynamic entities. As self-assembly rests on spontaneous reactions between materials, matter is no longer viewed as a passive receptacle upon which information is imprinted from the outside. Molecules have an inherent activity, an intrinsic *dunamis* that facilitates the construction of a variety of geometrical shapes (helix, spiral, etc.).

Second, like early modern chemists guided by the affinity tables, supramolecular chemists consider molecules to be relational entities. They are defined by their mutual interactions rather than by the nature and proportion of their components. Their properties as supramolecular edifices result less from their composition than from the arrangement of and the relations between the atoms and the molecules. Jean-Marie Lehn, who coined the phrase "supramolecular chemistry" in 1978, even occasionally ventured the idea of a molecular sociology. Molecules are social entities, and their being together changes their properties and behaviors. A single molecule of water does not behave as a crowd of water molecules in a bottle do, and synthetic chemists relying on the relations and interactions between chemical agencies take advantage of their collective behaviors.

Third, this practice of chemistry presupposes that something emerges from the molecules "being together." The coupling process is not simply the expression of the information contained in each individual component. At each step of the synthetic process, a new organization emerges. Emergence here should be understood in thermodynamic terms as the production of higher out of lower order. Self-assembly is a process leading from less ordered to higher thermodynamically ordered ensembles of molecules or macromolecules. The resulting aggregates have new properties that could not have been predicted from the characteristics of individual components. Therefore, supramolecular chemists cannot rely on a uniform view of nature as being the same at all scales. While it is true that the laws of nature are universal, the chemists do not assume that they apply equally to all scales. There is a hierarchy of structures, from the large molecules that assemble at the nanoscale to form organelles to the cells, tissues, and organs that ultimately compose unique organisms. This basic observation has led Lehn to evolve an ambitious project for chemistry, to control the basic forces of self-organization and reproduce life structures (Lehn 1996).

14.7 Conclusion

The first two sections presented various understandings of analysis/synthesis prior to 1800. The lexical puzzle is fascinating, with two contrasting views of the relations between analysis and synthesis developed by eighteenth-century French chemists: either two gradual and successive operations of decomposition and recomposition wherein mental and material operations merge or two simultaneous faces of a complex process of chemical transformation involving separations and unions. I have argued that this contrast relates to a heuristic simplification of the view of chemical composition based on the assumption that nature operates along the same lines as human logic. Chemical compounds have been named and defined by the nature and proportion of their constituent elements when the detailed processes of chemical operations occurring in the vessels have been black-boxed. In balancing the inputs and outputs of chemical reactions, post-Lavoisier chemists often reduced chemical reactions to algebraic equations and composition to a mere sum of discrete and inert chemical elements. Analysis and synthesis have thus been primarily conceived as an intellectual pathway leading from the compound to the simple and returning from the simple to the compound. This two-way journey was praised highly throughout the nineteenth century for its cognitive and probative force while the pragmatic project of synthesizing new compounds with never-before-seen properties appeared to be marginalized. This does not mean that the pragmatist model had no future.

However, the contrast between the intellectual and pragmatic notions must be nuanced given that the two sometimes overlap. Berthelot maintained the intellectualist approach to synthesis, although he championed synthetic chemistry for its creative power and promises. His program of gradual synthesis could have favored the pragmatic view because he paid attention to the actual genetic process of synthesis of acetylene that he named triglycerin and opposed atomic chemists, such as Wurtz, Kekulé, and others, who named and defined organic compounds according to their alleged molecular structures. Quite surprisingly, however, Berthelot developed a hyper-intellectualist view of composition as the material implementation of the rational process from the simple to the compound. This rational design method is still recommended in nanotechnology and synthetic biology as a bottom-up process that assembles bricks and modules into systems.

Moreover, Berthelot's concept of organic synthesis has served as model for a non-intellectualist view of synthesis in Bachelard's epistemology. Although he adopted Berthelot's notion of synthesis as inventive power, Bachelard explicitly criticized his intellectualist view of synthesis as reverse analysis. He rather emphasized that synthesis both requires specific techniques and relies on interactions between atoms and molecules. Bachelard became the champion of a rationalist materialism that retains several features of the intellectual tradition of interpretation of synthesis.

The intellectualist and pragmatic visions are combined in a considerably different mode by twentieth-century synthetic chemists, who describe it as a complex strategy for monitoring the spontaneous behavior of chemicals.

Table 14.1 Comparative survey of the two models of analysis and synthesis

Intellectualist episteme	Pragmatist episteme
Symmetry between analysis and synthesis	Disconnection between analysis and synthesis
Argumentative power, probative force	Heuristic and creative power
Identification of mental and material operations	Emphasis on skills, tacit knowledge, and tours de force
Metaphysical assumption: nature proceeds like human logic from the simple to the complex	Metaphysical assumption: material agencies capable of interactions bringing about emergent properties

If it is clear that the intellectualist and pragmatic views of analysis and synthesis are two ideal types rather than alternative or rival models, if they overlap in most particular cases, what might be the point of contrasting them and treating them as typological units, as summarized in the table below? (Table 14.1). It is worth distinguishing them to emphasize the role of the tacit metaphysical visions underlying the chemists' epistemic choices. The metaphysical assumptions underlying the two models differ substantially. The intellectualist model relies on the assumption that nature works like human logic from the simple to the complex. This conviction made explicit by the Condillac–Lavoisier connection forms the basis of the probative power conferred on experimental analysis and synthesis. It proceeds from a conflation of the material operations of composition and decomposition with the mathematical notions of analysis and synthesis as methods of reasoning. The corollary is the denial of the dynamics of matter, whether it be affinity or self-assembly.

By contrast, the pragmatist view of chemical synthesis as a complex game with the spontaneous dynamics of chemical substances is based on the assumption of active matter and relies on the agencies of atoms and molecules as much as on the empirical knowledge, skills, and tours de force of the synthetic chemists. In this model, synthesis complements analysis as an exploratory method for discovery rather than as a method of proof to establish the truth value of a hypothesis. It has a heuristic power rather than a demonstrative function. The modern conception of synthesis, while reactivating the old view of analysis and synthesis as an art, introduces something novel—namely, the idea that you can create something new with the materials at hand (i.e., materials that are not obtained by analysis). "Synthesis," as an act of design or creation, thus becomes wholly decoupled from the process of analysis.

References

Albury, William. 1972. *The logic of Condillac and the structure of French chemical and biological theory (1780–1800)*. PhDdissertation. Johns Hopkins University.

Bachelard, Gaston. 1953. *Le matérialisme rationnel*. Paris: PUF. Quoted in the Reedition 2021.

Ball, Philip. 2003. *Philip Ball, Nanotechnology in the firing line. Nanotechweb.org,* December 23, 2003, http://www.nanotechweb.org/articles/society/2/12/1/1.

Baumé, Antoine. 1773. *Chymie expérimentale et raisonnée, par …* Paris: P. F. Didot le jeune.

Bensaude-Vincent, Bernadette. 1993. *Lavoisier. Mémoires d'une révolution*. Paris: Flammarion.
———. 1996. Between history and memory: Centennial and bicentennial images of Lavoisier. *Isis* 87: 481–499.
———. 2009a. Synthetic biology as a replica of synthetic chemistry? Uses and misuses of history. *Biological Theory* 4 (4): 314–318.
———. 2009b. Le mixte ou l'affirmation d'une identité de la chimie. *Corpus. Revue de Philosophie* 56: 117–142.
———. 2010. Lavoisier, lecteur de Condillac. *Dix-Huitième Siècle* 42: 49–65.
Bensaude-Vincent, Bernadette, and Ferdinando Abbri, eds. 1995. *Lavoisier in European context. Negotiating a new language for chemistry*. Cambridge, MA: Science History Publications.
Bensaude-Vincent, Bernadette, and Bruno Bernardi, eds. 2003. *Roussseau et les sciences*. Paris: L'Harmattan.
Beretta, Marco. 1993. *The enlightenment of matter*. Cambridge: Science History Publications.
Bergman, Torbern. 1784. Meditationes de systemate fossilium naturali. *Nova Acta Regiae Societatis Scientarum Uppsaliensis* 4: 63–128.
Berthelot, Marcellin. 1893. *La Synthèse chimique*. Paris: Alcan.
———. 1894. *Discours au banquet de la Chambre Syndicale des Produits Chimiques*, le 5 avril 1894. https://sniadecki.wordpress.com/1894/04/05/berthelot-01/.
Campos, Louis. 2009. That was the synthetic biology that was. In *Synthetic biology: The Technoscience and its consequences*, ed. Markus Schmidt, Kelle Alexander, Agomoni Ganguli-Mitra, and Huib de Vriend, 5–21. Dordrecht: Springer.
Clericuzio, Antonio. 2000. *Elements, principles and corpuscules: A study of atomism and chemistry in the seventeenth century*. Dordrecht: Kluwer.
Crosland, Maurice P. 1962. *Historical studies in the language of chemistry*. London: Heineman.
Dagognet, François. 1969. *Tableaux et Langages de la chimie*. Paris: Vrin.
Dam Henry, J.W. 1894. Foods in the Year 2000. Professor Berthelot's theory that chemistry would displace agriculture. *McClure's Magazine* 3 (4): 303.
Daumas, Maurice, and D.I. Duveen. 1959. Lavoisier's relatively unknown large-scale experiment of decomposition and synthesis of water; February 27–28, 1785. *Chymia* 5: 111–157.
de Condillac, Etienne Bonnot. 1780. *La Logique ou les premiers développements de l'art de penser* (1780). Paris: Vrin-Reprise, 1980. Bilingual edition with English transl. by William Albury *Logique/Logic,* New York: Aboris, 1980.
Debus, Allen. 1967. Fire analysis and the elements in the sixteenth and seventeen centuries. *Annals of Science* 23: 127–147.
Diderot, Denis, and Jean Le Rond d'Alembert, eds. 1751. *Encyclopédie ou dictionnaire raisonné des sciences des arts et des métiers*. Paris.
Drexler, Eric K. 1986. *Engines of creation*. New York: Anchor Books.
Endy, Drew. 2005. Foundations for engineering biology. *Nature* 438: 449–453.
Fourcroy, Antoine F. 1800. *Système des connaisances chimiques et de leurs applications aux phénomènes de la nature et de l'art, par...* Paris: Baudouin, an IX.
Guyton de Morveau, Louis-Bernard. 1782. Mémoire sur les dénominations chimiques, la nécessité d'en perfectionner le système, les règles pour y parvenir, suivi d'un tableau de nomenclature chimique. *Observations sur la Physique, sur l'histoire naturelle et sur les arts* 19: 370–82. Also published as a brochure in Dijon, 1782.
Guyton de Morveau, Louis-Bernard, Antoine Lavoisier, Claude-Louis Berthollet, Antoine-François Fourcroy. 1787. *Méthode de nomenclature chimique*. Paris: Cuchet. Reprint Institute for the History of Chemistry, Philadelphia, 1987.
Hoffmann, Roald. 1995. *The same and not the same*. New York: Columbia University Press.
Holmes, Frederic L. 1971. Analysis by fire and solvent extraction. *The Metamorphosis of a Tradition. Isis* 62: 128–148.
———. 1989. *Eighteenth-century chemistry as an investigative enterprise*. Berkeley: University of California Press.

Kahn, Didier. 2007. *Alchimie et paracelsisme en France à la fin de la Renaissance (1567–1625)*. Geneva: Droz.
Kant, Immanuel. 1787. *Critic of pure reason*. Preface to the 2nd edition 1787. Available at https://www.gutenberg.org/files/4280/4280-h/4280-h.htm#chap02
Kim, Gyung Mi. 2001. The analytic ideal of chemical elements: Robert Boyle and the French didactic tradition in chemistry. *Science in Context* 14: 361–395.
———. 2003. *Affinity. That elusive dream. A genealogy of the chemical revolution*. Cambridge, MA: MIT Press.
Klein, Ursula, and Wolfgang Lefevre. 2007. *Materials in eighteenth-century science: A historical ontology*. Cambridge, MA: MIT Press.
Klein, Joel A., and Evan R. Ragland. 2014. Introduction. Analysis and synthesis in medieval and early-modern chemistry. *Ambix* 61 (4): 319–326.
Lavoisier, Antoine. 1789. *Traité élémentaire de chimie*. Paris: Cuchet. Quoted in the English transl. by R. Kerr. *Elements of chemistry*. New York: Dover Publications, 1965.
Lehn, Jean-Marie. 1996. From matter to life: Chemistry?! *Resonance* 1: 39–53. https://link.springer.com/article/10.1007/BF02835621.
Levere, Trevor. 1992. Balance and gasometer in Lavoisier's chemical revolution. In *Lavoisier et la révolution chimique*, ed. M. Goupil, 313–331. Palaiseau: Ecole Polytechnique.
Macquer, Pierre-Joseph. 1753. *Elémens de chymie théorique....* Paris: J. T. Hérissant.
Malouin, Pierre Jacques. 1751. Entry 'analysis (chemistry)'. In *Encyclopédie ou dictionnaire raisonné des sciences des arts et des métiers,* vol. 1, ed. Diderot Denis and Jean Le Rond d'Alembert, 403.
Newman, William R., and Lawrence M. Principe. 2002. *Alchemy tried in the fire: Starkey, Boyle, and the fate of helmontian chymistry*. Chicago: Chicago University Press.
———. 2005. Alchemy and the changing significance of analysis. In *Wrong for the right reasons*, ed. Jed Z. Buchwald and A. Franklin, 73–89. Dordrecht: Springer.
Pagel, Walter. 1982. *Paracelsus. An introduction to philosophical medicine in the era of the renaissance*. 2nd ed. Basel: Karger.
Powers, John C. 2014. Fire analysis in the eighteenth-century: Herman Boerhaave and scepticism about the elements. *Ambix* 61 (4): 385–406.
Rappaport, Rhoda. 1958. *G.F. Rouelle, his Cours de Chimie and their significance for eighteenth-century chemistry*. Unpublished master's thesis, Cornell University.
———. 1960. G.F. Rouelle: An eighteenth-century chemist and teacher. *Chymia* 6: 68–101.
Rocke, Alan J. 2001. Chemical atomism and the evolution of chemical theory in the nineteenth century. In *Tools and modes of representation in the laboratory sciences*, ed. U. Klein. Boston Studies in the Philosophy and History of Science, vol. 222. Springer: Dordrecht. https://doi.org/10.1007/978-94-015-9737-1_1
Rousseau, Jean-Jacques. 1999. *Institutions chymiques* (undated). First published by Maurice Gautier, *Annales de la Société Jean-Jacques Rousseau*, 12, 1918–19 and 13, 1920–21. Republished in 1999. Paris: Fayard.
Seeman, Jeffrey I. 2015. Woodward Hoffman's stereochemistry of electrocyclic reactions: From day 1 to theJ ACS receipt dtae (May 5, 1964 to November 30, 1964). *Journal of Organic Chemistry* 80: 11632–11671.
Siegfried, Robert. 2002. From elements to atoms: A history of chemical composition. *Transactions of the American Philosophical Society, New Series* 92(4): i–iii, v–x, 1–263, 265–278.
Siegfried, Robert, and B.J. Dobbs. 1988. The chemical revolution in the history of chemistry. *Osiris* 4: 34–52.
Simon, Jonathan. 2002. Analysis and the hierarchy of nature in the eighteenth century. *British Journal for the History of Science* 35: 1–16.
———. 2005. *Chemistry, pharmacy and the French revolution*. London: Routledge.
Smalley, Richard E. 1996. *Discovering the fullerenes*. Nobel Lecture December 7, 1996, http://cnrst.rice.edu.
———. 2001. Of chemistry, love, and nanobots. *Scientific American* 285: 76–77.

Smeaton, William A. 1954. The contributions of P.J. Macquer, T.O. Bergman and L.B. Guyton de Morveau to the reform of chemical nomenclature. *Annals of Science* 10: 144–167.
Venel, Gabriel-François. 1753. Chymie. In *Encyclopédie ou dictionnaire raisonné des sciences des arts et des métiers,* vol 3, ed. Denis Diderot and Jean Le Rond d'Alembert, 1055–.
———. 1754. Distillation. In *Encyclopédie ou dictionnaire raisonné des sciences des arts et des métiers,* vol. 4, ed. Denis Diderot and Jean Le Rond d'Alembert, 1055–1059.
———. 1765a. Menstruelle analyse chimie. In *Encyclopédie ou dictionnaire raisonné des sciences des arts et des métiers*, vol. 10, ed. Denis Diderot and Jean Le Rond d'Alembert, 342.
———. 1765b. Mixte & Mixtion. In *Encyclopédie ou dictionnaire raisonné des sciences des arts et des métiers*, vol. 10, ed. Denis Diderot and Jean Le Rond d'Alembert, 585–588.
Whitesides, George M. 1995. Self-assembling materials. *Scientific American* 1995: 146–149.

Bernadette Bensaude-Vincent is emeritus professor at Université Paris 1 Panthéon-Sorbonne. She specializes in philosophy and history of science and technology. She is a member of the French Academy of Technology. She received the Dexter Award from the American Chemical Society in 1994, the Sarton Medal in 2021, the Franklin-Lavoisier Prize in 2024. She recently published *Between Nature and Society. Biographies of Materials* (2022).

Chapter 15
Contesting the Musical Ear: Hermann von Helmholtz, Gottfried Weber and Carl Stumpf Analyzing Mozart

Julia Kursell

Abstract Music analysis emerged in the late nineteenth century as an occupation in its own right, independent from the education of composers. This chapter uses three case studies to describe how and with what aims three authors, in analyzing music by Mozart, produced samples to which to compare his music. It thereby traces how a pairing of the notions of analysis and composition increasingly created room for a concept of synthesis that eventually replaced the notion of the "fine ear" for music with procedures devised by the analyst that take into account a genuine malleability of hearing and listening. The cases are Hermann von Helmholtz's charts for indicating the harmoniousness of its component chords in Mozart's *Ave verum corpus*, K. 618; Gottfried Weber's fabricated alternatives to the famous dissonances opening the string quartet C major, K. 465; and, finally, Carl Stumpf's comments on his listening of the Serenade B-flat major, K. 361, after having extensively used the first technical analysis and synthesis of the sound of musical instruments.

Keywords Hermann von Helmholtz · Gottfried Weber · Carl Stumpf · Wolfgang Amadeus Mozart · Analysis · Music · Hearing · Listening

15.1 Introduction

"Mozart is certainly the composer who had the surest instinct for the delicacies of his art," Hermann von Helmholtz (1821–1874) wrote in his groundbreaking book *On the Sensations of Tone as a Physiological Basis for the Theory of Music* (Helmholtz 1863, 366; 1885, 225). To demonstrate that music corroborated his assumptions regarding the functioning of the ear, Helmholtz performed an analysis of Mozart's choral piece *Ave verum corpus*, K. 618. Composers such as Wolfgang Amadeus Mozart (1756–1791), Helmholtz argued, had access to that which in

J. Kursell (✉)
Musicology, Faculty of Humanities, University of Amsterdam, Amsterdam, The Netherlands
e-mail: J.J.E.Kursell@uva.nl

W. R. Newman, J. Schickore (eds.), *Traditions of Analysis and Synthesis*, Archimedes 73, https://doi.org/10.1007/978-3-031-76398-4_15

music teaching remained unexplained: “It was left to the musician himself to obtain some insight into the various effects of the various positions of chords by mere use and experience. No rule could be given to guide him” (224).

Whether Helmholtz’s notes on Mozart may be considered as amounting to a “musical analysis” is debatable. The historiography of the scholarly study of music generally considers musical analysis to have emerged in the late nineteenth century—after Helmholtz and after the foundation of the discipline of musicology, which, for the German speaking realm, is generally considered to have occurred in the mid 1880s, heralded by the foundation of the journal *Vierteljahrsschrift für Musikwissenschaft* in 1885.[1] However, the first examples of full-fledged musical analyses mentioned in scholarly literature stem from the first half of the nineteenth century. Gottfried Weber (1779–1839), an author and music teacher of private means, is typically included among the first to have published an analysis of music in the form that would subsequently become standard for the discipline: an extended discussion of a single composition according to explicit criteria. His analysis of the slow introduction to Mozart’s string quartet in C Major, K. 465 was first published in Weber’s own journal *Caecilia* in 1831 and later inserted into his three-volume teaching manual. Weber explicitly addressed his own procedure as such, using the German equivalent to analysis—namely, “taking apart” (*zergliedern*) as well as the Greek borrowing “Analyse.” More significantly, he did not assume the perspective of a composer wishing to learn from Mozart but that of a listener who wished to understand the composer’s music. It was this perspective—that of the informed listener—that would later go on to define the target of musical analysis.

While Helmholtz did not explicitly frame the discussion of the *Ave verum corpus* as an “analysis,” he did embed it into what he called “a correct and careful analysis of a mass of sound” (1885, 227). As such, it formed part of the core of his book, the testing of knowledge about hearing that may be in contradiction to his hypothesis about the functioning of the ear. He was aware that he would be unable to prove in vivo his claim that small bodies in the inner ear reacted selectively to the frequency components in sound. However, while confirming that his theory of selective resonance in the organ of Corti had “no immediate connection” to his investigation into music, he insisted that the theory could be said to gather “all the various acoustical phenomena with which we are concerned into one sheaf,” giving a “clear, intelligible, and evident explanation of the whole phenomena and their connection” (227). What mattered most was that his observations of Mozart’s piece did not contradict his findings on hearing.

The present chapter, which discusses three examples of nineteenth-century analyses of Mozart’s music, confronts two key approaches: listening to music versus hearing music. Both Helmholtz and Weber, who instantiate these two approaches respectively, used methods that are said to be analytical and implemented these

[1] For a more recent general reference on the history of musicology, see Melanie Wald-Fuhrmann’s, entry *Musikwissenschaft, Zur Fachgeschichte* in: *MGG Online*, ed.by Laurenz Lütteken, New York, Kassel, Stuttgart 2016ff., published June 2022, https://www.mgg-online.com/mgg/stable/421611 (accessed June 29, 2024).

methods as such, but their emphases on listening and hearing afford different degrees of agency to the subject. While Weber emphasizes that his analysis empowers the listening subject to understand what Mozart was doing, Helmholtz's analysis empowers those to whom musical training was not accessible but nevertheless could not but hear music in the manner in which he demonstrated it to affect the ear. Helmholtz's instruments assisted "a researcher without any music training," as noted by Franz Joseph Pisko (1827–1888), author of a popular introduction to Helmholtz's new apparatus of acoustics, in 1865. "Even one who is hard of hearing can undertake acoustic studies, in which weak tones that are covered by a number of simultaneous sounds are supposed to be perceived" (1865, 7).

Marshaling his knowledge of mathematics, physics, anatomy, physiology *and* music, Helmholtz not only proposed a theory of hearing—which was later disproven by György Békésy, who was awarded the Nobel prize for his explanation of the mechanics that performed the selective analysis in the inner ear—but he also designed an *experimentum crucis* in which his analyses of sound were subject to a synthesis to test the validity of his hypothesis of the ear's capacity to discriminate sound. The coupling of analysis and synthesis within the domain of sound also exerted a considerable impact on musical discourse, as I shall argue in a third example in this chapter—a remark by philosopher and experimental psychologist Carl Stumpf (1848–1936) on Mozart's Serenade, K. 361, from his book *Speech Sounds* (1926) will be used to investigate whether and how the paired notions of analysis and synthesis in hearing music interfere with those of composition and analysis in listening to music. Stumpf prominently introduced sound analysis and synthesis into his psychological laboratory in the 1910s. Yet, his remark on Mozart, which concerns the role of timbre, reveals a trajectory of experiences that the individual listener—Stumpf, in this case—must have undergone before being able to analyze the features of a composition at a given moment. As the remark on timbre demonstrates, this experience may take any form, and the analysis, accordingly, may be just as well informed by sound analysis and synthesis.

Musical analysis will thus be discussed in this chapter from the following three perspectives. In the first part, which focuses on Helmholtz, I shall demonstrate that the criteria for analysis stem from an attribution of properties that operate on several levels while taking apart the matter at hand. The notion that sound can be analyzed and synthesized offers a new basis for arranging charts of the properties under discussion. In the second part, I shall trace how Weber was first obliged to instruct his readership about what they might expect. They first had to acquire the rules of composition. Once this stage was assumed, the process of "taking apart" could address a given piece. This section will emphasize how the object for analysis must be constituted, including not only the authority of the canonized piece but also the trajectory that leads to its understanding. The third part concerning Stumpf's remark on Mozart discusses how he takes the analysis and its object to be co-constituted both through listening and hearing. The working hypothesis at stake in this chapter is that the analysis of music always co-constitutes its objects. In observing how actors in three different fields use different strategies for this co-constitution, the chapter traces how the notion of musical analysis became detached from that of composition.

15.2 Experimental Sensory Physiology: Helmholtz Analyzing K. 618

Helmholtz's experimental physiology of hearing made extensive use of music as an object of experimentation. Like its counterpart, the *Physiological Optics* (1867), the book *On the Sensations of Tone as a Physiological Basis for the Theory of Music* (1863) has a tripartite structure that covers the physical, physiological, and psychological conditions of sensory—in this case, auditory— physiology. However, while the *Optics* discussed stereoscopic vision as the main object for discussing the highest level of sensory processing in physiology, the *Sensations of Tone* turned to music instead. Music permeated all three parts of the research presented in the book. Periodic sound was the privileged object not only of Western tonal music but also of acoustic science, and Helmholtz's work fostered its status as an aesthetic premise of music composed in the nineteenth century.

In the first part of the *Sensations of Tone*, which discusses the physical and anatomical preconditions for audible sound entering the ear up to the nerve endings, Helmholtz's presented his resonance theory of hearing. According to this theory, minute bodies along the inner ear's basilar membrane were capable of selectively resonating, each in its respective eigenfrequency, with incoming frequency components. This theory and, more specifically, the claim of resonating bodies whose eigenfrequencies covered the range of hearing could not be proven. The minute dimensions of the inner ear, located within the hardest bone of the human body, prohibited observation, and no resonance was demonstrable by post-mortem anatomy. It was only in the twentieth century that the basilar membrane became visible in action. For Helmholtz, the workings of the inner ear remained inaccessible to verification by autopsy.

The wealth of experiments and knowledge accumulated in Helmholtz's book centered on his desire to fill this gap. Anything known about sound and music that was within reach—from the sound of musical instruments through speech sounds and compositions to music history and non-European theoretical writing—was used in the interest of excluding contradictions to the main hypothesis. The second part—on the physiology of hearing proper—focused on distortions that could reveal the ear's functioning by exposing its limitations. Again, music provided most of the phenomena on which he reported, as the distortions in question were those that occurred from simultaneously given periodic sounds. Helmholtz proposed, for instance, that the beats resulting from the superposition of periodic waves could be correlated with the musical notion of dissonance, and he speculated that the happy or gloomy impression conveyed by major and minor modes, respectively, was partly due to the emergence of combination tones that occurred when the sound's amplitudes exceeded the ear's dimensions. The non-linear distortion that he was able to calculate had not previously been described for hearing. Its discovery, published in *Annalen der Physik und Chemie* in 1856, granted Helmholtz his entry into the field of auditory physiology.

Helmholtz's new explanation of combination tones—difference tones, to use today's terminology—turned the field of auditory physiology upside down. If musical tones had previously been held to be a figment of the imagination—namely, as that which the mind makes up from the physical vibration—the understanding of the ear as a distorted auditory channel showed that the body produced what would be perceived in the mind. The notion of acoustics as a branch of physics was thus unnecessary, according to Helmholtz, as it did not relate to any properties of vibration that might not hold to inaudible vibration. The bodily conditions of sound production now had to be integrated into the physical study of sound. This also entailed that physics did not stop before the ear to leave perception to the mind: rather, physics entered the body. In the ear itself, two tones could produce a third, and the mind—or, for that matter, the music listener—could not but hear it.[2]

Two tones or periodic sounds producing a third meant that two musical notes would produce a third note that was unwanted by the musician. That this phenomenon was of the same kind as the culturally produced musical tones made music an interesting object of inquiry. Mozart's choral piece, *Ave verum corpus*, K. 618, which, Helmholtz writes, was praised for its beauty and simplicity, could serve his purposes for an examination of whether a composer such as Mozart could avoid the combination tones or use them artfully.

Prior to Helmholtz's discussion of the piece, the composer and author Hector Berlioz (1803–1869) had mentioned the *Ave verum corpus* in his treatise on instrumentation, the *Grand traité d'instrumentation et d'orchestration modernes* (1844). There, it served to exemplify the use of the human voice in choral music. One feature that Berlioz noted as particularly remarkable was the indication "sotto voce"—that is, with a soft voice—that applied throughout the entire piece. Berlioz acknowledged the difficulty of singing softly in a controlled way over a lengthy period of time and remarked that it was therefore advisable to use the range in which the singers would feel most comfortable. The composer, he wrote,

> should use only notes of the medium [range] in an Andante with soft and sustained sounds; those alone can possess the suitable quality of tone, dwell with calm and precision, and be sustained without the least effort in a pianissimo. This is what Mozart has done in his celestial prayer: "Ave verum corpus." (1858, 179)

Such soft sound production would be unlikely to produce distortion of the kind that interested Helmholtz. Nevertheless, Mozart's piece was one of two compositions that he scrutinized in a laboratory setting, the other being a *Stabat Mater* by

---

[2] On Helmholtz's research into combination tones, see Pantalony 2005; Kursell 2009, 2015; Hiebert 2014. In Kursell 2018a, I attempt to link the research on combination tones, e.g. in Johannes Müller and Jan Purkyne, to that on "subjective phenomena." On the latter, especially Johan Wolfgang Goethe's notion of subjective phenomena, see Crary 1990; Vogl 2007; Schimma and Vogl 2009; Schäfer 2011.

Renaissance composer Giovanni Pierluigi da Palestrina (ca. 1525–1594). On a keyboard instrument that Helmholtz had constructed for his experimental acoustic work—a harmonium with a steady sound that was tuned according to his instructions—he investigated an aspect of the combination of notes that he had found to be neglected in harmony textbooks:

> In musical theory, as hitherto expounded, very little has been said of the influence of the transposition of chords on harmonious effect. It is usual to give as a rule that close intervals must not be used in the bass, and that the intervals should be tolerably evenly distributed between the extreme tones. And even these rules do not appear as consequences of the theoretical views and laws usually given, according to which a consonant interval remains consonant in whatever part of the scale it is taken, and however it may be transposed or combined with other. They rather appear as practical exceptions from general rules. (1875, 339)

Helmholtz himself completed an extensive study of these distributions, as he expected them to differ with respect to their degree of distortion. These distortions are prominent, for example, in the piercing sound of recorder ensembles who need not play falsely to sound so: the disturbing additional pitches emerge in the listeners' ears. The measurements and experimental verifications that Helmholtz undertook using his harmonium resulted in a list of the "best sounding combinations" (1875, 229) of triads and tetrads in major and minor. He then turned to actual compositions to determine whether his conjectures were borne out by any composer. The *Ave verum corpus*, he wrote,

> is particularly celebrated for its wonderfully pure and smooth harmonies. On examining this little piece as one of the most suitable examples for our purpose, we find in its first clause, which has an extremely soft and sweet effect, none but major chords, and chords of the dominant Seventh. All these major chords belong to those which we have noted as having the more perfect positions.[3]

He then enumerated the frequencies of chords as they were ranked in his chart of harmoniousness, disregarding their melodic and harmonic sequence in the piece. However, he discussed the larger formal units again, continuing,

> It is very striking, by way of comparison, to find that the second clause of the same piece, which is more veiled, longing, and mystical, and laboriously modulates through bolder transitions and harsher dissonances, has many more minor chords, which, as well as the major chords scattered among them, are for the most part brought into unfavourable positions, until the final chord again restores perfect harmony. (339–340)

The status of the chords in this investigation is remarkable. To define them, Helmholtz did not resort to harmony teaching manuals, at least not directly, but rather created his own chart of building blocks. Chapter 12 on "chords" presents a table that aligns the intervals he discusses two chapters earlier with regard to their consonance and dissonance (Fig. 15.1). In the top left corner, the letter C indicates the lower of two notes, the second of which is aligned vertically and horizontally

[3] The German original is even more affirmative than Ellis' translation: "Alle diese Durakkorde gehören den von uns als vollkommen wohlklingend bezeichneten Akkorden an" (366).

322 Zweite Abtheilung. Zwölfter Abschnitt.

C	*G* $\frac{3}{2}$	*F* $\frac{4}{3}$	*A* $\frac{5}{3}$	*E* $\frac{5}{4}$	*Es* $\frac{6}{5}$	*As* $\frac{8}{5}$
G $\frac{3}{2}$						
F $\frac{4}{3}$	Grosse Secunde $\frac{9}{8}$					
A $\frac{5}{3}$	Grosse Secunde $\frac{10}{9}$	G r o s s e T e r z $\frac{5}{4}$				
E $\frac{5}{4}$	K l e i n e T e r z $\frac{6}{5}$	Kleine Secunde $\frac{16}{15}$	Q u a r t e $\frac{4}{3}$			
Es $\frac{6}{5}$	G r o s s e T e r z $\frac{5}{4}$	Grosse Secunde $\frac{10}{9}$	Uebermässige Quarte $\frac{25}{18}$	Kleine Secunde $\frac{25}{24}$		
As $\frac{8}{5}$	Kleine Secunde $\frac{16}{15}$	K l e i n e T e r z $\frac{6}{5}$	Kleine Secunde $\frac{25}{24}$	Verminderte Quarte $\frac{32}{25}$	Q u a r t e $\frac{4}{3}$	
B– $\frac{7}{4}$	Verminderte Terz $\frac{7}{6}$	Falsche Quarte $\frac{21}{16}$	Kleine Secunde $\frac{21}{20}$	Verminderte Quinte $\frac{7}{5}$	Falsche Quinte $\frac{35}{24}$	Grosse Secunde $\frac{35}{32}$

Fig. 15.1 Chart of consonant intervals that produce the most consonant triad. (Helmholtz 1863, p. 322, cf. 1885, 212)

with the letter C. These are positioned in such a way as to present intervals that decrease in consonance. The ratio between C and G is the simplest next to self-identity or a doubling of the frequency—that is, an octave in musical terms. C–G, or the "fifth," indicates a ratio of 2–3 between the fundamental frequencies of two sounds. The next, "F" or the interval with a ratio of 3–4 is slightly less consonant, producing more beats, as Helmholtz calculated two chapters earlier, where he famously proposed the explanation of dissonance as the presence of interferences between two notes' spectra. A, following Helmholtz, is the next-best note to sound together with C, followed by E, E-flat, etc. He now fills in the fields that result from the horizontal and vertical alignments, indicating the interval that results between the notes that denote the further lines and columns in the chart. Thus, self-identity is left empty; below, a column of resulting intervals follows it. If the resulting interval is consonant, its name is rendered in spaced print.

From this, he obtains a list of just six consonant combinations of three pitches (Fig. 15.2).

To the reader who is familiar with musical harmony teaching, the result is rather banal. Helmholtz, however, explains it again, adhering strictly to the logic of his own deduction:

> The two first of these triads are considered in musical theory as the fundamental triads from which all others are deduced. They may each be regarded as composed of two Thirds, on major and the other minor, superimposed in different orders. The chord C E G in which the major Third is below, and the minor above, is a *major triad*. It is distinguished from all other major triads by having its tones in the closest position, that is, forming the smallest intervals with each other. It is hence considered as the *fundamental chord* ("Grundakkord oder Stammakkord") or basis of all other major chords. The triad C E-flat G which has the minor Third below, and the major above, is the *fundamental chord* of all *minor triads*. (1885, 212)

In the next step, the rankings of the most harmonious triads and tetrads result from distributing the intervals over more than one octave and observing how strongly the distortions come to the fore when played on a harmonium in just intonation. On this instrument, the approximation to the integer ratios that are taken to define the intervals is closer than on a piano. Helmholtz even had a harmonium built for laboratory purposes according to his own instructions as to how its double keyboard should be tuned. Mozart's "fine ear" was thus replaced in the experiment by an instrument. Figure 15.3 shows the triad ranking with the triads themselves in hollow notes (half note or minim) and the distortions in black notes (quarter note or crotchet).

Next, Helmholtz expanded this procedure to more than three simultaneous notes, ranking "the most perfect positions of major tetrads within the compass of two octaves" (Fig. 15.4) as well as minor tetrads and less favorable positions.

The second position, encircled in Fig. 15.4, equals the position of the choir's voices in the beginning of Mozart's piece (see Fig. 15.5).

Fig. 15.2 Combinations of three tones, resulting from the chart in Fig. 15.1 (Helmholtz 1863, p. 322)

1) *C E G* 2) *C Es G*
3) *C F A* 4) *C F As*
5) *C Es As* 6) *C E A.*

Fig. 15.3 The most perfect positions of major triads (Helmholtz 1863, 332, cf. 1885, 219)

Fig. 15.4 Best positions of major tetrads within the compass of two octaves. Here, the distortions are not indicated. (Helmholtz 1863, 337, cf. 1885, 223)

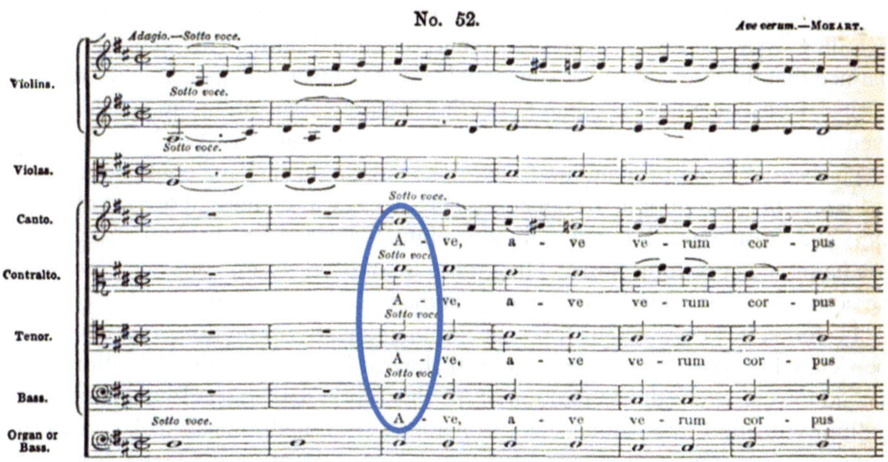

Fig. 15.5 The opening of Mozart's K. 618. Encircled: the entry of the choir in Helmholtz's "position 2"

The total ranking of Mozart's choices is, according to Helmholtz, as follows: "Position 2 occurs most frequently, and then 8, 10, 1 and 9. It is not till we come to the final modulation of this first clause that we meet with two minor chords, and a major chord in an unfavourable position" (1885, 225). Together with his description of the more "veiled" atmosphere of the second part and the bright conclusion, this results in the following distribution, given here for clarification (Fig. 15.6).

This idea of describing the overall harmonic sphere of a piece rather than following the voices and describing the transitions from chord to chord was wholly new. Rather than being built upon the rules of harmony and counterpoint, it had its roots in Helmholtz's new approach to the analysis of sound. To quote Pisko again, Helmholtz's major contribution to research in acoustics was the "analysis and

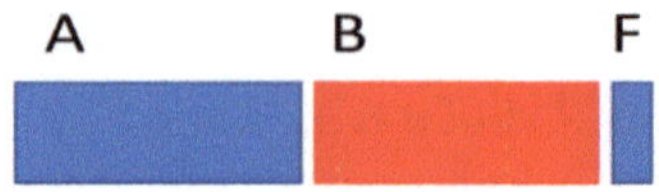

Fig. 15.6 Schematic depiction of the result of Helmholtz's analysis of Mozart's *Ave verum corpus*: A: the harmonious first part; B: less harmonious second part; F: the harmonious final chord

synthesis of sound" (1865 passim).[4] The "apparatus for the artificial composition of vowels" (*Apparat zur künstlichen Zusammensetzung der Vokalklänge*, 1912, v) that Helmholtz had designed for his acoustic research was based on his assumption that periodic sound could be analyzed into sinusoidal components in integer ratios. Following Jean-Baptist Fourier's theorem, Helmholtz further assumed that the results produced by such analyses were relevant to the ear. If two periodic sounds—or musical tones—of the same pitch and the same overall loudness could still be distinguished as stemming from different instruments or as instantiating two different vowels of the German language, then the ear would require some information regarding what made these sounds different. Based on Fourier's theorem, Helmholtz claimed that the strength of the sinusoidal components was responsible for that difference. Consequently, the organ of hearing required some ability to detect the frequencies in a sound selectively and according to their intensity. While the inner ear was inaccessible to observation, its abilities to distinguish sound could be tested.

To test this assumption, the apparatus (Fig. 15.7) was composed of a set of tuning forks and resonators, all tuned to the first eight—later the first twelve—frequencies in a harmonic spectrum. Surrounding these tuning forks with coils and connecting the coils to an "interrupter fork" that was tuned to the same frequency as the lowest of the forks, he could set all forks electromagnetically in motion. The interrupter then served to move every vibration of the lowest fork, every other vibration of the second lowest, every third vibration of the third fork, and so on. To manipulate the audibility of the forks' sound, the resonators were tuned to each fork's main frequency. The combination of fork and resonator would make the fork's soft sound audible while simultaneously eliminating unwanted additional components from its sound. As the—inharmonious—spectra of the forks and the resonators only converged in one frequency, what became audible in this experimental setup were single frequencies or "simple tones," to use Helmholtz's terminology. The resonators, in turn, could be opened and closed with movable lids, and their distance from the forks could be changed, as could the current that passed through the electromagnets with which they were surrounded. The tones would become audible when the lid of the adjacent resonator was open and almost inaudible when the lid was closed.[5]

---

[4] According to the state of Google's Ngram Viewer while this chapter was written, Pisko is the first to speak of an analysis *and synthesis* of sound. Helmholtz himself does not speak of sound synthesis but of the artificial composition (*künstliche Zusammensetzung*) of vowels (translated as "apparatus for the artificial construction of vowels" by Ellis in Helmholtz 1885, vi).

[5] The manipulation of phase will not be discussed here. To my knowledge, Helmholtz was the first to use the term "phase" consistently to refer to acoustic waves (Kursell 2018a, 269).

Fig. 15.7 Apparatus for the artificial construction of vowels after Helmholtz with twelve sets of tuning forks and resonators, an interrupter fork and a keyboard, taken from Koenig 1889, p. 26

Using this apparatus, Helmholtz re-instantiated the data he had gathered from analyzing sung vowels with sets of resonators of his invention. These were tuned to one Fourier series of frequencies and could be held to the ear. If a component of the series in question was prominently present in the sound under scrutiny, the resonator would single it out from the overall sound. For the synthesis, the components of the Fourier series of frequencies would then be adjusted to the values obtained in this way. Although the sounds that the 8 or 12 forks produced did not resemble any vowel sound in particular, a faint resemblance seems to have been possible when the settings were quickly switched for individual vowels. For this, a keyboard was added to the apparatus, with each key connected to one of the resonators' lids. That there was any difference at all, however, was sufficient to prove that the results of the Fourier analysis were relevant to the ear. Helmholtz termed the sound differences that he obtained in this way "musikalische Klangfarbe" or timbre.[6]

All these experiments presupposed the possibility of producing tones with a single frequency component. If such a sound was audible, an investigation of tones that were not equivalent to a musical note but to a mathematic symbol might be successful. This affected the basic concepts used in musical discourse. While German-language music theory continued to speak of one "tone" when one note was perceived, Helmholtz consistently exchanged "tone" with "tones," given that every note potentially contained multiple audible periodic components.[7] His description

[6] See Kursell 2013, 2017, 2018a, 2018b.

[7] The literature on Helmholtz tends to overestimate the relevance of the equivalence between note and periodic sound, even today. Helmholtz apparently was not reluctant to let go of an assumption that could be dropped without running into contradictions. By contrast, he took the greatest pains to eliminate the impact of phase on hearing. As he did not know about the early connection of both ears' nerves, he assumed that spatial hearing happens individually and is thus not subject to his

of Mozart's music is based on this renewed terminology. In two notes, all the components could potentially produce distortions, such as beats or combination tones, though his chart could only consider those that were most likely to occur. Yet, this new approach to defining a note explains the great care he took to explain the basic terminology of music from scratch — and why Pisko would call this an investigation even for those lacking musical education or being hard of hearing. Only those components that could be safely predicted would be taken into account. A phenomenological approach to the piece's perception, while not irrelevant, was only given in the final stage: the alternating harmonious and veiled character of the *Ave verum corpus* ultimately yielded a correlation with the predictable distortions in the sound.

15.3 Music Theory: Gottfried Weber Analyzing K. 465

Reflecting on his investigations of harmoniousness, Helmholtz summarized his findings on distortions in the conclusion to the second part. There, he attributes joy to the acknowledgment of order, also in music listening. He writes,

> A combination of tones will please us when we can discover the law of their arrangement. Hence it may well happen that one hearer finds it and that another does not, and that their judgments consequently differ. The more easily we perceive the order which characterizes the objects contemplated, the more simple and more perfect will they appear, and the more easily and joyfully shall we acknowledge them. But an order which costs trouble to discover, though it will indeed please us, will associate with that pleasure a certain degree of weariness and sadness (*tristitia*). (1885, 230)

A case in point to instantiate this claim may be found in a discussion of another piece by Mozart that caused early nineteenth-century music scholars to suffer from a heavy dose of such *tristitia*. A rather fierce debate had taken place between the Belgian music theorist and scholar François-Joseph Fétis (1784–1871) and his German counterpart, Gottfried Weber, between 1829 and 1832. Fétis found the slow introduction to Mozart's string quartet in C major, K. 465 so bold that he assumed the piece needed a correction, while Weber opposed the notion that Mozart made mistakes but felt the need to state,

> As regards *my own* ear, I frankly confess that it does *not* receive pleasure from sounds like these;—on this subject I can freely speak as I think, and, in defiance of the silly and envious, dare even take up the haughty words and say: *I know what I like in my Mozart*. (754)

The article that Weber wrote in answer to Fétis counts as one of the first examples of musical analysis before it became consolidated only in the late nineteenth century. Reference works mention Jérôme-Joseph de Momigny's *Cours complet d'harmonie et de composition* (1806), E.T.A Hoffmann's essay about Ludwig van

own terms of sensory physiology. Moreover, given his experimental means and the fact that he was first to consider phase at all, the concept of phase as a factor in spatial hearing did not enter the scope of his investigation.

Beethoven's *Fifth Symphony* (1810) next to Weber's article on Mozart's string quartet K. 465. Weber first published his analysis of this quartet in the music journal *Caecilia* in 1831 and republished in his multi-volume treatise *Versuch einer geordneten Theorie der Tonsetzkunst* (1832).[8] These texts are considered "milestones" (Gruber 1994) and "classic examples" (Moreno 2004, 19) of music analysis, as they proposed new approaches to describing music with specific sets of criteria that are expounded in the texts themselves.

Ian Bent's definition of music analysis as "that part of the study of music that takes as its starting-point the music itself, rather than external factors" in the music encyclopedia *The New Grove* (Bent and Pople 2001) is poised within a Western notion of music theory that builds upon centuries of text production on music. Since the early Middle Ages, a corpus of texts had developed from mainly expounding rules for singing toward an extensive prescription of how to compose any kind of music within the symbolic system of musical sound and eventually up to what has mostly been addressed in music theory as the analysis or "interpretation" of the classical music repertoire. A break occurred in this developmental trajectory when the bulk of theoretical writing became descriptive rather than prescriptive. One of the explanations for this break has been seen in the rise of the bourgeois middle class and its concept of the "masterwork" (*Meisterwerk*) that was set forth in opposition to aristocratic representational aesthetics (Gruber 1994). New types of literature divulged the canon of masterworks and fostered a discourse on music for the non-expert who mainly listened to music rather than producing it (Thorau and Ziemer 2019).

Weber's contribution to this development is situated at a point when the composer or "Tonsetzer" was still the main addressee. While Weber himself has been described as a *Bildungsbürger* and dilettante (Holtmeier 2007), the style of the analysis that he demonstrated in the journal *Caecilia* presupposed a high level of musical education. Weber had founded the journal himself, and he frequently used it as a platform from which to convey his own standpoint within musical discourse. In the case of Mozart's string quartet K. 465, the discussion surrounding the first movement's slow introduction had been ongoing. The debate had begun immediately after the publication of Mozart's six quartets dedicated to Joseph Haydn in 1785, earning K. 465 the nickname "Dissonance" (*Dissonanzenquartett*), and flared up when the Belgian music theoretician François-Joseph Fétis published a note on Mozart's alleged "mistakes" in 1829. The note appeared in the *Revue musicale*, founded by Fétis himself.

The bone of contention was the abundance of so-called "unintroduced dissonances" and "false relations" with which the piece opens. Experts invoked these

[8] See the entries on "Analysis" in, e.g., *MGG* (Gruber 1994), *Grove Music Online* (Bent 2001), *Lexikon der systematischen Musikwissenschaft* (Utz 2010). De Momigny provides reductions of opera settings, in which those adhering to the most practiced method of musical analysis in the English-speaking realm of scholars, so-called "Schenkerian" analysis, could see a precursor. E.T.A. Hoffmann, in turn, provided an early example of a hermeneutic interpretation with a similar impact on those who continued to work in this vein.

terms to refer to a particularly daring and sometimes even unruly use of the rules of counterpoint. In many ways, however, Mozart's piece alluded to exactly these rules and eighteenth-century musical learnedness by stretching the rules as far as possible. The first movement's slow introduction combines one of the oldest techniques in polyphonic setting—namely, an "imitation" (whereby a voice "imitates" the melodic steps of the preceding part)—with a feature of baroque basso continuo setting, whereby the bass repeats a single note like a foot on an organ pedal, above which a harmonic progression unfolds (with the bass note adding to the intricacy of that harmonic progression into which it does not fully fit). Furthermore, a slow introduction was not a regular feature of string quartets but rather was more characteristic of symphonies. The slow introduction to a symphony can often be seen to present features of the piece's chosen key, exploring, as it were, the tonal realm in which the piece is going to take place. This was a frequent feature in Haydn's symphonic music, for example. Mozart thus made a bow before his dedicatee. In K. 465 in addition, there is a witty contrast between the difficulty of the slow introduction and the bright and happy character of the main part after the introduction. The slow introduction reminds the addressee—listener and dedicatee—that C major was also the key in which Mozart is known to have composed his most adventurous explorations of tonality. C major normally needs no sharps or flats, yet Mozart's explorations of the limits of tonality in this key abound in so-called "accidentals." As such, this slow introduction showcased difficulty.

Fétis, in any case, was certain that Mozart had committed errors. He was prompted to publish his note on the passage in question by the opportunity to examine Mozart's original manuscript, owned by a London-based harp maker. Fétis noted,

> My first worry was thus to check this quartet where I hoped to find a confirmation of my conjectures; but I had to convince myself immediately of my own error. The passage that had received so much critique was written by Mozart unambiguously and without any sign of hesitation as it had been engraved in all editions, and the inconceivable dissonances without any aim that disrupt the ear are taken down by his hand. Thus, there is no longer any doubt about this error of a great artist. We may trust the evidence, but we should stick to saying, as did Haydn, that he had his reasons for writing this way; because mistakes of this kind hurt reason, senses, and taste. (606, my translation)

The conjectures that Fétis mentions here concerned an example that he gave in the same brief article. Including the printed score of the piece's beginning, he added an illustration of his own musical conjectures in an alternative to the quoted opening. Fétis reassured the reader that his alternative contained only minimal interventions, which, for him, made it all the more mysterious that Mozart would not have preferred not to use them immediately. He writes:

> On examining carefully this harmony that had been the object of so much astonishment and conjecture, I was struck by how easy it would have been to remove its defects, without changing either the main phrase or the form of the accompaniments, and even to make the imitation which seemed to be the cause of the gross errors that one notices in it more exact and more in keeping with the rules of all schools. (602)

Thus, the difference between the original and the alleged correction is rhetorically downplayed to render Fétis' astonishment about the faulty original even more prominent.

Weber opposed the notion that Mozart had made mistakes. He accepted the piece as a given and spoke of choices rather than mistakes. His own treatise of harmony spelled out the rules Fétis appeals to in the several volumes of his treatise on composition. An extended appendix on Mozart in the final part provided him with the occasion to discuss the breaching of the rules. However, he refrained from a "judgment on the frequently disputed theoretical allowableness and irregularity of the passage in question" (737/200), proposing instead to "observe" and "analyze" what Mozart did and how. He prepares the reader by distinguishing music from science:

> Once for all, music is not a science endowed with mathematical deduction and completeness; it is not a system presenting us with absolute rules of permission or prohibition, the adoption of which can in all cases determine — like "twice two are four"—the value or worthlessness, the accuracy or inaccuracy, the lawfulness or unlawfulness of this or that combination or succession of tones; and all the pretensions of those who have imagined they could found the theory of music on mathematics, and from such an assumed foundation deduce and establish absolute precepts, appear on the slightest examination as empty and ridiculous dreams, the fallacy of which can be clearly proved by the first best example. (Weber 1851, 737)[9]

Weber framed his activity as analyzing or "taking apart" (*zergliedern*), and he called the result an analysis (*Analyse*). His main method in doing so was to describe, note by note, not only the rules that explained the notes' correctness and contextualized the alleged breaches but more specifically the effect that each note might have exerted on a potential listening instance. This instance, he addressed as "Gehör" (ear), occasionally "Gehörsinn" and "Ohr"—that is, the sense and the organ of hearing—and presented it as having internalized the rules of composition and their possible application. In so doing, he also appealed to his readership to carefully study the book so as to be able to follow him now in applying the rules for analysis.

In musicological research, Weber has been said to track "an idealized listener's perception of the passage chord by chord. The result is an analysis that is historically noteworthy for its elegant descriptive language and its quasi-phenomenological awareness of musical harmony as it unfolds in time" (Bernstein 2002, 787). Most frequently, commentators have pointed to the excessive number of alternatives that Weber provides (e.g. Christensen 2019). Similar to Fétis' conjectures, these alternatives demonstrated what Mozart could have done, and they also instantiate what

[9] Die Tonkunst ist nun einmal keine mit mathematischer Consequenz und Absolutheit begabte Wissenschaft, kein System, welches uns absolute, verbietende oder gebietende Regeln darböte, aus deren Anwendung auf jeden vorliegenden Fall sich, wie 'Zweimal zwei ist vier' der Werth oder Unwerth, die Richtigkeit oder Unrichtigkeit, Erlaubtheit oder Verbotenheit dieser oder jener Verbindung und Zusammenstellung von Tönen bestimmen liesse, und alle Anmasungen derjenigen, welche träumten, die Tonsatzlehre mathematisch begründen und aus solcher anmaslichen Begründung absolute Präcepte ableiten und aufstellen zu können, zeigen sich bei der leichtesten Prüfung als leere, nur belachenswerthe Träume, deren Trüglichkeit sich durch das erste beste Beispiel handgreiflich zeigen lässt. (Weber 1830–1832, 202).

Weber takes to be the ear's—or the listener's, for that matter—expectations of how the music could, but does not, continue. Musicologist Jairo Moreno argues that Weber was intrigued by the ambiguity of Mozart's music, attributing that interest of Weber's to "Romantic irony."[10]

Other authors have confirmed that the emergence of the listener in the first decades of the nineteenth century relates to Romantic subjectivity (e.g. Dahlhaus 1988; Johnson 1996). In the context of a history of musical analysis, Weber's conjectural examples are particularly interesting in that they demonstrate how the pairing of the notions of "composition" and "analysis" is juxtaposed with a production of conjectural samples by the analyst. These samples do not have the status of compositions, as the author himself calls them conjectural and introduces them using a conditional clause: *if* that note in Mozart's setting were to be understood in such and such a way, *then* the next notes could have been the following (204, 208). This method spells out the ambiguities: the note in question is conjecturally explicated and heard "as" a particular symbol in the tonal system, which, in turn, is made understood by the provision of a continuation that pins down the ambiguity to one of its meanings. Weber describes the attitude of "the ear" as awaiting "further instruction and conformation" about the still underdetermined tonal key (205). One sample is integrated into a question—"etwa so?" (211; "for instance, like this?"; not translated in 1853/5).

The samples instantiate not only unrealized possibilities arising from the ambiguity of the setting but also alternatives that could mitigate harsh effects or alleviate the ear's responsibility for fitting what is heard into the tonal system. Where Fétis gave one alternative, Weber provides six alternatives to Mozart's imitation, introducing them with yet another conditional clause—"That the strangeness principally arises from the union of the above circumstances will be evident, if we so alter the passage as to omit them" (747). However, where eight bars were printed in the *Revue musicale*, Weber's readers were presented with only the first two bars and were obliged to complete the rest on their own. Inviting the reader to participate in a series of musical thought experiments, he challenges them to follow his suggestions based on the knowledge of the rules, "the comprehension of which will now present no farther difficulty to the reader of all that precedes" (746). After all, this remark appears in the third volume of a textbook on harmony.

The mode that the reader is intended to pursue in the explication of unused possibilities is to play the alternative samples on a piano. Almost all the examples are presented in piano reduction—that is, in the usual notation for piano players on two

[10] In spelling out the ways in which Weber's text continually renews that ambiguity, Moreno (2003) is inspired by a remark, communicated to him in a letter from his colleague Kevin Korsyn: "Weber generates an almost absurd proliferation of detail, taking more than 15 dense pages to analyze four or five bars and giving almost every pitch a series of multiple and contradictory interpretations. One could relate this profusion of detail to the trope of irony, in the extended sense proposed by Hayden White. Irony is the trope that sanctions multiple linguistic perspectives on reality, because one realizes that language is not adequate to capture experience. One searches for multiple linguistic redescriptions of events." (99) In an exemplary gesture, the author Moreno publishes this private note, which contains the kernel of his argument.

Fig. 15.8 Diagrammatic overview of the harmonic progression in measures 3–4 of K.465 (Weber 1851, 749)

musical staves rather than the system of four staves used for a string quartet. The music notation for pianists provided an overview of all simultaneous notes. In addition to saving space on the printed paper, it allowed the reader to walk to their piano and hear what was written by playing it. Weber even speaks of "*anschlagen*"—hitting the keys of an unmentioned piano—where he addresses the sound effect more specifically. This is not how string players produced sound. Instruments of the violin family are not struck, as pianos are; typically, rather, the bow is drawn across the strings. However, Weber also does not address actual sound production. This is most conspicuous in one of the rare examples in which he uses a diagram to give an overview of the harmonic skeleton of the passage in question (Fig. 15.8). Even there, he writes, "the fundamental third B in the bass is again struck anew." (749)

It is important to note that Mozart's piece no longer appeared as an example for other composers to emulate in the context of Weber's analysis. Mozart's mastery was not intended to be within the reader's reach, including his eccentric application of the rules. The fabricated alternatives instead spelled out the educated listener's attempts to cope with ambiguities. They unfold the hesitation among possible interpretations into sequential conjectures. Rather than merely explaining the rules and breaches, these fabrications cause the reader to feel them. This transcends mere rhetoric, as it also fulfills a knowledge-making function. Although Weber guided the readers' listening, he explicitly refrained from imposing any "correct" application or interpretation of the piece, leaving the last word to the listener, among which he counted himself.

15.4 An Experimental-Experiential Trajectory: Stumpf's Analysis of K. 361

The most prominent attempts to follow up on Helmholtz's research into the analysis and synthesis of sound—before electronically produced simple tones became a standard in psychoacoustics laboratories—happened under the guidance of philosopher and experimental psychologist Carl Stumpf at Berlin University. Beginning in 1913, Stumpf had a two-part structure built at the university's Institute for Psychology for the analysis and the synthesis of sounds by interference. When the construction began, Stumpf had already turned 65. His interest in Helmholtz's research, however, dated back 40 years, to the moment when Stumpf had assumed his first professorial position at Würzburg University, where he followed his former mentor Franz Brentano as professor of philosophy in 1873.

In 1883, when Stumpf had moved from Würzburg to the German University at Prague, the first volume of his book *Tonpsychologie* (Tone Psychology, 2020) appeared. This, together with the second volume of 1890 in addition to his philosophical work, earned the reputation of an experimental psychologist, which eventually brought him via Halle and Munich to Berlin. In the foreword to *Tonpsychologie's* first volume, Stumpf gave credit to Helmholtz's "classic work." Notwithstanding the wealth of inspiration that psychologists could gain from it, they had left the greatest share still to do (1883, v; 2020, lxi). If Stumpf considered his first volume to already be bulky, he likely did not expect that it would keep him busy for several decades to come. The interference apparatus turned his interest in sound perception and cognition in Helmholtz's wake into cutting-edge experimental research.

The experiments with musical instruments in particular surprised Stumpf himself, who was well aware that their spectra exceeded the range of possibilities with his interference device. That device's operation principle worked as follows: it used interference to cancel out select frequency components in periodic sound that traveled through a system of tubes. To produce interference, spikes of various lengths were inserted into the main tube. A spike would cancel any wave with a wavelength of four times that of the spike, projecting the reverse pattern of rarefication and densification on that particular wavelength, thereby canceling it out. When all spikes were inserted, which covered a wide frequency range rather densely, any periodic sound should ideally disappear and gradually reappear when the spikes were taken off again.

The device worked best for vowel sounds, whose spectra are situated in the rather small range of the greatest frequency resolution in human hearing. Speech sound, however, traveled so well through the complicated design of the amassed tubes that some of the problems that the apparatus posed remained unnoticed. For the sounds of the instruments, by contrast, the deformation of the sound through the system of the tubes became audible and disturbing. For instance, funnels had to be added at the points where musicians produced the sound, so that it could enter the system of the interference tubes without loss. Where possible, narrow tubes were replaced with wider ones. Any experiment on the frequency components present in the sounds of instruments had to begin by testing whether the instrument could be recognized through the system of tubes at all. If this was not the case, "then one has to renounce this way of researching" (382). If the study could continue, however, it proved "extremely instructive" (383) even beyond the mere description of the formants. Joined by Curt Sachs, an eminent expert on musical instruments, the researchers observed the gradual composition of the sounds from their fundamental frequency up to the full (transmittable) spectrum and discovered several expected and unexpected phenomena.

As was to be anticipated, for example, the fundamentals stripped from the higher partials resembled one another for all instruments. However, they were often found to be rather weak. Given the fact that musical notes were understood as representing the pitch that equaled the fundamental frequency, the fundamental's weak presence in the spectrum was a surprise.[11] For many instruments, even the perceived pitch only emerged once the fundamental had been joined by at least one, sometimes several, of the higher partials—by removing the canceling spikes and thereby reversing the procedure of analyzing the sound. Again, the appearance of the perceived pitch based on several partials raised questions. It was unclear, why the ear would nevertheless depend on the fundamental for determining the overall perceived pitch. In other cases, however, the opposite occurred, and the ear was found to follow one of the partials for determining the pitch of an instrument rather than hearing the fundamental as its pitch. Sometimes, dissonant chords suddenly appeared in the spectrum, at other times, beats among partials could be perceived, or the likeness of the timbre to a vowel sound was observed.

Although many instruments were analyzed using the interference apparatus, only two— the clarinet and bassoon—are extensively described. A third, the French horn, added to the curious phenomena that Stumpf and Sachs observed: its second partial could not be isolated in analysis. The analysis of the French horn's sound, Stumpf added, had to remain incomplete, as it exceeded the technical means. The horn's sound, however, was also emulated in a series of sound synthesis experiments that used the second part of the structure composing the interference apparatus, which eliminated the overtones of continuous pipe sounds until only their fundamental remained audible. This resulted in a range of simple tones—that is, tones with only one frequency each. They build a Fourier series as had been the case in Helmholtz's apparatus, but a far greater number of them was available in this device. The simple tones could be selected, their strength modified, and the resulting modified selection be merged into a single sound in the control room of the interference device. In this way, the patterns obtained from the various modes of analysis could be re-instantiated.

The resulting synthetic sounds were compared to the natural sound of the respective instruments. For this, some concessions were made with respect to the situation of such observation. "Only those partials that come up for the characteristic features of a sound at some distance were observed, even though more partials can be found at close vicinity when using the resonance method" (1926). That is, the nearness or distance at which a sound was heard significantly altered the spectrum. Stumpf and his team chose to work only on those spectra that withstood propagation. In the experiments, the distancing was achieved by placing the musicians and observers in

[11] Here and elsewhere, I use the notion of the "spectrum," although this was introduced for acoustics only in the later 1920s, when it had become too complicated to address the frequency composition of sounds otherwise.

opposite rooms across a corridor, or the door to the control room was simply opened with the musician staying on the corridor while playing the test notes.

Experiments with the analysis and synthesis devices, in turn, were particularly instructive when single partials were omitted by interference or by excluding them from the re-synthesis. The second partial of the horn sound, for instance, which had been inaudible in the interference experiments, proved decisive in the synthesis experiments. When it was omitted from the synthesis, the sound resembled that of a clarinet. Joined by other experts, including Georg Schünemann, also a professor of musicology, and Emil Prill, first flutist of the Staatskapelle Berlin and professor at the music academy (Musikhochschule), Stumpf and Sachs listened to the synthetic and natural sounds in this way. The experts judged the synthetic sounds to be "very good," as Stumpf reports with some amazement.

As an example of Stumpf's methods of sound analysis, the description of the experiments on the bassoon's sound is particularly instructive. According to Stumpf, four regions could be discerned in that instrument's sound with the help of the interference device. Even when it reached the upper end of its main formant, it still remained closer to a trombone. Above, a buzzing sound resulting from superimpositions of the higher partials joining became audible. Adding yet higher partials, the sound acquired one of its characteristic features, which Stumpf called "nasal" (385), until finally, with the last partials that could be addressed with the interference, he made the typical "furry" sound of the bassoon reemerge. He added,

> These four stages of the development can also be discerned in the full sound, but this is equivalent to the discrimination of single partial tones: one hears them in all degrees of clearness, depending on one's attitude. If one concentrates fully on one zone [meaning a bandwidth of partial frequencies], however, then, of course, the impression of the timbre itself disappears. If one concentrates somewhat less on that alone, then the timbre remains present next to the partial impression of the zone. (385)

With the interference apparatus, the bassoon's sound was transformed into a sequence of otherwise unnoticeable frequency compounds. In reconstructing the spectrum from the fundamental back to full transmission, the compounds were added to one another like building blocks. Stumpf rounded off the research on the sounds of musical instruments with an excursion in small print that described his impression from a concert in which he had heard Mozart's Serenade no. 10 for 13 instruments, the so-called *Gran Partita*, K. 361. He wrote,

> Observations concerning the impact of the register on the sound of wind instruments in musical performance can be particularly fruitful, when several of them come together in a *concertante* style. I noted much to that matter in an execution of Mozart's Serenade for 13 wind instruments by members of the Berlin State Opera Orchestra (Staatskapelle). For instance, the pitch of the French horn playing *piano* or *mezzoforte* often appeared to be an octave lower due to the softness and expanding width of its sound, as this is also the case for simple tones; the oboe in its low register (c^1–g^1) sounded like a cornet, very beautiful and closer to the vowel A then E, which pointed to stronger partials in the 2-lined octave, etc. (1926, 391f.)

The analytical hearing that was used and trained in these experiments continued to shape the experience of the concert with the Staatskapelle. That the object of the study was now a piece by Mozart is not accidental. The serenade is a genre that was originally played in the open air and therefore uses wind instruments, whose sounds carry further than strings. Traditionally, a serenade consisted of several movements, and the examples known from earlier times often had single instruments excel as concerting soloists, while simultaneously seeking an equilibrium between all instruments in the ensemble. This also holds for Mozart's K. 361, in which some of the movements feature select smaller constellations of the instruments while in others sometimes one or two instruments serve as soloists of the ensemble.

Mozart's music was not the main target of Stumpf's description. The music served as yet another sample for study that entered into the context of the experimental research that Stumpf engaged in both before and after. Nonetheless, the listening on which Stumpf reported explicates the *concertante* style of this particular composition in an original fashion. Stumpf freely associated previous experiences with the sounds that instruments produced, referring as much to his own trajectory as a listener as to the experimentation. Thus, experimental and experiential knowledge were combined, undergirded with the acquired knowledge that was presented in the book's preceding argument.

15.5 Musical Analysis in Three Different Contexts

Musical analysis—in the context of Western composition, which is relevant here—relied on established categories. Where expounded at greater length, analyses served the purpose of providing models for composers—that is, those who put together what was taken apart in the analysis. A layer of a guided subjectivity was added in the nineteenth century to analysis, when listening became the predominant mode of engaging with music. Weber's proposal that Mozart's alleged blunders be reconsidered is exemplary in this respect. His analysis informed the way in which his readers would listen to the piece, enabling them to appreciate each note's complex relationship to the rules of composition and to grasp all the options that were virtually present, as Weber claimed, but that were not selected by the composer. This approach to educating his readers would not mitigate the cognitive dissonance that resulted from the superimposed instances of bending the rules of composition but, rather, the opposite. Weber's avowal that he disliked the passage he analyzed is telling in this respect: The fabricated examples summarize what the readers were supposed to have learned by the time they read the analysis. They presented artificial results of the application of the rules distilled in the book. Although these were "correct" alternatives, they were intended not to substitute Mozart's composition but rather to enrich its listening.

Fétis's alternative version, in turn, shows that a "correction," even when introduced as a substitute, by the same token was not a synthetic counterpart of analysis. The relationship to the analyzed item being one of authority, there was no way of producing similar alternatives.[12] The same holds true for Weber's conjectural examples: they are syntheses in the sense that they synthetically using the rules of composition to produce correct music. However, they re-instantiated the analysis of Mozart's piece only in so far as the piece was already instantiated in the listener's mind. Musical analysis, as it was practiced in the nineteenth century, still took composition as its point of departure. Analysis was possible only to the extent that its object was constituted by composition. No demand yet existed for a concept of synthesis. What was new in the nineteenth century was that the addressee of the analysis was no longer the composer but the listener, who was initiated into the rules while learning to analyze at the same time. Any synthetic activity would, then, occur in the listener's mind. As Weber's analysis shows, this eventually resulted in a new way of perceiving the analyzed piece.

Stumpf's setup of analysis and synthesis refers to both this context and the new notion of analysis and synthesis that Helmholtz had brought from the natural sciences into the study of hearing. As Helmholtz had in his work on vowels, Stumpf used the interference device to generate data from analyses of the frequency composition of speech sounds and the sounds of musical instruments. Contrary to Helmholtz, whose attempts to synthesize sound were restricted to the crucial question of whether the ear *distinguished* the synthesized sounds among themselves, Stumpf used synthesis systematically to verify the *quality* of his analyses, apparently with some success, and he even reports that his analyses found some sounds of musical instruments to be very good. What the remark on the sounds of musical instruments in Mozart's serenade reveals, then, is that Stumpf concluded from both the learning trajectories that musical analysis designed for its new readership *and* from Helmholtz's appeal to harmony teachers that they shape their listening in the concrete material setting in which they undergo such trajectories. Most significantly, however, Stumpf integrated his own trajectory of learning from his experimental work into his analytical listening. His description of Mozart took into account the fact that his hearing had changed. The music as it is synthesized in the listener's mind was his point of departure for analyzing the listener's previous trajectory.

These three examples thus demonstrate that musical analysis was co-constituted in three different ways. In Weber, the reader underwent a trajectory of learning rules—to follow Mozart and to follow Weber analyzing Mozart. In Helmholtz, the analysis of vowel sounds seems to relate to a stable object. This stability, however, is precarious. The distinction was more than ephemeral: it depended on the quick change between preset choices of the partials' strengths. Helmholtz' analysis of Mozart's *Ave verum corpus* reaches stability via a different route. The confirmation the analysis of the moods in the piece's two parts and conclusion is in line with the analysis is achieved by extrapolating a stable set of chords, whose actual sequence is not considered. The stability of a statistics of harmonious or less favorable chords is, in addition, based on features that are extrapolated in a new, experimental setting, notwithstanding the fact that they are unlikely to appear in a performance. Stumpf,

[12] Perhaps the closest to this would be the slow introduction to Hyacinthe Jadin's (1776–1800) string quartet in E-flat major op. 2, No. 1, which is closely modelled after Mozart's but reverses the direction of his melodic lines.

finally, not only co-constitutes his object—namely, an experience of listening to Mozart that is informed by his previous encounters with the sounds he hears, but he explicitly refers to this as a trajectory of learning that co-constitutes his analytical listening. It would not be until later in the twentieth century, when—for instance, in ethnomusicology or cultural musicology—new constellations of listening to other communities' music became the subject of new modes of analysis, that attempts to reconstitute the communities' rules of compiling music would become a synthesis that was up for discussion but now in a social act of communication.

References

Bent, Ian D., and Anthony Pople. 2001. Analysis. *Grove music* Online https://www.oxfordmusiconline.com/grovemusic/view/10.1093/gmo/9781561592630.001.0001/omo-9781561592630-e-0000041862. Accessed 2 Feb 2024.

Bernstein, David W. 2002. Nineteenth-century harmonic theory. The Austro-German legacy. In *The Cambridge history of western music theory*, ed. Thomas Christensen, 778–811. Cambridge: CUP.

Christensen, Thomas. 2019. *Stories of tonality in the age of François-Joseph Fétis*. Chicago: University of Chicago Press.

Crary, Jonathan. 1990. *Techniques of the observer. On vision and modernity in the nineteenth century*. Cambridge, MA: MIT Press.

Dahlhaus, Carl. 1988. *Klassische und Romantische Musikästhetik*. Laaber: Laaber.

Fétis, François-Joseph. 1829. Sur un passage singulier d'un quatuor de Mozart. *Revue Musicale* 5: 601–606.

Gruber, Gerold W. 1994. Analyse. In: *MGG Online*, ed. Laurenz Lütteken. New York/Kassel/Stuttgart: Bärenreiter, Metzler, RILM. https://www.mgg-online.com/mgg/stable/12832. Accessed 2 Feb 2024.

Helmholtz, Hermann von. 1863. *Die Lehre von den Tonempfindungen als physiologische Grundlage für die Theorie der Musik*. Braunschweig: F. Vieweg.

———. 1865. Ueber Combinationstöne. *Annalen der Physik und Chemie* 175: 497–540.

———. 1885. *On the Sensations of Tone as a Physiological Basis for the Theory of Music*. Translated by Alexander J. Ellis [after the German edition of 1877]. London: Longmans, Green & Co.

Hiebert, Erwin. 2014. *The Helmholtz legacy in physiological acoustics* (Archimedes: New Studies in the History and Philosophy of Science and Technology, 39). Cham: Springer.

Holtmeier, Ludwig. 2007/2016. Weber, Gottfried, WÜRDIGUNG. In *MGG Online*, ed. Laurenz Lütteken. RILM, Bärenreiter, Metzler, 2016–. Accessed 11 Jan 2023. https://www.mgg-online.com/mgg/stable/393810

Johnson, James H. 1996. *Listening in Paris: A cultural history*. Berkley/Los Angeles: University of California Press.

Koenig, Rudolph. 1889. *Catalogue des Appareils d'Acoustique construits par Rudolph Koenig*. Paris.

Kursell, Julia. 2009. Wohlklang im Körper. Kombinationstöne in der experimentellen Hörphysiologie von Hermann von Helmholtz. In *Resonanz. Potentiale einer akustischen Figur*, ed. Karsten Lichau, Viktoria Tkaczyk, and Rebecca Wolf, 55–74. Munich: Fink.

———. 2013. Experiments on tone color in music and acoustics. In Music, sound, and the laboratory, 1750–1980, ed. Alexandra E. Hui, Julia Kursell, and Myles W. Jackson (*Osiris* 28), 191–211. Chicago: University of Chicago Press.

———. 2015. A third note: Helmholtz, Palestrina, and the early history of musicology. *Isis* 106: 353–366.

———. 2017. Klangfarbe um 1850. Ein epistemischer Raum. In *Wissensgeschichte des Hörens in der Moderne*, ed. Netzwerk "Hör-Wissen im Wandel," 21–40. Berlin: De Gruyter.
———. 2018a. *Epistemologie des Hörens*. Paderborn: Fink.
———. 2018b. Klangfarbe. In *Handbuch Sound: Geschichte – Begriffe – Ansätze*, ed. Daniel Morat and Hansjakob Ziemer, 57–62. Stuttgart: Metzler.
Moreno, Jairo. 2003. Subjectivity, interpretation, and irony in Gottfried Weber's analysis of Mozart's 'Dissonance' quartet. *Music Theory Spectrum* 25: 99–120.
———. 2004. In *Musical representations, subjects, and objects: The construction of musical thought in Zarlino*, ed. Rameau Descartes and Weber. Bloomington/Indianapolis: Indiana University Press.
Pantalony, David. 2005. Rudolph Koenigs's acoustical workshop in nineteenth-century Paris. *Annals of Science* 62: 57–82.
Pisko, Fr. Jos. 1865. *Die neueren Apparate der Akustik: Für Freunde der Naturwissenschaft und der Tonkunst*. Vienna: Carl Gerold's Sohn.
Schäfer, Armin. 2011. Goethes naturwissenschaftliche Kunstauffassung. In *Goethe-Handbuch. Supplemente*, ed. Andreas Beyer and Ernst Osterkamp, vol 3: *Kunst*, 183–196. Stuttgart, Weimar: Metzler.
Schimma, Sabine, and Joseph Vogl, eds. 2009. *Versuchsanordnungen 1800*. Berlin/Zurich: Diaphanes.
Thorau, Christian, and Hansjakob Ziemer. 2019. The art of listening and its histories: An introduction. In *The Oxford handbook of music listening in the 19th and 20th centuries*, ed. Christian Thorau and Hansjakob Ziemer, 1–36. New York: Oxford University Press.
Utz, Christian. 2010. Analyse. In *Lexikon der systematischen Musikwissenschaft*, ed. Helga de la Motte-Haber, Heinz von Loesch, Günther Rötter, and Christian Utz, 38–41. Laaber: Laaber.
Vogl, Joseph. 2007. Der Weg der Farbe. In *Räume der Romantik*, ed. Inka Mülder-Bach and Gerhard Neumann, 157–168. Würzburg: Königshausen & Neumann.
Wald-Fuhrmann, Melanie. 2022. Musikwissenschaft, Zur Fachgeschichte. In *MGG Online*, ed. Laurenz Lütteken. New York/Kassel/Stuttgart: Bärenreiter, Metzler, RILM. https://www.mgg-online.com/mgg/stable/421611. Accessed 28 June 2024.
Weber, Gottfried. 1830–1832. *Versuch einer geordneten Theorie der Tonsetzkunst, Dritte, neuerdings überarbeitete Auflage*. Vol. 3. Mainz/Paris/Antwerp: Schott.
———. 1851. *Theory of musical composition, treated with a view to a naturally consecutive arrangement of topics*. Trans. James F. Warner, ed. John Bishop, vol. 2. London: Robert Cocks and Co.

Julia Kursell is Professor of Musicology at the University of Amsterdam. Her research interests include twentieth and twenty-first-century composition, the history of musicology and the relation between music and science. She is co-director of the UvA's Vossius Center for the History of Humanities and Sciences (http://vossius.uva.nl).

MIX
Papier aus verantwortungsvollen Quellen
Paper from responsible sources
FSC® C105338

If you have any concerns about our products,
you can contact us on
ProductSafety@springernature.com

In case Publisher is established outside the EU,
the EU authorized representative is:
Springer Nature Customer Service Center GmbH
Europaplatz 3, 69115 Heidelberg, Germany

Printed by Libri Plureos GmbH
in Hamburg, Germany